MW01437114

ENGINEERING DESIGN
From Art to Practice

ENGINEERING DESIGN
From Art to Practice

Joseph W. Walton

WEST PUBLISHING COMPANY
ST. PAUL • NEW YORK • SAN FRANCISCO • LOS ANGELES

COPYRIGHT © 1991 By WEST PUBLISHING COMPANY
50 W. Kellogg Boulevard
P.O. Box 64526
St. Paul, MN 55164-1003

All rights reserved

Printed in the United States of America

98 97 96 95 94 93 92 91   8 7 6 5 4 3 2 1 0

Library of Congress Cataloging-in-Publication Data

Walton, Joseph W.
   Engineering design: from art to practice/Joseph W. Walton.
     p.   cm.
   Includes index.
   ISBN 0-314-76551-4
   1. Engineering design.   I. Title.
TA174.W264   1991
620′.0042—dc20

90-40234
CIP

Production Credits
Copyediting: Michael Michoud
Illustrations: Scientific Illustrators
Text Design: Geri Davis, Quadrata, Inc.
Cover Design: Pollock Design Group
Cover Image: FPG International
Composition: Trigraph Inc.

Photo Credits

14, H. Armstrong Roberts; **15**, H. Armstrong Roberts; **16**, H. Armstrong Roberts; **17**, H. Armstrong Roberts; **18**, Stock-Boston; **19**, H. Armstrong Roberts; **28**, H. Armstrong Roberts; **28**, H. Armstrong Roberts; **29**, Camerique; **30**, Camerique; **31**, H. Armstrong Roberts; **32**, Camerique; **33**, H. Armstrong Roberts; **34**, H. Armstrong Roberts; **35**, H. Armstrong Roberts; **36**, Camerique; **37**, H. Armstrong Roberts; **38**, Paul Fortin, Stock-Boston; **40**, Courtesy of Numonics Corporation; **42**, Courtesy of Deere & Co.; **51**, H. Armstrong Roberts; **52**, H. Armstrong Roberts; **53**, H. Armstrong Roberts; **174**, **255**, Courtesy of Deere & Co.; **256**, Courtesy of Metalcut; **259**, Courtesy of Deere & Co.; **260**, Courtesy of Deere & Co.; **261**, H. Armstrong Roberts; **262**, Courtesy of Minster Machine Co.; **263**, H. Armstrong Roberts; **264**, Courtesy of Minster Machine Co.; **275**, Courtesy of Minster Machine Co.; **276**, Courtesy of De Vlieg Machine Co.; **277**, Courtesy of De Vlieg Machine Co.; **278**, Courtesy of De Vlieg Machine Co.; **279**, H. Armstrong Roberts; **280**, Courtesy of Deere & Co.; **286**, Courtesy of Deere & Co.; **288**, H. Armstrong Roberts; **289**, H. Armstrong Roberts; **438**, Courtesy of Xavier Park Unit of Mercy Health Center, Dubuque, Iowa; **446**, H. Armstrong Roberts.

Contents

Preface XIV
Introduction to Student XVIII
List of Symbols XX

SECTION ONE ENGINEERING PROBLEM SOLVING PROCEDURES 2

1 Introduction To Engineering 4

1.1　Engineering Experience　5
1.2　Sources of Design Failure　10
1.3　Unsolved Problems　13
1.4　References　19
1.5　Exercises　21

2 Engineering Activities 24

2.1　Primary Job　24
2.2　Necessary Skills　25
2.3　Engineering Developments and Advancements　26
2.4　Responsibilities　38
2.5　Day-to-day Concerns　44
2.6　Constraints and Limited Resources　46
2.7　Project Planning　52
2.8　Communication　57
2.9　Personal Characteristics and Abilities　64
2.10　References　65
2.11　Exercises　67

3 Problem Solving 74
- 3.1 Sources of Problems 74
- 3.2 Creativity 76
- 3.3 Roadblocks To Creative Problem Solving 76
  - **CASE STUDY #1:** Door Holder-opener 79
- 3.4 Problem Solving Activities 80
- 3.5 References 83
- 3.6 Exercises 84

4 Problem Definition 88
- 4.1 Methods of Problem Definition 88
  - **CASE STUDY #2:** Parts Moving Problem; basic problem statement 89
- 4.2 Related Activities 96
- 4.3 Cautions 102
- 4.4 System Problems 103
- 4.5 References 104
- 4.6 Exercises 105

5 Problem Solution Idea Generation 108
- 5.1 Specific Considerations 109
- 5.2 Physical Laws 117
- 5.3 Brainstorming 119
  - **CASE STUDY #3:** Idea Generation 121
- 5.4 Work Simplification 124
- 5.5 Feasibility Study 125
- 5.6 Summary 127
- 5.7 References 128
- 5.8 Exercises 128

6 Refinement and Analysis 134
- 6.1 Levels of Analysis 135
- 6.2 Related Activities 138
  - **CASE STUDY #4:** Drawing table adjustment 141
  - **CASE STUDY #5:** Room cooling problem 147
- 6.3 Summary 154
- 6.4 References 154
- 6.5 Exercises 155

7 Decision and Implementation 166

7.1   Idea Selection   168
      CASE STUDY #6:   Baseball field revision   171
7.2   Implementation   174
7.3   References   176
7.4   Exercises   177

Design Problems 182

Group I–Incomplete Problem Identification   182
Group II–Preliminary Problem Solution Given   186

SECTION TWO   DETAILS OF REFINEMENT AND ANALYSIS   196

8 Materials 198

8.1   Metals   198
8.2   Polymers   202
8.3   Ceramics   207
8.4   Wood   207
8.5   Material Properties   208
8.6   Other Selection Criteria   233
8.7   Composite Materials   235
8.8   Additional Study Topics   237
8.9   References   238
8.10   Exercises   240

9 Manufacturing 246

9.1   Process Selection   246
9.2   Material Forms   252
9.3   Secondary Operations   267
      CASE STUDY #7:   Special Machine   287
9.4   Tooling   288
9.5   Design Considerations   290
9.6   Summary   297
9.7   Additional Study Topics   297
9.8   References   298
9.9   Exercises   300

10 Probability and Statistics 318

- 10.1 Probability 318
- 10.2 Normal Distribution 327
- 10.3 Z Test 330
- 10.4 Decision Error 331
- 10.5 *t* Test 332
- 10.6 Effect of Sample Size 333
- 10.7 *F* Test 334
- 10.8 χ^2 test 335
- 10.9 Normal Distribution Test 338
- 10.10 Binomial Distribution 340
- 10.11 Poisson Distribution 342
- 10.12 Weibull Distribution 344
- 10.13 Process Capability 346
- 10.14 Table Data Analysis 352
- 10.15 Additional Study Topics 354
- 10.16 References 355
- 10.17 Exercises 356

11 Failure Analysis 366

- 11.1 Design Interferences 368
- 11.2 Measure of Reliability 369
- 11.3 Safety Factors 373
- 11.4 Probabilistic Design 374
  CASE STUDY #8: Probabilistic Design Analysis 376
- 11.5 Material Strength Theories 377
- 11.6 Impact Loading 383
- 11.7 Fatigue 389
- 11.8 Column Buckling 402
- 11.9 Fracture Mechanics 407
- 11.10 Creep 411
- 11.11 Corrosion 412
- 11.12 Fretting 412
- 11.13 Hydrogen Embrittlement 413
- 11.14 Vibration 413
- 11.15 Summary 415
  CASE STUDY #9: Hydraulic Cylinder 415
- 11.16 Additional Study Topics 419
- 11.17 References 417
- 11.18 Exercises 421

12 Engineering Models 436

- 12.1 Types of Models 436
  - **CASE STUDY #10:** Tree Age Analysis 440
- 12.2 Boundary Conditions 443
- 12.3 Full-Scale Models 444
- 12.4 Scale Models 445
- 12.5 Dimensional Analysis 447
  - **CASE STUDY #11:** Wind force modeling 450
- 12.6 Experimental Data Analysis 454
  - **CASE STUDY #12:** Empirical formula refinement 460
- 12.7 Experimentation 467
- 12.8 Additional Study Topics 471
- 12.9 References 471
- 12.10 Exercises 472

13 Optimization 480

- 13.1 Calculus Techniques 481
  - **CASE STUDY #13:** Tool Life Optimization 486
  - **CASE STUDY #14:** Pipe Packing Optimization 491
  - **CASE STUDY #15:** Cost Ratio Analysis 493
- 13.2 Non-Calculus Techniques 496
- 13.3 Linear Programming 508
- 13.4 Additional Concerns 512
  - **CASE STUDY #16:** Balcony Seating Optimization 512
  - **CASE STUDY #17:** Spherical Storage Tank Optimization 516
- 13.5 Experimental Optimization 519
  - **CASE STUDY #18:** Experimental Optimization Performed on Three Variables 523
- 13.6 Additional Study Topics 527
- 13.7 References 528
- 13.8 Exercises 528

14 Financial Analysis 538

- 14.1 Simple Interest 539
- 14.2 Installment Loans 540
- 14.3 Compound Interest 540
- 14.4 Mortgages 542
- 14.5 Annuities 543
- 14.6 Taxes 546

14.7	Depreciation & Depletion 545
14.8	Overhead Charges To Operations 547
	CASE STUDY #19: Misapplied Overhead 549
14.9	Return On Investment 551
14.10	Cash Flow 552
	CASE STUDY #20: Cost Reduction Proposal 552
	CASE STUDY #21: New Process Proposal 553
14.11	Forecast Uncertainty 555
14.12	Summary 556
14.13	Additional Study Topics 556
14.14	References 556
14.15	Exercises 557

15 Ethics 562

15.1	Definition 562
15.2	Sources of Problems 563
15.3	Codes 564
15.4	Specific Situations 569
	CASE STUDY #22: Personal Conflict 570
	CASE STUDY #23: Corporate Conflict 571
	CASE STUDY #24: Government Ethics 572
15.5	Study Cases 573
15.6	References 573
15.7	Exercises 574

Appendices 577

A-1	Basic quantities and units. 578
A-2	Basic quantities and units. 579
B	SI Prefixes and multiplying factors. 580
C	Selected conversion factors. 581
D	Energy values of common fuels. 582
E	Element data. 583
F-1	Properties of selected materials. 584
F-2	Properties of selected materials. 586
F-3	Properties of cast iron. 588
F-4	Properties of selected aluminum alloys. 590
F-5	Properties of selected magnesium alloys. 591
F-6	Properties of selected brasses. 592
F-7	Properties of selected bronzes. 593
F-8	Properties of selected nickel alloys. 594

F-9	Properties of selected titanium alloys.	595
F-10	Properties of selected ceramics.	596
F-11	Properties of concrete.	597
F-12	Selected values of the modulus of elasticity.	598
G	Normal distribution.	599
H	t distribution.	601
I	F distribution.	602
J	Chi-squared distribution	606
K	Correlation coefficient.	607
L	Stress concentration factors.	608
M	Stress intensity factors.	612
N	Copper wire data.	617
O	Shear, moment, and deflection equations for cantilever beams.	618
P	Properties of selected geometric shapes.	625
Q-1	Wide flange structural section properties.	625
Q-2	Standard flange structural section properties.	627
Q-3	Standard channel structural section properties.	628
Q-4	Standard angle structural section properties.	629
R	R-values of insulating materials.	630-631

Answers To Selected Exercises 632

Index 660

Preface

Engineering Design: From Art to Practice is an overview of fundamental methods and procedures for solving engineering design problems. It includes discussion of and emphasis on performing the essential, though sometimes routine, details of analyzing and implementing design solutions while stressing design flexibility and adaptability. It emphasizes analysis and synthesis techniques, for these allow an engineer to tackle new unsolved problems. Examples that illustrate various concepts are drawn primarily from my personal experiences in engineering. During the 13 years I was a manufacturing engineer in industry, I saw more problems caused by a low concern over how to manufacture a product than by poor product ideas. One of the reasons the United States is falling behind other industrial countries, Japan most notably, is not a lack of good ideas, but a lack of efficient manufacturing and financial implementation. Engineers don't spend most of their time solving earth-shaking revolutionary design problems; day-to-day mundane-seeming problems occupy most of their time. It is not usual for an engineer to solve calculus integrals after graduation, the more important problem being to determine how the best ideas can be manufactured economically into reliable products.

This book is my solution to a problem: to have a text for an engineering design course that contains a variety and quantity of problems without suggesting that there is a final word on any problem. This book has been used successfully in a two-semester, eight credit, senior-level engineering design course for Engineering Physics majors. The course is the last chance to put the pieces of many different courses together, many of which have been taught by professors who are experts in their own fields, but who do not combine the concepts into a coherent whole. Technology and state of the art changes require that engineers be prepared to keep up with the pace of change and be aware of the various interrelationships that exist in such areas as electronics, atomic physics, and material science.

During my 18 years of teaching, one of the most common student comments has been that textbooks are hard to read. I have tried to write in a style that reads easily and also keeps the reader aware of details. Many authors write for the experts, not for the students. The result is that students often do not read for learning or understanding. I also think that students learn as much from example as from theory, so I include examples where others would include theory.

This book has material for two, 15-week semesters meeting four hours a week. The amount of time spent on extended design problems, semester-long or year-long design projects, will influence how much time is spent on the exercises in each chapter.

Multiple, shorter projects can serve the same purpose as a semester long project, depending on personal objectives. Projects should be undertaken to broaden the experience of the student and the teacher, and to facilitate accurate and detailed analysis on the part of the student. Detailed formal reports, both written and oral, are part of the project work. Suggested technical prerequisite courses for this text are graphics, calculus, chemistry, physics, statics, strength of materials, material science, and computer programming.

The text is divided into two sections. Section I, Chapters 1-7, surveys the engineering problem solving procedure. Chapters 1 and 2 illustrate the range of problems and activities in which engineers are involved. Chapter 3 presents an overview of problem-solving procedures, and Chapters 4 through 7 present the details of the tasks engineers perform while solving design problems. It is assumed that student projects will be assigned and developed as the topics of the design procedure are studied. Some engineering students have never been exposed to the range of activities that complement the problem-solving task. Chapters 4 through 7 cover these activities and offer examples for illustration. Chapter 5 lists questions an engineer should ask to insure an optimum final design solution.

Section I contains over 200 exercises, ranging from those appropriate for group discussion, to those a student could spend a week or more investigating. The exercises stress the common, but essential, considerations necessary for good design. Many exercises have no single answer, which is a true engineering predicament. Answers the students provide will vary depending on their background, personal experience, and available references. Class discussion of the exercises is important so that ideas and experiences can be shared and the assumptions and compromises needed for solutions can be agreed on. Cooperation, understanding, and communication between specialty areas of engineering is a real engineering requirement.

A variety of design problems are proposed at the end of Chapter 7 for student work. The problems can be used for a month-long, semester-long, or year-long projects depending on the depth of analysis desired, how many students work together on a project, and whether working models are built. Students can start a design project from Chapter 7 at the beginning of the semester, and solutions can be developed as Chapters 4 through 7 are studied. To make the exercises on experimentation meaningful, solutions must be proposed, experiments run, and data collected and analyzed. Project analysis and refinement can be continued while students discuss the ideas and develop proficiency with the concepts discussed in the chapters in Section II.

Section II, Chapters 8-15, covers details necessary for thorough project analysis. Exercises in Section II are more apt to have one answer because they are intended for single concept discussion and practice. They are small details of the overall design process. Chapters 8 through 15 are mostly independent and can be studied as need and interest dictate. Some engineering students may have had separate courses on some of these topics, and if so, the chapters can be skipped over or used as a review. I have learned that students need review on a regular basis. There are a few exercises and case studies that are chapter-sequence dependent. The alignment of exercises with text material may at times appear random, and this is intentional. Real engineering problems do not come in chapter/topic order, they occur randomly and without hints. Over 450 exercises in Section II range from short answer to the nearly impossible. Time and talent will dictate which to use. Each chapter in Section II has additional study topics listed to encourage further study, provide topics for student research papers, and emphasize the seemingly inexhaustible nature of engineering concepts. The topics are purposely left unreferenced, and in the spirit of unstructured problems, the student must start from scratch. As a guide for the study topics, and as support for chapter discussions, references are listed at the end of each chapter and in Section 4.2, neither of which is meant to be exhaustive. Parenthetical notation, (2), used in the text refers to the numbered references at the end of the chapter. English to metric conversions were done using Theodore Wildi's *Units*, 2nd ed., published by Volta Inc., Canada, 1972.

Acknowledgements

I am grateful to the Literary Executor of the late Sir Ronald A. Fisher, F.R.S., to Dr. Frank Yates, F.R.S. and the Longman Group Ltd, London for permission to reprint Table VII from their book *Statistical Tables for Biological, Agricultural and Medical Research* (6th Edition 1974).

I want to thank the following reviewers who have offered constructive criticism during the development of this text.

Joseph R. Baumgarten, Iowa State University

George Bibel, University of Akron

James L. Blechschmidt, Arizona State University

Larry Clemen, Iowa State University

Annette Fetter, Howard R. Green Consulting Engineers

Gary A. Gabriele, Rensselaer Polytechnic Institute

R. A. Gardner, Washington University in Saint Louis

Charles Reif Hammond, Dover, NH

Timothy K. Hight, Santa Clara University

Paul P. Lin, Cleveland State University

K. Ross Johnson, Michigan Technological University

Barry W. McNeill, Arizona State University

Steven P. Nichols, The University of Texas at Austin

John V. Perry, Jr. Texas A&M University

George Saliba, University of Pittsburgh

A. Samson, University of California at Los Angeles

Robert F. Steidel, Jr., University of California at Berkeley

Donald H. Thomas, Drexel University

David G. Ullman, Oregon State University

John R. Zimmerman, University of Delaware

Introduction For Students

For students of engineering, or any discipline, understanding the levels of intellectual activity needed to enhance learning will assist with developing proficiency in using the problem solving procedures. A summary and brief discussion of these main intellectual activities will aid in your understanding of the role they play in your development as an engineer. The key words for the activities that affect the intellectual thought process are:

1. Know
2. Comprehend
3. Apply
4. Analyze
5. Organize
6. Synthesize
7. Evaluate
8. Create

Basic knowledge is important. Facts and figures must be *known*, but knowing is not enough. Knowing that the yield strength of a material is 30,000 psi is not of much value unless one *comprehends* what yield strength is. Once a concept is known and comprehended, it can be *applied* to a specific case. Example: A physical member made from a material with an allowable yield strength of 30,000 psi will withstand a load of 15,000 lbf before deforming permanently if it has a cross section of 0.50 in^2.

A problem can be *analyzed* by breaking it into component parts, each of which can be studied separately, as well as in the context of assembly. Fracture and other failure modes for each part are studied and the components designed accordingly. If the analysis results are *organized*, then the problem solver can *synthesize*. Synthesis includes the prediction of what will happen to the overall design solution when a change is made in a component. If the location of a component is changed, or its method of attachment to the next member is altered, or its material is changed, the size and shape of other components and the life of the product may also be affected. Synthesis is not complete until the prediction is *evaluated*, perhaps with a computer model, a scale model, or a full size model. If the evaluation results do not meet expectations, then alternative solutions must be *created* and the test process repeated. Iteration of ideas, analysis, and test is continued until the results satisfactorily solve the design problem.

The purpose of this book is to provide you with ideas, suggestions, and exercises to help develop your skills so you can perform design problem-solving activities in an efficient way. Reading about how to solve problems is only the beginning; practice is essential. Exercises are provided so you can practice the concepts and improve your problem-solving skills. Some of the exercises will be so easy you may wonder what the catch is. Others will be so complex you may wonder if a solution is possible. Open-ended design exercises at the end of Chapter 7 give you the opportunity to practice the steps required to solve a problem, including how to go about obtaining necessary support information. There is no single best answer to a design problem, only a compromise after you consider relevant and restricting factors. Discussion with your instructor and your fellow students is good learning; reading theory and examples is not enough. Example and practice are as valuable as theory.

Make good use of what you have learned in previous courses: calculus, differential equations, chemistry, material science, statics, dynamics, strength of materials, or computer science. Engineering problems don't appear with a footnote telling you what solution technique to use. Be creative and develop new solution techniques along the way.

The real problems of day-to-day engineering cannot be itemized in a checklist. Your ability to work and communicate with people, combined with your resourcefulness and creativity, will determine the success or failure of your projects. Do not be discouraged with failures. Learn from them and make the next solution better.

List of Symbols

A	= Area, generally perpendicular to applied force; Annuity beginning amount.
α	= Type I statistical error.
B	= Light absorptivity; Intercept on semilog and log-log graph paper.
b	= Intercept on linear coordinate graph paper.
β	= Type II statistical error.
D	= Ductility; Circle diameter.
δ	= Beam deflection.
e	= Column eccentricity.
ϵ	= Normal strain.
E	= Modulus of elasticity.
E_t	= Tangent modulus of elasticity.
F_e	= Static force equivalent to the effect of an impact load.
f_n	= Natural frequency.
Φ	= Shear strain.
γ	= Thermal expansion coefficient.
G	= Modulus of rigidity.
G_t	= Tangent modulus of rigidity.
g	= Gravitational constant, 32.17 ft/sec^2 (9.81 m/sec^2).
ν	= Poisson's ratio.
I	= Light intensity; Area moment of inertia.
I_r	= Reflected light.
I_t	= Transmitted light.
i	= Period interest rate.
k	= Electrical resistivity; Spring constant.
K_c	= Fracture toughness.
K_t	= Stress concentration factor.
LCL	= Lower process control limit.
l_e	= Column effective length.

List of Symbols

m	= Refraction index; Slope of a line on linear coordinate graph paper; Mass of an object.
μ	= Absolute viscosity of a fluid.
P	= Loan principal.
P_c	= Critical load for column buckling.
ρ	= Material density.
R	= Modulus of resilience; Thermal resistance coefficient; Annuity periodic withdrawal.
r	= Radius of gyration; Annual interest rate; Radius of a circle; Coefficient of correlation.
σ	= Normal stress; Standard deviation of the universe.
S	= Standard deviation of a sample of the universe.
S_e	= Reverse bending fatigue strength.
S_e'	= Adjusted reverse bending fatigue strength.
S_{su}	= Modulus of rupture.
S_{ut}, S_u	= Ultimate tensile strength.
S_{uc}	= Ultimate compressive strength.
S_{us}	= Ultimate shear strength.
S_{yt}, σ_y	= Yield tensile strength.
S_{yc}	= Yield compressive strength.
S_{ys}	= Yield shear strength.
τ	= Shear stress; Natural period of oscillation.
T_g	= Glass transition temperature.
UCL	= Upper process control limit.
W	= Applied weight force.
ω	= Circular velocity.
x or **y**	= Mean of a set of data.

SECTION ONE

ENGINEERING PROBLEM SOLVING PROCEDURES

Engineering design and problem solving procedures are the lifeblood of practicing engineers. Problems are encountered at every stage of an engineering project and engineers must know how to deal with them. Some problems repeat themselves, resulting in specialized fields of engineering. Engineering has become so specialized that different courses of study and disciplines are available. Engineers become trained to solve a variety of problems and needs of society: nuclear, computer, chemical, mechanical, electrical, agricultural, and oceanographic, to name just a few. Some problems, because of constant changes in the state of the art, do not repeat themselves, but still must be solved. Finding a way to repair or replace a distorted space telescope mirror, and finding a reliable, economical source of non-polluting energy are examples of problems yet to be solved as this book is written.

Regardless of which speciality a student engineer wishes to master, or which uncharted problem area is going to be searched for answers to the world's problems, there is common ground to cover, common information to learn, and common procedures to guide in the quest for acceptable solutions to problems. Section I of this text provides procedures, study material, and learning exercises for problem solving. These procedures can then be applied to any engineering (and also non-engineering) problem. The guidelines for problem solving presented and discussed do not guarantee that a utopian solution will be found, only the "best" within your limits. Specific data on material and material failure theories, as well as solutions to problems that have already been solved, are of little value unless an engineer uses the information in a logical and systematic way, hence the need for guidelines to the problem-solving procedure. Section I includes hints, examples, and illustrations on how to sort and sift through information, whether it comes from handbooks or from your own creativity.

All the guidelines and data in the world, however, will not be of value to you unless they are tempered with experience, unbiased analysis, and good judgment. Engineering provides many stories of peculiar, surprising, and sometimes unfortunate events. These events teach us to be careful because they are unforgiving if we make a mistake. The bridge fails, the ship sinks, or the building collapses in the storm. Even as engineers work within their own area of specialty they must be constantly alert for the unusual and the unexpected. Following the guidelines for problem solving will only help reach an optimum solution, they will not provide a guarantee of success.

Examples and exercises in Section I are specific, even though the procedures which are presented are universally applicable and generally can be used on any problem. Section II also contains specific examples and exercises which use specific procedures. However, these procedures are not universally applicable.

1 Introduction To Engineering

"The trouble with the future is that it usually arrives before we're ready for it."
◆ ARNOLD H. GLASOW

Engineers are professional people who work to solve the problems of today's society and also plan ahead to prevent problems in the future. The purpose of this book is to help develop the skills necessary to perform problem solving activities effectively. This is done by using theory, examples, and exercises to illustrate that engineering problems in general have alternate acceptable solutions. Solutions are a compromise, a balance between conflicting restrictions that are overcome using personal judgment based on experience and theory. It is unfortunate that so much of engineering training is connected with "getting the right answer," because the conclusion many students have is that a problem has only one correct answer. For some exercise problems this may be so. Single answer problems are required so that students can study new concepts and become familiar with elementary procedures and methods. Engineers polish their skills by practicing simple techniques and then applying them to more complex issues. Few, if any, real problems have only one answer. What, you ask, can be required of the problem $2x = 4$ other than solving for the answer $x = 2$? The difference is that $2x = 4$ is not an engineering problem. It is a mathematical model that might be used to help solve an engineering problem. The model $2x = 4$ may be but one solution to dividing up a space of 4 feet, namely into 2 equal pieces. Perhaps a better model to help solve the real problem is $3x = 4$, dividing the space into 3 equal pieces. More generally the model $aX = 4$ is applicable, using the parameter a and investigating many solutions.

1.1 Engineering Experience

Engineering decisions are based on theories studied in classes, tempered with the experience of making theories work, and adjusted for the actual restrictive conditions encountered. This is one reason why state of the art and related procedures are built up from past successes and failures. Some of the best learning an engineer will ever get occurs when a proposed solution is tried and it does not work as planned. It is hoped that the failure will not cost lives or too much money, and only personal pride will suffer. On the other hand a design may be successful for a reason other than the one assumed critical by the engineer. History has taught the engineering profession many things. Some of these are obvious in today's technical world; others have been learned more recently and were not obvious at all. Following are summaries of cases that illustrate some of the lessons that have been learned the hard way.

Case #1: In the 1850s, when cast iron railroad bridges in Pennsylvania were failing, the order went out to replace them with wooden bridges. Wood held up under the same conditions, but no one knew for sure why the cast iron did not. Today we know why cast iron was not a good choice for the bridges; the combination of tension loads at stress concentrations is more than brittle cast iron can withstand. When the proposed change in the state of the art from wood to cast iron was attempted and failed, the engineering decision was to revert to a tried and proved method, wood bridges.

Case #2: In December, 1928, one leaf of the bridge over the Hackensack River in New Jersey failed after less than two years of service. The bridge was a two-leaf drawbridge that used a 770-ton counterweight on each leaf. The moving counterweight, accelerating and decelerating, coupled with component friction, caused forces and vibrations that the supporting structure could not withstand. The dynamics of the bridge had not been properly understood by the designers. Strain gage tests and transit measuring verification of the problem were made on the leaf that had not collapsed. The bridge was strengthened with additional tower supports and was returned to service. The theoretical model analysis had not been substantiated with real time testing.

Case #3: In March, 1937, a high school in Texas, just four years old, converted from burning methane city gas to natural gas obtained less expensively from a nearby oil field. Shortly after conversion, the school exploded, killing 455 people. Leaks in the system, which apparently were there prior to the fuel switch, caused the natural gas, denser than the methane, to accumulate in the basement and explode when an electrical switch was used. Investigation concluded that the area through which the pipes ran was inadequately ventilated. The investigation report also recommended that malodorants be added to all natural gas so leaks can be more easily detected. Lessons are often learned the hard way, but learning and reacting are essential for technological advancements.

Case #4: On November 7, 1940, the Tacoma Narrows Bridge failed during a moderate 40 mph cross wind. Other bridges, notably the Golden Gate Bridge, had been designed with the same basic state of the art, but had not failed. Analysis done after the bridge collapsed showed that the bridge was not deep enough for its length. It was aerodynamically unstable and simply "flew" apart. No model wind testing for this phenomenon had been done prior to construction. Aerodynamic instability had never before been recognized as a cause of bridge failure.

Case #5: In 1943, Liberty ships were failing when used in the cold environment of the North Sea. Liberty ships were transport vessels, fabricated with welded rather than riveted seams, and used during World War II to ship soldiers and supplies. Many of the ships were cracking, and some sank because of the cracks. Engineers realized that welded construction did not allow for the suppression of crack growth as did riveted construction. The failures confirmed stress concentration theory because many of the cracks started at the square corners of hatchways. The failures also demonstrated that stress concentrations that are not critical when metal is above the transition temperature, become critical below the transition temperature. The materials react brittlely, and cracks grow rapidly and unexpectedly. Fracture mechanics analysis is now used regularly by design engineers.

Case #6: In the early 1950s, a de Havilland Comet airplane failed by exploding in flight. The prototype had received a test load at twice the expected load pressure, and all seemed safe; the source of the failure was a puzzle. It is now known that the overload testing resulted in a residual compressive stress at stress concentration points. A component containing residual compressive stresses will withstand more cycles of tensile loads than will a part that does not contain residual compressive stresses. Comet airplanes which had not been tested exploded in flight when a crack formed at the corner of a window. The internal cabin pressure caused the crack to grow at a high rate which in turn caused the fuselage to explode.

Case #7: In August, 1953, a fire in the General Motors transmission plant in Lavonia, Michigan made clear the need to design large buildings with fire walls, protection for steel roof framing, and sprinkler systems, even though the contents are not considered a fire hazard.

Case #8: In 1963, the Vaiont Dam in Italy withstood unusual forces in an impressive, yet disastrous way. A landslide caused a wave of water to pass *over* the dam and kill more than 2000 people in the valley below. The dam was strong enough to withstand the high forces, but the project was a disaster. The dam remains filled with 300 million cubic yards of earth and is useless for storing water or generating electricity. A design engineer must always consider the effect and consequences of designs on the surrounding environment; analyze the overall system.

Case #9: In 1966, a 14-foot long, earth-filled pocket was discovered in an eight-foot diameter concrete shaft during the construction of the John Hancock Center in Chicago. The shaft's top had shifted during construction, and an inspection uncovered the problem. Worker failure? Sabotage? Ignorance? Good design won't succeed with poor construction techniques.

Case #10: A rule of thumb occasionally heard is "If the building holds together while being built, it will hold together when it's complete." One reason behind this statement is that structures often have stresses placed on them during construction that exceed the stress placed on them after construction is complete. One such concern is the load caused by wet concrete and the low strength of uncured concrete. This was made clear in a dramatic fashion in March, 1973 when a 26-story building under construction in Washington, D.C. collapsed killing 14 workers. Support shores had been removed before poured concrete had become strong enough to hold up the ensuing construction. The collapse started on the 22nd floor, and all 22 floors collapsed to the ground.

Bridge construction is another area where construction loads and stresses may govern the sizes of many components. Many members hang in space as cantilever beams until other members are built up to them and properly attached.

Case #11: In May, 1976, the South Carolina Canadian Bridge fell, and two semitrailer truck drivers were killed. An expansion joint had locked, and all the bridge contraction due to temperature change occurred at the one joint that would move. One joint did not have enough safe travel to accommodate the contraction, and it pulled itself off the supporting structure. Poor maintenance on the joints? Poor design on the joint surfaces? Poor design for not considering the contraction accumulation at one joint? All lessons learned the hard way.

Case #12: In January, 1978, the roof of the Hartford Civic Center collapsed under the weight of snow. The roof structure had been designed with the help of computer analysis. The program, however, had been written with assumptions that did not "hold up." Even during construction it was noted by contractors that the roof was sagging. No concern was given to this observation for the designers felt confident the design was correct because a computer was used for the analysis. Computers can't make right what engineers don't start right.

Case #13: In 1979, a DC-10 airplane lost an engine in flight and crashed. Design and test data were reviewed, and no design problems were uncovered. After more than one accident occurred, further checking discovered that a normal engine service operation was causing the problem. The engines were periodically dismounted, serviced, and remounted. Mounting holes were becoming elongated during the service procedures and were the source of failure.

Case #14: In 1980, the Alexander L. Kielland, an offshore oil rig, broke up in normal North Sea weather. When components of the failed structure were analyzed, a three-inch crack was found in a part near a weld joint. Paint in the crack revealed that the crack had been there before the unit went into service. Poor material? Poor inspection? Tragic results.

Case #15: In 1981, the Kansas City Hyatt Regency Hotel had a skywalk collapse during use just after opening. The skywalk had been designed to hang from the overhead structure so that the open area below was not cluttered with support columns. After the accident, analysis revealed that a change had been made during construction. The change had not been checked out with proper calculations and had placed more load on a hanging joint than it had originally been designed to carry. When the live load of use was added, the total was more than the structure could carry and it collapsed. The Hyatt lobby was rebuilt using a proven design, that of placing columns under the structure.

Case #16: Also in 1981, Grumman Flexible buses used in New York started developing fatigue cracks in their frames. The design had used a low safety factor because of design restrictions. The buses had to be fuel efficient, smooth riding, maneuverable, lightweight, accessible for the handicapped, air conditioned, and low in cost. The number of conflicting requirements resulted in a compromised design that was too fragile for the use environment.

Case #17: In 1983, the Mianus River Bridge in Connecticut collapsed after 38 years of service when a section hanger cracked and failed. How many cycles had the hanger absorbed in 38 years? Post-failure analysis indicated that a combination of questionable design and poor maintenance caused the problem. Periodic bridge inspections had been made. Was the design done in 1945 consistent with the state of the art in that year? Don't judge an old design by today's standards, judge them by standards of the time they were built. If state of the art changes are significant, a structure may require strengthening or replacement.

Case #18: In 1986, Joe Walton built a log rack in Iowa to store firewood off the ground. He built the rack from standard two-inch water pipe and water-resistant treated wood. The pipes were driven into the ground for legs. The rack fell over during the first rainstorm because the low compressive strength of wet soil had not been accounted for.

Case #19: In 1987, a private Iowa college allocated money to update and rebuild its outdoor track facilities. Track widening and resurfacing were the primary objectives. As work on the project progressed, it was discovered that the original track surface was not level. In fact the enclosed football field and track area sloped nearly three feet from one end to the other. The project reportedly ran 33 percent over its budget of $150,000 just to fill in and make the track level. Why was the critical project step of defining the problem and collecting adequate data for planning and scheduling omitted?

Case #20: In April, 1988, the lesson of metal fatigue was again brought to attention when a Boeing 737-200 lost its cabin roof during flight. A preliminary report indicated that six metal fatigue cracks were found emanating from rivet holes in the aluminum skin. Inadequate test? Inadequate inspection? Improper use? More problems for the engineer.

Case #21: In the summer of 1988, beaches in New York and New Jersey were closed to the public when hospital waste washed ashore. Tests on the waste showed evidence of the AIDS virus. Accidental? Illegal dumping? Is the world population to succumb to its own waste? Pollution prevention and control is a full time job for an environmental engineer, and a concern of every engineer.

Case #22: In November, 1988 an arm on a carnival ride in Florida broke off and one person died and seven others were injured. The ride had been recently inspected and no problems were found. The arm failed at a crack that had been painted over and was not visible. The crack had been there for some time as demonstrated by the rust found under the paint. Poor inspection procedures? Painted to hide flaws? Fatigue continues to control the life of physical structures.

Case #23: Lest the impression is made that all engineering problems are structural and material oriented, consider the collision of two trains in August, 1988. The collision caused the evacuation of 1500 area residents because one of the trains had two derailed cars of denatured alcohol that caught fire. Three human errors contributed to the collision and the dangerous conditions.

1. The collision occurred 16 minutes before one of the trains was scheduled to leave the train yard. Someone made a bad decision on starting time.
2. The two cars containing alcohol were the second and third cars of the train, though company rules required them to be at the end of the train. Someone violated company rules.
3. The tracks where the collision occurred had no wayside signals because the tracks were regional tracks with limited traffic. Someone had decided the cost of the signals was not warranted.

It is unfortunate that some things have to be learned from mistakes. Mistakes and ignorance cost money, time, and often lives. That is why efforts between engineers with different specialties are commonplace today. There are so many aspects to large commercial projects that it would be impossible for one person to be able to be responsible for all of them. Even seemingly small projects such as creating a strain gage pressure-sensing system for roller bearing races requires the joint efforts of mechanical, electrical, metallurgical, and manufacturing engineers. A project such as a large building certainly needs the cooperation of many specialists such as structural, electrical, safety, civil, and mechanical engineers.

Similarly, the design of nuclear power plants requires the joint efforts of nuclear engineers and material engineers. Nuclear power plants use pipe made from stainless steel because of its resistance to corrosion. However, salt from any source in the presence of moisture will corrode stainless steel. Salt can come from the air or from the sweat of the workers who weld the pipe. Pipes are routinely inspected for cracks so they can be repaired or replaced before serious leakage occurs. Materials used in nuclear power plants are also exposed to neutron bombardment. When the plants were first designed, it was not known what effect this would have on the structural materials. It is now known that continuous exposure to neutrons will raise the transition temperature of the metals. The result is that sudden cooling of the material can cause it to fail in a brittle manner.

The more complicated a problem, the less likely it is that direct application of any one theory will solve it. The restrictions and assumptions made during theory development are one source of possible conflict. The problem at hand may have variations and options present that were not known or considered during theory development. The following examples illustrate how poor understanding of a new situation and improper application of a theory may also contribute to inappropriate analysis.

EXAMPLE 1.1:
Theory development assumptions.

During the development of many material deformation theories, a mathematical simplification about angles is usually made. If the angle of deformation is small enough, the angle in radians is equal to the tangent of the angle. This simplification is fine for "small" angles, but is inadequate as the size of the deformation increases. If you forget the theory development assumption you may incorrectly apply the theory.

EXAMPLE 1.2:
Temperature variation affect.

Recall that the strength of materials decreases as the temperature increases. If you forget this, trouble can occur. An unplanned temperature increase can occur if an object is placed in a confined space under the sun's rays. Temperature problems can also occur to a part that is ordinarily convection cooled if the cooling fan stops or the air flow passages become blocked. Unplanned events may happen if an object made from a thermoplastic material is left too close to an incandescent light bulb.

Your engineering training has placed emphasis on theories and known scientific principles. It is up to you to see that they are applied correctly. Many problems often arise because of decisions that "seemed like a good idea at the time." Always consider the total effect of what you do, not just present conditions.

Engineers have learned many lessons the hard way, but they have also achieved great successes. Following are some of the notable engineering achievements that are often taken for granted by many people:

1. The Brooklyn Bridge, which opened for service in 1883, was the longest suspension bridge in the world until 1931. The bridge is still in service, safely carrying traffic from Brooklyn to Manhattan.
2. The Panama Canal, a 51-mile navigable link between the Pacific and Atlantic Oceans, was built in the period 1905-1914, and required the movement of 268 million cubic yards of dirt.
3. The Mackinac suspension bridge, a four-lane, 26,444 foot connection between the upper and lower peninsulas of Michigan, replaced car ferry service in November, 1957.
4. The series of 29 locks and dams built on the upper Mississippi River controls water flow to maintain a nine foot channel depth necessary for river barge traffic from the Ohio River to Minneapolis-St. Paul, Minnesota.
5. San Francisco buildings built after 1906 to be earthquake resistant survived the October, 1989 earthquake.
6. The spaceship Voyager 2, launched in August, 1977, is powered by the heat from the radioactive decay of plutonium oxide. The craft sent images and data on the makeup of planets back to earth during its 12-year trip, culminating with a close up look at Neptune in August, 1989.

1.2 Sources Of Design Failure

There are many reasons why products fail to live up to expectations. Failures can range from a catastrophe like a bridge collapsing, to an inconvenience such as a door opener that reacts too slowly. Engineers must be aware of the reasons why these things happen and try to avoid the problems. Following is a summary of 14 of the more common sources of engineering errors, omissions, and failures.

The *problem to be solved may not be understood* because the originator of the problem did not convey information correctly. Associated with a possible poor problem description may be a lack of questioning. Even if you think you know what the problem is, ask questions. Review your impression of the problem with the originator to be sure you understand it. Added discussion can only help clarify the situation. Don't be in too big a hurry to get started on the solution. A bad start to a design problem almost always assures a bad ending.

The design may be based on *erroneous data*. If the design of a dam is based on a high water level of 15 feet, but the measurement was taken from an undefined reference point, the final design probably won't do the intended job. For example, a problem with a surveying job was solved by clarifying a reference point. Two different surveys located the boundary of a lot in different places. The difference was 40 feet, half a lot width. Additional checking showed that the two surveys were started at opposite ends of the section of land, and that the section was too big. An original survey had made the section half a lot too big.

Design errors can be made because of *invalid or overextended assumptions*. Errors of invalid assumptions are very difficult to trace because often a designer doesn't record why a particular design decision was made. Get in the habit of recording all your work, not just the conclusions. If a storm sewer design was based on average rainfall, but this was not documented, it would be hard to verify why the system flooded from time to time.

Faulty reasoning from good assumptions can also cause problems. Bad reasoning can occur when the premise is wrong, or if the premise is applied to the wrong condition. Assuming that 1010 steel is always ductile is a bad assumption. If heated and quenched, the 1010 steel might be harder than desired; if an impact load is applied, the material might react in a brittle fashion. Concluding that the sine $0.25_R = 0.25$ could be a bad conclusion depending on the accuracy required. See Exercises 1.7 and 1.8.

An *incorrectly stated principle* can lead to many design problems. The statement "A feather and a ball bearing will fall to the ground from equal heights in equal time" is true if the surrounding medium is a vacuum, but it is not true in normal atmosphere.

Improper experiments are often a source of problems. An experiment can be set up so that preconceived results will almost be guaranteed. If no control sample is run at the same time, the results may be attributable to the wrong variable. It might be possible that a variation was present that was not recorded, like line voltage variation in an electrical experiment, or temperature variation in a chemical experiment.

Poor data collection from a valid experiment can also result in invalid conclusions. Data recorded in experiments are subject to errors. Errors include the incorrect reading of instruments or incorrect recording of correct readings. Included in this can be the transposition of figures or illegible handwriting. Errors in copying can also occur when data are collected from handbooks or charts. The wrong row or column can be entered, or a critical footnote might be overlooked. In very critical conditions reading and recording by more than one person is appropriate. Often errors can go unnoticed for a long time as in this example.

EXAMPLE 1.3:
Data error.

Isaac Newton recorded and used an incorrect value for the distance from the earth to the sun and calculated a mass for the earth inconsistent with the value used in other of his calculations. The error was not discovered until 1987, 300 years later.

Good data can be made bad with *errors in calculation*. Incorrect use of a calculator or computer program will give errors, and the possibility is strong that an error will not be obvious. Practice with orders of magnitude review will help. Repeating calculations or verifying them graphically is suggested for critical designs.

Once a design has been determined, drawings of the parts to be manufactured or system flow charts must be created because drawings are the communication link between the design engineer and manufacturing. A *drawing error* that causes a breakdown in communication can create many problems, as the following example illustrates.

> **EXAMPLE 1.4:**
> **Drawing error.**
>
> A layout drawing of a tank to hold liquid detergent and parts to be washed was made at half size by engineer A. Engineer A went on vacation and engineer B was called in to complete the drawings. Engineer B was not familiar with the project and dimensioned the drawing as if it had been drawn full size. The result was a nice half-size model of the wash tank, but the tank was of no value to the production shop.

Many other drawing errors can occur: incorrect threads to fit a purchased part, wrong sizes because of a layout measuring error, incorrect tolerances for function, and errors in basic view representation.

Parts drawn correctly must be manufactured, and *incorrect manufacture* can cause a good design to turn bad. A part that was machined to size for test purposes, but flame cut to size for production, failed because of surface irregularity stress concentrations. Manufacturing control is just as important as using the correct design failure criteria during product refinement and analysis. *Incorrect assembly* can also cause properly machined parts to fail. A part that was spread and forced into assembly over two dowel pins failed because of the pre-use tensile load placed on a critical part of a casting.

A well made part must be shipped to the customer and protected from damage. *Errors in packaging and shipping* can cause even the best made products to fail. Parts being shipped are often subjected to loads and vibrations that are not present when the product is in real use. Springs and other structural members can accumulate many cycles of stress before the product ever reaches the customer.

> **EXAMPLE 1.5:**
> **Shipping problem.**
>
> One problem related to shipping was an axle bearing that became out of round due to transportation-induced vibration. The vibration peened a flat spot on the bearing. The load was being applied to the same place on the bearing with each cycle of vibration because the vehicle was tied down. The bearing joint design was changed, and modifications to the shipping procedure were made. Air pressure in the rubber tires was increased after tie down to reduce the spring effect of the tires.

Products that do not go directly to the customer must be stored safely so they do not deteriorate. *Incorrect storage* can cause products to deteriorate and fail when placed into service. Unprotected steel parts must be stored in low humidity areas, or coated with rust inhibitor to avoid rust. High temperature can also cause materials to deteriorate. Parts must be stored in such a way that creep distortion does not occur because of inadequate support.

A successful product must be used properly by the consumer, but *improper use by the consumer* is very difficult to prevent. Good owner manuals and safety labels are minimal protection against improper use. Over time, many products have become laden with safety features because of improper owner use. Examples include lawn mowers with blade-stopping interlock mechanisms, and snow blowers with auger-stopping interlock mechanisms. Improper use by consumers includes failure to perform regular maintenance. Engineers try to design around this problem by creating "maintenance-free" products such as cars that do not need chassis lubrication, motors with self-lubricating sintered bronze bearings, and house siding that never needs painting.

Designing and building products that meet all these potential problems and still allow for a profit is indeed a challenge. It points out the large responsibility that engineers have; engineering is not a profession to enter into lightly.

Engineering design principles do not change when problems are solved in areas other than your specialty. Principles don't change; only the problems and the applications change. Every job and profession has its own set of problems. Although there are some very specific considerations made in this book, such as the fatigue life of rotating shafts, a new problem may require that a known procedure be used in a new way. Whatever your job, and whatever type of business you find yourself in, the problem-solving approach can be used. You should have the potential to solve any problem that comes along. Even if you don't know anything about underwater agriculture or making parts on the moon, if the problem-solving procedure is followed and analysis is done properly, problems in these areas are not insurmountable. This is not to say that someone more knowledgeable than you won't solve the problems faster and arrive at more efficient solutions, but problem solving and engineering design activities will provide an acceptable answer if you perform them with purpose and desire.

1.3 | Unsolved Problems

Many unsolved problems require creativity and inventiveness to solve. As technology advances, new levels of performance are required. People and products are extended. Rotating parts need to spin faster, resistance to higher operating temperatures is required, and equipment needs to operate in the weightlessness of outer space. Supersonic flight has its own set of problems, as does nuclear power. Larger and stronger components are needed for taller buildings, longer bridges, and higher-payload airplanes. Other products need to be smaller, such as electronic components and artificial human implants. Systems to control the latest innovations have to be more responsive to people and the variations that can occur.

The problems you work on as an engineer may be less conspicuous than those mentioned. You may work to make a failing part function as it was intended, or simply make it more reliable. Perhaps the problem is that one part of a system is interfering with another part of the system. Recall that some of the first computer systems interfered with radio and TV signals when units were too close together. You might have to redesign parts to make them lighter or smaller, or so they can be assembled more easily.

There are many areas of science and industry that have problems that have not been solved satisfactorily, or solved at all. If you are searching for problem areas in which to try out your problem-solving skills, you will have many opportunities.

Environmental pollution is one of society's largest unsolved problems. As the world population grows, it places demands on the supplies of food, clothing, and shelter,

FIGURE 1.1

Worn-out and discarded product disposal pollution.

and it also creates waste disposal problems as illustrated in Figures 1.1, 1.2 and 2.17. Some of the by-products of our society that must be handled include human body wastes (sewage systems), product use waste (garbage, both biodegradable and non-biodegradable), manufacturing wastes (coal ashes, spent nuclear fuel rods, and waste chemicals), and old buildings. Environmental problems must concern engineers as well as the general public, and require combined efforts to solve. These problems include human wastes, mining wastes, worn out motor vehicles, industrial heating and power needs, water supply, radioactivity (natural and human-made), power plant by-products (heat, air pollutants, and fuel waste), animal waste, foundry waste, home heating and ventilation, refuse disposal, use of pesticides and weed killers, fertilizer residue, detergent residue, carbon monoxide, sulfur oxides, hydrocarbons from combustion processes, and

FIGURE 1.2

Garbage disposal pollution.

nitrous oxides. Some of these problems affect our water supply, as shown in Figure 1.3, while others affect the atmosphere. Breathable air can be in short supply in areas where industry and people crowd together and cause air toxicity. An average person needs 1000 gallons of breathable air per day. Where does it come from? Figures 1.4 and 1.5 are examples of environmental air pollution.

Other real and immediate problems exist in the urban areas of the world. Food, water, clothing, and shelter are not always available to the people who need them. Engineers are involved in the systems and technology required to meet the demands.

Problems concerning **energy** need attention. Fossil fuels such as coal, oil, and natural gas are being used up and sources depleted. Coal, which is the most abundant fossil fuel, is higher in by-product contaminants than the less abundant gas and oil.

FIGURE 1.3

Industrial water pollution.

Special equipment is required to burn coal and control air pollution. Special safety precautions are required for transporting and storing natural gas and oil products. Research is necessary to make other forms of energy more convenient and efficient to use. Many solar energy products are on the market; maybe you can improve on their efficiency or reduce their cost. Solar energy is received at the rate of one calorie/cm^2/minute; how can a high percentage of it be captured efficiently?

Wood is a replaceable resource that is used for construction, heating, and paper products. How can wood's use be better engineered so less will do more?

Wind energy is studied and has been developed with little commercial success. The payback time for the required investment is too great for most people's thinking, while others think the rotating blades are too dangerous, or that windmills are too noisy. Can consumer fears be overcome while reducing cost and increasing efficiency?

Another problem facing our technological society is the storage and retrieval of **information**. People learn more, know more, write more, and want access to more information. Computers have made some of this job easier. Computers can search through compiled data. How does it all get compiled, and in what fashion so it can easily be found? The stock market is a good example of the use of computer storage and retrieval, but even there the time-honored paper stock certificate is still used.

Good communications systems are required as a result of the mass of information available and the variety of problems and projects people undertake. All forms of communication, oral, written, and pictorial, need to be considered in any system. Short term temporary communication is needed as well as for long term permanent records.

FIGURE 1.4

Industrial air pollution.

Society has come a long way with the telephone, television, radio, printing, telefax, and other methods of communication; but problems of accuracy, timeliness, availability, and permanence still exist.

Transportation still has problems, whether it is a short 0–100 mile trip or a long 1000–5000 mile trip, (see Exercise 7.16). Connections between types of transportation need improving. A one-hour, 500-mile plane ride can end in frustration during the one-hour, 5-mile taxi ride to the final destination. Even connections within a given transportation mode can be bad as anyone who has had to wait at O'Hare airport in Chicago five hours between "connecting" flights can attest.

As airways, highways, and waterways get more crowded, transportation safety becomes more important. Safety involves more than having a buoyant seat cushion available.

Medicine still has unsolved problems. Tools and instruments are needed for the delicate and sensitive work performed on the human body. Existing artificial body parts

FIGURE 1.5

Transportation air pollution.

can always be improved, and additional artificial body parts are needed. Studies on the effects of medicine on humans and animals need to be done. Will a person be safe over a period of years after taking combinations of a variety of medicines? The design and use of laboratory testing equipment are problems for the engineer.

Much work needs to be done to gain access to the products of the **ocean and other waterways**. The waters of the world are more than just a place to go fishing or a place to dump wastes. Fresh water supplies are a great concern to many parts of the world. Potentials of the ocean include mineral extraction and underwater agriculture. The world water supply may hold some of the answers to energy shortages. Figure 1.3 illustrates a growing problem with our water supply, industrial waste dumping.

There are also many problems of **product distribution** faced by manufacturers associated with supplying products to the market place. Products must be reliable and sell at a reasonable cost. Automation can increase the consistency of products manufactured, but the quantity supplied may be more than is needed. Versatility and alternate use of facilities need to be considered.

There are also many problems for humans caused by **natural events** such as earthquakes, floods, spontaneous combustion and fires, tornadoes, hurricanes, volcanic eruptions, soil erosion, natural atomic radiation, avalanches, landslides, or a tsunami. Figure 1.6 is but one example.

There are many problems and too few trained people and funds available to solve them. The solutions require a combination of present and new theory, and new applications of acceptable solutions. This means answers are not found in textbooks; new thoughts and efforts are required.

FIGURE 1.6

Erosion problem.

1.4 References

1. Alm, Rick and Thomas G. Watts. "Critical Design Change Is Linked to Collapse of Hyatt's Skywalk." *Kansas City Star*, July 21, 1981.
2. Bennett, Gary F., Frank S. Feates, and Ira Wilder. *Hazardous Materials Spills Handbook*. New York: McGraw-Hill, 1982.
3. Bolt, B.A. et. al. *Geological Hazards*. New York: Springer-Verlag, 1977.
4. Boraiko, Allen A. "Earthquake in Mexico." *National Geographic,* Vol. 169, no. 5, May 1986, 654-675.

5. Bronikowski, Raymond J. *Managing the Engineering Design Function*. New York: Van Nostrand Reinhold, 1986. Chapter 16 on product liability.
6. Chiles, James R. "The Ships That Broke Hitler's Blockade." *Invention & Technology*. Vol. 3, No. 3, Winter 1988, 26-41.
7. Constance, John. *How to Become a Professional Engineer*. New York: McGraw-Hill, 1978.
8. Cooper, Sol E. *Designing and Structures: Methods and Cases*. Englewood Cliffs, N. J.: Prentice Hall, 1985.
9. Cottrell, Alan. *How Safe Is Nuclear Energy?* London: Heineman, 1981.
10. Cross, Jean and Donald Farrer. *Dust Explosions*. New York: Plenum Press, 1982.
11. Dawson, Gaynor W. & Basil W. Merces. *Hazardous Waste Management*. New York: John Wiley & Sons, 1986.
12. Hopf, Peter S., editor. *Handbook of Building Security*. New York: McGraw-Hill, 1979.
13. Howard, Ross. *Acid Rain: The North America Forecast*. Toronto: The University of Toronto, 1980.
14. Kohl, Larry. "Wall Against the Sea." *National Geographic*, Vol.170, no. 4, October 1986, 526-537.
15. Lindgren, Gary F. *Guide to Managing Industrial Hazardous Waste*. Stoneham, Mass.: Butterworth, 1983.
16. Martin, Edward J. and James H. Johnson Jr. *Hazardous Waste Management Engineering*. New York: Van Nostrand, 1987.
17. Melaragno, Michele. *Wind in Architectural and Environmental Design*. New York: Van Nostrand Reinhold, 1982.
18. Monakhov, V.T. *Methods for Studying the Flammability of Substances*. New Delhi, India: Amerind, 1986.
19. McDowell, Bart. "Eruption in Colombia." *National Geographic*, Vol. 169, No. 5, May 1986, 640-653.
20. Ormerod, Richard. *Nuclear Shelters—A Guide to Design*. London: The Architectural Press, 1983.
21. Petroski, Henry. *To Engineer Is Human*. New York: St Martin's Press, 1985.
22. Petroski, William. "Eastbound Crew Claims Train Left Yard Too Early." *Des Moines Register,* August 3, 1988, 3A.
23. Putnam, Palmer Cosslett. *Putnam's Power from the Wind*, 2nd ed. New York: Van Nostrand Reinhold, 1982
24. Ramson, W.H. *Building Failures—Diagnosis & Avoidance*. New York: E & F.N. Spon, Ltd., 1981.
25. Rogers, Michael. "After Oil; What Next?" *Newsweek*, June 30, 1986.
26. Ross, Steven S. *Construction Disasters: Design Failures, Causes, and Prevention*. New York: McGraw-Hill, 1984.
27. Sykes, Jane, editor. *Designing Against Vandalism*. New York: Van Nostrand, 1980.
28. Wanielista, Martin P. et al. *Engineering and the Environment*. Monterey, Cal.: Brooks/Cole, 1984.
29. Zackrison, Harry B. *Energy Conservation Techniques for Engineers*. New York: Van Nostrand Reinhold, 1984.

1.5 Exercises

1. Find samples from your own life or that of family or friends in which the first solution to a problem did not adequately solve the problem and necessitated additional tries.

2. Research and write a paper on any of the product failures mentioned in the chapter, or on any product failure for which data are available. Include discussion of the failure, what happened, why, the "cost" to society, and how the problem was solved.

3. Research and write a paper on a "successful" design, such as the Brooklyn Bridge, the Mackinac Bridge, the Panama Canal, the Alaska pipe line, or the 1986 Voyager airplane.

4. Research and prepare a paper on the social and environmental implications for one of the following waste disposal projects.
    a) Sanitary land fill
    b) Garbage incineration
    c) Refuse resource recovery
    d) Separation and concentration of hazardous wastes
    e) Biological degradation of wastes
    f) Chemical treatment of hazardous wastes
    g) Waste recycling (paper, metals, glass, plastic)
    h) Radioactive wastes

5. Research and prepare a use proposal on one of the following energy sources. Include reasons for using and reasons for not using the source. Consider environmental effects.
    a) Nuclear
    b) Ocean thermal
    c) Ocean tide
    d) Solar
    e) Geothermal
    f) Wind
    g) Synthetic fuels
    h) Hydroelectric
    i) Coal
    j) Coal gasification
    k) Petroleum

6. Give examples from nature that illustrate the following concepts.
    a) Tubular structural members
    b) Levers and fulcrums
    c) Squeeze pump
    d) Aerodynamic streamlining
    e) Water repellent surfaces
    f) Ultrasonic communications
    g) Light sources
    h) Light sensitivity

7. What is the largest value of $\emptyset$ in degrees such that $\tan \emptyset = \emptyset_R$ to the nearest 0.01? 0.001? 0.0001?

8. Repeat Exercise 1.7 using the sine function. Compare the results.

9. Write an algorithm that determines the $\sin \emptyset$ if the $\tan \emptyset$ is known, but does not determine the value of $\emptyset$.

10. A 10 megawatt power plant burns coal for its energy source. How much ash is released into the atmosphere if the coal used is 8% ash, 65% of the ash reaches the electrostatic ash precipitator, and the precipitator is 96% efficient?

11. Find real life problems that are examples of the 14 points made in Section 1.2 about types of product failure.

12. If the coefficient of linear expansion of brick masonry is 0.0000031 in/in·°F, and that of concrete is 0.0000067 in/in·°F, what concerns must be designed for in normal concrete-masonry buildings? Quantify the problem of a 100 foot length wall and 100°F temperature change.

13. A 3000 x 4000 mm piece of 5 mm thick plate glass is used in the construction of a building.
    a) Consider what will happen when part of the glass is in the shade in the winter, and another part of the glass is in the sun and is warm to the touch, $\Delta T = 60°C$.
    b) Will thicker glass solve any potential problem?
    c) Consider two possibilities:
       (1) The glass is well grouted and free to move in the flexible grout.
       (2) The glass is oversized, the frame is undersized, and there is no grout between them.

14. The design for the center cable on a suspension walk bridge is to have an in-place span of 75 feet and a sag height of 17 feet. The cable is made from a steel alloy, $E = 30 \cdot 10^6$ psi, and supports an assembled load of 500 pounds per foot of walk-bed. The engineer in charge claims a 0.75 inch effective diameter cable, with an allowable tensile strength of 20,000 psi, will be sufficient for the job. Do you agree?
    Explain your conclusion. Reference equation: $T = \{wL/2\}\{\sqrt{1 + (L/4h)^2}\}$, where T = tensile force in the cable at the support, L = distance between supports, w = weight load/unit of length L, and h = sag height.

15. A 95 foot high cylindrical water tower is designed to hold 80,000 gallons.
    a) What is the minimum diameter of the tank?
    b) Is it likely that the diameter is larger than that calculated in part (a)? Why?
    c) What minimum compressive load is placed on the ground by a full tank?
    d) Using the answer from part (a) what amount of uneven settling will cause the tower to lean four feet out of line with vertical?
    e) The condition in part (d) happened to a water tower in Charter Oak, Iowa in 1988. Discuss the likely cause(s) of the leaning tower and why it should not have occurred.

16. In Exercise 1.15, part d, does it may make a difference if the four feet is measured on the surface of the ground or from the corner of the tank that has sunk below the ground? Calculate the amount of sinking using this assumption and compare with Exercise 1.15.

2 Engineering Activities

"Technology is a function of societal selection based on perceived needs—social and economic....Who perceives the needs and shapes the technology? And to whom are they accountable?"
◆ MARTY STRANGE, FAMILY FARMING

The purpose of this book is to investigate those things that are important for an engineer to consider while designing solutions to problems. The professional engineer practices the art of applying scientific theory and principles to the efficient conversion of natural resources for the benefit of humans, to satisfy their perceived needs and desires. Engineers link theory and technology to the needs and desires of the consumer. Part of the engineer's work during the linkage is to convert laboratory-scale operations to the industrial level, from test tubes and beakers to barrels, pipes, and pumps. Problems and restrictions arise during these transitions, and the engineer tries to find optimum solutions to the problems of delivering a product or service to the marketplace. The procedures used to find optimum solutions while solving these problems make up engineering design. Optimum does not mean the utopian best, only the best within the restrictions placed on the problem environment. Optimization techniques will be covered in detail in a later chapter. Engineers are also responsible, either directly or indirectly, for converting natural resources to usable form. Engineers may not dig the iron ore from the ground, but by specifying 1020 HR on a drawing, they are requiring that the digging be done.

2.1 Primary Job

The engineer's primary job is to solve problems relating to the marketing of a product: its creation, manufacture, distribution, safe use, and disposal. The product can be a physical object such as a building or a diesel locomotive, or it can be a system such as a scheduling procedure for using the locomotives, cars, and track of a train network. In the process of getting a product to market, an engineer may be responsible for a product's original design, the proverbial idea person. It is more likely, however, that day-to-day work will be related to many other activities required to get the product to the customer, such as the following:

1. Determine sales forecasts using marketing studies.
2. Determine product specifications as a result of market studies, consumer requests, and competitor product analysis.
3. Determine product cost, required capital, project profitability, and return on investment.
4. Determine project plans including personnel requirements and time schedules.
5. Establish acceptable levels of reliability, maintenance, and repair criteria.
6. Test products to verify levels of performance and reliability.
7. Determine processes and procedures to manufacture the product in a consistent and economical way.
8. Design the machines and processes needed to manufacture the product.
9. Design the systems that will use the product.
10. Write the product maintenance and repair manuals, or conduct the service training school sessions for product dealers and their personnel.
11. Assist with promotion plans and procedures for timely market introduction.
12. Be involved with 101 other necessary details to make a good product and a successful company.

College courses cannot include all the details of the necessary activities to carry out an assignment. Be prepared to learn new things when you begin nonacademic engineering problem solving in business and industry. Learning will include working with practicing engineers, reading current literature, attending seminars and other training programs, taking advanced classes in your specialty, taking classes in new technology, and trying out your own ideas. The designs of the engineer must keep pace with the expansion of technology and the state of the art.

As you read this book, keep in mind that lists, such as the one above, will not necessarily be complete; indeed, I hope you will be able to add to all lists that appear. Secondly, the order of items in a list does not imply an order of importance. The most important items depend on the situation in which you find yourself.

2.2 Necessary Skills

Engineers need many skills and talents to get the required jobs done. Among the talents engineers need are:

1. The ability to express problems clearly, both orally and on paper, using written and graphical representations. Communicating with others is essential, both in the activity of collecting information to solve a problem, and in the dissemination of results. See Section 2.8 for additional discussion on communications.

2. The experience and knowledge to propose ideas for the solution of problems. The range and possible success of ideas is enhanced as your experience and knowledge grow, because you will have more information to apply to the problem.
3. The knowledge and ability to analyze possible solutions for their true merit. This knowledge must include the ability to model problems using mathematics, to test ideas experimentally, to draw proper conclusions from the results, and not be influenced by the prejudice of others.
4. The ability to search out information, study it, and apply it to a new situation. Problem solving is made more efficient, and is directly proportional to the ability to use data sources efficiently.
5. The ability to project ideas from what is known to what is not known. Projecting ideas and state of the art to new applications requires creative thought, one of the most difficult things to learn or teach. See Section 3.2 for additional discussion on creativity.

Some people have a preconceived notion that all engineering work is directed toward the generation of inventions and new products. That is not true, and in general more of the engineer's time and effort is spent making an existing idea work as originally intended than developing new products. The primary job of a research engineer is to develop working prototypes of a problem solution. The primary job of a development engineer is to convert a working prototype of a problem solution into a market-ready, mass producible product. The work of any engineer involves trying new approaches, using old approaches on new problems, testing an idea to see how it works, making changes, trying again, trying a different approach, starting over, revising, reiterating, retesting, and compromising. Only then can a final decision be made which is the best under the given conditions and constraints.

2.3 Engineering Developments And Advancements

Engineers and scientists have built a vast storehouse of data and knowledge to deal with design problems. This knowledge consists of such items as:

1. Theories of the physical universe, many of which have been verified by experiments and observations. Refer to Section 5.2 for a summary of some of the most commonly used physical theories of the physical universe.
2. The data collected from experiments and observations that may lead to new theories and more experimental tests. The appendices of this book contain often used data derived from experiments and theory.
3. The mathematical models that are used to quantify and qualify the results of experiments and observations. Mathematical models are discussed in a later chapter, and are used extensively for quantitative manipulations.
4. The procedures used to control the behavior of products, processes, and people for the desired results. Efficient procedures are necessary to capitalize on theory, data, and models.

5. Rules of thumb, preliminary estimating procedures, orders of magnitude of estimating, and heuristics. A heuristic aids to the direction of solution, but it is not necessarily justifiable.

Engineers are not often thought of as experimenters and while some people think of scientists only in this way, both groups do experiments, however with different emphasis. Engineers are concerned with solving a problem that is of immediate concern, or at least is a known concern. A scientist may be working on an experiment and may not be involved with a current consumer problem. This difference causes an engineer to be under more pressure. Someone—the public, the government, or an employer—is waiting impatiently for a solution.

Engineers and scientists do share concerns. Both are dependent upon information and data that have preceded them. Developments are made, changes are implemented, and the wealth of knowledge expands. What is known today is more than yesterday and less than tomorrow. As knowledge expands, specialities are developed. Engineers specialize to keep up with the knowledge expansion. Civil engineering may be the oldest field of engineering because the first problems to be solved on a need basis were related to water supply, land division, sewage, and roads; today however, nuclear engineering and electrical engineering are disciplines in themselves because the need for these specialties has grown as technology has advanced and consumer needs have expanded. As knowledge and data accumulate, it becomes more apparent that divisions of effort are necessary. No one can remember all the data and knowledge available, so engineering becomes more and more a team effort.

The development of products, processes, and procedures follows the expansion of knowledge. The first self-propelled automobiles were similar in design to the horse-drawn carriages of the time. No one was concerned about suspension, tires, exhaust fumes, or driving in the snow. Nevertheless developments came, sometimes slowly, sometimes dangerously, but always as a result of trying to satisfy the needs of the consumer, or at least the needs the manufacturer thought the consumer had.

During the evolution of the automobile, for example, many developments occurred. Among these were progress in the development of metals and their alloys, plastics, electronics, rubber, glass, manufacturing processes, petroleum refining, paints, fabrics, heat transfer, storage batteries, safety, noise control, pollution control, road construction, bridge construction, and traffic control devices. Figures 2.1, 2.2, and 2.3 are samples illustrating some of the most obvious changes that have occurred in automobile design.

Developments that occur in one industry, such as the automobile industry, carry over to other industries. Like the automotive industry, the aviation industry is concerned with materials; they both need materials that improve their product's performance. The aircraft industry was delighted when a manufacturing process was improved to reduce the thickness tolerance of the airplane skin material to a smaller value. The smaller tolerance made possible planes with a higher guaranteed payload because engineers could calculate more accurately how much the plane would weigh. The automotive and aviation industries both make use of the high octane fuels from the petroleum refining process. Petroleum refining gives products other than gasoline, so other petroleum-based research and products, such as polymers and medicines, were developed. The successful nine-day global nonstop flight of the airplane Voyager in December, 1986 illustrates a technological advance in the use of composite polymer materials.

Many of the advances of modern technology are interrelated. When more iron is

FIGURE 2.1

Early vintage Cadillac.

FIGURE 2.2

1955 Ford.

FIGURE 2.3

1986 Mercedes.

required for ships, cars, bridges, and buildings, steel mills need equipment to produce more steel in less time, and to produce new alloys and new shapes. Mines need machinery to extract more ore in a shorter period of time. The ships that carry the ore to the mills need to be larger and faster. Figures 2.4, 2.5, 2.6, and 2.7 show some of the equipment that has been designed and built as a result of the increased need for iron and steel. Advances bring about needs, needs bring about problems, and problems challenge engineers. The problem range is almost limitless, from product development to efficient production methods.

    The engineer spends time developing products, making them easier to use, stronger, more efficient, less expensive, safer, more reliable; or tries to adapt a product to a different application or environment. Consider how automobiles, airplanes, ships, buildings, and bridges have evolved since their first conception. Engineers are always trying to make things better. They push materials, processes, and procedures to the limit, often learning the hard way that a limit has been reached and that a different approach is needed. When the speed of airplanes exceeded the speed of sound, new forces, reactions, and design solutions had to be considered.

Engineering Activities

FIGURE 2.4

Equipment for underground coal mining.

Sometimes theory comes before application, while in other cases a physical phenomenon can occur and be used long before theories are proposed as to why it works. Electricity is an example of using a phenomenon for practical uses without "seeing" what is happening. When Edison worked on the design of an incandescent light bulb and the electric dynamo, he made improvements with limited knowledge of theory. He just knew his design was getting better, because he compared input and output of previous models to new designs. Although he could not see electricity, he designed to control the effects of it efficiently.

In metallurgy, it is known that when hot steel is cooled rapidly it is harder and can be manufactured into an edge that holds its sharpness. In ancient metallurgy practice, however, this hardening was thought to be a special event that was only successful when a hot sword was thrust through a slave; not a very humane process, but one that worked. Today we know why rapid cooling is effective in hardening steel, and we have developed efficient and practical methods of performing the process.

Product development often follows interesting paths. A major change in chain saw blade design came in the early 1940s when Joe Cox observed the way a timber beetle, *Ergates Spiculatus*, ate through wood. The blade improvement made a chain saw more efficient, and other improvements soon followed. Saw weight, for example, was reduced from 72 to 16 pounds for an equivalent machine between 1938 and 1966. Other

FIGURE 2.5

Equipment for open pit mining.

modifications were also made, such as spark arrestors to prevent brush fires, and design improvements to reduce vibration, noise, and exhaust fumes.

As engineers make technological changes, they must be alert to the side effects of the changes. For example, nonstick surfaces make cleaning of cooking utensils easy, but rob people of a source of iron in their diet. Cast iron cookware supplies trace amounts of iron to the food prepared in them. Today many people, women in particular, suffer from iron deficiencies and need to take iron supplements because they no longer use hard-to-clean cast iron cookware. The next example illustrates how complex project side effects can be.

FIGURE 2.6

Equipment to transport ore to refineries.

EXAMPLE 2.1:

Aswan Dam.

Side effects of the Aswan Dam, built in Egypt at a cost of $1 billion to control the Nile River, may have been overlooked or ignored. The primary purpose of the dam was to provide electric power and water for irrigation purposes. Side effects, which became apparent after the dam was built and placed in operation in 1964, include the following:

1. The dam prevents the annual flooding of the farm lands beside the Nile, which had brought nutrients to the soil, kept ground water levels high, and flushed salts from the soil. The farmers now have to purchase and add fertilizer to maintain the same crop level at a cost of $100 million per year, and the increased salt level reduces the overall productivity of the soil.

2. The dam prevents the nutrient laden silt from reaching the Mediterranean Sea, where it had supplied food for sardines, mackerel, shrimp, and lobster. The harvesting of these "crops" has about vanished from the area.

Engineering Developments And Advancements   *33*

FIGURE 2.7

Equipment required to process the ore.

3. The clear water that now flows is washing out the river's bed and bridges downstream from the dam. Solutions to the washing out problem include the proposal to build additional dams between the Aswan Dam and the Mediterranean Sea at an estimated cost of $250 million.

4. The lack of silt is causing the delta areas to be eroded by the sea, and farm land is being lost.

5. The higher water levels behind the dam were coming close to washing out the 3000 year old Ramses Temple at Abu Simbel, and the temple was dismantled and moved to save it.

6. The lake and irrigation distribution of water has improved the habitat of snails that carry schistosomiasis, a disease that affects the strength of humans. The disease has spread, and there is no known cure.

7. It was projected that the dam would be full by 1970, but it was only half full in 1985 because of underground seepage and surface evaporation.

Engineers must be aware that the wealth of information available to them cannot be held in constant readiness to be applied the instant it is required. It does not take too long to become aware there is more to know than one person can keep up with; that's why specialties are developed. It is also apparent that good reference sources, such as those listed in Section 4.2, are necessary. Each area of engineering has its own emphasis, its own jargon, its own realm of expertise. Regardless of the area of engineering that becomes your specialty, the primary objective of your work will be to work with others to solve problems.

An engineering problem is often the direct result of trying to adapt to a new situation. Consider the Liberty ships that had been used satisfactorily in some waters, but began breaking up in the North Atlantic during World War II. In this case an existing design solution, a ship that carried men and equipment, had performed satisfactorily; but the new need, to move men and equipment in the cold North Atlantic sea, required a modified design. What was learned from this situation led to the specialty area of fracture mechanics, the study of cracks in metals and of crack growth under load-time-temperature variations. Engineers now know that cracks can grow at low temperatures, and at stresses below ultimate or yield strength levels.

Another problem that arose due to technological advances is the effect of neutron radiation on the materials used in nuclear power plants. Neutron radiation increases the

FIGURE 2.8

Truss-designed Queensboro Bridge, New York City.

temperature at which a metal changes its characteristics from ductile to brittle. Until nuclear reactors were used regularly, the rates of reaction to neutrons and the limits of change were not known.

The aerodynamics of bridges was not considered a design analysis criterion until after the Tacoma Narrows Suspension Bridge failed due to wind-induced oscillations. Bridges built before this time had been designed to withstand aerodynamic effects, but designers were not aware they had accounted for them. The Golden Gate Bridge, for example, was built on design principles similar to those of the Tacoma Bridge, but it had a different depth/width/length ratio and was safe from aerodynamic effects, something the engineers weren't aware of at the time of design. The variety of bridge designs shown in Figures 2.8 through 2.12 illustrates that no one design is always the best, and that good designs must be adaptable.

The development of automobile springs is a case where the best attempts at designing a more reliable spring, even with more expensive steels, did not solve the problem. When engineers decided to revert to less expensive steel, they discovered that the less expensive steels made the springs last longer. Later they learned that the lower grade steels were not susceptible to hydrogen embrittlement. At the time it was just not known what the operating characteristics and reasons for failure were, just that something else worked better. Today we know of other improvements to make springs more

FIGURE 2.9

Draw bridge, St. Joseph, Michigan.

FIGURE 2.10

Arch bridge in Salzburg, Austria.

reliable, such as shot peening the surface which was introduced to commercial production by the Associated Spring Corporation in 1929.

Additions to the information available to engineers may come as slow developments and improvements to existing products, or as breakthroughs in technology. Transistors, for example, made vacuum tubes obsolete, and are one of the biggest contributors to the practicality of small computers. Computers themselves made obso-

FIGURE 2.11

Walt Whitman suspension bridge, Philadelphia, Pennsylvainia.

lete such items as the slide rule and the mechanical calculator. Internal combustion engines made the horse-drawn carriage and buggy whips obsolete, but helped make engine-powered airplanes a possibility. Engines helped advance the conversion of energy forms including the capacity to generate more electricity. Electricity gave us the potential to design the telephone, telegraph, radio, television, and radar. Many technological advances are interrelated this way.

Some engineering designs are less sensational, but still important for our way of life. The design of a fire escape procedure from a skyscraper may never be put to the test, but if the procedure is not designed correctly, a fire can cause the loss of life, and may result in a lawsuit against the design engineer. The life boat shown in Figure 2.13 may never be used, but if needed, the launch system must function as intended. There won't be a second chance.

Designs used today are the result of many trials and failures. Edison tried more than 30 different materials, and in many different forms, in his search for an appropriate filament for the incandescent light bulb before he succeeded in finding one that lasted long enough for practical use. He proceeded with tests by trying alternative forms of the material, different filament geometries, and alternate filament support systems, always trying to improve the efficiency, output, and life of the light bulb.

Sometimes, failure is the result of not having enough knowledge, not making use of existing knowledge, or not having the right equipment. An early experiment performed to determine the speed of light is an example. Two men with lanterns stood on facing hills. The first man uncovered his lantern and started timing. When the man on the other hill saw the light, he uncovered his lantern. The first man stopped timing when he saw the light from the second man's lantern. By using the recorded time and the distance between hills, they thought the speed of light could be calculated. As we are now aware, this experiment was doomed to failure because the apparatus and technique were not capable of doing the job.

FIGURE 2.12
Arch truss bridge, Cape Cod Canal, Massachusetts.

2.4 Responsibilities

As engineers design and problems are solved, responsibilities are undertaken and engineers are involved in a wide variety of tasks. Some engineers work specifically on new product development, which arises when the consumer or a company recognizes a need that is not being met. New products range from a supersonic airplane to an

FIGURE 2.13

Life Boat Launch System.

automatic door opener for a building. The product can be in the public eye, such as the space shuttle, or as inconspicuous as a new coin sorter used at a toll booth.

Engineers probably spend more time improving existing products than doing any other design job, and many reasons account for this. A product may fail before its time; it may not perform as intended; or sales may be dropping due to competition. An improvement may be required to reduce cost, or a purchased part or material may no longer be available so a change in design becomes necessary.

Product modifications may be needed when a larger model or an adaptation of a product to a different environment is required. For example, an existing crane may lift 50,000 pounds (22,680 kilograms) to a height of 100 feet (30.5 meters) and someone wants to lift 70,000 pounds (31,750 kilograms) to a height of 130 feet (39.6 meters). This situation may be solved with an adaptation of a current product, or a new product may be required. The decision depends on whether an existing design can be modified and extended to meet the new objective, or whether a totally new concept is required. If a new concept is required to solve the problem, then the task belongs in the new product category. If changes in material, material size, or part configuration can be made to meet

Engineering Activities

FIGURE 2.14

Computer input tablet and specifications.

the new conditions, then a product modification or update is appropriate. The advantage of redesigning a current product is that existing parts, components, and design theory can be used. If a new product is necessary, additional time may be required to develop a new technology, capital expenditures may be required for new processes and facilities, and reliability and safety may become questionable.

Before serious design work is started, the design problem must be defined and specifications goals established. Product specifications may include requirements in any category including mechanical, chemical or electrical. A working capacity specification might be: dig 10 feet deep, lift 1000 pounds, travel at 10 miles per hour, withstand 2000 psi of internal pressure, process 1000 transactions an hour, operate at 120° C, consume

model 2207
SPECIFICATIONS

Physical Data:
Model/Size — Dimensions
2207-1212 — 15⅞" x 16⅞" x ½"
(3030) — (40.3$_{cm}$ x 42.9$_{cm}$ x 1.3$_{cm}$)

Weight — Active Area
6 lbs — 12" x 12"
(2.75 kg) — (30.5$_{cm}$ x 30.5$_{cm}$)

Interface:
RS232C—25 pin cable attached
Baud Rate—9600

Operating Temperature:
50-104 degrees F. (10-40 degrees C.)

Storage Temperature:
14-131 degrees F. (−10-55 degrees C.)

Humidity:
0-90% (non-condensing)

Resolution:
0.001" (0.025mm)

Accuracy:
± 0.02" (0.5mm)

Output Rate:
Up to 160 points/sec. HOST selectable
(rate dependent on options selected)

Standard Accessories:
Pen stylus and 4-button cursor
 (same connector)
Power supply
RS232C cable (integral)
Tablet riser
5¼" floppy disk (3½" for Mac)
 —mouse driver (Microsoft)
 —Windows driver
 —tablet set-up utility
 —diagnostics
 —instruction manual text

Optional Accessories:
Menu hold down

Output Formats:
Host selectable
x(n) and y(n) represent coordinate bits for each respective axis.
pB(n) represents the condition of each respective cursor switch (n).
Keys depressed = 1 and Keys released = 0

SummaSketch® and Bit Pad® are registered trademarks of Summagraphics Corporation.

SummaSketch MM Series, Absolute Coordinates, Packed Binary, 8 bits

CHAR. #	7	6	5	4	3	2	1	0
C₁	1	Px	T	Sx	Sy	pB4	pB2	pB1
C₂	0	x6	x5	x4	x3	x2	x1	x0
C₃	0	x13	x12	x11	x10	x9	x8	x7
C₄	0	y6	y5	y4	y3	y2	y1	y0
C₅	0	y13	y12	y11	y10	y9	y8	y7

Px = Proximity (0 = in proximity, 1 = out of proximity)
T = Tablet identifier, either 0 or 1 (for dual tablet configurations)
Sy, Sx = 1
pB3 sets flags pB1 and pB2 simultaneously

SummaSketch MM Series, Relative Coordinates, Packed Binary, 8 bits

CHAR. #	7	6	5	4	3	2	1	0
C₁	1	Px	T	Sx	Sy	pB4	pB2	pB1
C₂	0	x6	x5	x4	x3	x2	x1	x0
C₃	0	y6	y5	y4	y3	y2	y1	y0

Px = Proximity (0 = in proximity, 1 = out of proximity)
T = Tablet identifier, either 0 or 1 (for dual tablet configurations)
Sy, Sx Coordinate signs (1 = positive, 0 = negative) pB3 sets flags pB1 and pB2 simultaneously

Numonics Standard, Packed Binary, 8 bits

CHAR. #	7	6	5	4	3	2	1	0
C₁	1	pBA	pB4	pB3	pB2	pB1	x15	x14
C₂	0	x13	x12	x11	x10	x9	x8	x7
C₃	0	x6	x5	x4	x3	x2	x1	x0
C₄	0	0	0	0	0	0	y15	y14
C₅	0	y13	y12	y11	y10	y9	y8	y7
C₆	0	y6	y5	y4	y3	y2	y1	y0

(pBA = 1 when pB4, pB3, pB2 or pB1 = 1)

Bit Pad I, Packed Binary, 7 bits

CHAR. #	6	5	4	3	2	1	0
C₁	1	pB4	pB3	pB2	pB1	0	0
C₂	0	x5	x4	x3	x2	x1	x0
C₃	0	x11	x10	x9	x8	x7	x6
C₄	0	y5	y4	y3	y2	y1	y0
C₅	0	y11	y10	y9	y8	y7	y6

Bit Pad I, ASCII, 7 bits
 XXXX,YYYY,FCL
X = The X Coordinate BCD ASCII digits.
Y = The Y Coordinate BCD ASCII digits.
F = Indicates which key(s) are pressed.
C = Carriage Return
L = Line Feed

Default Start-up:
Summagraphics, SummaSketch (MM series, Absolute Coordinates, Packed Binary 8 bits, odd parity)
switch stream mode
500 ppi resolution
9600 baud

Power:
External: 115 VAC 60 Hz; 20 Watt max

Internal:
+12 VDC, 100 MA; −12 VDC, 100 MA; +5 VDC, 50 MA; (± 10%)

NUMONICS

101 COMMERCE DRIVE, MONTGOMERYVILLE, PA 18936. (215)362-2766. TWX 510-661-6585

21-79
0488

---

less than 5 kilowatts of power. Or they may involve physical capacities, a maximum or minimum weight or dimensions, color, and geometric shape. Specifications may change as a product is developed, and the final specifications may not be the same as the original. Specification goals are important however, so that development effort is focused in the right direction. Figure 2.14 is an example of a product and its final specification.

Some designs end up with "frill" specifications that do not contribute to the function of the product. Does a faucet have to be gold plated to function? No, but the contact points on an electronic component board might have to be gold plated so they won't corrode in use. Never confuse a well-designed product with an over-featured product. A Cadillac gets a driver to a destination as does a bicycle. The Cadillac just has more features, not the least of which is a roof to keep the occupants dry when it rains. Does a car need a computer to determine an estimated time of arrival, or is it just a nice feature?

Engineers spend much of their time determining part function, ranging from press fit assemblies to wind fit clearances. These decisions in turn determine part tolerances, manufacturing processes, and subsequently the cost. Deciding the function of a part so a product performs properly requires a combination of functional design tolerances, experience, and testing. An inexperienced engineer may specify small tolerances which are safe but expensive, while an experienced engineer will specify larger tolerances, adequate to make a part function, but at a lower cost.

A design cannot be manufactured unless drawings of the parts are made, as illustrated in Figure 2.15. The drawing includes sizes, tolerances, the material to be used, and other data that the engineer feels is necessary to insure a functional part.

The details on part drawings determine how parts are manufactured. Parts should be designed so they can be made with the least expensive processes available. The cost of tooling, gages, and machine tools as well as the amount of time to manufacture can be reduced with thoughtful component design. Parts are put together in assemblies and

FIGURE 2.15

Production part drawing.

Courtesy of Deere & Company.

documented using a product bill of material. A bill of material includes not only the manufactured parts, but also the purchased parts such as nuts and bolts, nails, glue, washers, paint, and special packaging. Part names and required quantities are included. The bill of material is used for assembly purposes as well as for determining final product cost, and often includes assembly instructions. Figure 2.16 is an example of a subassembly bill of material.

Engineers also are responsible for the life of a product. Establishing life limits of a product is not done on an independent basis; input from the customer, edict from management, concern for the safety of the consumer, and legal restrictions must all be considered. Should the part last a week, a year, 100 years, 1000 cycles, 1,000,000 cycles? These are decisions that must be made, because product design is based on these decisions, and the decisions affect cost.

Engineering decisions affect product cost. The cost of a product includes the cost of the material from which the parts are made, the cost of the manufacturing time required to make them, and the cost of supporting facilities. Information needed to determine cost includes process times; required process tooling (dies, jigs, and fixtures), determined by manufacturing engineers; wage and overhead rates, determined by the accounting

FIGURE 2.16

Sample Bill of Material, Courtesy of Deere & Co.

```
AT 107691 SLIDING FRAME INSTALLATION
    Quantity    Component Part Number
    1           AT 114649    Frame (Sliding with pistons)
    1           AT 106584    Sliding Frame with bushings
    4           R 81195      Ring, Back-up
    4           R 83469      Packing, O-Ring
    4           T 113162     Piston
    0.001       UN 6012      Gals of oil, JDM J20A Transmission: Apply to O.D. of
                             locking cylinder pistons before installing into sliding
                             frame.
    2           JD 7806      Fitting, Lubrication, 90 Degree
    1           T 106066     Bar (Sliding Frame Locking Bar)
    2           T 106398     Shim
    2           T 106399     Shim
    0.010       UN 5826      Lbs of JDM J13E3 Lubricating grease: Grease area of
                             sliding frame which contacts lower rail of backhoe frame.
    0.010       UN 6012      Gals of oil, JDM J20A Transmission: Apply to O.D.
                             of locking cylinder pistons.
    4           19M7369      Hex Screw
    4           24M7241      Washer, Hardened
```

Position sliding frame onto backhoe frame making sure it seats completely down on either the top rail or bottom rail. Place lock bar into position. Install as many shims as possible under the lock bar (need not have equal amounts on each side). Now remove 2 mm worth of shims. Note: Thick shims are 3 mm and thin shims are 1 mm thick. Install cap screws with washers and tighten.

department; shipping costs; required return on investment, determined by first level management; and administrative costs.

An engineer is also responsible for specifying service requirements necessary after the sale. Regular service and recommended maintenance of a product are important so that it will function and last as long as intended. The preparation of instructions and requirements for service activities is as important as any other step in designing a product for the customer. Good service is needed to back up the sale so that the customer remains satisfied. The original design should always take into consideration how a product will be serviced and repaired. There is no value in placing a grease fitting on a moving joint if a maintenance person cannot get to it. Periodic adjustments that are hard to make probably won't be made. Clearly specifying maintenance procedure in owner's manuals is also a legal requirement meant to protect consumers from injury caused by a malfunctioning product.

Engineers must be concerned with safe and efficient ways to get a product to the customer in usable condition, as well as safe use by the consumer. Sharp edges and corners must be avoided, and glass and other fragile parts must be packaged safely and protected during shipment. Surfaces that could rust while being stored must be protected. Openings must be protected from the weather if outside storage is required. Special packaging is required when hoisting or forklift equipment is used for moving the product. If the limit on width of railroad or highway shipping is 12 feet, then the product must be designed within that limit. Some products are designed so they can be easily disassembled for shipment and reassembled upon delivery. Many products cost more to package, protect, and ship than to manufacture.

Safe use requires that the consumer be protected from potential dangers. In the age range of 1 to 44 years, more people are injured or die from accidents than from any other cause. Engineers cannot prevent all accidents from happening, but products must be designed to prevent normal-use accidents from happening. Consumers must be shielded and protected from very hot or cold objects, radioactivity, electric shock, chemicals, high sound levels, and excess vibration. Products must be designed to reduce the level of undesirable side effects. Ample and clear warning labels and instructions are a must.

2.5 Day-To-Day Concerns

Regardless of the role an engineer plays in getting a product from the conception stage to the customer, there are day-to-day concerns that are common to all engineering activities.

An engineer must be willing and able to accept an assigned responsibility, whether working for someone else, or self-employed. An engineer has to accept the challenge of the problem, make decisions, and accept the consequences.

Engineers must advance themselves and the company they work for by delving into areas in which they are not expert, as well as by probing deeper into their own area of expertise. Without taking some risk you and your employer can stagnate. As knowledge grows and new technology is developed, so must you grow and develop.

Engineers are as certain to make mistakes as the sun is to rise in the morning. In this respect engineers are no different from anyone else. To minimize mistakes engineers must check their work, test the possibilities, and check again, but mistakes will happen. An error may result in a catastrophic event, such as the failure of the Tacoma Narrows Bridge; or be as embarrassing as the wash tank that was drawn to half size, and then dimensioned and built as though drawn to full size. Two things are worse than making a mistake. From a personal development standpoint, worse than making a mistake is not learning from it. From a professional and ethical standpoint, worse than making a mistake is knowing it was made and not trying to correct it.

Trying to satisfy other people's needs and desires is the main source of engineering problems. Don't get trapped into thinking you are the only person a design has to satisfy. There isn't much profit to be made if the only customer of a product is the engineer who designed it.

All engineering projects are limited by the amount of money, time, material, or people that are available. A good engineer must make efficient use of available resources. Having resources left over at the end of a project is a sign of good management. Even if an overall budget is approved, it doesn't give the engineer a blank check for all the money so budgeted. Proper ethical decisions must be made. Each expenditure must be justified as supportive of the project objectives.

Engineering ideas need to be tested before the consumer has the opportunity to purchase a product. Testing may show up a problem that was not accounted for during the theoretical design of a product. Testing of a complete product or just a portion of a product may be appropriate. Frequently, testing will be used for design analysis because no combination of theories can be used to estimate an answer; the total design will be developed experimentally.

Engineers are also involved in training people who will use a new product; it may be different enough to require special instruction, or require the learning of new skills. Engineers may also have to train the people who will run product tests and experiments, or who are responsible for assembling a new product.

Engineers make decisions. Decisions are made on such details as material, material form, fabrication methods, and system controls. Final designs rarely satisfy everyone, and tradeoffs (compromises) among such items as cost, function, appearance, reliability, serviceability, and maintenance are the rule. Balance of goals is needed, and good engineers make judgments that lead to optimum, but user safe, solutions. Many designs have to be reviewed, tested, redesigned, and retested several times before satisfactory results are obtained. Don't be frustrated when an activity must be redone; knowledge and progress are taking place and the product is getting better.

All projects require the proper use and scheduling of people, machines, and money. Personnel support must be efficiently used or a project will fall behind schedule. Machines, whether they are testing apparatus or being built to manufacture the parts, must be coordinated so tests are run on time and production parts are manufactured on time. Money must be used efficiently, for if it runs out before the project is complete, an out-of-budget condition will exist, and a good project may be canceled.

Engineers also analyze project proposals so that resources are properly directed. Analysis includes activities that supply data so intelligent decisions can be made. This includes comparing designs, checking test data, running computer simulations, studying company warranty records, calculating critical sizes, and checking modes of possible failure.

Engineering projects and proposals must be communicated to others. Communication is one of the most used and most neglected skills an engineer has. Engineers must communicate to complete a project efficiently. How else can the details of what is needed be known, and how can a response to a design proposal be considered without good communication? Good communication is necessary as an engineer collects data and ideas; reports status; prepares work orders; reports test results; prepares drawings, specifications, and assembly procedures; or sells a completed project. See Section 2.9 for more discussion on communications.

2.6 | Constraints and Limited Resources

Design optimization requires finding a compromise that is the best solution within the constraints of the problem. Constraints include anything that in some way controls the outcome of a project including time, money, personnel, facilities, and laws. Some of the constraints imposed on engineering projects are within the engineer's control, others are not.

A schedule of expected events by dates or amount of **time** is almost always a constraint for an engineering project, and usually outside the engineer's control. For example:

1. Build a bridge before the tourist season starts.
2. Build a boat before the ice goes out of the river.
3. Overhaul the boiler system before the winter heating season starts.
4. Finish remodeling the student union building before fall semester begins.

The following example illustrates problem solving under the restraint of time.

EXAMPLE 2.2:
Time constraint.

How many Ping-Pong balls will fit into the room you are in? You can probably think of several methods of obtaining an answer. Now add the restriction that an answer must be given in 30 seconds. Perhaps none of the ideas you have will work in 30 seconds and you will simply have to guess. Any guess seems appropriate under the severe restriction of a 30-second time limit. An answer might be: Greater than 1000, 10 million, or some other random guess. Unless you have remarkable powers, a 30-second answer has to be a guess.

Now suppose 30 minutes are available to determine an answer. In 30 minutes you can measure the room, calculate its volume, and divide by the estimated volume of a Ping-Pong ball. A Ping-Pong room might be close by so you can measure the diameter of the ball to calculate the volume more closely. But the answer is still an estimate, although probably a better one than the 30-second answer, for you still don't know for sure how many Ping-Pong balls will pack into the room.

Suppose 30 days are available to obtain an answer. In 30 days you might buy enough Ping-Pong balls to fill the room, and then count them. You might also be creative and just do part of the room and calculate a total answer by proportion. Either way your answer is more accurate than in the two previous methods.

A **budget** is a necessity for the control of a project. A budget lists the cost estimates necessary to get a job done, and projects a profit based on estimated sales and an estimated sales price. After a company agrees to a proposal, the amount of money committed to it must be controlled so funds are adequate to complete the project. If not, a financial loss can result, and the company may be forced out of business. The solution to the Ping-Pong problem may be restricted by a budget constraint if there is not enough money to buy enough Ping-Pong balls to fill the room.

Another limitation is **support staff.** A project is generally assigned to a limited number of engineers and supporting staff. No project can afford the luxury of unlimited staff.

An engineer is often told what to design based on **consumer needs** and abilities. These needs come from current users of your product, or through market surveys. A product that will be used by an average person must be designed within the limits of human size, reach, and strength.

Available materials are always a constraint within which you must work. The combination of desired material properties and what is commercially available, generally forces a compromise between such factors as weight, strength, corrosion resistance, electrical conductivity, machinability, cost, and delivery dates. This restriction also includes standard purchased parts such as nuts, bolts, springs, keys, light bulbs, and pipe.

Available manufacturing facilities are as critical as available material. If your company only has casting facilities for sand casting iron, and you design an aluminum part to be die cast, a compromise will be necessary. The part needs to be redesigned, purchased from another company, or new casting equipment must be purchased. Purchasing a part rather than manufacturing it may add lead time to the project or idle existing capital equipment.

Products must be designed to be used in **different environments**. A product designed to be used in the salt water of the ocean might be designed differently than a product used on dry land. A structure built in the earthquake zone of California might be designed and built differently than one in Chicago.

Another restriction that engineers must take into account are **laws, codes, and government regulations.** There are many laws that place restrictions on designs. The following laws, codes, and commission regulations are a sample of sources of constraints that may affect your design work. The example following this list illustrates the details that codes force an engineer to consider.

- Local building codes
- Federal Trade Commission
- Flammable Fabrics Act
- Refrigeration Safety Act
- Federal Boat Safety Act
- Highway Safety Act
- Toxic Substances Control Act
- Occupational Safety And Health Act
- Federal Housing Authority regulations
- National Traffic And Motor Safety Act
- Federal Food, Drug, And Cosmetic Act
- Federal Coal Mine Health And Safety Act

- Local fire regulations
- Department Of Labour
- Poison Prevention Packaging Act
- Magnuson-Moss Warranty Act
- Department Of Transportation
- Federal Railroad Safety Act
- Federal Hazardous Substance Act
- Environmental Protection Agency regulations
- Consumer Product Safety Commission
- Radiation Control For Health And Safety Act
- Federal Metal And Nonmetal Mine Safety Act

EXAMPLE 2.3:

Code restrictions.

An existing city building code contains a section on the construction of parking lots that includes the following restrictions and requirements.

1. No building permit or occupancy permit shall be issued, and no construction, grading, or other land development activity listed below may be commenced on property unless a Site Plan has been submitted and approved for such activity as set forth in this code.
2. The Site Plan shall include one or more appropriately scaled maps or drawings of the property clearly and accurately indicating the following:
   a) Complete property dimensions;
   b) The location, grade, and dimensions of all present and/or proposed streets or other paved surfaces and engineering cross-sections of proposed curbs and pavement;
   c) Complete parking and traffic circulation plan, if applicable, showing location and dimensions of parking stalls, dividers, planters or similar permanent improvements; perimeter screening treatment, including landscaping;
   d) Location and full dimensions of all buildings or major structure, both proposed and existing, showing exterior dimensions, number and area of floors, location, number and type of dwelling units, height of buildings;
   e) Existing and proposed contours of the property taken at regular contour intervals not to exceed five feet, or two feet if the City Planner determines that greater contour detail is necessary to satisfactorily make the determinations required by this ordinance;
   f) The general nature, location, and size of all significant existing natural features, including but not limited to sidewalks or paths, tree or bush masses, all individual trees over four inches in diameter, grassed areas, surface rock and/or soil features, and all springs, streams, or other permanent or temporary bodies of water;

  g) A locational map or other drawing at appropriate scale showing the general location and relation of the property to surrounding areas, including where relevant, the zoning and land use pattern of adjacent properties, the existing street system in the area and location of nearby public facilities.

3. All Site Plans shall include a report or narrative containing the following:
    a) Legal description and address of the property.
    b) Name, address, and phone number(s) of the property owner(s).
    c) Name, address, and phone number(s) of the developer(s) or contractor(s), if different that the owner(s).
    d) Proposed use(s) for all non-residential buildings or structures.
    e) Data clearly identifying the following: total number and type(s) of dwelling units on the property; number and type of all structures or buildings, whether residential or non-residential; total area of the property; number of dwelling units per acre; total floor area of each building.
    f) Proposed landscaping schedule indicating plant types, number and timing for installation.
    g) Proposed construction schedule of all structures and physical improvements indicating the timing and sequence of each major structure and improvement.
    h) Present zoning classification(s) of the property.
    i) Present and proposed type and number of parking spaces on the property.
4. The Site Plan must show that a reasonable effort has been made to conserve and protect those natural characteristics that are of some lasting benefit to the site, its environs and the community at large.
5. Slopes which exceed ten percent shall be protected by appropriate measures against erosion, run-off, unstable soil, trees and rock. Measures shall be taken to stabilize the land surface from unnecessary disruption. The Soil Conservation Service shall be consulted for soil erosion control practices and allowable soil loss as permitted by the State of Iowa. Said erosion control and soil loss limitations shall be the responsibility of the property owner.
6. The placement of buildings, structures, fences, lighting and fixtures on each site shall not interfere with traffic circulation, safety, appropriate use and enjoyment of adjacent properties.
7. Adequate illumination shall be provided to parking lots, sidewalks and other areas for vehicular and pedestrians circulation. In no case shall illuminating devices be placed above fifteen feet in height in a residential district.
8. All parking spaces shall be clearly marked in accordance with Parking Lot Design Standards. Signs and pavement markings shall be used as appropriate to control traffic access and egress.
9. All areas designed for vehicular use shall be paved with a minimum of either an eight inch rolled stone base and two and one-half inch asphaltic concrete mat, a six inch Portland cement concrete pavement, or other equivalent pavement approved by the city. The paving surface must be so designed and maintained as to allow prompt and effective drainage of natural precipitation. No water drainage across sidewalks shall be allowed.

10. The minimum single car exit or entrance access lane width shall be fifteen feet.
11. The minimum combined exit and entrance access lane width shall be twenty-five feet.
12. Standard car stall depth shall be nineteen feet for 90° parking, twenty-one feet for 60° parking and twenty feet for 45° parking.
13. The minimum isle width for standard cars shall be twenty-four feet for 90° parking, eighteen feet for 60° parking, and thirteen feet for 45° parking.
14. Single-family, two-family dwellings, and parking lots shall maintain a minimum of twenty percent of lot area as a permeable and uncovered surface that contains living material.
15. All parking lots abutting a residential district or public right-of-way shall be screened from grade level to a height not less than three feet.
16. All commercial and industrial uses that abut residential, office, or institutional districts, shall maintain screening not less than six feet along the abutting property lines.
17. Screening will be equivalent to:
    a) Fences with at least fifty percent opaque construction; or
    b) Hedges, shrubs or evergreen tress of at least thirty percent opacity at the time of installation and fifty percent opacity maintained within three years of installation; or
    c) Berms or graded slopes of not less than three feet of mean height. Such berms or graded slopes shall contain at least fifty percent living material.

Product life is also a design constraint. Engineers must design a product to last an acceptable length of time. In some cases, such as a fire extinguisher, the life may be one use. In the case of an automobile valve spring, the life must be greater than 10^6 cycles. A typewriter ribbon may be designed for 10 cycles of use, oil in an engine crankcase for three months use. The life of a product can be measured either by the number of cycles used or a fixed period of time.

Product design is also influenced by the anticipated **disposal** after the product has served its useful life. For example, disposal of spent fuel rods from a nuclear reactor is an important concern. A product designed to be biodegradable might be better received by the consumer, and certainly makes for a better resource-use cycle than one that isn't. An engineer who is not concerned with the disposal of a used product is not concerned with the total solution of a problem. Figure 2.17 illustrates one major product disposal problem in the United States, junk cars.

The projected product **sales price** is also an engineering restriction. In most cases the sales price goal of a product is set before a project is complete, and it is up to the engineer to design a product whose cost meets the established goal. Cost invariably becomes the ultimate restriction, and engineering decisions affect product cost. If the cost is so high that no one will buy a product, it doesn't matter how well the product functions.

Another restriction, though sometimes nonfunctional, is one of **styling and appearance**. This restriction is more common in products that people will look at as well

FIGURE 2.17

Product disposal problem.

as use, as illustrated in Figure 2.18. Houses, cars, furniture, jewelry, and clothes fall into this category, and large companies employ engineers who spend the majority of their time on styling. On the opposite side, where styling and appearance have little consequence, are such products as drainpipes and electronic components that are rarely seen by the customer, and whose appearance is not critical.

An engineer might be told to design an **unnecessary feature** into a product, one that has nothing to do with the function of the product. Many of these features relate to appearance and styling just mentioned; however, they may also be features that are just nice to have, like automatic door locks and power seat adjustments on an automobile, or automatic re-dialing on a telephone. This does not imply that features aren't necessary or very useful for certain people; they just aren't needed by the majority of people. A remote control on a television, for example, is much needed by a bed-ridden invalid desiring to watch television, but it is not really needed by others.

Size and weight limits exist on the **shipping** and delivering of completed products. If a required product exceeds these restrictions, then final construction and assembly must be done at the use site, not at the factory. Prefabricated homes and offshore oil drilling platforms, Figure 2.19, are products which are usually assembled on site.

Engineers must specify all the **maintenance** required so a product remains functional. This includes specifying what must be done, and how often, such as greasing the bearings once a week, checking the fluid levels each day, or changing the oil every 2000 miles of operation.

FIGURE 2.18

Products which appeal because of styling

Company policy may also be restrictive. If the company you work for has a history of using hydraulic mechanisms, your superior might take a dim view of a pneumatic mechanism design.

2.7 Project Planning

Engineers are managers of time and other resources, and so they must plan, schedule, and follow-up on those plans. Plans for project implementation and logical schedules for events are crucial to the success of a project. Project planning starts with a statement of the intended goal. The more specific this statement can be made, the better the plan will be. An acceptable planning statement might be: Design a 100 horsepower, diesel powered bulldozer, and have it ready for sale in 19 months.

After the overall project plan is specified, the project is divided into major project activities. Completion times, personnel requirements, and cost estimates are made for each activity. The time limits and cost for each activity are based on similar completed projects, or from estimates by those who must do the work. Methods of making preliminary time and cost estimates include using a per part, per pound of product, per

FIGURE 2.19

Construction of an off-shore oil drilling platform.

floor of a building, or per cubic yard basis. Major product activities include first design, first build, first test and evaluation, first cost, redesign, retest, plant modification, build production tooling, establish production methods, final cost, and production. The actual list depends on the project (See Exercises 2.37, 2.38, and 2.39 for examples).

The sequence of project activities must be determined so an overall time plan can be developed and personnel, money, and plant resources scheduled. The sequence requires input from each department and work subgroup that has some responsibility for the project. This is an interdepartmental effort and includes product design, test and evaluation, purchasing, manufacturing, production, sales, service publication, plant layout, and accounting.

Most projects allow overlap of activities because design changes, redesign, rebuild, and retest can be done on component parts at the same time that overall product build and test is in progress. Parallel activities such as product cost, plant facility modifications, and sales literature design and promotional activities can be completed while product test and design are going on. Gantt charts are often used to map out the overall project schedule, showing the major activities and their overlap (see Figure 2.20).

Once a preliminary schedule of activities and times is presented, total period costs and personnel requirements can be estimated by passing down the Gantt chart for any interval of time. In Figure 2.20, for example, November, 1991 shows six activities occurring at the same time. This is only possible if the resources used for first build test are not required for second build test, as the two activities overlap. The resources include people as well as equipment. Either different resources must be made available, or the schedule of 19 months must be extended.

The advantage of the activity overlap of the Gantt chart is also one of its shortcomings. The overlap makes a critical time path difficult to determine. The general chronological order is down and to the right, but all of the preceding events do not have to be complete for the next level of activity to begin. In Figure 2.20, for example, six activities are shown to be active during the month of December, 1991. If a project is broken into smaller, more specific increments of work, or shorter elements of time, then progress can be checked more accurately; however, the chart can become large and cumbersome when more activities are added.

A network chart, often called a Critical Path Network (CPN), overcomes the problems of the Gantt chart because of the use of directed activity arrows. In a network chart it is assumed that an arrow activity must be complete before the next activity can begin. The time and effort required to conduct the detailed project analysis and determine activity prerequisites and logical activity order is time well spent. It forces attention on the details of a project and it may reveal a problem that otherwise might go unnoticed. Key personnel or critical test equipment may not be available at the time required to meet schedule. Plan ahead for special events such as holidays and vacations; schedule problems don't go away by neglect.

To use the network chart, you must break down activities into segments small enough to show their interdependence. Because of the sub-activity detail required, network charts quickly become complicated, and the use of a computer becomes helpful. Subdivisions of network charts are often based on subassemblies, material types, manufacturing method, or major functional groups such as engine, transmission, sheet metal, seats and upholstery for an automotive project. The extreme subdivision for a project development network is to list every part separately. This becomes necessary if specific parts require the use of new untried material or manufacturing processes.

Figure 2.21 shows an activity network and the interdependence and nondependence of the activities. Nondependent activities can be worked on at the same time. Vertical analysis of resources can be done on a network chart more efficiently if the activity arrows are drawn in proportion to their time estimates. If parallel activities do not require the same resources, then they can be completed anytime within the confines of the overall plan. Perhaps the greatest advantage of the network chart is that the critical

FIGURE 2.20

Sample Gantt Chart.

	1991	1992	1993
A S O N D	J F M A M J J A S O N D	J F M

- First design
- First build test
- First cost est.
- Second design
- Second build test
- Second cost est.
- Final design
- Release design for production
- Final design test
- Final cost est.
- Order material
- Build tooling
- Tooling tryout
- Pre-production build
- Production build
- Production cost
- First ship date →

FIGURE 2.21

Sample Activity Chart.

```
          Build patterns        Pour castings         Machine castings
            (25–30–35)    J      (15–25–30)    K        (35–40–50)

            Order          Receive       Fabricate      Weld            Line
            steel           steel         parts         assemblies      assembly        Ship
       A   (10–15–20)  B  (25–30–40)  C  (45–50–60)  D  (40–45–55)  E  (10–15–25)  F  (10–12–15)  G

          Order non-manufactured parts      Receive non-manufactured parts
                  (10–15–20)          H              (50–60–75)

              Create part drawings              Prepare repair parts manuals
                 (10–15–20)           I                (90–100–120)
```

Numbers represent working days. (Optimistic Estimate–Normal Estimate–Pessimistic Estimate)

path can be determined. The critical path is the chain of dependent events that encompass the entire project, and whose total time is a maximum. Major management and work force emphasis should be placed on the current critical path.

In Figure 2.21 using the normal time estimates, path A-B-C-D-E-F-G is the critical path and projects a project length of 167 days. Path A-J-K-E and A-H-E cannot be the critical sub-path because their most pessimistic times of 115 and 95 days are less than the most optimistic time of 120 days for the parallel path A-B-C-D-E. Path A-I-F could become critical because its pessimistic time of 140 days is greater than the most optimistic time of 130 days for path A-B-C-D-E-F.

A Program Evaluation and Review Technique (PERT) chart is a network chart showing earliest and latest completion dates for each activity based on the no-slack critical path of activities. PERT requires constant update when activities are completed, and the results require many calculations that are usually implemented using available computer software. PERT on computer was first implemented in 1958 by Admiral William F. Raborn on the Navy Polaris missile project.

The main problem with using activity networks and PERT is establishing the various sub-activities, their interdependence, and an estimate of their completion time. A well documented PERT chart will be complemented with a table listing the earliest start date, latest start date, and any slack time for each event. The chart must be updated on a regular basis so project status is always current. The earliest start date for an activity is the first date the activity can begin based on the earliest completion date of all preceding events. The latest start date for an activity is the latest date an activity can be started so all succeeding activities can be completed by the project due date. An activity has slack if there is a difference between the earliest start date and the latest start date. Slack time might be negative if a preceding event is late in being completed. Negative slack requires a change in plan, such as changing the due date or reducing future activity time. Activity time can be reduced by assigning more people to the task or working overtime.

2.8 Communication

Communication is vitally important to an engineer working to solve a problem. An engineer receives problem statements from people who may have essential details specified incorrectly or incompletely; indeed they may not even understand the situation. Many people are contacted to establish the proper project groundwork so that money, time, and talent are properly directed.

Communication is also important in gathering support information. An engineer spends time gathering data from books, periodicals, other engineers, production workers, and maintenance personnel. During the life of an engineering project, no resource can be excluded as a source of information. In the search for problem understanding or problem solution, there is no substitute for good communication. The majority of engineering tasks are complex and related to other fields and disciplines. The ability to work with the ideas of others, and the results of their work, is essential to the efficient completion of any project. Communications usually involve both written and oral activities, and both are important as the following example illustrates.

EXAMPLE 2.4:
Typical communication situation.

A meeting had two project proposals on the agenda that were to be presented, discussed, and voted on. The person reporting on the first project distributed a half-page proposal summary. After ten minutes of discussion and a few clarifying remarks, the proposal passed and work started on the project the next day.

The second proposal was no more complicated, but was done orally only. After 20 minutes of confused discussion concerning implementation, a vague motion for acceptance was made which included the phrase "...and will distribute a copy of the final proposal." The motion passed but it wasn't clear if anyone really knew what they were voting for, nor whether the final form of the proposal would be well received. Work on the project didn't start for several weeks.

2.8.1 Written Communication

The need for good report writing cannot be overstressed because from time to time all engineering activities require written communication. Reports are written in response to questions such as: What is the problem with the new control system? What ideas do you have to fix the problem with the control system? What is the status of the warranty replacement program on the control system? What is the progress of the design of the new computer control system? Reports are used to communicate the results of a feasibility study which lays the groundwork for a new project. Periodic reports are also required to communicate the status of projects that cover extended periods of time. Status reports are required at regular time intervals or at major scheduled completion dates, and are critical for determining whether or not schedules and budgets are on target and if adjusting is required. Major scheduled completion dates include those times when budget requests are made, when an important product development step is complete, or when approval is needed for the next phase of a project. Report writing should be thought of as a talent to improve and polish just like any other skill. Regardless of how good an idea is, a report is needed to get approval for its implementation. Reports are also required to maintain contact with management, and to insure project support from other operating departments.

Engineers must be ready to communicate on the many occasions when reports are required. Your management or your client will require a report for a contract bid proposal, a project status report, or a report on a meeting. A request may come to prepare instructions on a new procedure, a new product specification, or a new product release notice. Reports are required as a result of research and development, product testing, or customer reactions to new products. Reports are also needed when trips are made to other companies, when inspection or product manufacturing capability studies are made, and for personnel evaluations. In addition, if you work for a large company the only way other people in the organization will get to know you is through reports you write.

Data and related information required for reports may come from laboratory experiments, library resources, product vendors, field tests, warranty claim records, letters from customers, market surveys, cost accounting records, or product scrap records. Some of the ways information is presented in reports include charts, graphs, photographs, lists, quotes from other people, copies of official documents, or a diary of chronological events (of great value during patent negotiations).

The first phase of an engineering project is usually a **feasibility report** written to consider whether or not a project should be pursued. This report is written to clarify an existing situation, to suggest solutions that are practical, to consider the proper technology to use, and to estimate costs and schedules. This report is made after the problem is identified, ideas are suggested, and preliminary analysis has been made, but before money is allocated to begin long term design, model building, and product testing. Space in the report is allocated to an accurate and complete discussion of the problem, discussion of solution ideas suggested, reasons why certain ideas were rejected, and reasons why specific ideas are considered for further study. A feasibility report may include multiple possibilities, the most acceptable being selected only after extensive research, testing, and development. The report will also contain a discussion of related problems such as manufacturing capacity, material availability, personnel training requirements, and the impact on other products presently being produced. See Section 5.5 for more on feasibility studies.

A **project proposal report** is made to request implementation of a project that has come through the problem solving and feasibility acceptance procedure. This report includes the final recommendation and details that led to the decision to recommend pursuing the project. It is not necessary to discuss all discarded ideas or give them equal time in the report, but it is important to include those that were in the final running, and the reasons why they were discarded. Give enough background to support the recommended decision.

Depending on the complexity and magnitude of a proposed project, and the assignment originally undertaken, the following items are included in a proposal report:

1. An abstract of the problem and the proposed solution, sometimes included in a cover letter.
2. Details of the problem situation as known, its background, needs, and scope.
3. Details of the proposed solution including economic feasibility arguments, the solution which at the time appears to be the best, budget needs, personnel needs, and a time schedule.
4. An appendix with materials that may be important for backup, but not required to follow the logic of the proposal.

There are occasions when the format of a report is preset. This usually happens when a proposal or report is being made to a government agency, another company, or an associated support foundation. Reports of this nature must follow the established guidelines. If they are not, a request for project funding may be denied just because the proper format was not followed; a bid for a job may be rejected even though the bid price is the best; or legal action may be instigated because terms of a contract are not followed. If a report is required before next-phase funding is released, and the information is not complete or not in the proper format, delays in funding may occur, projects will get off schedule, and personnel may have to be released.

A **project status report** is not as lengthy as a project proposal report because only activity related to the original schedule and budget is reported. The main points to include are:

1. Work completed since the last report and any special results.
2. Schedule status, including time and money.
3. Work to be done next, and any anticipated problems.

Extra information need be included only when it is necessary to emphasize particularly good results, or problems that may require extra budget, testing, personnel, or a schedule change. If serious problems occur, then the status report will be supplemented with a new proposal, including much the same information as the original proposal, but extended to rationalize the necessity for a change. In the discussion of the reasons for a change other people or departments are not to be attacked. More important are facts supported by test reports from the field or other reliable information, letters from vendors of material or parts, letters and other documentation from marketing people, or information on changes in the state of the art of the technology being used.

Technical writing is simple and to the point. For clarity, sentences are short and words used are common to engineering and understandable by the reader. Be sure that verb tense is consistent throughout the report. Double check pronoun antecedents; if in doubt repeat the proper name rather than use a pronoun. Do not use the pronouns he, his, him, she, or her unless in reference to a specific person. Avoid opinion when presenting data. Do not use ambiguous words and phrases such as large, lightweight, mostly, higher priced than, and more significant than.

EXAMPLE 2.5:

Communicating with specifics.

1. Rather than saying the mixture has significant amounts of water in it, say the mixture has 35 percent water in it and the next highest ingredient, latex, is 30 percent of the mixture.
2. Rather than saying the group was made up mostly of dealer representatives, say in the group of 50, 35 were dealer representatives.
3. Rather than saying the part is worn out, say the bearing is worn so that the shaft rotates 0.10 inch out of line.

In many ways writing a report is just like any other problem that needs solving, and the steps of the problem solving procedure can be of value. The following discussion offers suggestions to help you write better reports. You need to develop your own style and at the same time cover the necessities. Engineers develop an individual style of writing as well as a report draft procedure that works best for them. There are, however, elements common to many types of writing.

Begin a report by stating what you want the report to do; make a proposal, report a status, ask for action from someone else, request money, or change a schedule. This statement may not be in the final report but it will serve as your objective as you write the report. Write a draft or outline of your thoughts without specific regard for sentence and paragraph structure, spelling, or consistent tense. The purpose of the first draft is to get ideas down, as in the design problem solving procedure. The first draft can include phrases, incomplete sentences, or a list of words. Some writers, for example, outline all the points that need to be covered, place them in preferred order, and then go back and fill in sentence details under each point. Still others start with the conclusion and work backward through the logical developments that precede the conclusion. As with geometry proofs, the end conclusion is known, and the thought process works backward to the beginning assumptions.

Read the first draft over, looking for the best logical sequence of the elements of the message you have to deliver. Rephrase and combine similar thoughts in the same paragraph. Eliminate expressions such as more than, better than, higher than, significant amounts, made up mostly of, superior to, and safer than, and replace them with quantitative data such as $350 compared to $400, no failures in 10,000 hours of test, or lasted 1000 cycles compared to 600 cycles. Eliminate unnecessary words and condense sentences to basic information.

Rewrite your marked up draft. Start polishing by checking for proper word usage, proper spelling, consistent tense, and correct punctuation. Use a dictionary, thesaurus, or a guide to grammar and punctuation for reference. Generally other changes will come to light as this second draft is being worked on, perhaps a change of order of topics, or the addition of more recent test results. Reworking your own draft helps refine thoughts and their logical sequence. Some engineers think that typing should only be done by secretaries, and their own time is too valuable to be spent on redrafting. With the advent of computer word processing, however, writing and editing is done efficiently, and with the ease of making changes, writing quality should improve. This redraft-reread cycle may occur several times; it is similar to refinement in the problem solving procedure.

When the final draft is complete, allow some time to pass, such as overnight, or from morning to afternoon if time is short. Rereading a draft after a time break will permit better judgment on the completeness and coherence of the writing. Additional thoughts and developmental ideas may come to mind and can be included. Request a co-worker to read the draft to help you decide if the message is clear. If a co-worker has to ask a question for clarification, it is a sure bet that other people reading the report will have questions too.

As you reread the draft yourself try to read as the receiver of the information rather than the sender. Be certain you know the background of your reader. Make it clear if you want action taken, want a response of acceptance or rejection, want information, or want an activity started. Specify dates, dollar amounts, and people's names. A good report often has more influence on the final approval for implementation than does a good idea.

2.8.2 Oral Communication

Daily engineering activities and project work require communication, and information is regularly passed between people orally, in a casual manner. Oral communication is usually used when instructing someone on a new procedure, explaining how to use a new piece of equipment, clarifying a new operating policy, or informing someone about an event of mutual interest. Oral communication, used more than written communication, has a problem that written communication does not: oral communication cannot be studied by the receiver for full and complete meaning. For this reason effective oral communications and presentations require some approaches that are different from those for written communication.

Oral communications require information to be passed in smaller increments; long complicated sentences that are written and can be reread, studied, and interpreted are likely to be confusing when spoken. A report that is read out loud requires too much work on the part of the listener, and is usually ineffective because too much information is coming in too short a period of time. Most listeners do not like to be read to; it makes them seem uninvolved and unimportant.

Oral communication is more effective if important statements are made shorter than when writing. Listeners, even if they take notes, need information in small pieces so it can be digested before more information can be received and processed. Since a good oral presentation will be easy to outline by the listener, prepare an oral presentation by outlining the points you wish to make. Use one note card or one piece of paper for each major point. Write larger than normal and leave space between lines so you can read easily.

The main advantage of oral communication is that it permits an exchange of ideas and information over a short period of time; clarifying questions can be asked and clarifying statements made. To be effective, good oral communication requires some feedback so the giver of the information knows if the information has been received correctly. This can be accomplished by asking questions of the receiver or by having the receiver write the information as it is heard, interpreted, and will be used. General discussion between the giver and receiver, along with examples, assist the oral transfer of information.

There are many times, however, that individual discussion and question and answer feedback are not practical, such as a presentation before a large group, or when the presentation is recorded for future use. For occasions like those it is better to be overprepared for a presentation than underprepared. Careful and complete preparation for a presentation, informal or formal, will result in a smoother and more effective transfer of information.

Some oral report preparation is the same as for a written report. The background and technological level of the audience must be known so the presentation can be given at the correct level; terminology must be appropriate, the common interest of the group must be pursued, and the purpose of the presentation must be clear. Is the presentation being made to inform the group of a new project proposal? A present project status? A request for additional time and money? Different purposes require different approaches.

Concern for the length of an oral presentation parallels that of a written report. A presentation that is too long will lose listeners just as a report that is too long will lose readers. Be well organized; get to the point, state the facts, summarize, and stop. A report over 20–30 minutes may be trying to cover too much information, but if more time is required be sure the presentation is straightforward and logical, and not a rambling sermon.

A strong introduction is important to attract and hold an audience's attention. When delivering a technical report do not use the typical after dinner speaker's opening joke. State concisely and precisely the purpose of the meeting. The following openings may be appropriate:

1. To office workers who may have to alter their parking procedures. "I want to explain to you what will happen next week when construction begins on the building addition."
2. To department members prior to the vacation season. "I am going to review the vacation policy because last summer there were several problems that occurred."
3. To engineers who will be designing parts and to engineers who determine production manufacturing procedure. "Manufacturing has installed a new metal spraying process. It is important we all understand the advantages and limitations of this process."

Oral presentations can usually benefit from visual aids, slides, videos, poster-size graphs and charts, and overhead transparencies. These should be prepared ahead of time and reviewed by practicing the presentation. The proper timing of visual aids is important. Don't show a slide and then say "I'll get to this in a few minutes." The visuals must keep pace with the presentation.

If overhead transparencies are used, do not turn your back on the audience to read the information to them. Stand near the projector and use a sharp pencil to point to key items as you discuss them. Limit the amount of information on each overhead to one idea or an outline of main points; too much information presented at one time will not be easy to absorb and understand. Be sure your overheads can be read from the back of the room you are using for the presentation. Practice ahead of time, and check to be sure a spare projector bulb is available.

The use of slides can also enhance a presentation. The point of each slide must be appropriate, and as with overheads, they must be visible from the back of the room. Try to avoid complete darkening of the room to project visuals, as many people like to take notes and a dark room will prevent this.

The use of video is often appropriate because it can show motion which is not possible with slides or overhead transparencies. Be sure the point of the video is relevant. The shortcomings of videos include:

1. Poor visibility because the television screen is too small to be seen from a distance.
2. Improper sound levels, too loud for those up front and too quiet for those in the back.
3. Inattention because watching television is usually considered entertainment, not information sharing.

Do not talk while a sound supported video is showing. If comments are required use a remote control and stop the action, or tell the audience ahead of time what to watch for. Video displays should not monopolize the meeting, only support it.

If charts and graphs are to be displayed without using slides or overheads, prepare them in advance of the meeting. Drawing graphs and charts during a presentation is not a skill most people have, and it takes time away from the meeting. Make the charts large enough to show necessary detail and to be seen clearly from the back of the room.

When appropriate, use real samples of the object under discussion. Samples must be large enough to be seen clearly or small enough to pass around. Scale models are also appropriate if the group is small, as at a meeting for a project proposal presentation.

If the information to present is highly technical, or there are many details, consider distributing a handout that contains the major points. Leave space on the handout for additional notes. The last page of a handout can be a questionnaire if you require meeting feedback.

Always practice an important oral presentation ahead of time, preferably to someone, or to a small group of people, who will give you honest feedback. Practice will give you a chance to check the length of the presentation and the appropriateness of visuals, as well as logical order and completeness. Practice in the room the meeting will be held in if possible. This will let you check the effect of visuals, and let you determine the proper voice volume to use. This is particularly important if a sound system is not available.

If a presentation is made as a result of a project feasibility study, or is a project proposal review in conjunction with a written report, be flexible; it is not uncommon for managers who have not thought about a particular problem for a while to change the boundary conditions and restrictions at the last minute. Try to anticipate the consequences of different conditions by considering alternate courses of action before the presentation.

Pace the presentation to a level you would use if talking one-on-one with a co-worker. An audience will tune out quickly if you talk too fast. Use a pace that allows you to emphasize the key points and give the audience a chance to absorb them. Do not fill in normal conversational pauses with "you know," "right," "um," or "uh." A momentary pause is normal; every second does not have to be filled with words. Maintain eye contact with the audience. A speaker who never looks up from notes quickly loses the communication link. Stand on both feet with your weight balanced; don't put your hands in your pockets and jingle coins or keys. Keep your hands away from your mouth so you do not mumble; articulate clearly and complete the endings on words; working, not workin'.

At the end of the presentation summarize the main points. If there is time ask for questions. If questions are asked and you are in a large room, repeat the question before answering so you keep the audience together. If the presentation is limited by time and time is up, stop. End with a "thank you."

2.9 Personal Characteristics and Abilities

Regardless of the problem situation an engineer is in, to get all the necessary jobs done and solve problems efficiently, many talents, favorable personal characteristics, and abilities are needed. The most important ones include:

1. *Technical competence.* Engineers must know the details of their specialty area and how their area interacts and works with other areas.

2. *An understanding of nature.* Nature's resources, laws, and governing principles must be used to advantage when an engineer designs an optimum solution to a problem. An engineer has to know how nature works. Section 5.2 lists some of the laws of nature to consider when solving engineering problems.

3. *Empathy for the requirements of other people.* An engineer who can stand in the shoes of another person can better reach an optimum solution to that person's problem.

4. *Active curiosity.* Curiosity may have killed the cat, but curiosity allows an engineer to seek out conditions that may influence the results of design work. Seeking answers to "What if?" questions is a sign of curiosity.

5. *Ability to observe without judgment.* Prejudging a condition restricts thinking and prevents an optimum solution from being chosen. Observation and recording of data and details in an impartial way lays the best groundwork for an efficient and economical solution.

6. *An innovative mind.* An innovative engineer will continually think of ideas and solutions that are better than average, easier to implement, and less expensive. Chapter 5 reviews methods which assist innovative idea generation.

7. *Motivation to design just for the pleasure of accomplishment.* This is just a fancy way of saying that an engineer must enjoy the work of design. Self-motivation is an important ingredient to job satisfaction. An engineer must want to get a job done, and get it done well.

8. *Confidence.* Engineers must be confident in their talents, education, and skills. They should feel confident to take on any task in their realm of expertise, even if it is beyond their present status.

9. *Integrity.* Engineers must be honest with themselves and those with whom they work. Talents must be used to support the goals of those who are paying the way. Decisions must be the best possible under the company restrictions. The chapter on ethics includes some concerns that an engineer must consider when making decisions.
10. *Willingness to take risks and assume responsibility.* Some risk is always present; nothing can be certain. At many times in the design process decisions will be required. Engineers have to take risks based on the data available, and be ready to accept the responsibility of their actions.
11. *Capacity to synthesize.* One of the most common methods of problem solving is to synthesize, that is, to combine principles and ideas that have worked before in a manner that will solve a new problem. Many problems are solved by using old solutions in new ways or in new combinations. Creativity is important when synthesizing; see Section 3.2 for hints on being more creative.
12. *Persistence and a sense of purpose.* Some problems require many attempts before a satisfactory solution is found. Keep trying. Iterations, retests, and rebuilds are common procedures in problem solving. Engineers must keep goals clearly before them so work is concentrated and focused toward goals.
13. *Patience with products and people.* Engineers who are patient can keep their mind clear to think of better ideas and are able to think of more possibilities. By being patient, work will be careful and meticulous. This will help prevent errors of judgment as well as errors in computation and data collection. Patience will also help when working with people. The result will be more cooperation and more useful communication.

2.10 References

1. Acret, James. *Architects and Engineers: Their Professional Responsibilities.* New York: McGraw-Hill, 1977.
2. Burstall, Aubrey. *A History of Mechanical Engineering.* Cambridge, Mass.: M.I.T. Press, 1965.
3. Chiras, Daniel D. *Environmental Science.* Menlo Park, Cal.: The Benjamin/Cummings Publishing Co., 1985.
4. Diesch, Kurt H. *Analytical Methods in Project Management.* Ames, Iowa: Iowa State University, 1987.
5. Drescher, Nuala McGann. *Engineers for the Public Good.* Buffalo: U.S. Army Corps of Engineers, 1982.
6. Florman, Samuel C. *The Civilized Engineer.* New York: St. Martin's Press, 1987.
7. Friedel, Robert and Paul Israel. *Edison's Electric Light.* New Brunswick, N.J.: Rutgers University Press, 1986.
8. Gaither, Norman. *Production and Operations Management.* Hinsdale, Ill.: The Dryden Press, 1987. Chapter 10, Resource Requirements Planning, and Chapter 15, Planning and Controlling Projects.

9. Garvey, William D. *Communication: The Essence of Science*. Elmsford, New York: Pergamon Press, 1979.
10. Gastel, Barbara, M.D. *Presenting Science to the Public*. Philadelphia: ISI Press, 1983.
11. Glorioso, Robert M. and Francis S. Hill Jr. *Introduction to Engineering*. Englewood Cliffs, N.J.: Prentice Hall, 1975.
12. Halcomb, James. *Project Manager's PERT/CPM Handbook*. Sunnyvale, Cal.: Halcomb Associates, 1966.
13. Hajek, Victo. *Management of Engineering Projects*. New York: McGraw-Hill, 1984.
14. Howell, James F. and Dean Memoring. *Brief Handbook for Writers*, 2nd ed. Englewood Cliffs, N.J.: Prentice Hall, 1989.
15. Kamm, Lawrence J. "On the Road to Success." *Design News*, March 27, 1989, 158-166.
16. Katz, Michael J. *Elements of the Scientific Paper*. 3rd ed. New Haven, Conn.: Yale University Press, 1985.
17. Kemper, John Dustin. *Engineers & Their Professions*, 3rd ed. New York: Holt Rinehart & Winston, 1982.
18. Kemper, John Dustin. *Introduction to the Engineering Profession*. New York: Holt Rinehart & Winston, 1984.
19. Koen, Billy. *Definition of the Engineering Method*. Washington, D.C.: American Society For Engineering Education, 1985.
20. Krick, Edward V. *An Introduction to Engineering and Engineering Design*, 2nd ed. New York: John Wiley & Sons, 1969.
21. Lannon, John M. *Technical Writing*, 3rd ed. Boston: Little, Brown & Company, 1985. Many good examples.
22. Lesikar, Raymond V. *How to Write a Report Your Boss Will Read and Remember*. Homewood, Ill.: Dow Jones-Irwin, 1974.
23. Lesly, Philip. *How We Discommunicate*. New York: AMACOM, 1979.
24. Markel, Michael H. *Technical Writing*, 2nd ed. New York: St. Martin's Press, 1988.
25. Martina, R.L. *Project Management*. Philadelphia: Management Development Institute, 1968.
26. Mayr, Otto, ed. *History of Science*. New York: Science History Publications, 1976.
27. Meredith, Dale D. et al. *Design and Planning of Engineering Systems*. Englewood Cliffs, N.J.: Prentice Hall, 1973.
28. Miller, Tyler G. Jr. *Living in the Environment*, 5th ed. Belmont, Calif.: Wadsworth Publishing Co., 1988.
29. Petroski, Henry. *To Engineer Is Human*. New York: St. Martin's Press, 1985. Easy reading.
30. Red, W. E. and Benjamin Mooring. *Engineering Fundamentals of Problem Solving*. Boston: PWS Engineering, 1983.
31. Ryckman, W.G. *What Do You Mean by That?* Homewood, Ill.: Dow Jones-Irwin, 1980.
32. Sandler, Ben-Zion. *Creative Machine Design*. Jamaica, N.Y.: Paragon Press, 1985.

33. Sherwin, Keith. *Engineering Design for Performance*. New York: John Wiley & Sons, 1982.
34. Thomas, Elain, et. al. *Safe Chain Saw Design*. Durham, N.C.: Institute for Product Safety, 1983.
35. Van Fleet, James K. *The 22 Biggest Mistakes Managers Make and How to Correct Them*. New York: Parker, 1973.
36. Vidosic, Joseph P. *Elements of Design Engineering*. New York: John Wiley & Sons, 1969.
37. West, Jerome D. and Ferdinand U. Levy. *A Management Guide to PERT/CPM*. Englewood Cliffs, N.J.: Prentice Hall, 1969.

2.11 Exercises

1. Consider each of the following items and outline a sequence of product development. Include technical advances and any products made less important or obsolete in the process.
    a) Television
    b) Computers
    c) Ships at sea
    d) Travel on snow
    e) Bridges
    f) Firearms
    g) Writing with ink
    h) Roads
    i) Airplanes
    j) Electric lights

2. Give examples of products for each of the following restrictions.
    a) Is to travel no faster than 25 mph.
    b) Must be rectangular in shape.
    c) Has a given maximum weight; specify the weight.
    d) Has a minimum weight; specify the weight.
    e) Has a required color; specify the color.
    f) Has a minimum width, length, or height; specify the dimensions.

3. Give an example of a new product on which an engineer might work.

4. Give an example of a product you are familiar with that has a functional problem, and for which a redesign or modification seems appropriate.

5. Give an example of an existing product that might be modified to fill a related but different need.

6. Find a product specification sheet for a product such as an automobile, backhoe, stereo, pump, or electric motor, and determine if the specifications are easy to interpret and whether they are usable to make comparison judgments.

7. Find specification sheets on two or more competitive products and chart a comparison of the items listed.
8. Prepare a bill of materials for a small product such as a chair, table, or pencil sharpener. Include part name, material, basic function, and whether you think the company manufactured or purchased the part.
9. Obtain and read a service, repair, or installation instruction. Evaluate how well the description is written. Is there a chance for misinterpretation or serious error?
10. Study how some product is packaged and evaluate its effectiveness.
11. Give examples of products that sell primarily because of their necessary features.
12. Give examples of products that sell primarily because of the features that are not necessary.
13. Analyze some different products and distinguish between necessary functions and unnecessary, though nice, features.
14. Select a product for analysis and judge which components were purchased by the manufacturer and which were not. Use a product such as a desk lamp, pencil sharpener, stapler or some other small product that can be easily taken apart or for which all parts can be seen without disassembly.
15. Make a list of possible filament materials Edison may have tried while he was developing the incandescent electric light bulb.
16. If an airplane has 5000 square feet of surface area covered with 0.100 inch thick aluminum, by how much could the advertised payload change if the process tolerance used to manufacture the 0.100 aluminum was reduced from ± 0.010 to ± 0.005?
17. Find physical examples of parts that are:
    a) Wind fits, where no physical object comes closer than the wind.
    b) Assembly clearance fits, where parts must clear each other so assembly is possible.
    c) Press fits so parts stay assembled.
    d) Close tolerance no-interference fits.
18. Give examples of products or parts that must work only once before failure.
19. Give examples of products or parts that must work millions of times before failure.
20. Give examples of products that have time life limits rather than cycles-used life limits.
21. Give examples of products that have cycles-used life limits rather than time life limits.
22. Give examples of parts that have performed their designed job if they do fail.
23. List some products that can be used equally well by right and left handed people.
24. List some products that can't be used equally well by right and left handed people and could use right and left hand versions. Suggest design changes on these products so they would be easier for both right and left handed people to use.
25. Discuss reasons why car manufacturers test cars by driving them into walls. Why is that test considered less expensive than other design analysis methods?

26. How would the task of finding how many Ping-Pong balls would fit into an automobile differ from that of finding how many would fit into the room you are in?

27. From a market data search, determine the cost of sugar in the following forms and estimate the cost of packaging from your data.
    a) In as large a container size as possible.
    b) In a five-pound package.
    c) In a two-pound package
    d) In cube form.
    e) In one-teaspoon package form.

28. Why do many foods have "Use before..." warnings on their packaging?

29. Consider and discuss some products, or main features of products that are very human-limit dependent. List the product and the human function limit involved. Example: The use of a car jack requires a certain level of strength in a person's arm muscles.

30. List some products that have been redesigned over the years to reduce maintenance. Specify the maintenance that used to be performed and what the design feature was that eliminated the need for maintenance.

31. List some products that still require maintenance and speculate as to why the maintenance needs have not been designed out of the product. Example: Lawn mower blades still require sharpening. The cost of sharpening is less than the cost of a new blade.

32. List some products that are well designed for after-use disposal.

33. List some products that are not well designed for after-use disposal. Suggest changes so they are.

34. Suggest ways that chain saw weight and exhaust fume levels have been reduced.

35. Compare food and beverage containers made from metal, glass, plastic, coated paper, and uncoated paper on the basis of raw material use, methods of fabrication, disposal by efficient means, reuse, litter life potential, litter cleanup ease, by-product effect on the environment and humans, ease of use by the consumer, and shelf life of the product.

36. For each of the codes, laws, and regulatory commissions listed in the chapter, specify a product you could be working on that would be affected by each one.

37. Draw a network chart for the project whose activities are listed in Table 2.1.
    a) Can this project be completed in 27 weeks?
    b) Determine which events have slack time, and how much.
    c) Assuming each activity is a two-person job, determine an efficient schedule of required personnel for the project.

38. Table 2.2 is a list of some of the preliminary steps used during a new product introduction. Determine a sequence of activities and an estimated minimum time to product introduction. *Note*: This may not represent any actual project you may work on.

TABLE 2.1
Project activities for Exercise 2.37.

Activity	Description	Time, weeks	Preceding Event
A	Remove old boiler and controls	4	—
B	Install new external steam lines	5	—
C	Install new internal steam lines	4	A
D	Construct new footings	5	A
E	Install new thermostats	6	A
F	Install new boiler	4	D, C
G	Install new boiler controls	7	F, B
H	Test boiler and lines for leaks	4	G, E
I	Check thermostat	3	H
J	Install system insulation	2	I
K	Test final system	1	J

TABLE 2.2
New product activities for Exercise 2.38.

Activity	Description	Time, weeks
A	Determine customer needs	4
B	Evaluate competition	3
C	Perform market survey	3
D	Prepare product specs	3
E	Project sales forecast	2
F	Evaluate marketing methods	2
G	Design product	5
H	Build test product	4
I	Test product	5
J	Plan marketing activities	4
K	Mount advertising campaign	4
L	Distribute sales literature	1
M	Print sales literature	4
N	Establish sales price	2
O	Build customer product	3

TABLE 2.3

Activities for building a small building, Exercise 2.39.

Activity	Description	Time, hours	Personnel
A	Draw plans	24	1 designer
B	Obtain building permit	3 (days)	1 designer
C	Clear site	12	2 workers
D	Purchase and deliver materials	6	1 worker
E	Excavate for footings	4	2 workers
F	Form and pour footings	8	2 workers
G	Allow footing cure time	overnight	0 workers
H	Form and pour frost walls	16	2 workers
I	Allow frost walls cure time	36	0 workers
J	Pour floor	8	2 workers
K	Allow floor cure time	36	0 workers
L	Erect walls	16	2 workers
M	Erect roof trusses	16	3 workers
N	Install siding	24	2 workers
O	Lay plywood roof	8	2 workers
P	Shingle roof	16	2 workers
Q	Install windows	6	1 worker
R	Paint siding, 2 coats	16	1 worker
S	Form and pour walkways	16	2 workers
T	Allow walkway cure time	3 (days)	0 workers
U	Install Doors	8	1 worker
V	Clean up site	6	2 workers

39. Prepare a project network chart for the construction of a small storage building using the activities listed in Table 2.3. Add any activities you think are missing
    a) Determine a critical path for an optimistic time schedule and the number of days required. Assume an 8-hour work day.
    b) Determine a pessimistic time schedule assuming rain every other day.
    c) Determine the required number of workers for each day.

40. An executive from a ski manufacturing company said, "Computer analysis defines the ski in theory, but determining the best combinations of characteristics requires field testing." Explain the meaning of this statement.

41. Explain the difference between the meaning of these two statements:
    a) Parts are also difficult to obtain.
    b) Also, parts are difficult to obtain.

42. Explain the difference in the meaning in the following three phrases:
    a) high priced method
    b) higher priced method
    c) highest priced method

43. Explain why each of the following statements is not well defined.
    a) Coors is the best beer on the market.
    b) Henry studied very hard for the exam.
    c) The car stopped running.
    d) The workers in the assembly department are overworked.
    e) This gear box is too heavy.
    f) The backlog of orders is unbearable.
    g) This is a good milling machine.
    h) The warehouse is crowded.
    i) This flame cutting operation is a dangerous operation.
    j) The finish on this car is beautiful.
    k) The new transmission tested better than the old one.

44. For each of the statements in Exercise 2.43 explain what detail or details you would want to clarify to improve on the accuracy and precision of the statement.

45. The following two sentences imply an understanding of economics but are poorly worded. Reword the sentences and list other information needed to clarify the statements even further.

    "When solar heating devices are installed, customers usually consider overall cost. This cost is offset by fuel savings and maintenance expenditures."

46. What is wrong with the logic of this statement? "The pipes are inaccessible, making repairs costly and time consuming."

3 Problem Solving

"One of the best things you can do when you have a problem is to talk about it."
♦ *MOSHE F. RUBINSTEIN & KENNETH PFEIFFER*

Engineering design is a process of using compromise (mutual concession and agreement) and iteration (repetition for improvement) to solve problems relating to the creation of a product or service to satisfy a real or perceived human need. The variety of problems encountered requires that an engineering design procedure be flexible and adaptive. Procedures must be flexible because problems rarely have the same conditions and restrictions, and will not be solved using the same activities. The engineer must be creative in the approaches used, and sensitive to the restrictions and the environment surrounding the problem, and must decide how to proceed with a problem by judging its complexity and considering various courses of action. A complex problem, such as designing a space station, will require the efforts of a team of engineers, each contributing knowledge and solutions. Experienced engineers recognize that there are many items that need to be considered when solving a problem, and they understand that design is not just following a grocery list of individual tasks. Design is a series of interrelated activities that lead to optimum use of available resources under the restrictions imposed. Many activities are done more than once. New ideas and possibilities may occur at any time, and feedback and reevaluation are important. Iteration and refinement are required to narrow the gap between what is, and what is desired.

3.1 Sources Of Problems

Where do the engineering problems come from? Whether you work for a large corporation or have your own consulting company, problems will come to you from many different sources.

One primary source of problems is existing products. An existing product or procedure may have an **observable deficiency**. Deficiencies may exist because a product was designed using older state of the art analysis, or because it is being used differently than originally intended, or because it was poorly designed or manufactured. Product deficiencies are encountered in many different places, and include such items as:

1. A signpost that bends over in the wind.
2. A conveyor that stalls when too many packages are placed on it.
3. A plastic tool handle that bends or shatters when the tool is used.
4. A copier paper feeder that feeds multiple sheets when only one is required.

Engineers also work on **unmet needs** and design new products or services such as:

1. A storm sewer system for a subdivision to control water runoff.
2. Desks and proper lighting so students can study efficiently in the library.
3. A new style drilling rig for offshore drilling in deep water.
4. A process to detect explosives in airline baggage.

An engineer may be assigned to search for a **product improvement**. Perhaps the present cost is too high and the job will be to reduce cost. A competitor's product may be selling better and a redesign is needed to meet competition. Examples include:

1. Redesign a furnace to increase its efficiency by 10%.
2. Redesign a computer printer to print twice as fast and to be 25% quieter while it is printing.
3. Redesign a lathe so that it will hold tolerances of ± 0.0003 on turned diameters.
4. Redesign a temperature sensing system so it will send a signal to the security office if the temperature is greater than 120°F.

A company's product or service may be influenced by a **change in a law.** Law changes occur in building codes, fire regulations, environmental constraints, or may be caused by newly discovered product side effects or increased liability for a particular existing defect. Some recent law changes require:

1. An average automobile mileage of 25 miles per gallon of gasoline.
2. An engine that runs efficiently on unleaded gasoline.
3. An automatic passive restraining system for airplane passengers.
4. An automatic blade brake on a power mower that activates when the mower handle is released.
5. A limit on the use and disposal of asbestos.

You may have a **natural curiosity** to solve a problem whether the problem has an immediate application or not. This is a common trait of engineers, for new and unique situations arouse interest and stimulate creativity. Curiosity can also create valuable experience to be used later when solving other problems as well as improvements or inventions that are patentable. The following example illustrates a natural curiosity problem.

EXAMPLE 3.1:
Nine dot problem.

Connect the nine dots shown in Figure 3.1 with four straight, sequentially connected lines.

Problems like Example 3.1 provide excellent opportunities to practice problem solving, and often make people aware of barriers that can interfere with creative problem solving. The difficulty in finding a solution is the restriction placed on the problem: four sequentially connected straight lines. The basic problem of connecting all the dots is no challenge; there are an infinite number of possibilities. And so it is with most engineering problems; the restrictions, whether they be weight, volume, cost, material, or time, are the real challenges you will face. See reference (1) for interesting solutions to Example 3.1, including some that use only one line.

3.2 Creativity

A productive, creative engineer has the capacity to propose innovative, workable solutions to problems. Creative people can visualize alternatives without being inhibited by outside factors, and have the ability to mull over in their mind present input and past experiences while searching for a logical problem structure that removes solution conflicts. Most engineers want to be creative, but often daily chores get in the way of creative work. Working on unstructured or open-ended problems is good practice in creativity. An unstructured problem is one in which any solution that is not precluded by the problem statement is acceptable. Finding a door-holder-opener is an unstructured problem. Locating a wedge to hold the door open is not, unless material selection is the unstructured search.

Not all people can be brought to the same level of creative problem solving, but practice will help, the more practice the better. One of the purposes of practicing creative problem solving is to change the way people look at the world around them. Engineers who become more aware of surroundings will be better problem solvers. Everyone can be creative, but some people are better prepared for creative thought than others. Various studies indicate that creative people tend to be adventurous and willing to take risks to

FIGURE 3.1

Connect the nine dots with 4 sequentially connected straight lines.

achieve high gain; are perceptive in observation rather than judgmental; and are emotionally sensitive, self-confident, and open-minded.

There are a number of personal traits that will help you in your problem solving activities:

1. Have an inquiring and questioning attitude. Ask questions to determine what is expected and what restrictions and limitations are present and to discuss options. Ask "What if?" questions to gain understanding and to open the door to new ideas. See Example 4.3 and Section 5.1.7 for sample questions to ask about a problem to get the creative juices flowing.

2. Develop communication skills. Information and ideas transfer from person to person most efficiently when communications are good. This applies to both spoken and written communications. Many people communicate orally with success because misunderstood points can be repeated. Not so with written words. Practice writing to sharpen your communication abilities. Reading other people's work is one aid in learning to write, because recognizing examples of clear writing will help you judge your own work.

   Improve communications with yourself. If an idea or question crosses your mind, make a note of it, as it may never again surface in your thoughts. Make lists of personal priority goals and objectives, about projects to undertake, and investigations to make. Know what you are thinking and how your thoughts relate to the problem at hand.

3. Be physically and emotionally healthy. Physical well being improves concentration ability and stamina to carry out experiments and studies. Emotional health allows you to put a lid on distracting problems such as sorrow, fear, and worry. Energy spent on emotional problems is energy that can't be used on productive problem solving.

4. Accept reality and be self critical. Accept the fact that conflict and tension will occur from time to time, but don't let frustration result. Frustration blocks the normal thought process, saps needed energy, and limits creative thinking.

5. Be willing to consider new ideas and not worry about your ego. Even if an idea conflicts with your present thoughts or past experience it must not automatically be discarded. An idea that didn't work before may work now. The circumstances may be different, or the acceptance levels of others may have changed. Don't ever wave away an idea by saying "We tried that once and it didn't work."

6. Be curious and observant. Notice what goes on around you. Take note of things you see that work well and also those that don't.

7. Allow time for subconscious thinking. Some people consider their problem-solving efforts a failure and become frustrated if a great idea doesn't occur to them after a brief period of time. Meditation and thinking time are important when a solution is elusive. Some suggestions that may improve your concentration and help ideas to flow are:
   - Take a walk away from disturbances.
   - Play music to force concentration on a problem.
   - "Sleep on it" and take a fresh approach the next day.
   - Read a book.
   - Work on another problem.
   - Do something physical: cut the grass; wash the car.

Everyone can improve innovativeness and creativity, but it requires practice. While personal intelligence may set an upper limit on how creative a person may become, high IQ does not guarantee high creativity.

Highly creative people get great personal satisfaction from solving problems in new and unusual ways. Their satisfaction comes from a sense of pride, loyalty, pleasure, joy, thrill, and competition. The chance of doing something creative favors the prepared person. Creativity will not come to you by revelation. For additional dialogue on creativity see references (1), (5), (6), (17), and (18).

3.3 Roadblocks To Creative Problem Solving

Roadblocks to creative problem solving can come from external sources, as well as from sources within ourselves.

Past experiences, and habits that result from them, may lead an engineer to think there is only one way something can be done. Habit may cause a problem solver to have difficulties such as:

1. A limited range of basic knowledge search. Many people feel that once they have something that works, there is no other choice: "This is the way it has always been done." Or if a shy person rarely offers an idea, some might think "Don't ask him, he never has a good idea."

2. Failure to consider all the details of a problem. Relevant details may be ignored due to preconception of the answer. Obvious details must always be considered in balance with details found through deep investigation. Example: The witness saw the smoking gun and the bullet hole in the chest, but the coroner found poison in the stomach.

3. Failure to distinguish between cause and effect. This can lead to wasted effort and time pursuing something that is not the real problem. Example: If I frequently run my car out of gas, and the engine stops running on my way to work, I will probably conclude I ran out of gas again. But there are many other reasons why an engine will stop running; no spark, loss of engine coolant and the engine "freezes," or a plugged fuel line.

4. Failure to see what is around us. We look, but we don't see. For example, how many people can correctly describe the face of a telephone, including the locations of the numbers and the letters? Creative people see more than noncreative people see. They also visualize images more easily and consider alternatives just by thinking about them. To increase your creative ability try practicing your visual imagery. For practice, try visualizing each of the following; a bullfight, a baseball double play (How many different ways can it happen?), an airplane flying upside down, a bird flying backward, hitting a golf ball made from a marshmallow, poured water flowing up rather than down, drinking cold coffee, drinking hot lemonade. Did you have more trouble with images that you have never seen before? Most people do and often comment "Why should I think of things I haven't seen?" Exactly the point. Creativity and innovativeness assist you to think of something new, or a new application of an old idea, not just something you have seen or done before.

Our **culture and society** pose roadblocks. The culture in which a person has been raised can have effects that limit creative freedom. Certain ideas or methods will not be a valid choice because of upbringing mores or social customs. Examples would include the different philosophies and life styles of the Mormons, Amish, Quakers, Eskimos, Japanese, Chinese, Hindus, or Africans.

Cultural restrictions prevent innovations and make large changes in day-to-day living difficult. A change may be ready before its time. Examples include the iron plow, which early opponents claimed would poison the ground, birth control for societies that consider wealth a function of family size, or nutritious food that is considered fit only to feed cattle.

Personal emotions can also be barriers that must be overcome. Emotional barriers include sadness, worry, and perhaps the biggest emotional barrier, fear. This includes fear of failure, and personal embarrassment; fear of criticism from those we respect; fear of ridicule, both public and private; fear of embarrassment; fear of loss of job; and fear of appearing uninformed or ignorant. Today's society places emphasis (sometimes too much) on knowledge. Students are pressured to know facts and to be able to regurgitate details. As a result, to ask a question (an attribute of a creative person) is to cast doubt on one's intelligence. However, once asking questions becomes a part of your way of life, a must for gathering data, the fear will disappear.

Associated with the fear of appearing stupid is the mental block of prejudging an idea before it is suggested or tried out. Prejudging, and not offering a suggestion prevents open discussion and often excludes the best idea from being considered. Even an idea that will not work may lead to an idea that will. Refer to Section 5.1.5 for more on the danger of prejudging ideas.

One method of overcoming potential embarrassing situations is to write down all the possible results of your actions you can think of. What is the worst thing that can happen if you try your idea or ask your question? Now write down the good things that can happen if you try your idea or ask your question. What is the best thing that can happen? If you have more to gain than to lose, then do it.

Roadblocks can also be imposed by supervisors and co-workers. If a co-worker responds to a suggestion for change with, "We tried that last year and it didn't work," or "That method won't work in our company," or "We don't have time to try that," creative effort can easily be stymied. The following case study illustrates how the problem definition can limit creativity.

CASE STUDY #1:
Door-Holder-Opener

Consider the day the Physics Department was setting up for a conference. The person in charge of the conference wanted the self-closing double doors that separate the lobby from the main hallway open, so guests would feel free to wander about. The person in charge said to me, "Find a wedge to hold these doors open."

A wedge is a logical door-holder-opener, but so are many other things: a brick, a chair, a table, or a person. So as I walked about looking for an efficient solution, I considered several alternatives. Although a student worker was nearby, I didn't think he would like to spend the day holding the door open, so I pursued other alternatives.

My solution to the problem was a rectangular block of wood 1.50 inch x 2.00 inch x 0.75 inch. I opened the door and placed the block between the hinge side of the door and the doorjamb. It held the door open all day, didn't get in the way of people, and saved a student from a most monotonous job. A wedge is a habit solution to the problem. I was sent after a wedge when I should have been sent after a door-holder-opener. If time had not been short, other more permanent and versatile solutions could have been considered. The name given to a problem can itself be a roadblock to a creative solution. The device could be called a door anti-closer, doorstop, or door-block to allow for more creative thought.

3.4 Problem Solving Activities

Chapters 4 through 7 contain specific details of problem solving activities, but an overview now will provide the framework for the details in those chapters. An engineer engages in many activities in the quest for the solution to a problem. No list of activities is complete, nor is a given list the best for every type of problem. Problems are rarely the same, either in terms of resources and talent available, or in potential reward. So the following activity outline is not intended to be a fail-safe or complete list, merely a reminder of general responsibilities and procedures.

3.4.1 Define the Problem

Before a problem can be solved, it must be determined what the problem really is. If the activity of defining, clarifying, and understanding the problem is properly done, subsequent work is more efficiently performed. Nothing is as frustrating as to work on a project for some time and then realize you're not working on the real problem. The problem statement usually includes two main categories:

1. The demand specifications, criteria that are essential.
2. Criteria that are nice to have if they are affordable and do not jeopardize the performance of the demand specifications.

The criteria specified must not be so specific that they determine the solution to the problem (use a maximum of 500 watts of *power* rather than use 500 watts of *electricity*). The focus of problem definition is directed toward *what* has to be achieved, and on the problem limits and boundary conditions. If the problem of Case Study #1 had been described as "We need a wedge," I might not have found one in the building and a student might have had a long tiresome day holding the door open. But by stating the problem as "We need something to hold the door open," many possibilities can be considered. Methods to assist in defining problems are discussed in Chapter 4.

3.4.2 Consider Alternative Solutions

After a problem is adequately defined, possible solutions must be presented and the focus shifts to *how* the demand specification criteria can be accomplished. As the number of alternative solutions is increased, chances for an optimum solution increases.

If time is short, this activity can be brief, and the first idea that comes to mind may be used. For example, if a door is swinging toward you, and your arms are full of groceries, the idea of sticking your foot out to stop the door may be the only idea you can come up with in the brief time you have before the door hits you. However, if time is available, considering only one possibility may lead to expensive and inefficient solutions. Considering alternate solutions can also be a complicated activity requiring research and investigation lasting several years. Major projects, such as designing a space station, will consider many alternatives following lengthy research.

As ideas are generated, the scope of the project may be modified. This requires redefining or refining the problem statement. For example, the original problem statement may have been to create a computer program to generate payroll checks. After deliberation, the problem may be changed to include the generation of monthly, quarterly, and annual IRS and workman's compensation reports. If the scope of a project is changed, be sure to adjust the resources for the job, such as staff support, money, and due dates. Suggestions for generating ideas are presented in Chapter 5.

3.4.3 Refine Ideas

One objective of idea refinement is to narrow the selection of possible ideas and weed out ideas that have little chance of succeeding. Remaining effort, time, and money are then used to develop those ideas with the best chance of being workable and profitable. Some ideas may be discarded simply because the technology is not available for implementation. In other cases high preliminary cost estimates will cause an idea to be dropped early in the project. An equally important objective of refinement activities is to concentrate effort on improving how a problem will be solved. Good ideas are made better, and sub-problem solutions are combined for overall performance improvement.

Refinement activities may also require additional project scope changes. A refinement in the problem statement can result in new ideas to consider. An original problem statement "Design a ten-horsepower, front-tined rotary cultivator," may be modified to "Design an eight-horsepower, front-tined rotary cultivator with forward and reverse power drive."

Additional ideas for solution can be generated from a problem statement refinement made during idea refinement activities. Any activity may result in iteration back to a previous activity. No activity is performed once and then forgotten; always consider the whole picture. Suggestions for refining ideas are presented in Chapter 6.

3.4.4 Analyze Ideas

Analysis includes recording and studying all the efforts being made to verify that an idea is workable and will function. The questions that need answering are:

"Will the idea work as desired under the restrictions imposed?"

"Can a weak idea be altered so that it is viable?"

"Can this idea be implemented under the cost and time constraints imposed?"

Verification of function can be done in many ways, such as mathematical modeling, computer modeling, or actual building and testing. Additional refinement is usually made, and more ideas may be generated as the results of the analysis. Keep in mind that most projects require iteration through the idea, refinement, and analysis sequence of activities. On some projects, such as the development of a new material, process, or product, the iteration process may last for several years. Consider superconductors, for example, whose electrical resistivity properties are favorable, but which have low power carrying capacity and are difficult to manufacture. Their development will continue for a long time.

Models or prototypes are built and tested to verify function, and can be a portion of the product or the complete unit. For example, during the development and analysis of large computer systems, a program is often broken down into subprograms that can be tested independently of the total system. This is possible by using subroutines and a common data base. Some models and prototypes will be experimental in nature and will use experimental optimization procedures.

Product testing can be done using complete assemblies, or components. A shaft in a transmission can be tested outside of the complete transmission assembly, or it can be assembled into the transmission and then tested. The transmission can be tested by itself or installed in the completed vehicle and tested. Suggestions for the analysis of ideas is combined with refinement in Chapter 6.

3.4.5 Decision

The time comes when a decision on the final direction of a project is required. This may happen during the problem identification or idea generation phase of a project, but usually it comes after some amount of refinement and analysis has been done. In the worst case, none of the ideas being refined will function to an acceptable level, and the whole project must be started over or discontinued. It may be that the problem definition is too restrictive or demanding, or that time and money are inadequate to continue. If the decision is made to cancel the project, then salvage the work done to date, record accurately what has been done, and move on to a new project.

In the ideal case, one idea will clearly be the best, will be the least expensive, have a shorter lead time to production, and will suit the fancy of 1,000,000 customers.

In the usual case, no idea will work as well as desired, and a compromise between such items as function, cost, sales potential, and time to production will be made.

Perhaps the problem will have to be narrowed in scope. The reverse feature on the tiller may be discarded to reduce cost; or the four-horsepower engine already in production will be used to reduce development time and lower sales price; or about this time the Sales Department will report that only rear-tined cultivators will sell. Any number of unexpected changes may be required. Methods to assist in decision making are discussed in Chapter 7. In today's high technology environment of computers and Computer-Aided-Design/Computer-Aided-Manufacture (CAD/CAM) systems, perhaps the expression "Back to the drawing board," should be updated to "Back to the computer terminal."

3.4.6 Implementation

Once an acceptable solution is chosen, and assuming profit potential still exists, the final drawings, production methods, construction of buildings, hiring of people, ordering material and other supplies, and installation of required processes and procedures can proceed. This does not mean that new ideas and additional refinement are halted. New ideas and refinement should never be terminated; if changes are extensive, however, they may not be implemented on the first production model. They may have to wait for subsequent models. Implementation activities are included in Chapter 7 with decision making.

3.5 References

1. Alger, John R. M. and Carl V. Hays. *Creative Synthesis in Design*. Englewood Cliffs, N.J.: Prentice Hall, 1964.
2. Andrews, James G. "In Search of the Engineering Method." *Engineering Education*. Oct. 1987, 29-30, 55-57.
3. Bailey, Robert L. *Disciplined Creativity for Engineers*. Ann Arbor, Mich.: Ann Arbor Science, 1978.
4. Beakley, George C. and Ernest G. Chilton. *Design, Serving the Needs of Man*. New York: Macmillan, 1974.
5. Beakley, George C., et al. *Engineering, An Introduction to a Creative Profession*, 5th ed. New York: Macmillan, 1986.
6. Brightman, Harvey J. *Problem Solving: A Logical and Creative Approach*. Atlanta: Georgia State University, 1980.
7. Cross, Nigel. *Engineering Design Methods*. Essex, England: John Wiley & Sons Ltd., 1989.
8. George, F.H. *Problem Solving*. London: Duckworth & Co., 1980.
9. Harrisberger, Lee. *Engineersmanship*: *The Doing of Engineering Design*, 2nd ed. Monterey, Cal.: Brooks/Cole, 1982; 143 pages of easy, but stimulating, reading.
10. Kaku, Michio and Jennifer Trainer. *Nuclear Power: Both Sides*. New York: W.W. Norton & Co., 1983.
11. Love, Sydney F. *Planning and Creating Successful Engineering Designs: Managing the Design Process*. Los Angeles, Cal.: Advanced Professional Development Inc., 1986.
12. Meadow, Charles T. *Applied Data Management*. New York: John Wiley & Sons, 1976.
13. Osborn, Alex F. *Applied Imagination,* 3rd ed. New York: Scribner, 1979.
14. Pearl, Judea. *Heuristics*: Reading, Mass.: Addison Wesley, 1983.
15. Polya, G. *How to Solve It*, 2nd ed. New York: Doubleday Anchor Books, Doubleday & Co, 1957.
16. Rubinstein, Moshe F. *Patterns in Problem Solving*. Englewood Cliffs, N.J.: Prentice Hall, 1975.
17. Rubinstein, Moshe F. and Kenneth Pfeiffer. *Concepts in Problem Solving*. Englewood Cliffs, N.J.: Prentice Hall, 1980.
18. Ruggiers, Vincent Ryan. *The Art of Thinking*, 2nd ed. New York: Harper & Row, 1988.
19. Smith, Karl J. *The Nature of Mathematics*, 5th ed. Monterey, Cal.: Brooks/Cole, 1987.
20. Spotts, M.F. *Design Engineering Projects*. Englewood Cliffs, N.J.: Prentice Hall, 1968. Chapter 1 on creativity.

3.6 Exercises

1. List as many ideas as you can that could answer the question, "How can I get from my dorm to a movie theater."

2. List at least one advantage and one disadvantage of each idea listed in Exercise 3.1.

3. Rephrase each of the following statements to allow more room for creative solutions:
    a) Get a towel to dry your hands on.
    b) Take the car to go get a loaf of bread.
    c) Find a pen to write down a message while on the phone.
    d) Find a table to set our lunch on.
    e) Get a boat so we can row across this river.

4. Discuss this definition of design: Design is the combination of basic principles and available resources to solve a problem.

5. Discuss the implied meaning of the phrase "Trial and error," often heard in relation to the design process.

6. Give an example of a constraint so severe that an acceptable answer would be impossible.

7. When the Chernobyl nuclear accident occurred in Russia in April, 1986, the "China Syndrome" question was proposed: Where in the world is directly opposite Kiev/Chernobyl?
    a) Estimate an answer in 60 seconds.
    b) Discuss how you would go about making a more accurate answer.
    c) Follow your procedure in part (b) and determine an accurate answer.

8. Two ropes hang from the ceiling of a normal room, but they are too far apart for one person to reach both at the same time while they are hanging straight down.. Discuss ways to get the ropes together so the ends can be tied. Try out your ideas.

9. What is the next letter in the following sequence of letters? OTTFFSSENTETTF Explain your reasoning.

10. In the morning going to work, a man rode an elevator alone to the tenth floor and then walked up five flights of stairs to his office. At night when he left work he rode the elevator down all 15 floors. List possible explanations for this behavior.

11. Discuss what are often called "contingency plans." How would they be used in setting up an original project schedule and budget?

12. Make a list of methods for holding a door open. Include pictures or descriptions of commercially available products. Visit hardware stores and building supply houses for ideas.

13. List design shortcomings of each of the following products. User interviews may be required.
    a) Home vacuum cleaner
    b) Home clothes drier
    c) Electric stove

    d) Lawn mower
    e) Gas stove
    f) Snow blower
    g) Dishwasher
    h) Popcorn popper
    i) Refrigerator
    j) Coffee maker
    k) Clothes washer
    l) Toaster

14. Give some examples of development projects that you feel could not be speeded up by adding more money and personnel compared to some that could. These need not be absolute examples, only relative to each other.

15. List some possible design deficiencies with the signpost that bends over in the wind.

16. List some possible design deficiencies with the conveyor that stalls when too many packages are placed on it.

17. List some basic criteria to consider in the need for a storm sewer system.

18. List some basic criteria to consider in the need for study desks and proper lighting.

19. List some basic criteria to consider in the need for a new offshore drilling platform.

20. What can be measured to determine the success or failure of the redesigned furnace?

21. What approaches would you consider in the solution to the problem of designing an automobile engine that averages 25 miles per gallon?

22. Discuss possible sources and reasons why the problem statement was changed from "Design a five-horsepower, front-tined rotary cultivator," to "Design a five-horsepower, front-tined rotary cultivator with forward and reverse drive."

23. List advantages and disadvantages of the differences in testing a transmission shaft by itself, in an assembled transmission, and in a transmission in an automobile.

24. Make a list of five to ten machines in order of their versatility. Example: sewing machine, low versatility; computer, high versatility.

25. Make a list of five machines that run themselves compared to five machines that must have an operator to control them. Example: An electric clock runs itself, a bicycle requires an operator.

26. For each machine listed in Exercise 3.25 that does not need an operator, describe a design or equivalent situation that would do the same function with an operator. Example: A clock that needs an operator would be an hourglass.

27. For each machine listed in Exercise 3.25 that needs an operator, describe a design or equivalent situation that would do the same function without an operator. Example: A "bicycle" without an operator is used on overhead cables to carry loads from one location to another.

28. Discuss this definition of problem solving: "To reach a least-cost method of achieving a transformation from one state of affairs to another."

29. A passenger in an airplane, properly seated and buckled in for takeoff noticed a sign

on the back of the seat in front of him. The sign said "This outboard seat back does not recline." Which seat doesn't recline, his or the one in front of him? Discuss the sign labeling.

30. In 1976, Kenneth Appel and Wolfgang Haken, professors at the University of Illinois, Urbana-Champaign, announced they had proven that any map could be colored with only four colors, and no border would have the same color on both sides. They used the computer to check "all possible combinations." Any exception, of course, disproves their hypothesis. Does the map in Figure 3.2 disprove their hypothesis because no color is available for the inside circle? Explain.

31. Select a simple product, such as a crank pencil sharpener, and list the verb-noun functions of all the parts. For example one part of the pencil sharpener must have the function of "catch shavings."

32. Analyze a procedure, such as installing a chain link fence, changing a car tire, or building a book shelf, by listing the steps and the necessary tools for each step. Consider alternate sequences of activities to do the same job.

33. Three switches control the lights in a particular hallway. Two are 3-way switches, one at each end of the hall to control the even numbered lights. The third is a single switch controlling the odd numbered lights. The single switch is up if the lights are on, and down if the lights are off. A 3-way switch can be up or down when the lights are on depending on the position of the other 3-way switch. One 3-way switch is mounted next to the single switch at one end of the hall. For each of the following cases only one set of lights is on, either the even ones or the odd ones, but you don't

FIGURE 3.2

Map for Exercise 3.30.

B is blue
R is red
Y is yellow
G is green

know which. Your problem is to change the position of *one* of the two switches and, based on the results, regardless of what the results are, determine which switch is the 3-way switch and which is the single switch.

Case 1: S_1 down and S_2 down

Case 2: S_1 down and S_2 up

Case 3: S_1 up and S_2 down

Case 4: S_1 up and S_2 up

34. Same situation as Exercise 3.33 except both the even and odd lights are on for each of the cases of switch settings.
35. Same situation as Exercise 3.33 except no lights are on for each of the cases of switch settings.
36. If four cars are stopped at each corner of a four-way stop, how many ways can the cars proceed, all going at the same time, without colliding with any of the other cars? Describe them.

4 Problem Definition

"The engineer's first problem in any design situation is to discover what the real problem is."
✦ GEORGE C. BEAKLEY

Problem definitions can range from a simple exclamation, "Help! My boat is headed for the waterfall!" to a book of performance specifications such as those for a proposed power dam. A preliminary problem statement may also be quite vague. A customer may say "I sure could use a good door-closer." A problem statement with that little detail is not clear enough for an engineer to proceed with a design. Does the person want a product that is light, colorful, rustproof, installs with a screwdriver, one that closes and locks the door, closes the door in less than two seconds, or costs less than $50? Is it possible that what the customer wants is already available? The purpose of the problem definition is to clarify the situation so the true problem or need can be worked on. Communication with the people affected by the problem is essential. Make contact with as many people as practical, and discuss the situation with them. Go to the problem site to observe and collect information. Seeing the problem firsthand will clarify potential questions, and the physical restrictions will be observable and measurable. Take good notes.

4.1 Methods Of Problem Definition

A carefully thought-out problem definition includes as much information as possible relating to the problem. There are five general categories in which to accumulate project information. Case Study #2, concerning the movement of material between production departments, will be used to illustrate the steps of a typical problem solving situation.

The case study will be interrupted several times in this chapter with general problem definition discussion, and will also be continued in Chapters 5, 6, and 7.

Begin with a **basic statement of need**. A basic statement of need or want may have vagueness and generality. This is acceptable in an opening statement and allows more creative thought. The opening statement should allow general discussion and further clarification without assuming a solution. The basic statement of need should not include a preconceived solution.

CASE STUDY #2:
Parts moving problem; basic problem statement.

The statement "We need a four-wheel dolly to move this load of parts to the next department" includes the solution and restricts creative problem solving. A better statement, one that allows more creative solution ideas, is a question "How can we most easily and efficiently move these parts to the user department?"

Start problem clarification by listing **basic requirements**. As the basic problem statement is considered and discussed, certain desired outcomes and performance criteria should become apparent. These specifications should be listed or clarified with detailed sketches. They eventually become the technical goals for the project. Figure 2.14 is an example of product specifications resulting from a completed project.

CASE STUDY #2 CONTINUED:
Requirements.

1. Only one person should be involved in the movement of parts.
2. The movement of parts should take five minutes or less.
3. The movement of parts must be safe for the person doing the moving as well as for other people working in the area.

Basic limitations are as important to consider as basic requirements. As discussion and research on the problem continue, features may emerge that are considered undesirable, things you don't want the product to do. These limitations should be clarified and recorded with detailed sketches and/or word descriptions. There may also be constraints over which you have no control, such as relevant laws or codes. They are real constraints, and should be identified early so generated ideas can take them into account. Ideally, of course, the fewer restrictions the better, though all real problems have some constraints. There are also indirect constraints with engineering projects. They are the constraints of the effect of a change on the present status. All design projects will affect some component of a present system or of a related system. For example: If a four-lane bridge is built to replace ferry service, queuing is eliminated, approach roads may be inadequate, a company may be forced out of business, and tall-masted sailing boats may not be able to pass by when the water level is high. These side effects may cause the scope of the project to broaden (rebuild the access roads), may cause special features to be added (make the bridge a draw, swing, or lift bridge), or require a business relocation subsidy. Recognizing and including the consequences of project implementation early in the planning process will result in wider acceptance of engineering projects.

CASE STUDY #2 CONTINUED:
Limitations.

1. The parts must not come in contact with each other while moving because the surfaces have been ground to a 16-microinch surface finish.
2. A solution must be developed within ten days.
3. The budget for the project is limited to $10,000.
4. The union contract limits the weight one person may carry to 50 pounds (22.7 kg).
5. There is a building wall doorway between departments that is 8 feet wide and 8 feet high (2.44 × 2.44 m).

Everything that comes up during discussion about the project should be noted. Anything not clearly a requirement or limitation is listed as **other data**. This includes information about company policy, styling considerations, or available plant facilities. Law or code restrictions might also be specified, as well as other restrictions beyond the control of engineers. An example of an uncontrollable restriction is the contour map of a lot being considered for a shopping center.

It doesn't really matter whether a condition is listed as a requirement, a limitation, or as other data. The main objective is to allow room for a variety of considerations so a good framework and basic understanding of the problem can be established.

CASE STUDY #2 CONTINUED:
Other data.

1. Part orientation while moving is not critical, parts do not have an up or down side.
2. A variety of parts will be moved. Their weight ranges from 5 to 25 pounds (2.27 to 11.34 kg).
3. All of the parts are shafts that range in length from 6 to 16 inches (152 to 406 mm), and in diameter from 0.75 to 1.50 inches (19.05 to 38.10 mm).
4. Quantities of any given part to be moved will range from 10 to 100.
5. The distance to the next department is 150 feet (45.7 m), on the same floor.
6. That department uses the shafts for assembly into gear cases.

Most of the time when a new problem or project is started, **questions** are raised that cannot be answered immediately. Make note of those things that need to be answered during the course of solving the problem. Questions about performance specification, unknown restrictions, or related products commonly occur. As the answers to the questions are found, the scope of the problem may expand or contract.

CASE STUDY #2 CONTINUED:
Questions and answers.

1. How much floor space is available in the sending department and the receiving department? This will affect how often during the day parts must be moved.
2. Is part inspection required before moving or at the second department when received? This influences the time involved to prepare for moving, or the time for processing after the parts are received, as well as floor space required.
3. Can holes be put in the wall separating the departments? There might be a shorter way to the second department.
4. Is the wall a bearing or a partition wall? This is a logical follow-up question after question #3.

Answers to questions.

1. Parts are presently placed on flat pallets in a move area, and are moved whenever the move area gets crowded or when parts are needed in the next department. Parts are moved an average of four times a day. The pallets are moved with a hand operated pallet mover by the department parts expediter. The volume of parts produced has been increasing because two new grinders have been added to the department, making a total of eight machines.
2. Part inspection is done before the parts leave the machines so the parts are ready to move when they reach the move area.
3. and 4. Holes can be put in the walls, but they are partition fire walls, and there must be a heat activated automatic closer installed if an opening is cut.

If a new product is being considered, questions that need answering include the following:

1. Does this product replace an existing product, or does it add to the product line?
2. Is the product an extension of an existing series of models so that outer styling must be matched?
3. Are there parts already manufactured or purchased that should be considered for use rather than starting from scratch?
4. Are there existing plant facilities for manufacture, paint, assembly, and storage?
5. What size limitations for manufacturing exist relative to material handling, tonnage of presses, and bed sizes of machine tools?
6. What basic manufacturing methods are available and do the work stations have time capacity remaining to make additional parts?

Keep in mind that any statements made at this point in the project are not final. Many conditions can, and most likely will, be altered as the project proceeds. Keep in mind that a specific requirement on one project: "Power shall be a five-horsepower engine," might be a question on another project, "What horsepower is required?" In the first case, five-horsepower is a preliminary specification that can be changed, whereas in the second case, not enough information is available to make a preliminary decision. Perhaps the market survey isn't complete, or the power requirements of a required cultivation width aren't known. Don't let a project stagnate because some pieces of data are not available. Proceed and let data catch up. Work on the data available and keep the project moving.

The following example illustrates the five areas of problem definition for a proposed new product.

EXAMPLE 4.1:

New product preliminary problem definition.

1. Basic Statement—Design a garden rotary cultivator.
2. Preliminary Requirements
   - Power will be a five-horsepower gasoline engine.
   - The tines will be in the front.
   - Forward and reverse power operation will be available.
   - All controls will be reachable from the normal standing operating position.
   - Cultivating width will be 18 inches (457 mm).
   - Movement under power must stop if the controls are released.
3. Preliminary Limitations
   - Total weight will be 110 pounds (50 kg) or less.
   - Maximum storage area will be eight square feet (0.74 m^2).
   - Selling price will be $350 or less.
   - Maximum sales will be 1000 units per year, 70 percent in the spring, 30 percent the rest of the year.
   - The cultivator must be market ready for April selling in two years.
4. Other Data
   - This product is an addition to the current line of home garden equipment; no other cultivators are manufactured.
   - Exterior appearance should be compatible with existing products.
   - The manufactured components of a garden tractor now in production are larger than the expected size of the parts of the cultivator, so the size of existing machine tools and other manufacturing equipment is probably large enough.
   - Design the cultivator with features found on other products in the line so common parts, such as power train components and electrical controls can be used, and existing methods of manufacture will be adequate.
5. Questions
   - Will five-horsepower be adequate for an 18-inch (457 mm) cultivating width?
   - Should the drive mechanism be belt, gear, chain, or something else?
   - Should the tines come in segments, so narrow cultivation is possible?
   - Should the cultivator be self-propelled?

- Where should the starting controls be located?
- What parts should be painted? What color?
- Should any parts be plated?
- Should multiple use be considered so that the power unit can also be used for lawn mowing or snow blowing?
- Should the engine be two-cycle or four-cycle gasoline, or electric?
- Should regular maintenance be necessary or should the unit be maintenance free?

Remember that the lists are not necessarily complete or exhaustive. You will be able to add to all of them, particularly the questions. The more questions asked, the more thorough your thought process, and the greater the potential for creative solutions. An important part of problem identification is to make a record of details as they occur. A good beginning is important so that during the other stages of problem solving activities no aspect of the design will be neglected or overlooked.

When a problem is well defined, it is identified with a statement that includes expected results. Problem identification requires an understanding of the situation, specifying known relevant data, knowing what additional data is required, establishing limiting assumptions, and making a decision on priorities and personal values. Problem identification statements must be done carefully so they are not too restrictive or too vague to be approachable.

The following example illustrates how a problem statement can change based on different personal and environmental circumstances.

EXAMPLE 4.2:
Alternate problem statements.

Consider the different problem statements, all coming from the same situation.

1. My car engine has stopped, how do I get to work?
2. My car engine has stopped, how do I get it started?
3. My car engine has stopped, how do I stop safely?
4. My car engine has stopped, what do I do with the car while I go for help?
5. My car engine has stopped, how do I get the car out of traffic?
6. My car engine has stopped, what repairs will have to be done?

Problem statement #1 indicates my biggest concern is getting to work. Choices can include hitchhiking, calling a taxi, walking, or taking a bus.

Problem statement #2 indicates a desire to get the car restarted and continue on. Choices can include checking to be sure there is enough fuel, trying to get a push, jump-starting if the battery seems low, or calling emergency road service for help.

Problem statement #3 indicates I was traveling at some speed, and if the engine isn't running, I will not have power steering or power brakes. Various choices include slamming on the brakes, coasting to a stop on the shoulder (if there is one), or shifting to a lower gear to slow the car down gradually.

Problem statement #4 suggests I have already stopped, but perhaps in the middle of traffic or where there is no safe place to park. I could just leave the car where it is and hope for the best, or get help to push it out of the way.

Problem statement #5 definitely means I am in the middle of things and want to get out of the way. Getting a push or calling road service are among my choices.

Problem statement #6 might indicate I have been having trouble with the car and have recently spent money getting it fixed. Having more trouble now, and worrying about repairs might suggest I am ready to trade in the car and don't want to spend much money getting it repaired.

The next example illustrates how questions can be used to develop and clarify a problem when the initial statement has no restrictions or guidelines listed. These kinds of problems are often referred to as open-ended problems, as no one knows for sure where they will lead.

EXAMPLE 4.3:

Question approach to problem identification.

Consider the general problem statement "Design a desk."

Getting more information is a key to properly defining this problem and making clear what the parameters are. Among the questions that could be asked about designing a desk are:

1. What is the purpose of the desk? Study? Writing? Drawing? Storing material? Typing? Interviewing? Collecting money?
2. Who will use the desk? An adult? A child? A handicapped person?
3. How will the desk be used? Sitting down? Standing up? Straight-back chair? Chair on wheels? Wheelchair?
4. Will the desk need drawers? What for? Pencils and small articles? Papers? File folders? Books? Products to sell?
5. If drawers are needed should they be lockable?
6. Should drawers have bottom or side guides?
7. Should drawer guides be slides or rollers?
8. What material should be used for the desk? Solid wood? Plywood? Particle board? Metal? Plastic? Some combination?
9. What style is desired? Modern? Early American?
10. Is there a maximum permissible size? Does the desk have to fit a particular space?
11. Is there a maximum or minimum weight?
12. Should the desk be movable by one person?
13. Should the desk be stackable for storage?
14. Is color a concern for matching a particular decor?
15. Should there be room for permanent use of a typewriter, word processor, or computer?

16. Is a roll top appropriate?
17. What cost range is appropriate?
18. How many desks will be made?
19. What is the time schedule for completing the project?
20. How many people will be working on the project?
21. Are any differences required for right- or left-handed people?
22. Is a privacy screen desirable?
23. What height is comfortable?
24. Is a swinging door for a storage area appropriate?
25. Should the desk be made to be assembled by the customer?
26. What desks are presently available that come close to satisfying the current need?

Getting answers to questions like these will get the project off to a good start. Be sure to document answers in writing, including the source. One possible set of preliminary specifications and project guidelines coming from the questions are:

1. The desk will be used by a computer programmer.
2. The user will be a average sized adult. Refer to anthropometric charts in reference (11) and (12) listed in Section 6.4 for dimensions of average adults. For example, 99 percent of adult males have a sitting hip width less than 16.2 inches (411.5 mm), while the maximum is 18.11 inches (460.0 mm).
3. The programmer will be sitting down, so the desk needs space for leg clearance.
4. The desk needs one drawer for stand-up storage of $8\frac{1}{2} \times 11$-inch file folders, and one for pencils, pens, paper clips, and a stapler. A shelf or other storage area is needed for 24 inches (610 mm) of reference books, 12 inches (305 mm) maximum height.
5. Security is a problem so each drawer must be lockable. Same key for all drawers.
6. Drawers must slide and not cause static electricity or loud noises, but guide location is arbitrary.
7. No preference to drawer slides or rollers unless static electricity generation is affected.
8. Wood is preferred material, it has less potential for stray magnetic fields.
9. No preference on style, function is main guideline.
10. The maximum size allowed is 32 inches (813 mm) deep and 60 inches (1524 mm) wide.
11. No weight limitation.
12. Desk does not have to be movable by one person.
13. Stacking storage not required.
14. Natural wood color preferred to blend with the general office decor.
15. Room for a computer terminal required; a terminal and movable keyboard require 20 inches (508 mm) width and 25 inches (635 mm) depth, minimum.

16. No roll top.
17. Budget has been set at $400 maximum per desk.
18. Department will have ten people, five now, five more in about one year.
19. Five desks must be available in 90 days.
20. One person is assigned to the design of the desk, construction may be contracted out.
21. No right-hand or left-hand problems anticipated.
22. Privacy screen required.
23. See reference in item 2.
24. Do not eliminate swinging doors from possibilities.
25. Desks should come ready to use.
26. Designer will have to check catalogs, and local furniture and office equipment stores.

4.2 Related Activities

There are many activities in which an engineer might engage while preparing a problem statement. No single source will always be the best, so a wide range of possibilities must be kept in mind.

Conduct market surveys to establish customer wants and needs, price limits, and sales forecasts. A market survey can include personal interviews, telephone surveys, or mail questionnaires. Important sources of statistical business information include:

- *Census of Manufacturers,* U.S. Department of Commerce
- *Statistical Abstract of the United States,* U.S. Bureau of the Census
- *Metal Statistics,* American Metals Market, New York
- *Minerals Yearbook, Vol. 1, Metals and Minerals,* U.S. Bureau of Mines

Check products and material available for purchase on the market, and current standards to see how well the need is already being met. Use references such as:

- Annual catalog, American National Standards Institute.
- *Annual Book of ASTM Standards.* The 1981 Book of Standards is in 48 volumes and contains over 6400 standards.
- Chemical engineering catalog: *The Process Industries' Catalog*, Van Nostrand Reinhold.
- *Index of Specifications and Standards*, U.S. Department of Defense.
- *Index of Federal Specifications and Standards*, U.S. General Services Administration.
- *MacRae's Blue Book* (Hinsdale, IL), a multivolume collection of manufacturers' catalogs.
- National Association of Manufacturers.

- *Sweet's Catalog File* (McGraw-Hill, New York), a compilation of manufacturers' catalogs with emphasis on machine tools, plant engineering, and building materials.
- *Thomas Register of American Manufacturers* (New York), a comprehensive directory of addresses of manufacturers arranged by product.
- *VSMF Design Engineering System,* a compilation of vendors' catalogs and product specifications for mechanical and electronic parts; on microfilm, updated monthly.

Check patent records to determine the extent of workable ideas and protected ideas, as well as to determine a progression of technology. Use the weekly publication, the *Official Gazette* of the U.S. Patent Office, for the latest patents. For reference by class and subclass, you must visit the Public Patent Search Facilities of the Patent and Trademark Office, 2021 Jefferson Davis Highway, Arlington, VA. Thirty major libraries in the United States also have patent records arranged in numerical order.

Read books written on a specific subject to determine its history and development. This will provide data and insight into what has been tried and was successful, and also what has failed and why. This activity might also uncover legal concerns that should be considered, such as previous lawsuits, court rulings, governing laws, and code restrictions.

1. Dictionaries
   - Ballentyne, D.W.G. and D.R. Lovett, eds. *A Dictionary of Named Effects and Laws in Chemistry, Physics and Mathematics*, 3rd ed. London: Chapman and Hall, 1970
   - Bennett, H., ed. *Concise Chemical and Technical Dictionary*, 4th ed. London: Arnold, 1986
   - McGraw-Hill editors. *Dictionary of Engineering.* New York: McGraw-Hill, 1985
   - McGraw-Hill editors. *Dictionary of Mechanical & Engineering Design.* New York: McGraw-Hill, 1985

2. Thesauruses
   - Engineer's Joint Council editors. *Thesaurus of Engineering and Scientific Terms*. New York: Engineers Joint Council (AAES), 1969
   - Roget, Peter M. *Roget's Thesaurus of English Words.* New York: St. Martin's Press, 1978
   - *Thesaurus of Metallurgical Terms*, 4th ed. American Society for Metals. Metals Park, Ohio: American Society For Metals, 1980

3. Encyclopedias
   - Besancon, R.M., ed. *Encyclopedia of Physics*, 3rd ed. New York: Van Nostrand Reinhold, 1985
   - Considine, Douglas M., ed. *Encyclopedia of Chemistry*, 4th ed. New York: Van Nostrand Reinhold, 1984.
   - *McGraw-Hill Encyclopedia of Science and Technology*, 5th ed. New York: McGraw-Hill, 1982

4. Handbooks
   - Baumeister, T., ed. *Marks' Standard Handbook for Mechanical Engineering*, 9th ed. New York: McGraw-Hill, 1987

- Beaton, C. F. and G. F. Hewitt, eds. *Physical Property Data for the Design Engineer.* New York: Hemisphere Publishing, 1988
- White, John A., ed. *Production Handbook*, 3rd ed. New York: John Wiley & Sons, 1986
- Crocker, S., and R.C. King, eds. *Piping Handbook,* 5th ed. New York: McGraw-Hill, 1967
- Cummins, A.C., and I.A. Given, eds. *SME Mining Engineering Handbook.* New York: Society of Mining Engineers, Inc., 1973
- Fink, Donald G., and H. Beaty. *Standard Handbook for Electrical Engineers,* 12th ed. New York: McGraw-Hill, 1986
- Fink, D.G., and D. Christiansen, eds. *Electronics Engineers' Handbook,* 3rd ed. New York: McGraw-Hill, 1988
- Harper, Charles A., ed. *Handbook of Electronic Systems Design.* New York: McGraw-Hill, 1979
- Hewitt, Geoffrey F., ed. *Hemisphere Handbook of Heat Exchanger Design.* New York: Hemisphere Publishing Corp., Harper & Row, Inc., 1989
- Higgins, L.R., and L.C. Morrow: *Maintenance Engineering Handbook,* 44th ed. New York: McGraw-Hill, 1987
- Karassik, I.J., et al. *Pump Handbook.,* 2nd ed. New York: McGraw-Hill, 1985
- Maynard, H.B., ed. *Industrial Engineering Handbook*, 3rd ed. New York: McGraw-Hill, 1971
- Merritt, Frederick S., ed. *Standard Handbook for Civil Engineers,* 3rd ed. New York: McGraw-Hill, 1983
- Parmley, R.O. *Standard Handbook of Fastening and Joining,* 2nd ed. New York: McGraw-Hill, 1989
- Slote, Lawrence, ed. *Handbook of Occupational Safety and Health.* New York: John Wiley & Sons, 1987
- Tapley, Byron D. *Handbook of Engineering Fundamentals,* 4th ed. New York: John Wiley & Sons, 1989

5. Reference Books
    - Blake, L.S., ed. *Civil Engineer's Reference Book,* 4th ed. Boston: Butterworth, 1989.
    - Siemens Teams of Authors. *Electrical Engineering Handbook* . New York: John Wiley & Sons, 1982
    - Parrish, A., ed. *Mechanical Engineer's Reference Book*, 11th ed. Boston: Butterworth, 1973

Read technical journals to learn about applications and developments that are current. No attempt will be made here to select key journals from a list that includes over 50,000 titles. Reference to them, however, is guided by abstracting and indexing services such as:

- *Applied Mechanics Reviews*
- *Applied Science and Technology Index* (formerly *Industrial Arts Index*)
- *Building Science Abstracts*
- *Ceramic Abstracts*

- *Chemical Abstracts*
- *Computing Reviews*
- *Corrosion Control Abstracts*
- *Energy Index*
- *Engineering Index*
- *Environmental Index*
- *Fuel Abstracts*
- *Highway Research Abstracts*
- *Instrument Abstracts*
- *International Aerospace Abstracts*
- *Metals Abstracts*. Combines *ASM Review of Metals Literature* (U.S.) and *Metallurgical Abstracts* (British)
- *Nuclear Science Abstracts*
- *Public Health Engineering Abstracts*
- *Science Abstracts A: Physics*
- *Science Abstracts B: Electrical and Electronics Abstracts*
- *Solid State Abstracts*

Computerized data bases for abstracting and indexing are listed in Table 4.1.

Attend trade shows to see current products, collect literature, and ask questions.

Check your own company records for any research work that has already been done on your project. Include discussions with those who might have worked on the project before you.

Contact component product suppliers. They are generally eager to talk about their products and will contribute specification sheets and offer product demonstrations.

Contact government agencies. There may be policies, codes, or laws that will affect your project. This would be particularly important if your project involved waste by-products. The Government Printing Office is the source of all federal government publications. A Monthly Catalog of U.S. Government Publications lists all available publications. The following list is a brief guide to government technical literature:

TABLE 4.1

Abstracting and indexing databases

Data Base	Supplier	Number of abstracts
Metadex	American Society for Metals in cooperation with The Metals Society of London	555,000
CA Search	American Chemical Society	4,000,000
Compendex	Engineering Index, Inc.	766,000
NTIS	National Technical Information Service, U.S. Dept. of Commerce	725,000

- Directory of Information Resources in U.S., Vol. 1, Physical Sciences Engineering, 1971. National Referral Center, Sciences & Technology Division, Library of Congress
- Monthly Catalog of U.S. Government Publications, Government Printing Office
- Government Reports Announcements and Index (semi-monthly), (prior to 1971 was U.S. Government R&D Reports), National Technical Information Service, U.S. Dept. of Commerce. References R&D reports for government-sponsored reports plus translations of foreign material. Copies of reports may be purchased from NTIS (formerly DDIC and ASTIA).
- Scientific and Technical Aerospace Reports, (STAR), Abstracts compiled by NASA, Purchased from Government Printing Office
- Nuclear Science Abstracts, Abstracts compiled by Department of Energy, purchased from Government Printing Office

The government has also established specific information centers, which are necessary for the large amount of specific technical data available, including:

- Metals and Ceramics Information Center, Battelle Columbus Laboratories
- Mechanical Properties Data Center, Battelle Columbus Laboratories
- Plastics Technical Evaluation Center
- Shock and Vibration Information Center
- Machinability Data Center, Metcut Research
- Thermophysical and Electronic Properties Information Analysis Center
- Thermodynamics Information Center
- Atomic and Molecular Processes Center
- Metal Matrix Composites Information Analysis Center
- The Smithsonian Science Information Exchange (SSIE). The SSIE is the only source of information on all federally funded research in progress. The data are in machine-readable form and can be searched by computer.

Specific government agencies may also be of help. Some of the better known agencies are:

- U.S. Chamber of Commerce
- Interstate Commerce Commission
- Federal Communications Commission
- National Aeronautics and Space Administration
- U.S. Department of Agriculture
- U.S. Department of Commerce
- U.S. Department of Health, Education and Welfare

Contact trade associations for literature. Trade associations produce and review voluntary standards, and these may affect your project. Those that have produced a substantial number of standards include:

- American Petroleum Institute
- Association of American Railroads
- Electronics Industries Association
- Manufacturing Chemists Association
- National Electrical Manufacturers Association

Contact product testing organizations for information concerning specific products. These organizations include:

- National Fire Protection Association (NFPA)
- Underwriters Laboratories, Inc. (UL)
- American Gas Association
- American Society for Testing and Materials
- American Society of Heating, Refrigerating and Air Conditioning Engineers

Contact consulting firms that deal with the general area of your project. There are many specialty companies whose experiences would be most helpful. The telephone yellow pages may list companies to contact.

Contact finance and insurance companies. They may have price index, market money rates, bond discount data, product liability data, and other financial data that would prove helpful.

Contact professional and technical societies. They have also made important contributions through standards activities. Those societies that have contributed to standards development include the following. See reference (4) for addresses and details.

- Aerospace Industries Association
- American Concrete Institute
- American Institute of Steel Construction
- American Society of Agricultural Engineers
- American Society of Mechanical Engineers
- American Welding Society
- Building Officials and Code Administrators International
- Electronic Industries Association
- Institute for Interconnecting and Packaging Electronic Circuits
- Institute of Electrical and Electronics Engineers
- Instrument Society of America
- National Electrical Manufacturer's Association
- National Safety Council
- Prestressed Concrete Institute
- Society of Automotive Engineers
- Technical Association of the Pulp and Paper Industry

4.3 Cautions

Be aware of several things at this point. First, realize that for a given project, you will not use all the sources mentioned; in fact, you may not use any. The project could be handed to you with most of the preliminary work done, or you may already be familiar with the situation and will add details and solutions from experience. If you know the river's depth from previous encounters, a quick solution can be, "Jump out of the boat and wade to shore; the water is only two feet deep." The possible activities are just that, possible. Use them if you feel unsure, if you need more data, or if you simply want to gain additional insight into the problem.

Second, realize that you should not wait until a problem is handed to you to search out information. Whether your company deals primarily in office equipment, electronic equipment, computer services, conveyor equipment, or space shuttles, you should study the product and the related industry on a regular basis. Read books and periodicals, attend seminars, go to product shows, visit with customers and dealers. Keep informed and up to date; technology and the times change and you must change and develop with them.

Third, realize that projects in most companies are team projects, and expertise will come to the project from other people and other departments. There is too much for one person to know, so make good use of the knowledge of your co-workers. Check with others even if they are not assigned to your project; no source of information should be neglected.

Fourth, realize the difference between a statement of fact and a problem statement. Some people get themselves trapped by issuing a statement of fact and not defining a problem, as the following example illustrates.

EXAMPLE 4.4:

Circumstances and problem statements.

1. "My pen is broken," is not a problem definition; it is a statement of fact. The problem may be to find a service center capable of fixing the pen, or to find something with which to write.
2. "My car won't start," is a statement of fact. The problem to be solved could be "How do I start my car?" or "How do I get to work?" or "How do I get the car out from in front of the garage so I can use the car inside the garage?"
3. "The furnace won't run," is a statement of fact. The problem to be solved could be "How do I get the furnace running again?" or "How do I keep the pipes from freezing?"

Your first introduction to a problem may be preliminary specifications established by your company or your client. Also included might be basic restrictions such as sales price, sales volume forecasts, and a time schedule of development and market introduction. If this preliminary information is not adequate, or you feel there are too many unknowns, ask questions. The earlier clarifications are made, the earlier serious work can proceed. To do a good job of solving a problem, you must first do a good job of identifying the problem. This requires knowledge and understanding of the background of the project.

During the problem identification activity, ideas for solution generally come up as requirements and limitations are discussed. This is normal and should not be suppressed. In a separate record, start listing any ideas that come to mind during the identification process. Do not dwell on solution ideas, and do not spend time now trying to think of the solve-all problem solution. By concentrating on the identification process, you will get the boundary conditions clarified and be in a better position to generate ideas later. The main thing to do now is to think of the problem and record all the items that can possibly have an effect on subsequent activities. Getting the basis of the project clear will make the other steps of the procedure easier. None of the activities in the design process should be done without due regard for all the other steps.

4.4 System Problems

System problems can also be handled with the problem solving methods described. The main difference is that the questions asked will have to deal with discovering the present inputs, procedures, and expected outputs. System problems take a lot of detail work to find out what the users of the proposed system want. Communication skills are most important and can make or break a project. You must ask the right questions, and at the same time not offend people about the way they do their job. Systems involving people have habit built into them. People get comfortable with what they do, and resist change. For this reason it is often hard to find out what a system is either supposed to do, or how it does it. People may feel protective and will not answer completely or truthfully, or a particular event may not happen regularly and is easily forgotten. Later when this event occurs the engineer might be asked why it wasn't considered in the system design. Always maintain contact with the users during a system design. They should have numerous ideas, operational data, good judgment, and they can help evaluate your ideas. The following example illustrates the five areas of problem identification for a system problem.

EXAMPLE 4.5:
Preliminary problem definition for a system.

1. Basic Statement—Design a system to monitor and report on production material inventory.
2. Basic preliminary requirements
    - System must provide a weekly updates printout.
    - Updates must include breakdown of purchases, returns, factory use, scrap, and include costs associated with each category of material.
    - System must be on the central computer so any office can inquire about material status.

3. Basic limitations
    - System must be installed and operating in 6 months.
    - Computer terminals are not available in all departments that handle material.
4. Other data
    - Categories of material include steel sheets, steel plates, iron castings, electrical components, and rubber products.
    - The present system is manual and updated only once per quarter.
    - Shipments of material are received every day.
    - The last complete inventory was done three months ago.
    - The next complete inventory is scheduled to be done in nine months.
5. Questions
    - What support is available from the computer programming staff?
    - How much do computer terminals cost? What is their availability?
    - Does the present computer have connector ports available for additional terminals?
    - Does the present computer have enough storage capacity for added programs and data files?
    - Does the system have to be on line, or are daily update entries acceptable?
    - Is there room to locate computer terminals in all departments.
    - Will data entry be made by people experienced with computers or will training sessions be required?
    - Are any of the data already on the computer in accessible files: material deliveries, schedules, uses, or cost?

System designs are no less important to a profitable and well-run company than product design. Keeping track of the records of a company allows management to make timely and well informed decisions concerning purchases, investments, loans, new product introductions, personnel requirements, new markets (including foreign countries), and product pricing.

4.5 References

1. Caplan, Ralph. *By Design*. New York: St.Martin's Press, 1982.
2. Evans, Powell, Talbot, ed. *Changing Design*. New York: John Wiley & Sons, 1982.
3. Faupel, Joseph and Franklin E. Fisher. *Engineering Design: A Synthesis of Stress Analysis & Materials Engineering,* 2nd ed. New York: John Wiley & Sons, 1981.
4. Koek, Karin E. and Susan Boyles Martin, eds. *Encyclopedia of Associations*, 22nd ed. Detroit, Mich.: Gale Research Company, 1988. A guide to over 25,000 National and International Organizations.
5. Middendorf, William. *Engineering Design*. Boston: Allyn & Bacon, 1969.
6. Robinson, Judith Schiek. *Subject Guide To U.S. Government Reference Sources*. Littleton, Col.: Libraries Unlimited, Inc., 1985.

7. Ross, Steven S. "Rules' Role." *Graduating Engineer*, March 1985.
8. Smith, Ronald. *How To Plan, Design & Implement a Bad System*. New York: Petrocelli Books, 1981.
9. Whyte, R.R., ed. *Engineering Progress Through Trouble*. London: The Institute of Mechanical Engineers, 1975.

4.6 Exercises

Items 1–10 are potential problem statements. List at least five questions that you would ask to clarify each situation so that the real problem can be identified and solved. Some of the statements may result in more than one problem.

1. As I was driving to work, the hood of my car popped open.
2. As I was cleaning fish, the knife slipped and I cut myself.
3. This door won't stay open while I carry packages into the house.
4. When I graduate, where will I go to look for a job?
5. Next spring I want to plant a garden.
6. Next summer I want to add a bedroom to my house.
7. What will I do with my income tax refund?
8. If I fail my math class, will I be able to continue my engineering studies?
9. Where is that smoke coming from?
10. I can build a better lawn mower than that one.
11. Consider the following problem identification statement. Do you feel the statement is satisfactory? Why or why not? Outline the next series of activities to be followed on the project. Be specific.

    Design a writing desk that has at least one drawer for pencils and small office supplies, one drawer for letter size file folders, and one drawer to store necessary sales books in. Additionally, the desk will be used by a person sitting down. There should be ample room on top of the desk for a lamp, a telephone, an open loose leaf binder, and room to write out quotation sheets. A typewriter will not be used. Customers will be seated opposite the desk, so a privacy screen is required, and the desk must be attractive so the customer will feel at ease and comfortable.

12. Compare the advantages and disadvantages of the following three methods of conducting a market survey.
    a) Personal interview
    b) Telephone interview
    c) Mail questionnaire

    Often a statement of fact will generate many different possible problems and courses of action. For Exercises 13–15, list different possible problems that could be the result of the statement. After listing possibilities, rank them in order of most probable. Give reasons for your ranking.

13. I flipped the light switch and the light did not come on.
14. I tried to unlock the car door and I could not.
15. I pulled the rope on the lawn mower starter and the engine did not start.
16. Give a more complete problem definition for each of the following:
    a) Design a lamp
    b) Design a bike rack
    c) Design an apartment building
    d) Design a computer
    e) Design a mechanical heart
17. Which of the problems listed in Exercise 4.16 would you feel most confident in undertaking? Why?
18. A camping trailer, a powered vehicle to pull it, and a trailer hitch constitute a system problem because the components can be purchased, but they must match each other for safe and efficient use. List some considerations that would be necessary for this system to work effectively.
19. Make a list of as many components or subsystems you can think of for a bicycle. Each item listed may have a unique or multiple function.
20. As an example of code restrictions on design, consider floor loading levels required in buildings. Residential homes can be designed to 40 lbf/ft^2, but public buildings require 100 lbf/ft^2 design. Discuss the reasons for the differences.
21. A TV situation comedy was being acted out in a building during a rain storm, and someone entered the scene and said, "There is five feet of water on the roof." Even if the roof had a five foot high edge to contain the water, would you expect a building still to be standing if the water depth on the roof did reach five feet? Quantify your answer.
22. Many design problems have conflicting control conditions, that is, restrictions that cannot be met simultaneously. For example, suppose electric lights had to produce light and not reach a temperature higher than normal body temperature. Those would be conflicting restrictions, because the increase in temperature is a result of the light producing phenomenon. Give other examples of conflicting restrictions in design.
23. List one problem situation that each of the following objects helps solve if used properly. Then give an alternate solution to the problem. Compare both solutions.
    a) Airplane
    b) Electric motor
    c) Television
    d) Elevator
    e) Radar
    f) Computer
    g) Linotype machine
    h) Electric heat
    i) Phonograph
    j) Table lamp
    k) Automobile
    l) Car bridge

- *m*) Telephone
- *n*) River dam
- *o*) Electric water pump
- *p*) Traffic light
- *q*) Rifle
- *r*) Car signal lights
- *s*) Non-stick cooking utensils

24. A participant at a conference in Washington D.C. looked foreign and was wearing a turban. Another participant commented, "I didn't realize foreigners could attend this seminar. After all, it is paid for by the U.S. Government." Discuss any implied or stated restrictions, assumptions, generalizations, boundary conditions, limitations, or controls that were made by the person making the comment.

25. How were metals welded together before the invention of arc and acetylene welding? Are any methods similar to the first method of welding still being used? Describe them.

26. Many times people try to use machines beyond the range of use they were designed for. Example: A sewing machine designed for sewing cloth is used to sew leather, and the needle frequently breaks. Give other examples of attempts at using machines beyond their design limits and the possible consequences of that use.

27. Studying only the current solution idea is often neglecting the real problem. This frequently occurs and the situation is called "treating the symptom instead of the cause." In each of the following cases discuss what the real problem might be.
    - *a*) A student was often late for class and blamed rush hour traffic, saying, "I tried different routes, but they're all crowded."
    - *b*) A bicycle rider continually got the leg of his pants caught between the chain and the sprocket and blamed the chain guard design.
    - *c*) A student always had errors in computer programs and used poor typing skills as an excuse.
    - *d*) A bracket on a machine was breaking with high frequency under warranty. The engineer in charge wanted to increase the cross-sectional area of the bracket.

28. Complete the problem identification steps for one of the following design problems or select one of the problems from those listed in Section 7.6.
    - *a*) A reserved seat system for a passenger train
    - *b*) A taxi service system
    - *c*) Retractable window awnings
    - *d*) An automatic manufacturing and assembly line for picture frames
    - *e*) An automatic brick manufacturing process
    - *f*) An automatic ball bearing sorting machine
    - *g*) A device to control ambient temperature for a light bulb life experiment

29. List and discuss some related concerns that are indirectly controlled or influenced by each of the following engineering projects.
    - *a*) A dam on a river
    - *b*) A subway transportation system
    - *c*) A skyscraper office building
    - *d*) A downtown high-rise apartment complex

5 Problem Solution Idea Generation

"The more ambitious plan may have more chances of success."
♦ G. POLYA

The next activity on the road to successful design and problem solving, after adequately defining the problem, is to generate solution ideas. Generating ideas will exercise your creativity. Creativity is always important, but when you are starting with a blank page or a blank computer screen, originality and inventiveness are essential. Generating ideas does not cost much at this point, just the thinking time and a few pieces of paper or some computer time. Unencumbered thinking allows you to approach the problem with diversity and ingenuity.

The more ideas you formulate, the greater the chance you have to arrive at a great solution instead of just a mediocre one. Concentrate your effort on the generation of an ample quantity of ideas, as well as quality ideas. The first idea that comes along should not be judged the best solution. Accepting the first solution idea that comes along is one of the most frequent causes of failure for beginning engineers. Of course if time is very short this guideline becomes less critical. As your boat nears the waterfall you aren't going to get out a pad of paper and start listing alternative solutions to your predicament. In most engineering projects, however, time is needed to contemplate various approaches to a problem; sometimes several days, weeks, or months are required.

If you have done a thorough job of problem identification, you already have a start on the generation of ideas. The activities used to research the background of a project during problem identification in Section 4.2 are also sources of ideas.

Another source of ideas is the series of questions that were asked during problem identification. Solutions are certainly being suggested when someone asks if a mechanism should be driven by a belt, gears, or a chain. Recall that there are no sharp dividing lines between the activities used to solve problems. The generation of ideas is discussed as a separate activity in this chapter, but all the activities in the problem solving procedure must be meshed to achieve optimum results.

Search existing products for solution ideas to avoid making the same errors that others have made. Remember that a shortcoming in an existing product design is one source of problems, and one of your objectives is to avoid design shortcomings. Whether examining a competitor's product or your own, you can determine if there are still problems to be solved, or if the same job can be done less expensively or more safely. There is no single answer to a design problem. Consider, for example, all the styles of lawn mowers. They all answer the same basic problem question: "How can we cut the grass?"

A second reason to search existing products is to find out the current state of the art. Don't reinvent the wheel conducting research and development in an area that has already been adequately studied and developed. Review the basic concept, the theory, and the details of an existing technology if the need exists to apply it differently, but try to use the work of those who have come before you in an efficient and effective manner.

Engineers must also be able to call upon their own experience and creativity for ideas. Experience comes with time, and each engineering project you work on will increase your exposure to new ideas and techniques. Living through an experience will deepen your understanding, broaden your expertise, and enable you to draw on personal experience as you undertake more advanced problems.

5.1 Specific Considerations

Many people have trouble at the idea stage of a project; the paper is blank and they have difficulty getting started. Drawing the first line, or thinking of the first idea is often a roadblock to problem solving. Generating ideas requires more creativity than other activities performed during the problem solving procedure. Review the roadblocks that were discussed in Section 3.3.

Ideas don't cost money, and the more ideas that can be generated at the beginning of a project, the better will be the chance of finding an acceptable solution. A good idea generated too late may be costly in time and money spent developing a poorer idea. An engineer can improve the success rate for generating good ideas if some basic guidelines are followed.

Review the original problem statement, including all the questions, to refresh your mind about the specific task at hand. This is particularly important if you are working on more than one problem at a time, if some time has elapsed since you have worked on the problem, or if it just seems right to take a fresh look at the situation. Review is a reminder of the restrictions, limitations, and the boundary conditions within which you must work. Reviewing the unanswered questions will remind you of data that need to be gathered.

5.1.1 Break Problem Into Parts

Consider breaking a large problem into smaller problems. This will not be necessary if your problem is simple, such as "Find something to put in front of the tire so the car doesn't roll downhill.'' In the case of a project of great complexity, however, breaking the problem into sub-problems often makes idea generation easier. One way to break a problem into sub-problems is to address all functional inputs and outputs that can be identified. List specifically what the design result is to accomplish. One important sub-problem of all physical solution problems is material selection. Review materials and their properties to promote idea generation. The following two examples illustrate possible sub-problems of a larger problem.

EXAMPLE 5.1:
Design sub-problems.

Consider a primary problem of developing a surface vehicle for moon exploration. Sub-problems to consider include:

1. Power source; possibilities include solar, fossil fuel, or batteries.
2. Steering system; possibilities include manual, hydraulic, gears, or electric motor.
3. Motion control; possibilities include manual, mechanical linkage, radio control, umbilical cord from a control station, or on-board computer control.
4. Terrain contact; possibilities include rubber tires, driving tracks, moving "feet," rolling balls, or rolling cylinders.
5. Suspension system; possibilities include air cushion, hydraulic, springs, or direct with no damper.

EXAMPLE 5.2:
Design sub-problems.

Consider the design of a commercial building. Sub-problems would include site preparation, water supply and distribution, electric supply and distribution, the foundation, the exterior walls, the interior walls, windows and doors, people flow patterns, elevators, escalators, furniture, interior decorating, landscaping, heating, cooling, the roof, vehicle access, and parking. Each of these sub-problems generates its own set of concerns. On large scale projects, a different group of engineers might work on each sub-problem. Each group is required to coordinate its work with that of the other groups to be sure the solution to each sub-problem fits together with the solutions to all the other sub-problems. Before the final solution is determined, all the sub-answers must be meshed together to create the best overall solution, requiring more compromise.

5.1.2 Value Analysis

Value analysis is a technique that analyses a product or solution idea by breaking the problem into sub-problems using verb-noun functional expressions. Each verb-noun sub-problem expression is considered independently and ideas are generated for each one. Then the sub-problem ideas are joined for a total solution. This method assists in determining the essential functions and also the nonessential functions. Ordinarily, value analysis is used to evaluate an existing design, but the procedure is equally valid for helping to describe projected functional results.

The basic emphasis of value analysis is to consider the cost of providing a given function. Functional evaluation is aided by describing requirements in simple verb-noun statements. Phrases such as "support weight," "contain fluid," and "punch hole" are used. An overall project function might be more general, such as "sort mail," compared to a specific component function such as "weigh envelope" or "check stamp."

Each product, part, system, or procedure can be broken down into cost/function relationships. Individual functions must support the overall project goal. If a part has no function relative to the overall goal, it should be eliminated. There are no tables of minimum or maximum cost/function ratio values. The values are used only for comparison between ideas. Judgments about the maximum allowable cost for a function will be made, and the objective is to improve on that value. Most companies have standard cost procedures that use the resources of process engineering, plant layout, tool design, purchasing, and the accounting department to determine the necessary costs.

Value analysis places emphasis on creativity to determine other methods of performing the same function at less cost. Higher value is placed on a solution that accomplishes the same function with less cost. Guidelines such as cost per horsepower, cost per mile/hour, cost per pound, cost per inch-pound of torque, cost per unit of area covered, or cost per unit volume stored are often used. Understand that to make for less cost does not mean to make less reliable. The required task must be performed satisfactorily or the solution is not considered acceptable. The following example illustrates the verb-noun approach of value analysis to generate ideas, even though some of the identifying phrases are longer than two words.

EXAMPLE 5.3:
Environmental problem.

Consider the environmental problem of contaminated ground water and some broad objective functional sub-problems.

1. Raise ground water level.
2. Reduce natural contaminants (salts, sulfur).
3. Reduce human caused contaminants (pesticides, fertilizer residue, runoff from livestock yards).
4. Control use of water.

When these broad objectives are broken down in more detail, sub-problems and sub-problem solution ideas are more apt to come to mind.

How can ground water levels be raised? Increase precipitation. Reduce water runoff. Reduce water use. Reduce population. Increase water absorbing vegetation. Reduce evaporation.

How can natural contaminants be reduced? Reduce irrigation because it promotes salt accumulation in the ground. Pump water only from "clean water" areas.

How can human-made contaminants be reduced? Eliminate the use of septic tanks. Reduce the use of agricultural chemical fertilizers. Reduce the use of pesticides. Encapsulate all wastes to avoid leakage. Reduce the size and number of livestock yards. Collect animal and human wastes and process them for natural fertilizers. Eliminate land fills. Stop the manufacture of dioxins (chlorinated hydrocarbon compounds), organic solvents, and PCBs (polychlorinated biphenyls). Send all waste to outer space.

How can water use be controlled? Control flooding and store water for dry periods. Use pump-sprinkler irrigation rather than flood irrigation. Build more dams. Melt icebergs for a fresh water supply. Place automatic volume control shutoffs on home and business water supplies.

Notice that some of the ideas will cause additional problems to be considered and the scope of the problem to become larger. Many engineering problems have implications that prevent an arbitrary boundary cutoff. The related implications cannot be ignored. For the interested reader, reference (7) has an excellent discussion of the ramifications of water use control.

5.1.3 Combine Ideas

Another method to use, when generating ideas, is to try different combinations of the separate sub-function ideas, and then think of ways to put the parts together. Placing sub-function ideas together usually leads to new ideas. The number of possible combinations increases rapidly as the number of sub-functions and the sub-function solution ideas increase. If three sub-functions each has five ideas then there are $(5)(5)(5) = 5^3 = 125$ combinations to consider. The next example illustrates the combination of sub-problem procedure.

EXAMPLE 5.4:

Sub-problem combinations.

Reconsider the problem of designing a rotary cultivator. Some of the sub-problems and related ideas are:

1. Power source ideas
    - *a*) Electric motor with cord plug in
    - *b*) Electric motor with battery
    - *c*) Gasoline engine, four-cycle, vertical shaft
    - *d*) Gasoline engine, four-cycle, horizontal shaft
    - *e*) Gasoline engine, two-cycle, vertical shaft
    - *f*) Gasoline engine, two-cycle, horizontal shaft
    - *g*) Gas turbine engine
    - *h*) Diesel engine
    - *i*) Air motor with compressed air from a compressor
    - *j*) Hydraulic motor
2. Control ideas
    - *a*) Push-pull controls similar to bicycle brake cables
    - *b*) Solid rods, push or pull

      c) Turning crank
      d) Electric solenoid
      e) Hydraulic, similar to actuating automobile brakes
3. Ideas to power the drive and power tine rotation
      a) Belt drive, single or multiple
      b) Chain drive similar to bike chain, single or multiple strands
      c) Gear drive similar to motor bike
      d) Air motor
      e) Centrifugal clutch
      f) Flexible shaft
      g) Hydraulic motor
4. Ideas for a support frame
      a) Weldment from bent and formed steel.
      b) Cast iron or cast aluminum
      c) Molded plastic
      d) Bolted hardwood (oak, hickory)

The total number of combinations for the four sub-functions = (10)(5)(7)(4) = 1400. One of the combinations is a gasoline engine, two-cycle, vertical shaft, using push-pull cable control, chain drive on a molded plastic frame. Additional ideas must now be generated to put these pieces together. If tines rotate about a horizontal axis, and the power shaft is vertical, then the connection between them must change the direction of motion. Possibilities include a gear box to use with the chain drive, a friction wheel disk drive, or belts to change direction and chain connect before or after the change in direction.

5.1.4 Research

Research available sources of information for ideas, many of which are listed in Section 4.2. The more sources you use, the more ideas you should be able to generate. Don't let false pride (another roadblock to creativity) get in the way of obtaining ideas from others. It's what you do with the ideas that is important. Of course you will not steal a trade secret or infringe on patent or copyright protected information. If these sources yield the best workable ideas, then proper license and payment contracts can be worked out; obtain adequate legal counsel.

The use of a dictionary or thesaurus should not be overlooked as a source of ideas. Alternate meanings and methods of expressing thoughts will assist the idea generation activity as the following example illustrates.

EXAMPLE 5.5:

Alternate word idea generation.

Supposing one of the verb-noun generalized objectives generated using value analysis guidelines was "protect surfaces." Ideas are always needed for protecting products. In Roget's *Thesaurus*, protect leads to preserve, which has the following list of possibilities: embalm, deep-freeze, cold pack, refrigerate, boil, dry, dehydrate, ensilage, can, tin, prophylaxcize, moth-ball, pickle, salt, life belt, gas-mask, chair cover, bottle, marinate, smoke, paint, coat, cover, whitewash, kyanize, waterproof, and shore up. Granted that the thesaurus does not include electroplate, varnish, or anodize, but the point is to use a source to get a fresh perspective for ideas.

5.1.5 Record Ideas

Make a record of all ideas. The form of the ideas is not important. Use words, sketches, photos, xerographic copies from books or journals, or supplier catalogs. The recording of ideas cannot be over-stressed. It is from recorded data that future work will be done. A good idea, poorly recorded, or not recorded at all, may be lost due to ambiguity, oversight, or poor memory. When personnel are moved from project to project, as often happens in larger companies, good records are essential for project continuity. Many times a project will be handled by several people, or someone on a project may be reassigned. If the work done by one person is not properly documented, then the next person on the job, a co-worker or replacement employee, may have to repeat work already done. The new person may not think of as many ideas or as many good ideas, and time and effort will be lost.

Proper documentation, including dates and witnesses, is also critical for patent applications and legal protection. Recording of all project activity is a good habit to get into, whether a patent is your goal or not. If a product later causes property damage, or injury to humans, a well-documented trail of design activities and decisions is essential for a defense at court trials.

One of the largest roadblocks to creative design occurs during the idea generation activity. This roadblock is the mental mistake of judging an idea before it is recorded. Too many times an idea is discarded before it is recorded because it is judged to be unworkable. Do not fall into this trap. No idea is perfect, there is always something wrong with every idea. The point of generating ideas without judging them is to get as many ideas as possible. During subsequent activities, adaptations and combinations can be made to improve the ideas. *Do not judge ideas at this time*.

5.1.6 Answer Questions

During the problem identification activity, questions were asked. During the current activity of generating ideas some of the questions should be answered, and the product specifications clarified and quantified. Must all questions be answered? No. In fact, new ones may be added. If air or hydraulic power is considered as a possible solution, for example, then it must be determined whether such power is available at the site of use, or what the cost is to add it.

Some questions may require lengthy model or lab testing to answer. Some questions cannot be answered yet, but a variety of ideas should be considered to account for variations that may be found. Plan ahead for alternative possibilities.

If material selection has not been made, then the strength and availability of alternate choices must be checked out. This also may not be possible until after many tests on prototypes are complete. Tests will determine stresses on components, but preliminary material choices need to be considered.

An answered question may also change the scope of the project. If it is decided to make the rotary cultivator power unit also usable for snow blowing and grass cutting, ideas will have to be generated for connect and disconnect, and power takeoff concerns. It is not at all unusual to loop back to problem identification at this point in the procedure and repeat portions of activities previously done. Remember that no activity in the problem solving procedure is ever done and forgotten.

5.1.7 Ask Questions

Occasionally you will be stymied and unable to begin generating ideas. It is often helpful to ask questions to get over the starting hurdle and to start the ideas flowing. Even if ideas have been generated, answers to additional questions can broaden the range of ideas. The following questions are suggested to assist in the generation of ideas.

1. What materials could be used? Consider ferrous and nonferrous alloys, polymers, wood, and ceramics.
2. What safety features should be incorporated? Consider the possibility of fire, explosion, electrical shock, chemical burns, and cuts from sharp edges.
3. What methods of support can be considered? Consider support from underneath, overhead, or from an adjacent member. Consider welded, bolted, glued, friction forces, or riveted construction.
4. What sizes and weights seem appropriate? Consider human as well as equipment limitations.
5. Are there features that might only be used some of the time? Should there be more than one model? Is there more than one way to use a given model?
6. Are there features that should be attachments rather than part of the main structure? Some features can be added on for special cases, such as reach extenders, extra load brackets, or heavy duty shock absorbers on cars used for pulling trailers.
7. Can an idea or feature be enlarged or reduced? Large fuel tanks allow more mileage but add weight, while smaller electronic computer components allow faster processing time.
8. Can something else be substituted? Use a snap ring in place of a cotter pin; a plastic part for a metal one; a self tapping screw in place of a bolt and nut.
9. Can two or more ideas be combined? Design one part to do the job of two.
10. What are the advantages and disadvantages of other designs that have been considered? Consider features that make an existing product good or bad.
11. How can a particular disadvantage be overcome? Focus on a small detail and make improvements.
12. How can an advantage be improved? Make something good even better; faster, lighter, smaller, or stronger.
13. What is the scientific basis for the ideas to date? What other basis could be considered? Consider making gravity, centrifugal force, or electromotive force work for you.
14. How can reliability be improved? Work to increase the life of a part.

15. How can appearance be improved? Good appearance won't hide a bad product, but it can enhance a good one. Consider painting, plating, anodizing, or special fabrics.
16. How can a specification be improved? Make your product dig deeper, run with less power, have a greater payload, or operate faster.
17. Can an idea be put to another use? A different application of an idea may prove beneficial.
18. Can an idea be modified? Studying one idea carefully may lead to other related ideas.
19. Can components be rearranged? Rearrangement often allows combinations and alternatives not considered before.
20. Will reversing, or using the opposite work? Inside out? Upside down? First last? Last first?
21. Can the shape of a component be changed to fit better, to be made less expensively, or to do multiple tasks?
22. What happens if a part is made longer, shorter, thicker, or thinner?
23. Can some other power source be used? Electric, hydraulic, pneumatic, chemical, solar, wind, or nuclear?
24. Can a part be left out? Does it actually have a function related to the problem to be solved or is it an add on?
25. Can extra value be added at low cost? Better looking, easier to use, or more reliable?
26. What happens if a concept is taken to an extreme? Very short, very long, very light, very heavy?
27. Can the product be made more compact? Portability or storage requirements improved?
28. Can motion be added or subtracted to make an operation more efficient?
29. Can a feature be located at either end? In the middle?
30. Will a moving part work better sliding or rotating?
31. How can assembly be made easier? Nuts & bolts, rivets, welding, cotter pins, keys, splines, glue, Velcro™, or staples?
32. How can an operation be made safer and either faster or slower? A coin-operated toll gate may need speeding up whereas an automatic door closer may need slowing down.

You should also review your own feelings about the problem by asking questions about your basic assumptions and the boundary limits you have established. Questions should include:

1. What am I assuming?
2. What experiences and facts support my assumptions?
3. Is wishful thinking getting in the way of critical thinking?
4. Am I confusing chance and probability with cause and effect?
5. Will my assumptions pass a rigorous logic test?

5.2 Physical Laws

Another activity to consider when trying to generate ideas is to review the physical laws of the universe and determine if any are applicable to the problem at hand. The properties of materials and the principles of the physical universe have been studied and developed into a coherent set of relationships by the work of many people. The following list includes a brief description of many of these common laws and principles. Additional self-study is warranted depending on the problem under investigation. For example, the Piezoelectric effect in crystals might be a substitute for strain gages in a pressure transducer problem.

Ampere's law: Force relationship on a current-carrying conductor in a magnetic field.

Archimedes' principle: The buoyancy force on a floating object is equal to the weight of the displaced fluid.

Avogadro's law: There are $6 \cdot 10^{23}$ molecules per mole of gas at standard temperature and pressure.

Bernoulli's theorem: Conservation of energy in fluids, relates changes in normal pressure with axial velocity.

Biot-Savart law: Magnetic field produced by an electric current.

Biot's law: Linear conductivity of heat in a material.

Blagdeno law: Depression of freezing temperature due to impurities.

Boyle's law: Pressure times volume is a constant for a given gas at constant temperature.

Bragg's law: X-ray beam reinforcement as a function of wave length.

Brewster's law: The tangent of the polarized angle for a substance is equal to the index of refraction.

Brownian movement: Random motion of solids in a colloidal solution.

Charles' law: Pressure times temperature equals a constant for a given gas at a constant volume.

Christiansen effect: Refraction of powders in a liquid.

Corbino effect: Considers current flow induced in a metallic disc while rotating in a magnetic field.

Coulomb's law: Relates forces between electric charges.

Curie-Weiss law: Considers the temperature for transition from ferromagnetic to paramagnetic behavior.

Dalton principle: In gas mixtures, the total pressure is the sum of the partial pressures.

Debye frequency effect: Relates the increase of conductance of an electrolyte with frequency.

Doppler effect: Shift of received frequency due to motion of sound source.

Edison effect: Thermionic emission from heated metal in a vacuum.

Eötvos effect: Relates change in body weight with east-west motion at the surface of the earth.

Faraday electrolysis: Chemical change is proportional to electricity flow across an interface.

Faraday induction: Generation of electromotive force by change of magnetic linkages.

Gauss effect: Considers the increase of resistance of magnetized conductors.

Gibbs phase rule: Considers the equilibrium conditions in mixed systems.

Gladstone-Dale law: Relates change in index of refraction with density.

Graham's law: Osmotic flow of gases and solutions through a porous wall.

Hall effect: Voltage is generated perpendicular to the current in a magnetic field.

Henry's law: Behavior of a dissolved gas in a liquid.

Hertz effect: Effect of ultraviolet light on spark discharge.

Johnsen-Rahbek effect: Increase of friction of viscosity with applied voltage across interfaces between conductor, semi-conductor, and insulator.

Joule's law: Heat produced in an electrical resistor by current flow and time.

Kepler's laws: Planetary orbits are ellipses, equal areas are swept by radius vector in equal times.

Kerr effect: Optical polarization due to electric field applied to a substance.

Kirchoff's laws: Sum of currents into a circuit node is zero, algebraic sum of voltages around a circuit is zero.

Kohlrausch's law: Limiting conductance of an electrolytic.

Lambert's cosine law: Illumination of a surface varies with the cosine of the angle of incidence.

Le Chatelier's law: Equilibrium adjustments of a system under stress.

Lenz's law: Current and force induced so as to oppose motion of a conductor in a magnetic field.

Leidenfrost phenomenon: Non-wetting of a hot surface by a dancing liquid drop.

Meissner effect: The absence of a magnetic field inside a superconductor.

Nernst effect: Electromotive force generated by heat flow across force lines of a magnetic field.

Newton's laws: Acceleration is proportional to an applied force, actions are opposed by equal reactions.

Ohm's laws: Current through an impedance or resistance is proportional to the voltage drop.

Paschen's law: Sparking voltage of gas between electrodes.

Peltier effect: Absorption or liberation of heat by electric current flowing between different materials.

Piezoelectric effect: Dimension changes of electrified crystals, generation of electricity by stressing crystals.

Pinch effect: Reduction of cross section of liquid conductor carrying an electric current.

Pyroelectric effect: Charges developed between opposite crystal faces when the temperature is changed.

Purkinje effect: Color sensitivity change under dim light in the human eye.

Raoult's dimension effect: Change in electrical resistance with a change in length.

Snell's law: Relations between incident, refracted, and reflected rays at an optic interface.

Stark effect: The split of spectral lines when source is in a strong electric field.

Stefan-Boltzmann law: The heat radiated from a black body is proportional to the fourth power of its temperature.

Stokes' law: The wavelength of light emitted from a fluorescent material is always longer than that absorbed.

Tribo-electricity: Electromotive force produced at the interface of two dissimilar metals by relative motion and friction.

Weber-Fechner law: The minimum change of stimulus necessary for a response is proportional to the level of the stimulus already existing.

Weidemenn-Franz law: The ratio of thermal to electrical conductivity is proportional to absolute temperature.

Wien effect: The increase in the conductance of an electrolyte with applied voltage.

Wien's displacement law: As the temperature of a body is increased through incandescence, the color of its emitted spectrum shifts toward the blue.

5.3 Brainstorming

Another method used for the generation of ideas is brainstorming. Brainstorming is an activity in which a group of people get together specifically to think of ideas. A group of people can usually generate more and better ideas in a shorter period of time than can a single individual. When people hear one idea it often stimulates another idea: an offshoot of the idea just mentioned, a similar but different idea, a combination of ideas, or just a thought made conscious by the group's verbal exchange.

The following guidelines help to make a brainstorming session work effectively.

1. Arrange for a group of 4–12 people with various backgrounds, but relatively equal status levels. Try to get a few who have brainstormed before so the session will be easier to start. People in a group that is too small may not interact well, and people in a group that is too large may be difficult to control, and some may not participate.

2. Communicate with the people selected about a week in advance so they can prepare for the activity. Ask for their help, list the conditions of the planned session, and give them the topic to be brainstormed.

3. Select a leader who will:
    - Keep good notes and records of ideas generated. A tape recorder or stenographer may be useful. This is important for evaluation and analysis at a later time, usually done by a different group of people.

- Prevent judgment of the ideas presented so that ideas will flow naturally. Judging must not be permitted. If any idea or comment is judged, embarrassment and/or turf protection may result and idea generation will be reduced. The purpose of the session is to generate ideas, not to evaluate them. Non-judgmental questions are permitted. Often a clarification of words will trigger additional ideas.
- Keep participants on the topic so efforts will be concentrated on the specific problem presented. Side stories and other problems must not interfere with the specific problem being brainstormed.
- Encourage participants and help them feel free to contribute ideas.
- Have questions ready to ask to stimulate ideas in case participants' output lags. A leading question may set the group off on more ideas. Refer to the list of 32 questions in Section 5.1.7.
- Keep participants' responses short. Ideas are the main concern, not flamboyant rhetoric.
- Limit time so the session ends on a positive note. A session that drags on is not productive and may leave a bad impression on participants.
- Follow up with a written summary to the participants, thanking them for their help and requesting additional ideas that may have come up since the session. Include a list of the ideas that were generated for their reference.

Brainstorming can be used for a wide range of problem statements. A specific topic such as, "How can parking congestion on campus be alleviated?" or a general topic such as, "What can be done to make our campus more attractive?" can both be approached with brainstorming.

If a brainstorming session becomes nonproductive, whether it be by intimidation or other cause, consider the written brainstorming method. Give each person a sheet of paper and ask each one to write down three ideas, but not to include their name. Each sheet is then passed to another person; each person reads the ideas written, and adds three ideas. This continues until each person has read the ideas from each of the others and has added three ideas to each sheet of paper. Each sheet of paper should end up with three ideas for each person in the group; five people should conclude with 75 ideas, though some ideas may be duplicates because each person has only seen each piece of paper once. If this method stirs interest, the leader can read ideas from the sheets and participants can add ideas orally.

CASE STUDY #2 CONTINUED:

Ideas to consider for moving parts from one department to another.

1. Conveyor belts
2. Roller conveyor
3. Overhead conveyor with hanging hooks
4. Gravity chute
5. Fluid in which to float parts
6. Pneumatic pressure tube

7. Hand carry
8. Tote pans
9. A wheelbarrow
10. A sled
11. Two wheel hand dolly (like those used for moving refrigerators)
12. Three or more wheel push-by-hand carts
13. A forklift
14. A robot train on wheels
15. Catapult (similar to a hay bale handler)

The next case study presents a problem and shows the actual examples that resulted from the idea generating procedure.

CASE STUDY #3:
Idea Generation

A group of student engineers was given the problem, shown in Figure 5.1, of moving point C 3 inches (76.2 mm) vertically by moving a 24 inch (609.6 mm) hand lever through a maximum angle of 80°. Points A and B are available as pivot points. The resistance force at point C is known to be 50 lbf (222.4 N).

The students were prolific with idea proposals and properly recorded their ideas with words and sketches. Some of the ideas are shown in Figures 5.2–5.5. Figure 5.2 shows a preliminary list of five categories of ideas and a sketch of a specific idea for a sliding bar linkage. Figures 5.3 and 5.4 show sketches of ideas considering gears and cams. Figure 5.5 shows sketches of other ideas, including a consideration of the competitors. The ideas shown are not meant to be exhaustive, only a sample of ideas and the format used for recording them. The sketches also include written comments made during subsequent idea refinement and analysis.

FIGURE 5.1

Lever problem, Case Study #3.

FIGURE 5.2

HELICOPTER LINKAGE DESIGN
FELDMAN, TODD M   LAB 103   3-25-82

1. PROBLEM IDENTIFICATION

   IDENTIFIED SUBPROBLEMS: (USE WHERE APPLICABLE)
   I. TRANSMISSION OF POWER FROM HANDLE TO CONNECTING MEMBERS.
   II. TRANSMISSION OF POWER BETWEEN CONNECTING MEMBERS.
   III. TRANSMISSION OF POWER FROM CONNECTING MEMBERS TO POINT "A".

2. PRELIMINARY IDEAS
   LIST:
   a. LINKAGE SYSTEM
   b. CABLE SYSTEM
   c. HYDRAULIC SYSTEM
   d. GEAR/CAM SYSTEM
   e. ELECTRICAL SYSTEM

IDEA I a 1
- USE EXISTING LINKAGE
✓ NOT LINEAR!

IDEA I a 4
- VARY "θ","L","D"
- REVERSE POSITION
✓ TRIAL & ERROR

IDEA I a 2
- VARY "L"
VARYING "L" MAY AFFECT A CHANGE IN AMOUNT OF LINEARITY.

POSSIBLE
IDEA I a 3
- CHANGE SHAPE
✓ ON θ DEPENDENCE DOES SMALLER θ ⇔ MORE LINEAR?

FIGURE 5.3

HELICOPTER LINKAGE DESIGN
FELDMAN, TODD M.   LAB 103   4-1-82

2. PRELIMINARY IDEAS:

IDEA I d 3
- WORM AND WORM-WHEEL SYSTEM TRANSFORMS RADIAL MOTION TO ANOTHER PLANE OF RADIAL MOTION
POSSIBLE
✓ EXPENSIVE PRODUCTION!
✓ NOISY
✓ NEEDS SUPPORT.
✓ WEAR

IDEA I d 4
- TWO GEAR SYSTEM REVERSES DIRECTION OF RADIAL MOTION
VARY RADII TO ACHIEVE RATIO CHANGE
WHERE?
✓ COMPLEX CALCULATION
✓ EXPENSIVE!

IDEA I d 5
- CAM PROVIDES "UNIFORM" LINEAR MOTION.
✓ WEAR   ✓ SPRING NEEDED!
✓ EXPENSIVE

IDEA I d 6
- "POSITIVE ACTION" CAM LINEAR MOTION W/O. NEED FOR SPRING RETRACTION
EXTENDS INTO HANDLE.
POSSIBLE
✓ FIXED POSITIONING"

IDEA II a 1
- LOCATE POINT "X" IN FRONT OF POINT "F" & VARY "θ"
WEIGHT
✓ TRIAL & ERROR!

IDEA II d 1
- BEVEL GEARS INVESTIGATE "θ"
POSSIBLE RADIAL → RADIAL

IDEA II a 2
- VARY "θ" TO ALLOW FREEDOM OF MOVEMENT

IDEA II a 3
- "UNIVERSAL JOINT" INVESTIGATE "θ" RADIAL → RADIAL
✓ WILL NOT WORK @ 90°
DIFFICULT TO MOVE @ VERY ACUTE ANGLES.

Brainstorming

FIGURE 5.4

FIGURE 5.5

5.4 Work Simplification

Work simplification is a specialty study used to create ideas to eliminate wasted motion and effort while doing a particular job. The objective is to eliminate unnecessary work effort while getting a particular task performed. The usual method of using the work simplification approach is to break a procedure or process into its smallest component operations and determine the best sequence of operations, as well as the best way to perform each component operation. A product assembly line is a good example of a process that has been broken down into component operations, and each one made efficient relative to the whole. The product design itself must take the assembly procedures into consideration, as a design that cannot be efficiently and reliably assembled is not a good design.

Work conditions must be arranged to get the most done with the minimum effort. This includes positioning work so both hands can be used, or designing and supplying special tools or holding fixtures. Questions are asked about the sequence of operations being done and the tools and techniques used. Sometimes just changing the sequence of events will make a job more efficient. Work simplification analysis is most often used on a job already being performed, but the principles should also be used during the planning stages of a project. Efficient operation includes having the right tools available and in good condition, a work area large enough and at the right height, adequate lighting, heating, ventilation, and noise control.

When analyzing the work to be performed in any system it is important to know and understand the differences between the abilities of people and of machines. Humans have a greater capacity for sensory functions such as smell, taste, and sight, but in the areas of hearing and feel there are some tasks that machines can do better than humans. Fixing and matching sound frequencies, and sensing surface microfinishes are but two examples. Humans are very efficient at comprehending a complex situation, perceiving conditions, and deciding on a course of action. This is particularly true if the situation cannot be planned for in advance. A planned situation, such as a rocket flight, or an engine fuel mixture, can be controlled faster and with greater reliability by computers. In this same general area, humans are usually more alert for what is coming and can act on their hunches based not only on what is immediate, but what is anticipated, such as selecting a part from a load of jumbled parts. This makes humans more flexible than machines. A machine can be created to perform a single task and anything out of its range will be a total failure. Humans can at least suggest what to do if their ability range is exceeded. This also gives humans the edge when it comes to reasoning and judgments. Will machines ever be trial judges?

Machines are valuable for many reasons. They can operate faster and more consistently than a human. Most machines perform tasks that might otherwise be slow, dangerous, and difficult, if not impossible for a human to do. Machines perform at a constant pace and do not become bored, lose interest, and make errors because of lack of concentration. Robots are used extensively for applications where repetition of a task dominates a procedure, such as the handling of parts from work station to work station, as well as for applications that are dangerous to humans such as the handling of hot, cold, or radioactive objects. Computers are used extensively to make repeated calculations because they can make them faster and more reliably than humans. It does, however, take humans to control and interpret the results of the calculations. Computers can store and retrieve large amounts of data faster from memory, without error, than can a human. Well designed machines, including the new generation of parallel processor computers, can also perform tasks simultaneously.

5.5 Feasibility Study

The term feasibility study often causes confusion, because a feasibility study creates a special case of generating problem solution ideas for the engineer. Too often people use the term to imply that the project will be implemented as soon as the study is complete, and that can cause misunderstanding and difficulties for the investigating engineer. Often a feasibility study is the link that connects idea generation with analysis because only the feasible ideas, in terms of such concerns as time, money, and state of the art, should be pursued.

Feasibility studies usually cover the situation when someone has an idea and wants to know whether to proceed. The person may be capable of making the study, but will call an engineer in as a consultant to get a fresh outlook and new ideas. Be wary of the person who asks you to "Gather data so I can show the board that this project should be done." You may end up being used as a scapegoat. Always prepare a feasibility study in an honest and fair manner. A biased study may look only at the advantages and not consider the disadvantages, or vice versa. Taking sides will make a study biased, and that would be unethical.

A feasibility study should define the problem, identify factors that limit the scope of the project, and evaluate anticipated difficulties. The study may validate or invalidate the stated need, determine initial direction of a solution investigation, produce an initial set of possible solutions, and suggest a preliminary cost objective and schedule of activities. A feasibility study will use the same problem identification and idea generation procedures as have been discussed. The purpose of the feasibility study is to determine if a particular idea or scheme has enough merit to justify further study and development. But this applies to any engineering project. All projects must remain feasible in order to continue. Never get so attached to a project that you cannot muster the courage to call it to a halt if an insurmountable stumbling block is encountered.

During a feasibility study ask the usual questions about consumer needs and the current solution to the problem. A market survey may be required. Costs of implementation will be estimated at the macroscopic level rather than the microscopic level. The number of component parts required will be estimated, as well as their cost. Any items that can be specifically identified, such as an electric motor or other standard items, can be costed accurately. If the product under study is similar to one already manufactured, a comparison cost may be possible if the current state of the art will work in the new application. Some large equipment cost estimates are made on a per pound basis. In the case of engines or motors a cost per horsepower basis may be used. Perhaps just a cost per part and the estimated number of parts will suffice. Any basis for estimating cost is permissible, but documentation is necessary so follow-up and continued study can be logically done.

Items to consider when thinking of ideas for a feasibility study should include:

1. Is present material handling equipment adequate? This is particularly important if the product being considered is larger than existing models. Overhead hoists,

door openings, paint booths, material storage, and available machine tools all have size limits that may be exceeded with a new product.

2. Can the new product be shipped as current products are shipped?
3. Will the increased work load in the factory replace some current work load, or will additional people be required to meet production schedules?
4. If the new product will involve a new technology, are present employees capable of implementing the technology, or will special training or new hirees be required?
5. Will the present computer have adequate storage and terminal access for the new system?
6. How will the implementation of this project affect other projects and related operations?
7. What other ideas will satisfy the need as defined?
8. Is the public ready to accept the proposed project?
9. Are there unusual environmental issues involved? The Environmental Protection Agency requires an impact study for projects such as roads, dams, bridges, or dam-created lakes.
10. Are there unusual safety issues involved?

During a feasibility study, creativity in component design analysis is not the primary skill necessary. What is needed is the ability to determine whether or not it is possible to do the task being studied and if so, develop a plan for doing it. Consider all ramifications, and determine if it is economical to proceed.

A proper feasibility study will result in a list of objectives and goals for the project, and includes such items as alternative ideas for implementation, target manufacturing costs, time schedules, market penetration, personnel requirements, capital requirement costs, production goals, projected return on investment, and a list of questions not answerable without additional study. These items become the framework within which the detailed design activities must operate. Estimates of all the items listed must be made, and this reinforces the need for a group effort. The sales department can make a market survey for information on consumer needs and products presently available, and propose market penetration and selling price goals. Manufacturing can estimate the cost and availability of tooling and machine tool requirements. Industrial engineers can estimate direct labor cost and assist plant engineers with estimating facilities requirements. The report will give standards to use as a basis for follow-up progress reports. A well-managed project will promote an atmosphere in which all departments will contribute information so the stated objectives will be met.

EXAMPLE 5.6:
Feasibility study.

Consider a feasibility study to investigate adding storm windows to a building. The analysis of expenditures and savings is fairly straightforward. Single pane glass storm windows change the R factor from one to two. The average number of heating degree days can be obtained from the weather bureau or a fuel distribution company. The cost of fuel and the heating system efficiency can be obtained from records or estimated, and fuel savings can be compared to the installation cost.

Other factors should also be considered. More uniform heating may make people inside the building work more efficiently. If the heating system does not have to work as hard, it may last longer before repair or replacement. If the storm windows come with screens, then summer air circulation will be possible without the annoyance of insects. Summer cleaning cost will increase due to the additional glass area to wash. The change in looks of the building may attract more customers. A feasibility study should answer all the questions possible and list the ones that need further study and analysis.

If by now you are asking, "How does a feasibility study differ from the activities of regular problem solving, they both seem the same?" you understand well. There is no difference in activities, perhaps only in the depth and degree of completeness, and the amount of time used. Questions that may lead to a feasibility study include:

1. Would this be a good corner on which to build a parking lot?
2. Should I sell trailers to complement my already established tractor sales?
3. Can our company build and market a sheep's foot soil compactor attachment for highway building contractors?
4. Would it be economical to purchase and install sensors that will turn off the lights in this classroom when no one is in it?
5. Should our free parking lot be converted to a pay-as-you-use lot with toll gates or some other type of fee collection?

5.6 Summary

Methods proposed in this chapter are designed to help generate ideas to solve a given problem. Different problems will suggest different methods of idea generation. This makes the method of idea generation a problem in itself; so why not come up with your own ideas on how to generate ideas. Try to think of a method to generate ideas that was not discussed in this chapter.

The following method of generating ideas, though perhaps not original, came to me as I was reading the section on brainstorming in reference (1). Write down one solution idea to the problem at hand. List one advantage and one disadvantage of the idea. Think of an idea to overcome the disadvantage. List one advantage and one disadvantage of that idea. Think of an idea to overcome the disadvantage. List one advantage and one disadvantage of that idea. Continue the process of thinking of advantages, disadvantages, and ideas to overcome the disadvantages until a good overall solution to the problem takes shape. Not only should you have an acceptable solution, but you will know its advantages and how it will handle potential disadvantages. This procedure can also be done in a small group, but it is not brainstorming in the usual sense because judgments are made at each step of the procedure.

5.7 References

1. Adams, James L. *Conceptual Blockbusting: A Guide to Better Ideas.* New York: Norton Publishing Co., 1980. Includes a 9-page discussion of guidance to other readings on creativity.
2. Beakley, George C., et al. *Engineering, An Introduction to a Creative Profession,* 5th ed. New York: Macmillan, 1986. Chapter 12.
3. Freedman, George. *The Pursuit of Innovation.* New York: AMACOM, 1988.
4. Harrisberger, Lee. *Engineeringmanship—The Doing of Engineering Design,* 2nd ed. Monterey, Cal.: Brooks/Cole, 1982. Chapters 4 and 5.
5. Jones, J. Christopher. *Design Methods.* New York: John Wiley & Sons, 1981.
6. Keefe, William. *Open Minds.* New York: AMACOM, 1975.
7. Miller, G. Tyler Jr. *Living in the Environment,* 5th ed. Belmont, Cal.: Wadsworth Publishing Co., 1988.
8. Nadler, Gerald. *Work Simplification.* New York: McGraw-Hill, 1957.
9. *Value Analysis Value Engineering.* New York: Value Analysis Incorporated, 1968.
10. Also those references listed at the ends of Chapters 2, 3, and 4.

5.8 Exercises

1. List at least three ideas as possible solutions to each of the following "problem" statements:
    a) I don't have a date for the dance.
    b) My car won't start.
    c) I want to cook out and it is raining.
    d) I am nervous about taking the test.
    e) The computer is down and my program is due in two hours.
    f) How can I get to class on time?
2. A float in a carburetor for a gasoline engine is used to control the fuel flow valve. When the fluid level is high enough the float rises and closes the valve. One design of such a float is made from two brass stampings soldered together. The solder joint was not good in one sample and the float filled with gasoline, caused it to sink, and kept the valve open.
    a) What symptom might a lawn mower exhibit as a result of this failure?
    b) What other failure might result in similar symptoms?
    c) When the repairer took the carburetor apart, what clue do you suppose led to the discovery of the actual problem?
    d) List methods to remove the gasoline from the float. Give one advantage and one disadvantage of each method.
    e) Evaluate the ideas from (d) and order them, best to worst.
    f) Suggest ways to solve the problem of this non-floating float.

3. List subsystems of each of the following products.
   a) Bicycle
   b) Lock used at a dam site for boats
   c) AM radio
   d) Automatic toll booth on a toll road
   e) A residential home
   f) An adjustable vise grip
   g) A drafting machine
4. Record ideas that could be used for connecting a single power unit to a lawn mower, a snow blower, and a rotary cultivator.
5. Make a list of all the ways you could use:
   a) A single brick
   b) Ten bricks
6. In a group, brainstorm ideas for Exercise 5.4 or 5.5, or some other problem.
7. Make a list of any function that a connected telephone could be used for other than transmitting voice signals.
8. List as many ideas as you can on ways to "cover up" the front of a kitchen cupboard.
9. List ideas to be used to get students to begin assignments on time rather than the night before the due date.
10. List methods that could be used for determining the directions north, south, east, and west.
11. Suggest uses for football stadiums, many of which are used only 20–30 times a year.
12. Suggest ways to control traffic flow at road intersections.
13. List ideas for ways to use empty gallon plastic milk containers.
14. List ways to collect tolls on a toll road other than the stop-pay-start system which often causes traffic flow bottlenecks.
15. List ways of identifying pages of paper so they can be returned to their original order if dropped and scattered.
16. List some uses of old editions of books after new editions come out.
17. Suggest some uses for worn out car and truck tires.
18. Compare the differences in security measures that you would take if you were in charge of, or were designing, each of the following.
    a) A bank
    b) Ship cargo storage
    c) Computer power supply
    d) Educational facilities
    e) Health care facilities
    f) High rise office building
    g) An industrial plant
    h) A library
    i) Surface transportation

j) Air transportation
k) A personal residence
l) Single owner business building
m) A public building such as a post office

19. List ideas for solving each of the following problems.
    a) A sleepy driver wake up alarm.
    b) A method of locating a dropped contact lens.
    c) A way of displaying miles per gallon in a moving vehicle.
    d) A method of locking all the exterior doors in a house from one location.
    e) A method of determining if all those on board an airplane have their seat belt on and fastened.

20. One of the problems with older homes is lack of insulation. Discuss alternate methods of increasing the R factor of the walls and ceilings of an older home.

21. Gamma ray radiation can be attenuated by adding mass to the path that the radiation must pass to reach an object. Discuss ways to increase the mass of the walls and the ceiling of a residential home to make the inside safer from gamma rays.

22. Discuss alternate methods of collecting and returning golf balls on a driving range.

23. Discuss various methods of improving the lighting on drafting boards in a room where illumination is lower than minimum.

24. Discuss methods of determining which way is vertical and which way is horizontal.

25. Figure 5.6 shows a circular object with a nonhomogeneous mass at a distance R from the center. If the axis of the object is placed on knife edges the off center out-of-balance mass causes a torque, and the object rotates until the mass is at a low point. Devise a method of determining the out-of-balance mass and its location from the center so actions can be taken to balance the object.

FIGURE 5.6

Out of balance object for Exercise 5.25

26. A dynamic damper, often called a dashpot, is a device that dissipates energy of motion to some other form so a moving system will come to rest without an impact load and without causing oscillation. A dashpot rating can be quantified by a constant C, with units of lbf·sec/ft, as the proportionality constant in the equation $F = Cv$. F is the force that resists motion and v is the velocity of motion. A damper is a common device to use when vibration is present. List some ideas for the design of such a damper.

27. Figure 5.7 is a layout of a traffic intersection. In the morning hours the majority of traffic comes from A going downtown to work, and from E following E'' to the high school. Some of the traffic from A follows A', but the significant amount follows A''. The traffic from B merges with that from A'' and is controlled by a yield sign. A little traffic comes from C and is controlled by a stop sign. Traffic that comes to D is also controlled by a stop sign. Any traffic coming from E can follow E' or E''. The traffic following E' is <u>not</u> controlled by any signs. The roads from E and A are wide enough for four lanes, but only as turning lanes. The lanes are not wide

FIGURE 5.7

Traffic intersection for Exercise 5.27

enough for four lanes of moving traffic. The speed limit is 25 miles per hour. In the afternoon most of the traffic comes from E with about one-third following E'. From E is uphill. Near-miss collisions occur between traffic from A'' and E'. A sidewalk and a six-foot high chain link fence follow the inside of the curve. Suggest alternatives to the traffic pattern to minimize the risk of accidents.

28. An accelerometer is a device that allows the measuring of acceleration. Various mechanisms are suited for such a device, but most of them make use of the basic relationship $F = ma$. The problem becomes one of determining the force applied to a known mass and calculating acceleration or calibrating the device so it reads acceleration directly. Sketch and discuss various devices that could be used for an accelerometer.

29. An extension ladder generally has a warning label on it that suggests it is unsafe to extend beyond a certain limit. This limit is to maintain a minimum overlap of the two sections. Most extension ladders do not have a device on them to actually prevent or deter overextension. Suggest ideas to assist in limiting overextension.

30. A yardstick is balanced on a fulcrum at the 12-inch mark by placing an object at the 1-inch mark. After some time has passed, and without being touched from an external source, the yardstick tips due to imbalance. What object could have have been placed at the 1-inch mark that would cause such an event to occur?

31. Perform a feasibility study on one of the following:
    a) Permit free downtown parking.
    b) Require each freshman college student to purchase a microcomputer.
    c) Change the law to allow polygamy.
    d) Number the parking places on campus and sell their use to the highest bidder.
    e) Allow one nuclear power plant per state.
    f) Market a push rotary lawn mower.
    g) Purchase and install sensors which will detect the presence or absence of a person in a room and turn the lights out if no one is present. Use your classroom as a sample room.

32. List all the verb-noun functions of something complicated, such as a bicycle. Note that some parts can have more than one function. A bike frame, for example, "supports axle" and "supports seat post."

The following exercises are intended for practice in creative and inventive thinking.

33. Design a device or method to allow an egg to be dropped from a third-floor window and reach the ground below without breaking.

34. Determine a way to light a match from a distance of ten feet.

35. Determine a way to fill a pop bottle located on the ground, from a third-story window.

36. Design a device to turn the pages of a book to be used by people who don't have hands.

37. Design a device to allow the opening of a flip-top pop can with only one hand.

38. Determine a method to join two pieces of string from a distance of ten feet.

39. Determine a way to drive a nail into a board from a distance of ten feet.

6 Refinement and Analysis

"Refinement is the first departure from unrestricted creativity and imagination. Practicality and function must now be given primary consideration."
◆ *JAMES H. EARLE*

Refinement and analysis are the next activities to undertake in the design problem solving procedure, and the first activities that many students consider to be real engineering. The stereotyped engineer is often pictured as an analyzer of data, and a user of formulas, handbooks, tables, calculators, and computers. Engineers do spend time performing these activities; however, engineers also spend time performing related activities—communication being perhaps the most important, but often the most neglected.

Refinement and analysis activities are performed with the goal of reducing the list of possible solutions to the ones that have a chance of being the most acceptable, and at the same time improving the workable ideas to be better than originally conceived. Decisions need to be made so that additional time and money can be spent developing the most promising idea. The problem solving activities to be discussed in this chapter are no different than activities previously suggested; they are intended to help you focus on the essentials. Collecting additional information through refinement and analysis is important so alternative ideas can be fairly judged and compared. Judgments will be clearer, and made with more confidence, when the best information available is being considered.

6.1 Levels Of Analysis

The ideas being considered for final problem solution must be analyzed and compared on many performance levels and will require input from many individuals. Keep the design team involved at all stages of project development. The emphasis placed on each performance level depends on the problem being solved.

6.1.1 Function Analysis

The first performance level to analyze is basic function. The basic function of each design idea must be analyzed and verified. The idea must meet the specifications and minimum requirements as stated under problem identification. If an idea doesn't meet minimum requirements but still has favorable attributes, consider modifying the idea, combining it with another idea to make it acceptable, or changing a basic requirement. A compromise between an otherwise good idea and a current specification may be required. Compromise means that the idea and the requirements shift toward common ground. An original performance goal may be too ambitious, such as a minimum weight requirement that cannot be met even by using plastics and aluminum. If an idea cannot be improved to meet a minimum standard, and the standard cannot be changed, then the idea must be dropped from further consideration.

Basic function also includes whether or not the product is strong enough. Will the design adequately resist the loads that might be applied? All loads and forces, such as static, dynamic, cyclic, impact, and vibratory, must be considered. Forces can be applied by the user as the product is being used, by the environment (wind, tides, water erosion), or by the manufacturing processes used to make the product. The analysis of forces can be done manually by using accepted design strength criteria, by computer simulation and strength analysis using finite element analysis, or by studying either full size or scale models. See Case Study #4 for an example of preliminary product strength analysis.

Basic function also includes the necessity to analyze reliability. A design engineer wants the final product to last an acceptable length of time providing it is used properly by the consumer. At some point during the product analysis, when the list of potential ideas is reduced to the most viable, reliability must be considered. An idea that is least expensive and has the shortest lead time may be the least reliable. Nothing will turn potential customers away faster than an unreliable product. Failure rate determination and reliability programs conducted during product development are essential. Physical model testing will produce information to help analyze reliability of products taken as a whole as well as component parts. Associated criteria for products and components derived from testing data are mean time between failures, those components most susceptible to failure, the effect of the failure of one component on the failure of other components, and the time it takes to return a product to operation after a failure. Reliability analysis uses statistical concepts such as mean, standard deviation, and coefficient of correlation. Specific situations are handled by using a normal, binomial, Poisson, or Weibull distribution.

6.1.2 Safety Analysis

Next to basic function, the most important level of analysis is safety. Safety must always be analyzed because performance without safety is risky at best. Safety issues include designing a product to protect the user and bystanders from such things as electric shock, cuts from sharp edges or moving parts, explosion and fire, burns from hot surfaces, injury from loose flying parts, entanglement in moving machinery, and exposure to radiation, noxious gases, air contamination, caustic fluids, and acids.

If a hazard such as a sharp edge can be eliminated, it should be. A sharp edge on sheet metal can be folded over. Sharp edges and burrs on parts that have been machined should be removed using processes such as grinding, tumbling in an abrasive slurry, or thermal deburring. Consumers can be protected from the edges of sheet metal and glass by using metal, rubber, plastic, or wood edging. If glass shattering is a possibility, then perhaps shatterproof glass or Plexiglas™ should be used instead of regular glass. Blades and other parts that need to be sharp to function, or machinery that has pinch points due to the movement of one part into and out of another, should be shielded to prevent accessibility unless power is shut off, not just covered with an easily removable shield. Examples include the power cutoff interlock on lawn mowers and snowblowers, and two-hand controls on hydraulic and mechanical presses. Electrical shock potential in electric power environments is reduced by using third wire grounding or plastic cases. Seat belts were designed, developed, and added to automobiles and airplanes to reduce the damage caused by an accident. Electric sparks are avoided by using ceramic or polymer components, as well as placing electrical contacts in airtight enclosures.

Some dangers to the users of products cannot be eliminated or physically shielded, as in the case of volatile fuels, dangerous chemicals (insect sprays and herbicides), and acids. These situations require warning labels to be placed in appropriate locations and specialized use instructions included. Chemical mixing and use instructions must be written clearly, including precautions and proper procedures as well as antidotal measures in case of emergency. When operation or use of a product can cause harm to bystanders as well as the user, as in the case of motor vehicles, rotary lawn mowers, and automatic door closers, special design problems exist. Product features such as signal lights, brake lights, seat belt buzzers, backup bells, wide turn signs on trucks, presence sensors, and automatic stops were designed to help protect people and devices from harm.

Care must be exercised when designing warning labels. As products become more complex, and the consumer develops a right-to-know attitude, there is a tendency to write warning messages that are not understandable, too long, too complicated, or too intricate. Additional problems arise when products are distributed and used by people with varying degrees of literacy, and who use different languages. Many companies now print and include instructions in different languages with their products, all in an attempt to avoid improper and/or dangerous use by the consumer. In very special cases licensing is required before a product can be used because of the specialized knowledge and skill needed to perform an operation safely, such as piloting an airplane or supertanker.

6.1.3 Human Factors

Safe use must not prevent a product from being comfortable for the consumer to use. Consider the human aspects of your design. Some areas of concern for user comfort are: convenience of use, noise levels, illumination (too much as well as not enough), light

TABLE 6.1

Comfort and discomfort levels of selected human environmental aspects (11,12).

Aspect	Comfort Level Range	Discomfort or Dangerous Range
Humidity	30%–70%	0–10% & 90–100%
Temperature	65–75°F	<30 & > 100°F
Pressure	10–20 PSIa	< 8 PSIa & > 22 PSIa
Acceleration	0–0.1 G	> 1 G
Electric Current @ 60 Hertz	0–1 mA	> 10 mA
Radiation	0–0.2 REM/year	> 15 REM/year
Illumination	20–100 Footcandles	10,000 Footcandles
Noise	0–85 db	> 94 db
Vibration	0–1 cycle/sec	3–10 cycles/sec
Shock Waves	0–2.5 PSIg	> 7 PSIg
CO Level	0–10 PPM	> 10 PPM
CO_2 Level	0–1700 PPM	40,000 PPM
Oxygen Level	15–20%	> 60%
Moving Air	13–20 ft^3/min	0–5 & > 50 ft^3/min

glare and reflection, vibration, temperature (too hot or too cold), humidity, air flow, oxygen and other gas concentration levels, and acceleration. Certain levels of these conditions are not only uncomfortable to us as we work, but can be dangerous to our health, as shown in Table 6.1 and references (11) and (12).

Comfort and environmental aspects of product design are also important on a community and country-wide level. These include the quality of air, land, and water as well as resource allocation and replacement. For an illustration of such a concern see Example 5.3.

Safe and comfortable use also relates to the size and shape of people. Reference (12) contains anthropometric charts that list data relating to the average size and shape of humans. A properly designed product can increase user comfort, accuracy, and speed of performance while decreasing fatigue, boredom, isolation, and frustration. Proper design includes good location of controls, readable dials, good instructions, ease of making adjustments, ease of maintenance, and warnings for emergency situations using red/green lights, blinking lights, horns, bells, buzzers or automatic shut-down mechanisms.

6.1.4 Service Requirements

A product that functions as required and is safe and comfortable for the user may have problems when service is required. A well designed product is easy to service. Often this aspect of product design is ignored until it is too late to make changes and still meet production schedules, so a hard-to-service product reaches the marketplace. Consider an electric motor on an air conditioner that requires annual lubrication, but is located inside the unit and requires the removal of a dozen screws to get at, and then requires an oil can with an unusually long access spout. How often will a consumer lubricate the motor? Consider the car designed with the spark plugs located in such a way that engine accessories must be removed before the spark plugs can be accessed, and even then only with a special wrench. This design adds to consumer cost. Sometimes the decision made about which parts will be normal replacement parts is made hastily and without regard for serviceability. Should a light bulb within a lamp assembly be replaceable? One particular manufacturer did not have the replacement bulb listed as a repair part, but listed only the lamp assembly.

6.1.5 Market Concerns

Engineers must also be concerned with the market and economic aspects of design. There is no point having a great product if it can't be marketed. Will the design meet the acceptance level of consumers? Does it fill their perceived needs? Is the cost acceptable to them? Market study prior to offering a product to the marketplace, and after introduction for follow-up analysis is appropriate so customer acceptance or rejection information can be fed back to the design environment. This is particularly important when a new product changes a custom or traditional pattern, such as push-button phones, or short-life, undersized spare tires. Designers are required to consider other market concerns such as those involving styling, for automobiles, houses, and buildings; potential health problems and community acceptance of the location of a land fill or hazardous waste disposal; a change in a banking procedure (instant access cards); a change from a hydraulic system to a pneumatic one, or vice versa; or a change in a mass transportation system.

6.1.5 Storage And Shipping

While designing a product, consideration must also be given to how it will be shipped and stored, and how to determine an acceptable shelf life. Though food most obviously has a finite shelf life, any number of products can deteriorate over time: colors fade in the sun, mixtures separate if left standing, metals corrode, wood warps, chemicals decompose, fuels lose volatility, grease may melt out of a critical joint, and rubber will deteriorate in the presence of ozone.

6.2 Related Activities

The activities performed at this stage of the project should include anything that will help you make good judgments and decisions about the ideas under consideration.

6.2.1 Evaluate Ideas

One of the first activities to perform is to discard unworkable ideas. "Far-out" ideas, those for which no method exists to make them realistic or feasible, should be discarded. Revolutionary ideas ordinarily need state of the art changes to make them workable. An idea that might be discarded is the idea of a solar powered snowmobile. This does not imply that a snowmobile will never be solar powered, only that with the current state of technology, it is not commercially feasible. State of the art change involves research and development, which are jobs for the research engineer.

Other ideas might be discarded because of possible patent infringement, small market potential, or the amount of development and test time required to meet the current time and cost schedule. Poor ideas should be rejected before time and money are spent on development. Ideas and revision proposals can be rejected or postponed for many reasons, such as:

1. An idea is too similar to the competition. If a design is too similar to a competitor's, then your design needs to be lower priced, more reliable, available with shorter delivery time, or have a better warranty to be competitive.

2. An idea requires too many parts. If there are more parts in one design than another, the cost is also probably higher than necessary. More parts mean more drawings, more material, more labor (making and assembling), more complicated servicing and repair, and more repair parts in inventory, all of which add cost to the product.

3. An idea has parts that are difficult to manufacture. Parts can be so complex that standard methods of manufacture will not work.

4. An idea has a high estimated cost, product material and labor cost, high tooling cost, high facilities (buildings and machines) cost. An idea that is expensive must have features that make it better than the competition. It must be easier to use (changing a truck to four-wheel drive without getting out of the cab), last longer (ball or roller bearings instead of a sleeve bushing), perform a task that the competition can't (automatic re-dialing telephone), or have some other advantage.

5. Product development time is too long and schedule cannot be met. Product implementation time will increases if new state of the art is studied, or if a new product is being developed.

6. The new product is too similar to an existing product and there isn't any reason for the customer to trade up to the new model. A new product must have some advantage, such as cheaper operation, more work output in a shorter period of time, fewer skill requirements to operate, safer operation, or higher reliability.

7. An idea is too revolutionary, requiring a large increment in the state of the art or in the customer's adaptability: a three wheel car, a car with lever steering in place of wheel steering, or solar heating.

As ideas are reviewed, make notes about details to check or data to verify. These notes might include questions about clearance, amount of movement, fastening methods, or use of standard parts.

There is always the danger of discarding an idea that can be developed to be acceptable, so care must be exercised. Experience can be helpful. As you become more knowledgeable about ideas that work and those that don't work, you will become more confident about preliminary decisions. However, experience sometimes puts habit in front of logic and creativity. Don't get in the rut of thinking that the same solution will always be the best under all conditions.

6.2.2 Scale Drawings

Make scale drawings of the better ideas. Scale drawings help determine if an idea will really work as imagined, and help to determine specific product features. The specific details will be used later to compare the various design ideas. A comparison will be made to determine which of the most promising ideas, if any, are worth pursuing to final implementation and production. Scale drawings are generally started as complete product layout drawings. Then as each critical area is identified, sub-component layouts and details are drawn. Select a scale suitable for the amount of detail that needs to be shown. Scale drawings will help to determine specific details about your design. Sometimes the added detail will result in new ideas for solution.

Drawings will also aid in determining critical lengths, areas, volumes, and weights. In Case Study #4, the length of the arm and the slot dimensions can be determined from the scale drawings by measuring, or by calculations once the geometry is determined. Volumes, areas, and weights can also be calculated from drawing measurements. Computer-aided drafting systems will assist in the details of design layout and in many area and volume calculations.

FIGURE 6.1

Scale layout of a tent.

EXAMPLE 6.1:
Layout analysis.

Figure 6.1 shows several views of a small tent; specific auxiliary views show the areas of material required. The area of surface $ABC = (BC)(AD)/2 = (3.65$ units$)(1.95$ units$)/2 = 3.56$ unit^2. Not only can the area of ABC be calculated from the second auxiliary view, but a full size enlargement would give a pattern that could be used for manufacturing. Likewise, the area of rectangle $BCEF = (BF)(BC) = (5.95$ units$)(3.65$ units$) = 21.72$ unit^2.

The next case study is an illustration of how drawings and mathematical analysis are often combined.

CASE STUDY #4:
Drawing table adjustment.

Consider a drawing table tilt mechanism design shown as a preliminary layout in Figure 6.2. The scale drawing shows the arm movement required, represented by line AB, and overall relationships so details can be accurately considered. One of

FIGURE 6.2

Table adjustment layout. Units are inches (millimeters).

FIGURE 6.3

Adjustment arm details. Units are inches (millimeters).

- 15.14 (384.6)
- 1.00 (25.4)
- 0.25 (6.35)
- 10.00 (254.0)
- 14.14 (359.2)

0.391 (9.93) wide slot for 0.375 (9.53) dia bolts
0.391 (9.93) dia hole

the ideas for infinite angle adjustment from horizontal to 30° from horizontal is a slotted arm as shown in Figure 6.3. The preliminary arm dimensions of 0.25 x 1.00 inch (6.4 x 25.4 mm) are selected because this is a material size readily available from steel suppliers. The layout also permits analysis of other arm notch designs such as shown in Figure 6.4.

Drawings also assist in the analysis of forces. Let the force on the edge of the drawing table from a person leaning on it equal 100 lbf (444.8 N) as shown in Figure 6.5. (The 100 lbf was determined by placing a bathroom scale on the edge of the table and leaning on it.) Vector analysis can be used to determine the force in the arm, and how much force must be supported at the clamp. If F equals the force in the arm, determine the sum of the torques about point P as:

$$100(15 \cos \theta + 1.50 \sin \theta) - F[10 \cos ((90 - \theta)/2)] = 0, \quad \text{and}$$

$$F = \frac{1500 \cos \theta + 150 \sin \theta}{10 \cos ((90 - \theta)/2)}, \quad \text{where } 0° \leq \theta \leq 30°$$

$F_{0°} = 210$ lbf (934 N), and $F_{30°} = 160$ lbf (712 N)

If a bracket is placed on both sides of the table, then the force in each arm is one-half the amount calculated, 105 lbf (467 N) and 80 lbf (356 N).

The size of some components can be determined by measuring the scale drawing. Other sizes can be calculated by using data from the layout and applying appropriate mathematical models. Consider the drawing table arm. Once the compressive force is determined, the cross section dimensions can be analyzed based on a material's compressive strength, and its modulus of elasticity using column analysis. This would be most critical at $\theta = 0°$ because the force is the largest when the length of the moment arm is the greatest, and column failure is most likely when the column length is greatest.

FIGURE 6.4

Alternate arm design.

If a mild steel with S_{yc} = 36,000 psi (248 MPa) and E = 30·10⁶ psi (207 GPa) is selected, then the Euler column model, $P_c = \pi^2 EI/A^2$, can be applied as follows.

$$105 \text{ lbf} = \frac{(\pi^2)(300 \cdot 10^6)(I)}{[(0.65)10\sqrt{2}]^2} \text{ and } I = 0.00003 \text{ in}^4 \text{ (12.5 mm}^4\text{)}$$

The minimum moment of inertia in the arm shown in Figure 6.3 will occur in the area where the slot material is removed; $I = 2(0.30)(0.25)^3/12 = 0.00078$ in⁴(325 mm⁴). This is a conservative value because the slot is not continuous throughout the

FIGURE 6.5

Table adjustment vector analysis. Units are inches (millimeters).

$D_2 = 10 \cos [(90 - \theta)/2]$ inches
$D_1 = 15 \cos \theta + 1.50 \sin \theta$ inches

Force F

whole arm. The preliminary dimensions of 0.25 × 1.00 inches (6.35 × 25.4 mm) are adequate from a column analysis standpoint, even with the slot material removed, because 0.00078 > 0.00003.

A scale layout may also reveal that parts and assemblies do not work as originally conceived. If the problem is bad enough the idea may have to be dropped. However, the problem discovered may be solvable by using a new idea or by modifying an existing idea. An idea previously considered and rejected might be usable in a modified application. Review ideas previously discarded to see if one will be appropriate now. You did write them all down, didn't you? Solving new smaller problems will go on throughout the project. New ideas will be needed at all stages of the project, right up to the day of first production, and beyond.

Equally good ideas sometimes occur, and by studying drawings it may be possible to offer a product with optional features. Products with optional features include radios that run on batteries or plug into a wall socket, automobiles with a choice of engines, a power drill that can be used as a sander or a grinder, and a boat designed to run with sails or an engine. Scale drawings may also point out features that are particularly favorable when compared to the competition, such as larger fuel storage, greater maneuverability, fewer parts to service, more standard parts, or easier servicing.

6.2.3 Material and Manufacturing

Component requirements that are obtained from the scale drawings lead eventually to the selection of material. Materials are selected based on:

1. Their mechanical properties such as strength, toughness, ductility, or hardness.
2. Their physical properties such as specific weight, melting temperature, heat conductivity, thermal conductivity, or thermal expansion coefficient.
3. Their chemical properties such as corrosion resistance, state of matter at a given pressure and temperature, or the atomic lattice structure.
4. Their cost and availability.

Materials and their properties are important to make developmental design decisions. From a practical standpoint, materials must also be selected based on what is available. The drafting table support arm is an example. Regardless of how small a cross section is possible from a strength standpoint, an efficient choice of material will be a size that is readily available and can be handled by the existing manufacturing capabilities.

Component sizes and required details, coupled with an available material, will lead to the selection of manufacturing processes. Decisions made during the design of a product have a great effect on the method of manufacture. Always consider available processes and their relative costs when designing component parts. Processes fall into the following main categories.

1. Casting and Molding: Confining liquid or plastic material in a mold during solidification.
2. Shaping forms: Moving metal by forming, hammering, pressing, and forging.
3. Cutting from larger pieces: Shearing, sawing, nibbling, flame cutting, plasma arc cutting, and laser cutting.

4. Machining: Material removed by the cutting action of a tool bit such as milling, drilling, boring, and grinding.
5. Joining: Welds, glue, staples, rivets, screws, and nuts and bolts.
6. Finishing: Plating, painting, anodizing, and metal spraying or sputtering.

CASE STUDY #4 CONCLUDED:

The drawing table arm in Figure 6.3 is 15.14 inches (384.6 mm) long. If the double radius cutoff shear blade wastes 0.50 inch (12.7 mm) between cuts, then a standard 192.00 inch (4876.8 mm) bar will yield 192.00/15.64 = 12 pieces as illustrated in Figure 6.6.

Alternative methods can be used to manufacture the slotted arm shown in Figure 6.3.

Method #1:

Operation 10	Shear to length with double radius cutoff blade from a bar.
Operation 20	Punch the slot using a special punch.
Operation 30	Punch the hole using a standard punch.
Operation 40	Flatten as required.

Method #2:

Operation 10	Shear length with double radius cutoff blade from bar.
Operation 20	Drill (2) holes, one for slot beginning.
Operation 30	End mill slot.
Operation 40	Pencil grind machining burrs.

FIGURE 6.6

Bar cutoff operation. Units are inches (millimeters).

192.00 (4876.8)
15.64 (397.3)
15.14 (384.6)

Product cost is determined from the material requirements; the time required for each manufacturing step; the hourly wage rates; the overhead rates of labor costs, product development, and material storage; an expected rate of return from the capital investments; and an adequate profit. Scale drawings are also the basis for this activity. The material costs are generally obtained through the efforts of the purchasing department, while the labor estimates are generally obtained through the efforts of a manufacturing or industrial engineer working with the accounting department. The manufacturing methods proposed for the support arm need cost analysis data so the least cost method can be selected. A judgment is also needed to determine which method gives the most consistent quality control of the parts. Low cost without quality is false economy. Drawings are needed to make parts, whether they are for full-size or scale models.

6.2.5 Models

Some designs may be so complex that there are no mathematical models to analyze stresses and deformations. These products will be checked by building full-size or scale models. Scale drawing details can verify the concepts that in the idea stage of development were idealized with sketches, but models must be built before final appearance design can be established. The looks of a product can also have an effect on its marketability. Automobiles, home appliances, furniture, and houses are examples of products for which appearance contributes a great deal to marketing success.

Models are also useful to verify clearances between members, lengths of hoses and wire, and room for safe and efficient assembly and repair procedures. If an awkward assembly or repair procedure cannot be simplified, the need for a special tool may result, leading to another design project for the engineer and more cost for the company and the consumer.

There are alternatives to consider when building models for product evaluation. One alternative is to build and test components and variations on them. When the best of the components have been selected, they are assembled together for the final product. The advantage of this method is that individual effort can be concentrated on a small portion of the design. The disadvantage is that the interrelationship of various components may be overlooked and the final assembly will not perform as desired. The best of each of the components may not be the best for the total assembly.

A second alternative is to build a complete product based on preliminary component test, and then test and improve the product as a whole unit. A disadvantage of this method is that a particular component may not be improved to its full potential because its performance is adequate and major emphasis is placed on the less than adequate components. Advantages are that the whole product will be working. Changes will be made for total product improvement, not just a single component improvement, and there will be less total development time.

A third method is to build a complete product without any component test and then make changes to "make the product work" and improve its performance. The disadvantage of this method is that a product may be built that will not work, or will take a long time to get working, and money and time will be wasted. An advantage is that work on the final product starts immediately and if it works, time and money will be saved. Each project has to be considered separately because the procedure that worked last time may not work this time. Stay flexible and adaptive, and be ready to compromise.

Regardless of the activities undertaken to model and analyze ideas, the computer is likely to play an important role. Computer Aided Drafting (CAD) is commonly used for product development. Existing parts can be easily changed or used as the basis for a new but similar part. A computer search for an existing part to meet a given condition is also possible if Group Technology software is used to categorize existing parts. Many CAD systems will also calculate weights and volumes of parts as they are designed. If the CAD system is connected to material and production reference information, computer cost analysis can be performed. Connecting a CAD system to a computer aided manufacturing system allows process routing, including tooling, machine settings, and direct computer control procedures.

The finite element method of structural analysis is also practical using commercially available software and computer modeling techniques. "The finite element method is a computer-aided mathematical technique for obtaining approximate numerical solutions to the abstract equations of calculus that predict the response of physical systems subjected to external influences." (1)

Many activities related to component design can be enhanced by using a computer, from report writing on a word processor to statistical analysis of experimental data. It is beyond the scope of this text to go into details on computer software available to assist the engineering design function. The reader should, however, remember one thing: just because the answer comes from a computer, and has eight-digit accuracy, the result is not necessarily correct or applicable. The engineer must still judge the validity of the results. The following case study is an example of implementing a design solution before completing adequate analysis.

CASE STUDY #5:
Room cooling problem.

A laboratory on the first floor of a two-story building is used as a physics laboratory. The room is approximately 30 feet wide x 46 feet long x 10 feet high (9.15 x 14.02 x 3.05 m). Two walls have exposure to the outside, are constructed of 8-inch (0.203 m) block faced with 4-inch (0.104 m) brick, and have eleven, 3 foot-6 inch (1.067 m) x 6 foot-0 inch (1.829 m) single pane windows in them. The room is directly over the boiler room, which supplies heat, steam, and other utilities to the building. The lab is uncomfortably hot during the summer semester and during many weeks of the fall and spring semesters. Students have difficulty concentrating and doing good work. Opening windows for ventilation does little good, as outside temperature and humidity in the summer are also high, and breezes insufficient for cooling.

The first solution many people would implement is to install air conditioning, a common cure for an uncomfortable environment. And indeed this was done. Two 13,000 Btu/hour window air conditioners were installed. The result was no reduction in temperature, a slight reduction in humidity, and stale air.

When the problem was presented to an engineering class, an analysis of the situation revealed that the two window air conditioners could not compensate for the heat transfer through the floor, let alone the walls and ceiling. The boiler room was a nearly constant heat source at 85°F (29.4°C), and to keep the lab at 70°F

(21.1°C) would require the removal of approximately 52,000 Btu/hour. *Note:* The concrete ceiling of the boiler room has an R value of approximately 0.4 ft²·°F·hr/Btu (0.4 m²·°C·hr/Btu.).

Students had two primary solutions to the problem.

1. Open the windows in the boiler room to let hot air out. Every degree below 85°F (29.4°C) in the boiler room means a reduction of approximately 3500 Btu/hr transfer to the lab above. This solution is practical because the windows open into a metal grated window well.

2. Increase the R value of the boiler room ceiling by installing insulation. Only then do the window air conditioners (the habit solution to the problem) have a chance to cool the lab.

Solution #1 was a no-cost solution and was easily implemented. Solution #2 has not been implemented because of lack of funds.

Case Study #2 is continued here to further illustrate the idea analysis procedure.

CASE STUDY #2 CONTINUED:

All parts moving problems are not the same because of variations such as the configuration of the parts, the number of parts to be moved, the protection required, and the facilities available. Each of the ideas for moving parts has advantages and disadvantages that must be considered based on the limitations and restrictions known. Following is a brief discussion of the advantages and disadvantages of the ideas listed in Section 5.3, followed by a judgment about whether the idea should be considered for the final solution based on the limitations listed in Section 4.1.

Idea	Advantages	Disadvantages
1. Belt conveyor	Continuous operation, load and unload from many locations, can handle large volume.	Takes up floor space, parts may fall off, parts are exposed to environment, parts may bump into each other, power drive is required.
2. Roller conveyor	Continuous operation, load and unload from many locations, can handle large volume, no power drive required.	Takes up floor space, parts may fall off, parts are exposed to environment, parts may bump into each other, parts may fall between rollers.
3. Overhead conveyor	Continuous operation, load and unload from many locations, will handle moderate volume.	Parts may fall off, parts are exposed to environment, parts may bump into each other, power drive required, dangerous to pedestrians.

4. Gravity chute	No power required, continuous operation, high volume capacity, operator not required.	Parts may fall off, parts are exposed to environment, parts may bump into each other, need difference in elevation great enough to overcome friction, chute material may damage parts.
5. Liquid float	Continuous operation, operator not required, moderate volume.	Parts may bump into each other, need difference in elevation, need to dispose or pump return of fluid, possible fluid corrosion on parts.
6. Pneumatic tube	Fast, operator not required, parts protected from environment.	Parts may bump into each other, air compressor required, need load and unload station.
7. Hand carry	Use on demand, each operator can move own parts, low floor space requirement.	Personnel fatigue, may drop parts, weight limit controls parts per load.
8. Tote pans	Use on demand, each operator can move own parts, low floor space requirement.	Personnel fatigue, may drop parts, low weight limit, need supply of tote pans.
9. Wheelbarrow	Use on demand, each operator can move own parts, low floor space requirement.	Personnel fatigue, may tip over and damage parts, low weight limit, need supply of wheelbarrows.
10. Sled	Use on demand, each operator can move own parts, low floor space requirement, pull force in place of lift force.	High friction, need supply of sleds.
11. Two wheel dolly	Use on demand, each operator can move own parts, low floor space requirement.	Weight limit, lift and push load.
12. Push dolly	Use on demand, each operator can move own parts, more floor space required.	Push or pull force, need supply of dollies.
13. Forklift	High weight capacity, faster than by hand, use for other applications.	Need forklift and operator, need maneuvering space, more expensive.
14. Robot	Continuous operation, no operator needed, reliable, high load capacity.	Most expensive, need electronic control of some kind.
15. Catapult	Fast.	Need catapult device, part contact damage, need clear path, flying parts dangerous.

Comments and judgments:

1. Conveyor belt: Parts can be protected by using expandable plastic mesh protective sleeves. If standard conveyor units are used, the cost should be within budget. Unless conveyor has side spurs to each machine, a person is needed to load and unload parts on and off the conveyor. Reject because a conveyor will create more parts moving capacity than required, and the conveyor uses up floor space. Reconsider if volume of parts to move increases.

2. Roller conveyor: Parts can be protected by using expandable plastic mesh protective sleeves, or by placing parts on small skids that will ride the rollers. If standard roller units are used the cost should be within budget. Unless conveyor has side spurs to each machine, a person is needed to load and unload parts. Reject because a conveyor will create more parts moving capacity than required, and the conveyor uses up floor space. Reconsider if volume of parts to move increases.

3. Overhead conveyor: Easier to pass spurs to each machine. Reject because highly finished parts are hard to hang from a moving conveyor, and because of the danger to the parts and pedestrians.

4. Gravity chute: Lower cost than ideas 1, 2, or 3. Reject based on elevation drop required to make idea work, as departments are on the same floor.

5. Liquid float: Reject based on possible fluid leak, possible part corrosion, and elevation difference needed to make fluid flow, or the cost of a pump and plumbing to return fluid.

6. Pneumatic tube: Easy to route tube to each machine. Unanswered question: How much air pressure or vacuum is required to move 25-pound part? Reject because of time limit of ten days and unanswered question. Could research idea for future applications.

7. Hand carry: Reject because it is less efficient than present method.

8. Tote pans: Reject because it is less efficient than present method.

9. Wheelbarrow: Reject because it is less efficient than present method.

10. Sled: Reject because it is less efficient than present method.

11. Two wheel dolly: Reject because it is less efficient than present method. More danger of loads tipping over.

12. Push dolly: Reject because it is less efficient than present method.

13. Forklift: This is an expansion of the present system that will increase speed and increase load limit. Will the increase in speed also increase part damage? Will a new job classification be required for a forklift operator? Are there other jobs the forklift can do when not moving parts in this department? Is there a forklift already available that can work at this location part of the time? How much time is available for this use? Where can a forklift be stored?

14. Robot: Reliable, route can be easily reprogrammed. Reject based on high cost and higher capacity than required. Consider at a later time for a plant-wide parts moving system.

15. Catapult: Reject because of danger to parts and people.

It is not unusual for the majority of ideas to be rejected, nor is it unusual for the idea that surfaces as being "best" to be an extension of a present design. Unless a present design is totally inadequate, or a state of the art change comes along, most design changes are modifications of existing designs.

In this case the real problem turned out to be one of excessive part damage and poor inventory control, caused primarily by parts coming in contact with each other on the flat pallet, and the few number of parts moved per pallet. After additional discussion and problem understanding, it was decided to concentrate efforts on protecting the finished parts and increasing the number of parts per load.

Parts protection can be improved by using plastic or cardboard protective sleeves, but several sizes are required, and they would have to be returned and reused to keep cost down. Increasing the number of parts moved per load can be achieved by putting sides on the pallets. But perhaps there is a better solution. Figures 6.7 and 6.8 illustrate two of the more promising design proposals that will protect the parts, allow for more parts per load, and allow for movement to the next department by hand, hand pallet mover, or fork lift.

Full size test models of both designs were built and used for a short time to see how they worked. The upright model was easier to load and unload and was selected as the idea to implement even though the layered method allowed more parts to be moved per load. The layer separators were awkward to handle and they

FIGURE 6.7

Preliminary solution idea for Case Study #2.

A: (6) 6" Spaces down to (2) 16" spaces.

FIGURE 6.8

Preliminary solution idea for Case Study #2.

B: Spacing set by operator stacking parts, 6 to 16 inches.
C: Wood or plastic dividers
  1. Separate set for each major diameter range, or
  2. Make *A* slightly larger than largest diameter and do not use every notch for large diameter parts.
D: Radius equal to largest part radius, or notch only one side of separators equal to largest part diameter.
Notch can be circular or rectangular.

didn't stay with the pallets for reuse. The upright model was modified to a vertical rack rather than slanted, and a center column with four sides was used to increase the number of parts per load (Figure 6.9).

The wheels were selected based on an estimated maximum weight as follows.

2	2.50 x 2.50 x 40 inch hardwood cross members (50 lb/ft^3)	15 lbs
54	2.50 x 2.50 x 24 inch hardwood cross members (50 lb/ft^3)	234 lbs
28	1.50 dia x 16 inch steel parts (490 lb/ft^3)	225 lbs
1	40 x 40 inch hardwood pallet	90 lbs
	Total	564 lbs

The wheel selected was Globe #308504, a 4 inch diameter by 1.50 inch wide swivel rubber wheel, 5 inch overall height, with a 240 lb per wheel capacity, $9.20 each.

The final cost of $300 per carrier allowed an adequate number to be built. The carriers were not all built and in operation in ten days, but the project was under way, and enough people had been involved in the trial so that no major problems occurred during final implementation.

Related Activities   *153*

FIGURE 6.9

Final solution, Case Study #2.

6.3 Summary

As refinements of the project ideas are made, and details of manufacture, assembly, storage, and shipment are clarified, requirements for machine tools, material handling, painting, assembly, shipping, and storage must also be determined. Planning for the construction or procurement of plant facilities is as important to a successful project as a good product design. These requirements take design time, money, and a schedule of their own. The money and time must be incorporated into the overall project plan and schedule.

While refining ideas, it is common to think of new ideas, to realize that original specifications are too restrictive and must be relaxed, or that the original specifications can be improved. Modifying goals is a common occurrence. An early change in a specification that cannot be met will save time and money. An improvement in a specification will help evolve the product into a better than expected design. New ideas occur because of the increased concentration that accompanies refinement. New ideas might be a combination of the ideas being refined, or result from an improvement that becomes apparent while creating accurate drawings or while building a model. Additional thoughts may also suggest a change in the basic problem statement and specifications. Maybe your layout will show that a digging depth specification can be increased because of the linkage design you have proposed.

Do not suppress ideas that may change the scope of specifications of your project, but be cautious. A change in scope can require additional time, money, or staff support. These must be considered before a change is requested or approved. On the other hand, you may propose a change that will use less time, money, and staff. Those changes are the best kind. If you suggest a project change, be sure you have the answers to the questions that will certainly come up regarding project resources and the effect on the project time schedule.

6.4 References

1. Burnett, David S. *Finite Element Analysis*. Reading, Mass.: Addison-Wesley, 1987.
2. Chow, William Wai-Chung. *Cost Reduction in Product Design*. New York: Van Nostrand, 1978.
3. DiLavore, Philip. *Energy: Insights from Physics*. New York: John Wiley & Sons, 1984.
4. Dreyfuss, Henry. *The Measure of Man*. New York: Whitney Library of Design, 1967.
5. Fisk, Marian J. and Anderson, H. William. *Introduction to Solar Technology*. Reading, Mass.: Addison Wesley, 1982.
6. French, M. J. *Invention and Evolution*. New York: Cambridge University Press, 1988.

7. Lucas, Henry C. *The Analysis, Design, and Implementation of Information Systems*. New York: McGraw-Hill, 1981.
8. Schenck, Hilbert. *Theories of Engineering Experimentation*. New York: McGraw-Hill, 1979.
9. Stefanides, E. J. "Piezo Film Reduces Cost of Single-axis Accelerometer." *Design News*. Sept. 4, 1989, 212-213.
10. Tuve, George L. and L.C. Domholdt. *Engineering Experimentation*. New York: McGraw-Hill, 1966.
11. Wilcox, Alan D. *Engineering Design: Project Guidelines*. Englewood Cliffs, N.J.: Prentice Hall, 1987. Emphasis on electronics.
12. Woodson, Wesley E. & D.W. Conover. *Human Engineering Guide for Equipment Designers,* 2nd ed. Berkeley: University of California Press, 1965.
13. Woodson, Wesley E. *Human Factors Design Handbook*. New York: McGraw-Hill, 1981.

6.5 Exercises

1. In his book *Invention and Evolution*, M. J. French says,

    A large part of design consists of choosing from a very large number of possible ways of performing the various functions required of the thing designed, a single way for each function. The choice is complicated by the interactions of one function with another and by the many different ways in which the functions may be combined.

    Discuss his comment and how it affects the design process.

2. Think of other ideas for a drawing table support arm other than those shown in Figures 6.3 and 6.4.

3. Think of an idea for a support arm that would more likely be in tension than compression. This would eliminate column buckling as a possible mode of failure.

4. Name a product or circumstance that rates favorably, and one that rates unfavorably, for each of the following items.
    - *a)* Sound
    - *b)* High illumination
    - *c)* Vibration
    - *d)* Low illumination
    - *e)* Radiation
    - *f)* Low temperature
    - *g)* Pressure
    - *h)* High temperature
    - *i)* Shock waves
    - *j)* Moving air

5. List products that have "failed to function" as advertised or have disappeared from the marketplace, and give a reason why.

6. Assume that 20-foot long cars maintain one car length space between themselves for each 10 miles per hour of their traveling speed.
   a) What speed allows the maximum number of cars per minute to pass a given point?
   b) What is the maximum number of cars per minute that can pass a given point?
   c) What speed will allow 18 cars per minute to pass a given point?

7. For a particular period of time, aircraft accidents were recorded and tabulated based on the time of day they occurred. The representative data follows. Does the data support the conclusion that it is safer to fly at dawn? Explain.

Time of Day	% of total accidents
Daylight hours	86.0
Dark at night	7.0
Light at night	1.6
Dusk–twilight	4.0
Unknown	0.4

8. a) Analyze the apparatus in Figure 6.10 and explain how it can be used for an accelerometer when placed in the plane of motion. Be specific.
   b) Do both vertical tubes have to have the same area? What happens if the area of the vertical tube is one-half the area of the horizontal tube?
   c) Discuss advantages and disadvantages of having different cross-sectional areas for the vertical and horizontal tubes.

FIGURE 6.10

Accelerometer for Exercises 6.8, 6.9, and 6.10.

FIGURE 6.11

Accelerometer for Exercise 6.11.

9. How does the fluid selected for the apparatus of Figure 6.10 affect its operation? Some possible fluids are: water, density of 1.0 gm/cm^3; glycerin, density 1.26 gm/cm^3; ethyl alcohol, density 0.806 gm/cm^3; and mercury, density 13.6 gm/cm^3.

10. Analyze and refine the design in Figure 6.10 for an acceleration range of 1–5 G's. Complete scale drawings so the device can be built.

11. Analyze and refine the idea for an accelerometer sketched in Figure 6.11. Complete scale drawings so the device can be built.

12. Analyze and refine the idea for an accelerometer sketched in Figure 6.12. Complete scale drawings so the device can be built.

13. The ideas for accelerometers shown in Figures 6.10 and 6.12 are directional in concept. If they are not lined up with the direction of acceleration, the reading will be too low. Make additional refinement and analysis so the principles can be used regardless of the direction of acceleration.

FIGURE 6.12

Accelerometer for Exercise 6.12.

FIGURE 6.13

Sports car acceleration curve, Exercise 6.15.

14. A particular jet plane can accelerate from 0 to 575 miles per hour in 65 seconds. The speed-time graph is nearly a straight line. What force in G's does the pilot feel during the 65 seconds?

15. A particular sports car can accelerate from 0 to 200 miles per hour in 40 seconds. Due to various factors it cannot accelerate at a constant rate (Figure 6.13). Determine the maximum G's the driver feels.

16. From the data and answers from Exercises 6.14 and 6.15, determine the length of a race that would end in a tie for the jet and the race car.

17. A particular sports car can accelerate from 0 to 60 miles per hour in 6.08 seconds, and brake from 60 to 0 miles per hour in 126 feet. Which causes the driver to feel the higher number of G's, accelerating or braking? Quantify your answer.

18. What force would a seat belt system have to be designed for so the driver of the sports car in Exercise 6.17 would be safe?

19. Repeat Exercises 6.17 and 6.18 for a 5.29 second acceleration and 158 foot braking distance.

20. Refer to Figure 6.14. Water wells are located at A, B, C, D, E, and F. The water from the wells has to be pumped to the treatment plant before distribution. The distances between each well, and between each well and the treatment plant, are shown in the figure. Pipe for a direct line costs P $/meter, pipe for a double size pipe to handle the water from two wells costs $1.5P$ $/meter, and pipe for a triple size pipe to handle the water from three wells costs $1.8P$ $/meter. Find an appropriate network of pipe for this system for lowest cost.

21. An engineer is working on a project to design a method of moving sand from bulk ground storage to an overhead truck loading and weighing system. During the

FIGURE 6.14

Pump-pipe network for Exercise 6.20. Distances are meters.

course of development the following items have been listed as important to consider. The object is to keep cost down to meet the system needs.

a) Analyze how each of the following items is related to cost. Consider how the items listed are related to each other. Add items to the list as required.

E, Energy cost, $/kwh
T, Time equipment is used, hr/yr
C, Capacity of one bucket, lb/bkt
F, Material flow rate, lb/sec
I, Annual interest rate, %/yr
BC, Bucket cost, $/lb/bkt

BS, Bucket spacing, ft/bkt
S, Structure cost, $
P, Power required, hp
N, Number of buckets
CS, Conveyor speed, ft/sec

b) List questions that should be answered before proceeding with a design proposal for this project.

22. There are many products that have been designed to avoid catastrophe in the case where something goes wrong. For each of the following items specify a potential problem for the customer, specify what commercial designs are available to prevent the danger, and then think of other solutions to the problem. Example: Electric hedge trimmers may be used outside when it is wet. This creates the potential for electric shock. Grounded (three-wire) wire and grounded plugs are

available on commercial models. The problem can also be avoided by using double insulated cases, battery operated units, an electric circuit protected with a ground fault interrupter circuit breaker, or by growing bushes that do not have to be trimmed.
- *a*) Rotary lawn mower blades
- *b*) Snow blower augers
- *c*) Automatic garage door opener/closer
- *d*) Barrier arm at a toll gate
- *e*) Manually operated paper cutters

23. The following products have inherent dangers while being used. What is the danger, and discuss some ideas to overcome the danger. Example: An empty light bulb socket invites the insertion of a finger and an electric shock is likely. A smaller diameter base would make this less likely. Another possible solution is to have a safety seal that rotates to fill the opening of the hole when the bulb is removed.
- *a*) Power steering on an automobile
- *b*) Power brakes on an automobile
- *c*) Electric toaster
- *d*) Electric stove
- *e*) Incandescent light bulb
- *f*) Chain saw

24. An engineer is designing a ladle hook for supporting a ladle of molten steel in the steel making process. Figure 6.15 shows a sketch of the problem and of the proposed design solution. The engineer specified the following items as part of the solution. Analyze each item by itself and in conjunction with all of the items listed. Comment on the validity of each item. Be quantitative when possible.
- *a*) The method of construction proposed is five layers of 1.40 inch thick steel.
- *b*) Each of the five pieces is to be flame cut from plates.
- *c*) Twenty-six, 1.00 inch diameter rivets, made from steel with S_{ys} = 15,000 psi, will be used to hold the plates together. The safety factor for the rivets is given as 2.
- *d*) In the critical areas around the 8-inch support pin and the 15-inch ladle shaft, the plates will be joined by butt welding.
- *e*) The engineer claims that if the 6-inch central web width is changed to 12 inches and the thickness changed accordingly, the hook will hang so the center of the ladle shaft hole is directly under the upper shaft hole.
- *f*) The safety factor for the hook design is given as 3 when using material with S_{yt} = 37,500 psi.

25. Draw a fully dimensioned scale drawing of the ladle hook as proposed.

26. Determine an alternate design for the ladle hook and make complete scale drawings of the idea.

27. If electric heat is 100% efficient, that is, all the purchased energy is converted into heat, and a natural gas furnace is only 65% efficient, why is a gas furnace still the "best" buy?

FIGURE 6.15

Data for ladle design, Exercise 6.24.

FIGURE 6.16

Data for Exercise 6.30.

FIGURE 6.17

Schematic for Exercise 6.31.

28. List advantages and disadvantages of the following methods of heating a house.
    a) Natural gas
    b) Wood
    c) Electricity
    d) Fuel oil
    e) Coal
29. The frequency of "small" oscillations, in Hertz (cycles per second), of a simple pendulum of length L is $f_n = \sqrt{g/L}$, and the time of cyclic motion $t = 2\pi\sqrt{L/g}$.
    a) If a pendulum controlled clock is running slow, should the length of the pendulum be lengthened or shortened? Explain.

FIGURE 6.18

Ideas for Exercise 6.32.

(a) Sliding joint with set screw

(b) Same as (a) but have threads for adjustment

(c) Hinge joint with bolt and wing nut

(d) Fixed length / Protruding dowels

(e) Chain or other flexible linkage / Add weight / Protruding dowels

(f) Wing nut clamp / Curved slot

b) Explain why the effective length of a pendulum can be made shorter by adjusting the mass along the length of a fixed length pendulum bar.

c) Explain why pendulum clocks with long pendulum arms are easier to adjust to an accurate time compared to short pendulums.

30. Refer to Figure 6.16. A 20-foot high post is embedded in the ground at point C and has a 1000-pound pull on the top as shown. The post is beginning to bend over and it is desired to add a support. Space limitations prevent any near side brace further than 5 feet from C, and far side braces are impeded by an 8-foot high fence 10 feet from the base of the post. Several suggestions have been made.

   a) Add a brace on the near side to some point A on the post.
   b) Add a brace on the far side within the 10-foot limit.
   c) Add a brace from point B or beyond that passes over the fence.

   Analyze these ideas and others that you may have for adding support to the post. Consider cost of material, total structure stability, and installation methods.

31. The concept of a machine to test the shear strength of materials is not particularly complicated. However, the simplest of single shear machines might induce bending stress in addition to shear stress. Analyze and compare the differences between a single shear area testing machine and a double or triple shear area testing machine. Refer to Figure 6.17.

32. Figure 6.18 shows some possible designs for an adjustable drawing table support arm. Refine each of the ideas to workable solutions, including drawings of all component parts.

33. Suggest other possible solutions for the problem discussed in Case Study #5.

7 Decision and Implementation

"Intangible factors such as social values, political influences, and psychological effects are extremely difficult to quantify and measure. Yet, as all practicing engineers soon learn, these factors often control the acceptance or rejection of a design."
♦ DALE D. MEREDITH

Decisions are made during all phases of project activities. Decisions already made include the details of specifications, market potential, and sales forecasts, as well as whether to reject or develop various solution ideas. The decision necessary at this stage of the problem solving procedure is to determine whether the project should be canceled, modified, or implemented, and if implemented, which idea is most acceptable.

Canceling a project means that there is no solution available that has a chance of being successful under the present restrictions. A number of possibilities can lead to the canceling of a project. Perhaps the cost cannot be reduced to an acceptable level and sales projections at a high price indicate that too few products will be sold to make a profit. Perhaps the market has changed since the project was started. Maybe a competitor's product solves the problem better, or at less cost. Maybe a technological breakthrough has occurred and the state of the art has advanced beyond the present design: a more efficient slide rule doesn't have much market potential. Other reasons for canceling a project include:

1. No money left, the budget has been depleted. Perhaps research and development costs were higher than anticipated, and no acceptable solution was developed.

2. No time left, the due date arrived without a solution. The timing of product completion is often a critical concern, and perhaps project bid data arrived late and no proposal bid was possible.

3. No trained personnel to complete the project. If critical state of the art development depends on a few key people, and they leave, become ill, or die, a project may have to be canceled or postponed.
4. The project is badly timed, there is low customer interest, there is a saturated market or even a disappearing market. When the price of oil on the world market fell in the mid 1980s, and the number of oil wells drilled was reduced, the market for new oil-well drilling equipment disappeared.
5. The state of the art has outdistanced the expertise of the company. When transistors, integrated circuits, and printed circuit boards were developed, companies had to cancel or modify projects that involved vacuum tubes.
6. The consumer isn't ready for the product. The project is ahead of its time. Chester Carlson invented the process of xerography in 1938 but struggled to find commercial acceptability, rejected by companies such as RCA, General Electric, and IBM. It wasn't until 1949 that a small company in Rochester, New York marketed the first commercial machine. That company became the Xerox Corporation in 1961.
7. Public safety may be at risk. Perhaps the side effects of implementing an idea are worse than the effects of not implementing it. Refer to the discussion of the Aswan dam in Section 2.3.

There is nothing wrong with canceling a project, no matter whose favorite it is. Many projects should have been dealt this fate and weren't. One person's position of power and lust for success should not precipitate ludicrous spending of time and money on poor projects. Philanthropists can do as they like, but competitive businesses can't afford to. Fortunately, as the Xerox story indicates, persistence and determination can often overcome many problems.

If there are still questions of meeting market competition, maybe the best design can be modified to make it viable. Perhaps specifications need to be changed and the problem solving activities repeated. A large capital investment may be involved to make a power unit available for rototilling, grass cutting, and snow blowing, and inadequate sales increase is projected to justify the expense. A design specification adjustment may be required. An idea previously discarded may now be the most acceptable solution to the problem, or perhaps research has come up with new technology to apply to the problem. The most acceptable solution will always include compromises between conflicting alternatives.

Many product development projects are evolutionary in nature, that is, the weaker ideas are dropped during analysis and refinement, and the time and money go toward developing the most promising idea. If that is the case, then at this point the decision needed is whether or not the product presently developed meets enough of the design criteria to continue. It is hoped that the idea that has been refined and analyzed as workable will meet all of the problem identification criteria, will be clearly better than the other ideas, and be profitable. If it passes the test, the direction of implementation is clear.

If there are still several ideas that qualify under all the conditions set forth, and one answer is not clearly the best, then a decision must be made on the direction of the project. This will also be the situation when a feasibility study has been done, because one of the goals of a feasibility study is to propose several methods of solution to the problem.

7.1 Idea Selection

At any stage of an engineering project, decisions are made about which idea(s), if any, are to be further developed. The list of feasible solutions to a problem needs to be reduced to the one idea that seems to have the best chance of succeeding, or reduced to the best three or four ideas, any one of which may turn out to be most acceptable. Reducing the list to just one idea allows all available resources to be concentrated on developing that concept. If the list is reduced to three ideas and they are all developed, then either more money or more personnel must be assigned to the project or the project will take longer to complete. The next example compares resource allocation if three ideas are in the running for further development work.

EXAMPLE 7.1:
Idea development schedule.

In Figure 7.1, the horizontal axis represents periods of time, and the size of each rectangle represents the proportion of money and personnel being used during each time period. Figure 7.1(a) represents the method of idea selection where one idea is kept and developed. The project is shown to be completed in five periods of time and five units of money and personnel. During the first period of time one-third of the money and personnel time is spent on each idea to determine which should be continued.

Figure 7.1(b) illustrates what happens if each idea is studied longer to collect more information to support the final selection. If the total amount of money and personnel for each time period is the same as in Figure 7.1(a), then the project will not be completed in the same amount of time, it will take eight periods of time.

Figure 7.1(c) illustrates the approach where the project gets done in the same number of time periods as in Figure 7.1(a), but requires more money and personnel, three times more during time period two and twice as much during time period three. Unless other projects are available for the extra people to work on at the conclusion of this project, employee utilization will be poor, and a stable work force hard to maintain. It may be possible to utilize the extra personnel on ideas #2 and #3 after idea #1 is discarded. Difficulties exist if the time frame for idea #2 includes some tests that are time dependent, such as fatigue or creep tests. The total time may not be reducible even with additional personnel.

The main advantage of the methods shown in Figures 7.1(b) and (c) are that alternate ideas are being developed, and can be used if the idea that was first thought to be the best fails to succeed. If all the time and money goes into one idea and it doesn't work, backing up to develop an alternate idea will cause a product introduction delay.

FIGURE 7.1

Alternate time-personnel schedules.

If several preliminary ideas have been developed and are still feasible, the time will come when the preferred idea must be selected for implementation. One method for making the selection is by using a matrix, weighted-value technique as follows.

The first step is to make a **list of all the criteria** that you think could be used to compare the design solutions. Though this list can include anything that comes to mind, to be effective the list must include those things required for a fair comparison of the various proposals and must include the original specification goals. Some criteria to consider are:

- weight specification
- size specifications
- operation specifications
- cost to manufacture
- appearance
- number of parts
- ease of assembly
- ease of repair
- return on investment

- cost of use by the consumer
- time needed for delivery
- environmental specifications
- ease of use by the consumer
- cost of tooling
- possibility of harm to the user
- possibility of market penetration
- shelf life
- estimated life before failure

When the criteria list is complete, **assign a priority value** to each item on the list. Assigning a priority identifies the items on the list that are most important, and those that are the least important. Use a scale of 1 to 10 or 1 to 100 and assign a numeric value to each criterion. Most people prefer to associate a high value as good, so assign a 10 to the most important criteria and a 1 to the least important. If appearance is valued 10 and cost 8, then appearance is considered 25% more important than cost. More than one item on the list can have the same priority value if they are judged to have equal importance, or decimal values can be used if close refinement is necessary. To do an unbiased job of assigning criteria values, do not consider any of the solution ideas when assigning the values. Criteria values can also be assigned by several people on an independent basis and the results averaged.

After the value of each criterion has been determined, **normalize the values** by dividing each of the individual values by the total of the values. Each item will then have a normalized value between 0 and 1, and the total of the normalized values will equal 1. Normalizing is not mandatory but it gives range values more suitable for intuitive inspection because the values are easily compared as percentages.

Now **make a chart** with the criteria items to be compared as row entries, in order of their normalized priority values, and the idea solutions to be compared as column headings. For each criterion on the list **compare and rank the ideas.** The ranking values can cover any range—0 to 10 or 0 to 5 are common. Use high values for good rankings. Zero is included because an idea may not meet a particular criterion at all. Ranking can also be done by several people and the results averaged.

When each idea has been ranked for each of the criteria, multiply the ranked value by the normalized criterion value factor. Add the products and get a total score for each idea. The idea with the largest total is the "best" solution, that is, the most acceptable with the criteria and ranking selected. Ordinarily several ideas will have totals that are appreciably higher than others. Any solution idea with a very low total indicates that it does not measure up to the required criteria and can be discarded.

This system of evaluation is a tool to help weed out the poorest ideas and to focus remaining effort on the better ones that remain. Choices still must be made and optimization continued. The idea with the highest score can be selected and implemented, or the chart can be studied for the highest ranking in each category and the idea with the majority of high rankings selected. Time and energy can also be focused on improving the low ranked criteria of the "best" idea. This is simply a way to be sure the most acceptable idea is used. Another approach is to take the highest ranked idea from each criterion and generate a new idea that incorporates all the better features.

Obtaining numeric values of the total scores is not the purpose of this method because assigning numeric values to design criteria cannot be done in a consistent way. It is the relative values that are important. The procedure is recommended for a project that has several solutions under consideration, all of which seem workable. The procedure does take time, and it is difficult to agree on values while doing the evaluation. The main purpose of the procedure is to focus attention on the features of the design that are really important and to help decide which idea best meets the design criteria. Don't be surprised if new concerns are proposed during the procedure, because concentrated thinking is still taking place, and creativity hasn't been shut off. It may also happen that other comparison criteria will come up during this idea analysis and a new comparison will need to be done. Recall that one of the tasks of an engineer is to iterate so improvement can continue. It is essential that the design team agrees that the idea solution implemented is the "best." Another method of selection optimization is discussed in Exercise 7.14. The following case study illustrates the criteria evaluation and ranking procedure just discussed.

CASE STUDY #6:
Baseball field revision.

An engineer was assigned a design problem involving a baseball field. There was an existing field, but it had a left field foul line of only 240 feet (73.2 m). An investigation was made and alternatives proposed that would increase the foul line to 300 feet (91.4 m). Table 7.1 lists the design criteria selected to compare the ideas. The table includes the criteria values assigned. The criteria values listed in Table 7.1 indicate that cost has been selected as the most important factor, and shower and locker facilities as least important. The normalized values are calculated by dividing each value by 54, the total of the assigned values.

Table 7.2 shows each criterion ranked for each of five possible solutions. The present field was included as a basis because nothing has to be done. The best solution may be to make no change. The present field and the off-campus field both rank highest in cost because there is no reconstruction cost associated with these two ideas. Idea #1 is most costly to implement. All the on-campus ideas are equally close to locker facilities, whereas the off-campus idea is not close to locker facilities.

If the chart results are taken at face value, no change will be made because the present field ranks highest with a 7.74. Idea #2 is the poorest and probably doesn't

TABLE 7.1

Criteria and assigned values for baseball field project.

Criteria	Priority Values	Normalized Values
1. Cost of implementation, moving dirt, fences, retaining walls, etc.	10	0.19
2. Room around the diamond for benches, warm up, and spectators	6	0.11
3. Sun in the eyes of the players	5	0.09
4. Houses and cars in the location of home runs and foul balls	5	0.09
5. Meet minimum requirement of 300 foot foul lines	8	0.15
6. Shadows on the field from buildings	3	0.06
7. Near student dorms, convenience of stopping by to watch the game	6	0.11
8. Close to shower and locker facilities	2	0.04
9. Ground suitable for good grass and infield skin maintenance	4	0.07
10. Interference with other sports fields	5	0.09
Totals	54	1.00

Underlined words are keywords used in Table 7.2.

TABLE 7.2

Ranking comparison for five ideas for the ballfield revision.

Criteria	Normalized Priority Values, N	Idea #1 On Campus R_1 NR_1	Idea #2 On Campus R_2 NR_2	Idea #3 On Campus R_3 NR_3	Idea #4 Off Campus R_4 NR_4	Idea #5 Present Field R_5 NR_5
1. Cost	0.19	2 0.38	3 0.57	4 0.76	10 1.90	10 1.90
5. Minimum foul lines	0.15	10 1.50	10 1.50	10 1.50	10 1.50	0 0.00
7. Student convenience	0.11	10 1.10	10 1.10	10 1.10	1 0.11	10 1.10
2. Room	0.11	7 0.77	4 0.44	7 0.77	8 0.88	8 0.88
3. Sun	0.09	9 0.81	3 0.27	8 0.72	5 0.45	9 0.81
4. Foul balls	0.09	9 0.81	1 0.09	8 0.72	8 0.72	8 0.72
10. Interference	0.09	3 0.27	3 0.27	4 0.36	10 0.90	10 0.90
9. Ground	0.07	8 0.56	6 0.42	5 0.35	9 0.63	9 0.63
6. Shadows	0.06	8 0.48	6 0.36	9 0.54	7 0.42	8 0.48
8. Locker facilities	0.04	8 0.32	8 0.32	8 0.32	1 0.04	8 0.32
Total scores		7.00	5.34	7.14	7.55	7.74

The values in the R_1, R_2, R_3, R_4, and R_5 columns are the comparative rankings. For example, ideas #4 and #5 have minimum cost, idea #1 has highest cost, and ideas #2 and #3 are nearly as expensive.

warrant further consideration. If idea #1 could be done with less money than idea #3, and earn a 5 ranking in cost, idea #1 would be second best.

The final solution proposed for the project was a compromise. Did you expect anything different? It was decided to reduce the cost of idea #3 by eliminating the excavation for a full 300-foot (91.4 m) foul line and settle for 285 feet (86.9 m). This compromise increased the ranked value cost to an 8, and reduced the rating for minimum foul lines to an 8. This resulted in a new total value of 7.60. One could argue that a 285-foot (86.9 m) foul line rates a 0 because it doesn't meet minimum requirements, but recall that a compromise has been judged the most acceptable alternative under the imposed restrictions, not the utopian best. To date no action has been taken on the project because funds have not been made available. The final recommendation for the ball field revision project is given in Figures 7.2 and 7.3.

Follow-up to Case Study #6. In the fall of 1988 the existing baseball field was converted into a soccer field, and baseball games are now played at an off-campus field managed by the city. This is an example (not an uncommon one) of a problem that was resolved in a far different way than the original data, study, and analysis suggested.

FIGURE 7.2

Report on Case Study #6.

```
To: Coach Appleton
From: Jim Student
Re: Changing the baseball field configuration
Date: May 10, 1986
    I recommend a change in the baseball field layout to the one
shown on the attached drawing. This proposal is an answer to your
request to find a reasonable cost method of increasing the length
of the left field foul line. This proposal increases the length
from the present 240 feet to 285 feet. The cost of the proposal
is estimated at $2500. This cost covers relocating the fences and
backstop, and re-sodding the present infield, which becomes part
of the outfield of the proposed change. No excavation is
required.
    Other ideas were considered. One idea would have given a full
300-feet foul line, but excavation would have added an estimated
$7500 to the project.
    There is also a field in town that could be used for a rental
fee of $100 per game, but there are no shower facilities there.
Players would have to come back to campus to shower and change
clothes. An off-campus field would also discourage students from
''stopping by'' to watch a game.
    Please call me after you have studied the attached drawing.
I want to answer any questions you may have as well as discuss
possible implementation. If I have not heard from you by May 24,
1986, I will contact you.

cc: Members of the Athletic Committee
```

FIGURE 7.3

Final ball field proposal.

7.2 Implementation

Implementation includes the activities that are required to convert the "best" idea into a functioning product in the hands of consumers in an economical manner. Once a project has the green light, in some sense the hard work is yet to come. Good communication, that necessary ingredient to keep everyone's effort concentrated toward the common goal, is just as important now as before. More people will be involved once the decision to implement is made.

The final version of the project requires **detail and assembly drawings**, but additional development may be required, so changes continue to be made to drawings and related bills of materials. Preliminary drawings will already have been created, either for a full size prototype to be built, or for an accurate and complete product cost to be calculated. A complete cost analysis requires material and manufacturing labor costs, both of which come from detailed drawings and assembly specifications. For implementation, the drawings must be reviewed to be sure they include all the latest changes and incorporate appropriate company standards. Ordering 10,000 of an old version of a part is not only costly, it is demoralizing.

Product testing will continue during implementation, and last minute changes may be proposed. Some changes are mandatory, as when a part begins to fail under test. Other changes may be just the nice-to-do type and will be incorporated only if they can be done without disrupting the schedule, such as a change in appearance. Some proposed changes are saved for the next version or for a different model. A proposed change that cannot be implemented within the present time schedule may be offered later as an option. For example, a trailer pulling package option for a passenger car (heavy duty shock absorbers, special mirrors, higher horsepower engine, heavy duty transmission), or a software update for a computer system.

During implementation the **schedule for product release** generally becomes the controlling factor. An exception is a problem that may cause the consumer harm or place others in a dangerous position. Don't compromise a quality product just to get it out the door and on the way to the customer. The cost of warranty claims and recalls can offset short gains in sales. Lawsuits might occur that could force your company into financial disaster or out of business. Good ethics requires you to keep management notified about any known or potential problems. For the record and for future litigation possibilities, put all such findings in writing and send them to the appropriate management levels. Read reports on the April, 1986 space shuttle disaster and subsequent hearings, which appeared in newspapers and magazines, to learn how critical and important communication and timely action are.

Many **production activities** must be completed using the details from final drawings. Proper tooling, including dies, jigs, fixtures, gaging, and machine tools must be provided. The cost and complexity of these items should have been estimated for budget purposes, and now they must be ordered, designed, built, and tested. The testing of processes and tooling is no less important than the testing of product components. Production tooling must be capable of making parts to drawing specifications in a consistent manner. Time for the testing of processes and tooling is particularly important if a new process is being used. An engineer may have to spend many days at the factory of the vendor supplying the new process. It is easier to check out a machine in the factory of the supplier than in the user's plant. For complex operations on a new machine or process, the first run of production parts will likely be made at the vendor's factory.

Plant facilities such as assembly lines, paint booths, and storage areas must be prepared and made ready for full scale production. Will the new parts fit through doorways and under overhead lights? Many problems can arise as production starts. Methods of assembly that were no problem on a custom model building scale now become major obstacles. Forklifts used during development to assist with large component assembly must be replaced with cranes and hoists. The special touch of a technician's assembly procedure must now be specified at a particular torque or pressure setting, assembly procedures confirmed, and proper assembly tools and equipment provided.

A project involving system changes requires **personnel training**. Perhaps new forms and methods of processing them need to be introduced. Set up workshops in advance of full scale use to train users properly. Because some learning curves have very low slopes, adequate training time is essential. Plan for parallel system operation if all the portions of a new system have not been completely tested, because bugs will occur. This is particularly important when converting from a manual system to a computer system, or changing computer systems. A computer will not make a poor system run smoothly, it will only generate errors and chaos faster. New systems and products also require **procedures or operating manuals**. An operator's manual must be accurate and specific, and if a license is not required for operation of the product, the manual must include a warning that operation in any mode not specified in a list of safe modes can result in injury to both people and property, and possibly death.

Before a product is released to the consumer final **product parts listing and service manuals** need to be prepared. Parts lists and repair parts listings are important for support of dealers who make repairs and sell service parts, as well as for customers who need to know normal service operations. Poor customer relations can result from poor service information. Imagine the frustration on the part of a consumer who, after requesting a new light bulb for a light generator system, was told that the entire light bulb-light shroud-lens assembly had to be purchased because that was what was listed in the service manual.

Prior to product release, **market advertising** must be readied with product specifications and features highlighted. This includes information on available options, accessories, and cost. If a product unveiling show will be held to publicize the new product to the dealers, all the preparations must be planned and scheduled. Sometimes first production units are loaned to preferred customers for their use, so shakedown testing can be done that more accurately reflects customer applications.

When the first product is shipped, or the new system is put on its own, personal emotions run high. Many questions run through the minds of project staff. Will the product work as tests have predicted? Will the production units be as good as the test units? Was the marketplace reached in time? Were enough tests run so the reliability and confidence levels are high? Is the selling price too high? Too low? The questions will be answered by the consumer.

CASE STUDY #3 CONCLUDED:

The special pallet racks for moving ground shafts worked so well that their use was expanded. Empty racks start at the lathe department, where the parts are turned and center-drilled. The operator loads the parts on the racks and then the parts go to heat treatment. From heat treatment they go to the finish grind and then to assembly. As the racks are emptied, they are sent back to the lathe department and the cycle is repeated. Inventory control is improved and the surface of the parts does not get nicked or scratched any more. More racks were needed with this approach, but the increase was justified because of the improvement in inventory control and the reduction of scrap. The parts are now protected at each stage of production, parts are not lost in transit, and a more accurate count is maintained.

7.3 References

1. Amos, John M. and Bernard R. Sarchet. *Management for Engineers*. Englewood Cliffs, N.J.: Prentice Hall, 1981.
2. Andreasen, M. Myrup et. al. *Design for Assembly*. New York: Springer-Verlag, 1985.

3. Bronikowski, Raymond J. *Managing the Engineering Design Function*. New York: Van Nostrand Reinhold, 1986. A summary of the steps of the design procedure from a manager's point of view; chapter on patents.
4. Golembeski, Dean J. "Struggling to Become an Inventor." *Invention And Technology*, Vol. 4, No. 3, Winter 1989, 8-15.
5. Hajek, Victor. *Management of Engineering Projects*. New York: McGraw-Hill, 1984.
6. Lentz, Kendrick W. *Design of Automatic Machinery*. New York: Van Nostrand Reinhold, 1985.
7. Lipson, Charles and N. J. Seth. *Statistical Design and Analysis of Engineering Experiments*. New York: McGraw-Hill, 1972.
8. Meredith, Dale D. et al. *Design and Planning of Engineering Systems*. Englewood Cliffs, N.J.: Prentice Hall, 1973.
9. Morris, George. *Engineering—A Design Making Process*. Boston: Houghton Mifflin, 1977.
10. Pahl, Gerhard and Wolfgang Beitz, edited by Ken Wallace. *Engineering Design*. London: The Design Council, 1984. Many mechanical design hints.
11. Papanek, Victor. *Design For Human Scale*. New York: Van Nostrand Reinhold, 1983.
12. Schmidt, Frank W. and Robert E. Henderson. *Introduction to Thermal Sciences*. New York: John Wiley & Sons. 1984.
13. Spotts, M.F. *Design Engineering Projects*. Englewood Cliffs, N.J.: Prentice Hall, 1968.
14. Tichy, H.J. *Effective Writing for Engineers-Managers-Scientists*. New York: John Wiley & Sons, 1967.
15. *Consumer Reports* magazine.

7.4 Exercises

1. List comparison criteria you would consider if you were going to purchase each of the following products and wanted to compare brands.
    a) Rowboat
    b) Bookshelf
    c) Card table
    d) Automatic clothes washer
    e) Microcomputer
    f) Bicycle
    g) Camping tent
    h) House
2. List comparison criteria you would consider if you were going to make a decision about each of the following items.
    a) College to attend
    b) Vacation to take
    c) Engineering job to take

3. Select the product from Exercise 7.1 with which you are most familiar, place values on the criteria you have listed, and then normalize the values.

4. Select at least two products that are currently on the market for the product selected in Exercise 7.3 and compare them. Complete the calculation and select the most acceptable. If the selection comes out contrary to a selection you or someone you know has made, discuss reasons why.

5. Discuss specific feedback data that indicate if a mass-produced product, such as a moped, has been successful.

6. Discuss specific feedback data that indicate if a new system, such as inventory control, has been successful.

7. Suppose a new machine tool, a key process in a new product, was delayed in shipment well past the proposed production date. Discuss ways to work around the problem and meet the original date.

8. The following list includes some products that have critical market introduction dates. List the products in order, from those that have the most critical market introduction dates to those with the least critical introduction dates. List reasons for your ranking.
    a) Christmas trees
    b) Lawn mowers
    c) Snow blowers
    d) Water skis
    e) New car
    f) A new watch
    g) New residential house light controls
    h) A new typewriter
    i) A new college textbook

9. Discuss the problems that occur if the products listed in Exercise 7.8 arrive too early or too late.

10. Consider a new graduate engineer who is faced with the following decisions all of which are viable choices:
    A) Going into the consulting business as a sole proprietor,
    B) Going into the consulting business as a partner with two experienced engineers,
    C) Going to work for a local engineering firm, or
    D) Going to work for a national engineering firm.

    Assume that the probability for success depends upon the economy, which could:
    1) Enter a boom cycle,
    2) Continue at a steady pace, or
    3) Enter a recession.

    The following chart gives the engineer's estimated relative earnings expected for each of the choices. For example, going into business with two partners in a steady economy will yield half the earnings of going into business alone in a boom economy.

	1	2	3
A	100	40	0
B	80	50	10
C	85	60	15
D	90	75	15

a) Which choice should the engineer make if the economy has an equal chance at a boom, a steady pace, or a recession? Why?

b) Which choice should the engineer make if the economy has a 50 percent chance at a steady pace, 25 percent chance at a boom, and 25 percent at a recession? Why?

c) What choice should the engineer make if recession is certain? Why?

11. Table 7.3 represents a comparison made on five different cars by automotive experts.

   a) Determine a criterion evaluation value for each category and determine which car best suits your need if you were going to buy a new car.

   b) List other items you would add to the list if you were doing the analysis for yourself.

TABLE 7.3

Comparison data for five cars, Exercise 7.11.

Car #	1	2	3	4	5
a. Styling	20	19	14	19	14
b. Quality control	18	21	18	20	15
c. Instrumentation	16	19	17	20	16
d. Ride comfort	14	12	19	19	17
e. Visibility	18	20	21	20	16
f. Load hauling	18	12	21	21	16
g. Handling	13	19	14	17	15
h. Fuel economy	5	20	10	15	25
i. Acceleration ability	13	6	17	21	14
j. Dollar value	13	18	14	21	19

12. Table 7.4 lists five materials that are being considered for the outside covering of a building. Three criteria are to be used for evaluation:

   1) High density, which is good for shielding from γ rays
   2) High thermal resistance, which is good for reducing heat loss
   3) Low cost, which is less expensive to install

   a) Assign importance values to the three criteria.
   b) Rank each of the five materials for each criteria.
   c) Select a material.
   d) List other criteria you might want to consider if you were in charge of this project.

TABLE 7.4

Comparison data for five materials, Exercise 7.12.

Material Number	Density lb/ft^3	Thermal Resistance R value/inch of thickness	Cost $/ft^2
1	169	0.0007	17
2	120	0.20	5
3	140	0.10	3
4	170	0.02	30
5	62	0.25	0.50

13. List and compare the advantages and disadvantages of:
    a) a three-legged chair
    b) a four-legged chair
    c) a five-legged chair

14. An alternate method of comparison to determine the "best" of many ideas is to compare all possible combinations of ideas two at a time. No detailed breakdown of selection criteria is done, each idea is treated as a whole. For example, if six proposals are viable, and they are coded A, B, C, D, E, and F, then 15 comparisons can be made. If you keep track of the number of times that each proposal is preferred, then the ranking of "best" can be done based on that number. In the example shown in Figure 7.4, proposal C comes out on top followed by E, F, A, D, and B. In the case of a tie for "best," then further analysis of those tied needs to be done. Use this method of comparison on Case Study #6 or any of the other comparison problems. The procedure can also be performed three ideas at a time to confirm the two at a time decision. Comparing six things three at a time results in $_3C_6 = 6!/3!3! = 20$ combinations, while comparing six things two at a time results in $_2C_6 = 6!/2!4! = 15$ combinations.

FIGURE 7.4

Example of idea evaluation by pairs.

The idea at the top of the second column is compared with each idea in the first column, and the most acceptable one selected. Then the idea at the top of column three is compared with each idea in column one and the best idea recorded. This is continued until all combinations have been compared.

	A vs.				
B	A	B vs.			
C	C	C	C vs.		
D	A	D	C	D vs.	
E	E	B	E	D	E vs.
F	F	F	C	F	E

Idea	A	B	C	D	E	F
Number of "wins"	2	1	4	2	3	3

15. Compare buttons, zippers, and Velcro™ as methods of keeping a jacket closed. Consider such features as:
    a) Original cost
    b) Ease of use by two-handed people
    c) Ease of use by one-handed people
    d) Ease of repair
    e) Appearance
    f) Reliability
    g) Noise

16. Compare four modes of transportation: automobile, train, bus, and airplane. Consider such concerns as:
    a) Cost
    b) Ease of access
    c) Comfort in transit
    d) Safety
    e) Taking children
    f) Sleeping on route
    g) Eating on route
    h) Taking luggage

    Make the comparison for:
    a) A trip of 10 miles
    b) A trip of 100–500 miles
    c) A trip of 1000–2000 miles

17. Compare the advantages and disadvantages of a wooden sharpenable pencil and a mechanical pencil.

18. An alternate approach to lists of comparison criteria is to create a checklist with very specific items for a very narrow use, such as a checklist for inspecting a house to purchase, a car to buy, or a college to attend. Select a specific task, such as buying a car, and list the specific items to check for such a project.

Design Problems

The following project problems are proposed to allow you to practice the design procedure. There is no limit to the type of problem that can be considered, and you should be capable of leading any problem solving investigation. You may not be an expert in the area of the specific problem, but you know how it has to be approached. The problems are separated into two groups. Group I contains problems that do not have many details given, which is typical of open-ended design problems. Designs should be complete, with detail drawings, complete specifications, and either a scale model or a full-size working model.

Group II problems represent another approach to design, as they begin with a solution idea. It is not unusual for an engineer to be given a preliminary sketch and told, "Make it work." Someone has an idea, a brainstorm if you will, and wants to see if the idea can be made functional. This beginning idea could be a sketch or a picture of a competitor's product. In fairness to the person giving you the idea, and by necessity if the person is your superior, an analysis will be made. This is a case where the problem solving starts with the idea, and loops to problem identification, analysis, and then back and forth as required.

While working on any of the problems, remember there is no single best solution. That can be frustrating if you have not had practice in solving open-ended problems, but that is the real world of engineering. Real problems have no single answer. Good engineers know this and stay flexible and imaginative in their thinking. If your work on any of these problems leads you to a fortune-making idea, all the better.

Group I—Incomplete Problem Identification

1. Design a saddle support winch system for removing boats from the water for dry dock work.
2. Design a coin sorter.
3. Design a system and components that will make coffee on a continuous basis.
4. Design an automatic garage opener when only one inch of clearance exists between the open door and the ceiling of the garage.
5. Street and parking lot repair companies often use a gasoline powered abrasive cutoff machine to cut out sections of pavement for repair. Some are pushed by hand. Design a self-propelled machine that will cut through the concrete or asphalt and have a variable feed rate so the power used for cutting can be held constant.
6. Design a 1000 gallon water tank.

7. An area 200 feet by 150 feet is available for a parking lot. The lot is on a corner and has an average grade of 10 percent parallel to the 200 foot dimension. The local code limits the grade in parking lots to 6 percent. Design a parking lot for this property.

8. Design a system to allow the pilot of an airplane to know if everyone is buckled up before takeoff.

9. Design a log splitter.

10. Design a portable traffic signal light.

11. Discuss a device and a system to monitor and control the direction of movement of objects (people, vehicles, etc.) for an application such as in-only and out-only traffic flow.

12. Develop a conceptual design for an automatic machine that will accept boards that are 0.750 ± 0.030 inches thick, 3.00 ± 0.50 inches wide, and 16.00 ± 1.00 inches long, cut the boards to 14.00 ± 0.03 long, and drill two dowel holes in each end spaced 1.00 ± 0.01 inches apart and on center ± 0.01 inches.

13. Blast shelters generally have valves in their air intake ducts to prevent shock waves from entering. Springs or hydraulic cushions are often used in these mechanisms. Design such a valve for a 24-inch diameter opening. Check appropriate literature to determine the pressures involved.

14. A preliminary design for a manually operated, bicycle pedaled air ventilator system has the following specifications: 55 rpm pedal speed, 785 mm diameter fan, 480 rpm fan speed, 75 watt output, 50 m^3/min air flow for a 300 meter duct, or 100 m^3/min for a 25 meter duct. A two operator design specifies 100 m^3/min air flow for a 125 meter duct. Determine if the specifications seem reasonable, and suggest a possible design. Include an estimate of the human effort required to operate the device.

15. An abandoned railroad right-of-way has been taken over by a group and they intend to convert the property into a bicycle and hiking trail. They are worried that motorcycles and other powered vehicles will go on the trail and tear up the surface. Determine a solution to their problem to insure that powered vehicles will not have access to the trail.

16. Design a system of filling bags with 50 ± 2 pounds of granular material coming from a bulk storage system. Determine how many bags per hour can be filled with the device you design.

17. Design a child's chair that can double as a two-step step stool.

18. Design a device for controlling the on-off cycle of a series of lights. The on and off time should be variable.

19. Design a bicycle rack for home use.

20. Design a vise that will automatically center a part clamped under a drill press spindle.

21. Design an experiment to verify the value of the modulus of elasticity of a material using one of the beam deflection models. Reference your strength of materials textbook for various equations.

22. Design a bimetal temperature sensor that will control an electrical circuit for a temperature difference of 10°F.

23. Design a fixture that aligns pipes for welding. Select appropriate size ranges from commercially available sizes.

24. Design an educational toy. Part of the identification activity will be to determine the educational objective desired.

25. Design an apparatus to test the torque needed to unscrew a bolted assembly.

26. Design a windmill to convert wind energy to another form of energy.

27. Design an apparatus and a system to measure the efficiency of various windmill blade designs.

28. An experiment is to be conducted on the life of incandescent light bulbs as a function of ambient temperature. Design an apparatus that can be used to control the ambient temperature between 100°F to 500°F for individual bulbs ranging from 25 to 100 watts.

29. Design an experiment to test the effect of vibration on the life of incandescent light bulbs. Vibration amplitude and cycles per second should be variable.

30. Design a machine to weigh objects up to 500 pounds. A "high" number of weighings per minute is desired.

31. Design a test apparatus to test the compressive strength of wood dowels up to 1.00 inches in diameter.

32. Design a test apparatus to test the shear strength of wood dowels up to 1.00 inches in diameter.

33. Design an experiment to test the time to set for a given bonding strength for a wood glue. Include in the experiment such variables as: face grain joints, end grain joints, combination grain joints, butt joints, lap joints, type of wood, clamping pressure, time, humidity, and applied tensile and shear loads.

34. Design an experiment to study the pullout force of a wood screw. Include in the experiment such variables as: type of wood, diameter of screw, length of screw, size of pre-drilled hole, and direction of wood grain.

35. Design an apparatus that converts rotary motion into linear motion. The length of the linear motion should be variable, but be able to be fixed at any desired length. The stroke should be faster in one direction than the other. One-direction travel time should be at no more than 40 percent of the total stroke cycle time: 4.00 inches $\leq$ stroke length $\leq$ 10.00 inches.

36. An experiment is to be conducted to determine water-caused soil erosion in various soils growing various plants. The samples of plants will be grown in boxes 2 feet x 2 feet x 8 inches containing 6 inches of soil. A wire mesh box liner will allow the soil-planting sample to be removed from the box for testing purposes. Design an apparatus to hold the wire mesh sample, allow a measured amount of water to be applied, vary the angle of the sample so runoff angles can be changed, and accumulate the runoff for study. Type of soil and variety of plants are also variables.

37. Design a single assembly apparatus to determine the percent grade of land over a 10 foot length. The range of grades to measure is 0 to 100 percent with an accuracy of ±1 percent.

38. Design an experiment to study the variation of cable pull force of a rope or wire cable over a pulley. Allow for changes in pulley diameter, bearing changes, the angle of contact between the cable and the pulley, and the weight being lifted. How can this experiment be expanded to study coefficients of friction?

39. Design an experiment to study the variation of tension force of a V belt running at various speeds over a sheave. Allow for changes in sheave diameter, bearing changes, the angle of contact between the belt and the sheave, and the tension in the belt. How can this experiment be expanded to study coefficients of friction?

40. Design a device that can be used by a carpenter to support and hold a door in position while hinge and latch locations are being determined and cut.

41. Design an apparatus for sorting individual ball bearings used in bearing assemblies, in size intervals of 0.0001 inch in the range 0.2500–0.2505 inches ball diameter.

42. Wood and some other materials have directional strength properties. If a load is applied at some angle θ from a primary strength direction, an adjustment for safe loading must be made. Design and conduct an experiment to determine the maximum allowable angular load F_n as a function of the angle θ, F_c, and F_{cp}; F_c is the maximum allowable strength of the material in the primary direction and F_{cp} is the maximum allowable strength perpendicular to the primary direction. It has been suggested that the form of the answer is $F_n = (F_c)(F_{cp})/(F_c \sin^m \theta + F_{cp} \cos^m \theta)$ where m is dependent upon the type of wood.

43. In October, 1988 the Consumer Product Safety Commission banned the game of lawn darts from the marketplace, saying, "What limited recreational value lawn darts may have is far outweighed by the number of serious injuries and unnecessary deaths." (*Des Moines Register*, October 29, 1988). Design a similar game that would be safer for the public.

44. Design a mechanism that will automatically feed bars to a bar lathe operation. Consider bars from 1.00 to 6.00 inches in diameter and 8 to 16 feet long.

45. Design a "vandal proof" trash collection receptacle for use in parks and other public places.

46. Design a device that will shut off the flow of a fluid after a given amount has flowed past the device.

47. Design a device for sampling rainwater that will activate when rain begins to fall and will collect a sample of rain at regular intervals. The interval can be timed to the amount of rain. This device would be helpful for monitoring the acid content in rain.

48. The packing of a number of equal diameter cylinders together inside a larger cylinder is often required in engineering problems. The question that needs answering is "What is the smallest large cylinder for n small cylinders as $n = 2, 3, 4, \ldots$?" When $n = 7$ or 13 the configuration is easy to calculate. However, when $n = 5$ or 8 and for many other values the most efficient layout is not obvious. Design a device that will allow the smallest large cylinder to be determined for a given

Design Problems

number of smaller equal diameter cylinders. The problem is to have a device with a variable large diameter within which a given number of small cylinders can be arranged.

49. Checklists are good to have so that important items are not forgotten. Any activity can be assisted by using a checklist. Create a checklist for items to be considered when undertaking a design project. The more general it is, the wider the scope it will have.

Note: Exercises 6.10, 6.11, 6.12 are also for design exercises.

Group II—Preliminary Problem Solution Given

1. Figure D1 is an idea for a bar clamp. Complete the details of such a design.
2. Figure D2 is an idea for a monkey wrench. Complete the details of such a design.
3. Figure D3 is an idea for a parallel jaw clamp. Complete the details of such a design.

FIGURE D1

Bar clamp concept.

0 – 24.00"
Swivel
1.00" × 1.00" pad
2.00" Min.
0.250" × 0.750" × ?

Group II—Preliminary Problem Solution Given

FIGURE D2

Monkey wrench concept.

1.50" Max

FIGURE D3

Parallel jaw clamp concept.

0.750" × 0.750"

0 – 2.50"

2.50" Min.

0.375" dia

187

188 **Design Problems**

FIGURE D4

Rotating control layout.

FIGURE D5

Sliding handle concept.

FIGURE D6

Hand punch concept.

4. Figure D4 shows the layout of a mechanism whose purpose is to move a pin located at point C vertically 3 inches while the handle moves through an angle of 70°. The location of points A, B, and C are fixed, but D and E are not. The resistance to movement at point C is 50 pounds. Design a mechanism that will efficiently satisfy the requirement.

5. Figure D5 is an idea for a sliding handle for tools so the handle can be offset in tight locations. The handle should slide freely back and forth, but not come off when used. Complete the details of such a design.

6. Figure D6 is an idea for a hand punch. The device should hold a belt and allow buckle holes to be punched in the center of a leather belt and spaced appropriately. Complete the details of such a design.

FIGURE D7

Ink roller concept.

7. Figure D7 is an idea for an ink roller for a small printing operation. Complete the details of such a design.
8. Figure D8 is an idea for a hand fruit masher. Complete the details of such a design.
9. Figure D9 is an idea for a child's wagon. Complete the details of such a design.
10. Figure D10 is an idea for a movable and adjustable lawn lounge. Complete the details of such a design.

FIGURE D8

Fruit masher concept.

192 Design Problems

FIGURE D9

Wagon concept.

FIGURE D10

Lawn lounge concept.

FIGURE D11

Gear puller concept.

11. Figure D11 is an idea for a gear puller. The arms have two holes to allow for a larger range of gears. Complete the details of such a design. Consider a two-arm and a three-arm puller.

12. There are differences between gear pullers and ball and roller bearing pullers. Can the design in Exercise 11 be altered to be used for both? If yes, modify the design. If no, design an appropriate puller.

13. Figure D12 is an idea for modular retaining wall construction. Consider other ideas and complete the details of this design.

FIGURE D12

Retaining wall concept.

Cut to length or provide standard lengths: 4', 6', 8', ...

Drive stake after each layer assembly for anchoring

Holes to bolt sections together and for fastening deadman supports

All bolted assembly

Exposed side

Dirt side

Deadman arms

Mounting brackets — stagger from row to row so driven stakes do not occur in the same vertical line

A: Standard widths — 4", 6", 8", ...

14. Complete the details of other ideas for a modular retaining wall design and compare it to the one in Problem 13.

15. Figure D13, dimensioned in inches, represents a conceptual design for a pin locator assembly. The purpose of the assembly is to locate the $\frac{1}{2}$ inch radius at five different locations $\frac{3}{4}$ inch apart. The slot is used to adjust an idler arm for five speed settings of a machine. Complete the design.

Group II—Preliminary Problem Solution Given 195

FIGURE D13

Machine adjuster concept.

SECTION TWO

DETAILS OF REFINEMENT AND ANALYSIS

Refinement and analysis activities, which were summarized in Chapter 6, are the link between creative ideas and the practicalities of making an idea work as intended. The objective of refinement and analysis is to improve the ideas to create a functional product, and to make the product do its intended job in an efficient way. Poor ideas have to be discarded and good ideas must be improved, refined to work efficiently, and detailed to allow economical manufacture.

Development of the final design through refinement and analysis must be done to ensure proper function and to ensure the stated problem is solved. For some projects development activities may take years, particularly if they involve a change in the state of the art. Developments in medicine, hybrid corn, or a space station are examples of long-term projects. Many iterations of ideas, tests, redesigns, retests, and compromises are required. Test criteria and procedures must be established. Data taken must be studied and the results incorporated into redesign. There are so many details to be taken care of, and so much knowledge to be applied, that it is impossible to discuss a specific sequence of events for all projects. This is not to say a plan is not needed; indeed, without a plan the most noble of projects will flounder. Unknown and unplanned things do occur, and engineers have to deal with them. The more details considered, and the better use made of existing data, the better will be the chance for success and an optimum design.

Activities of the engineering design problem solving procedure have been discussed in Chapters 4 through 7; the specific steps for any project, however, will depend on the circumstances that exist at the time the project is being worked on. Always keep in mind such constraints as budget, market condition, time, and the state of the art. Chapters 8 through 14 go into more detail on many of the specific tasks to perform related to the concepts presented thus far, particularly modeling, product strength and failure models, and manufacturing.

As technology changes and theories and methods are altered, you too must change and alter your approach. To be a good engineer, you must be ready to change with a growing technological society. State of the art changes must be incorporated in new products as well as in product revisions. Lifelong learning is a necessity for engineers.

The following chapters discuss specific criteria to be considered during the solution of a design problem. The criteria discussed are themselves not design problems, but they include the details needed to judge different ideas and make good choices during the design process. Do not confuse the single answer exercises included in Chapters 8-14 with design problems. The exercises are intended to acquaint you with the concepts, and are necessary for practice and understanding. It is up to you to apply the concepts during the refinement and analysis steps of the design activity. The order of topics is not to suggest a specific sequence to be used for any given project. It may not be the sequence you will use on any project, but the points discussed are common to a variety of engineering projects. Even if some of the ideas are never used, just knowing that something might be considered increases your value to a project and to anyone you work for.

If you work on a systems design, don't think that the study of parts design isn't helpful. Knowledge of physical parts gives you an advantage when designing systems for such things as product distribution, inventory control, material ordering, storage, personnel training, customer service, cost control, and resource investment. This knowledge enables you to analyze vendor proposals more critically and be more confident in asking questions and understanding the answers. Remember that service industries are among the most capital intensive businesses there are when considering dollars invested per employee. Consider all the sorting and delivery equipment that United Parcel Service uses, or the computer equipment used by Wall Street.

8 Materials

"...individual engineers and technicians cannot hope to be familiar with all the properties of all types of materials in their numerous forms....try to learn some principles to guide...in the selection and processing of materials."
♦ LAWRENCE H. VANVLACK

One of the things to consider when designing a physical product is the selection of materials to use. Materials come in a wide variety of shapes, sizes, and compositions, and include components that are available, such as computer hardware and software, preprinted forms, or components that just need assembly, such as standard hardware and parts manufactured by others. This chapter reviews items that need to be considered when selecting materials for design, including basic material properties and specification values. For general discussion the classification of materials will be metals, polymers, ceramics and glasses, and woods and organics.

8.1 Metals

Metals are characterized by an atomic structure that has free electrons available. Metals are commonly subdivided into ferrous and nonferrous alloys.

Ferrous alloys have iron as their base material, which is the major commercially available metal. Two main reasons account for this: iron is relatively abundant and it has favorable mechanical properties.

Cast irons are those alloys of iron with 2.0 to 4.5 percent carbon as the primary alloying element. Other alloys such as manganese, phosphorus, nickel, chromium, molybdenum, copper, and vanadium are also added in smaller amounts to help control mechanical properties. The mechanical properties of the most commonly used cast irons are shown in Appendix F-3. Iron is cast to shape when material is desired in a shape not easily obtained from standard bar or plate. Typical cast iron parts are engine blocks, and transmission and axle housings.

Steels are those alloys of iron with less than 2.0 percent carbon as the primary alloying element. Other elements such as manganese, nickel, chromium, cobalt, vanadium, molybdenum, and silicon are also added to help control material properties. Table 8.1 lists some of the common alloying elements and the effect they have on steel.

Because there are so many alloy combinations for steels, a numbering classification has been established as a standard for identification of the alloying elements present. This is outlined in Table 8.2. The amount of carbon in the alloy, in hundredths of a percent, is shown as the right hand two digits. A 2320 steel is a low nickel alloy with 0.20 percent carbon while a 2540 steel is a high nickel alloy with 0.40 percent carbon. There is a growing attempt to use a new unified numbering system (UNS) for alloy designations; reference *Metals and Alloys in the UNS*, 4th ed., Vol. 1, *Ferrous*, Vol. 2, *Nonferrous*, Society of Automotive Engineers and American Society for Testing and Materials, Warrendale, Penn., 1986.

Some commonly used steel alloys, because of their availability, cost, and useful material properties include:

1. 1010, used for sheet metal parts where ductility is required, auto and appliance sheet metal;
2. 1020, used for low strength applications, and where standard machining and welding processes are available, can be case hardened;

TABLE 8.1

Effects of alloying elements on steel (2, 7, 21, and others).

Element	Effect
Aluminum	Restricts grain growth
Boron	Increases hardenability
Chromium	Increases hardenability, resistance to corrosion (stainless steels), and strength at high temperatures
Cobalt	Improves magnetic properties, and strength at high temperatures
Copper	Increases strength and corrosion resistance
Manganese	Increases hardenability, counteracts sulfur's brittleness
Molybdenum	Increases hardenability and toughness
Nickel	Increases strength, toughness, impact strength, and elevated temperature properties of stainless steel
Phosphorus	Increases corrosion resistance and machinability
Silicon	Improves magnetic properties, oxidation resistance, and strength of low carbon steels
Tellurium	Improves machinability in leaded steels
Titanium	Improves microstructure stability in stainless steel
Tungsten	Increases hardness and strength at elevated temperature
Vanadium	Raises coarse grain forming temperature

Materials

TABLE 8.2

Steel alloy numbering system (15, 35, 38 and others).

AISI or SAE Number	Primary Alloying Elements, weight %
10cc	Plain carbon steels
11cc	Plain carbon, re-sulfurized
13cc	Mn, 1.5–2.0%
23cc	Ni, 3.25–3.75%
25cc	Ni, 4.75–5.25%
31cc	Ni, 1.10–1.40%; Cr, 0.55–0.90%
33cc	Ni, 3.25–3.75%; Cr, 1.40–1.75%
40cc	Mo, 0.20–0.30%
41cc	Cr, 0.40–1.20%; Mo, 0.08–0.25%
43cc	Ni, 1.65–2.00%; Cr, 0.40–0.90%; Mo, 0.20–0.30%
46cc	Ni, 1.40–2.00%; Mo, 0.15–0.30%
48cc	Ni, 3.25–3.75%; Mo, 0.20–0.30%
51cc	Cr, 0.70–1.20%
52cc	Cr, 1.30–1.60%
61cc	Cr, 0.70–1.10%; V, 0.10%
81cc	Ni, 0.20–0.40%; Cr, 0.30–0.55%; Mo, 0.08–0.15%
86cc	Ni, 0.30–0.70%; Cr, 0.40–0.85%; Mo, 0.08–0.25%
87cc	Ni, 0.40–0.70%; Cr, 0.40–0.60%; Mo, 0.20–0.30%
92cc	Si, 1.80–2.20%
93cc	Ni, 3.00–3.50%; Cr, 1.00–1.40%; Mo, 0.08–0.15%

cc = carbon content in hundredths.
A B preceding the carbon content signifies a Boron steel, an L signifies a leaded steel. Examples: A 1020 steel has 0.20 parts of carbon by weight. A 10B18 steel has 0.18 parts carbon and has Boron added. A 12L14 steel has 0.14 parts carbon and has lead added.

3. 1040, used for application where post machining hardening and heat treatment is required, forgings, gears, brake levers, and sprockets;
4. 1080, used for applications where high hardness and abrasive resistance is required, snap rings, clutch plates, and valve springs;
5. A36-70 (0.26 percent carbon) used for structural steel, I beams, channels, and angle iron shapes;
6. 4340 and 4140, used for high strength applications where machining is required, axles, crankshafts, gears, and shafts;
7. 11-- series, preferred for parts turned on a lathe, easily machined.

Ferrous alloys also have the useful property of being able to be magnetized. Some iron alloys magnetize and demagnetize easily as measured by a low coercive force, and are classified as soft magnets. Other iron alloys do not demagnetize easily as measured by a high coercive force, and are classified as hard magnets. Table 8.3 lists some of these alloy combinations. Residual magnetism is the magnetism left in the metal when the source of applied magnetic force is removed. The coercive force is the applied

TABLE 8.3

Magnetic alloys, in order of coercive force (21).		
Material	Residual Magnetization weber/m²	Coercive Force weber/m²
Soft Magnets		
Supermalloy, 79% Ni, 16% Fe, 5% Mo	0.50	$0.0020 \cdot 10^{-4}$
Hypernik, 50% Ni, 50% Fe	0.80	$0.05 \cdot 10^{-4}$
Puron, 99.99% Fe	0.90	$0.05 \cdot 10^{-4}$
4.25% Silicon steel	0.70	$0.40 \cdot 10^{-4}$
2.5% Silicon steel	0.80	$0.80 \cdot 10^{-4}$
1% Silicon steel	0.90	$0.90 \cdot 10^{-4}$
Ingot iron, 99.9% Fe	1.10	$1.00 \cdot 10^{-4}$
Perminvar, 45% Ni, 25% Co, 30% Fe	0.60	$1.2 \cdot 10^{-4}$
Hard Magnets		
Carbon steel, 0.9% C, 1% Mn	1.00	$50 \cdot 10^{-4}$
Tungsten steel, 5% W, 0.7% C, 0.3% Mn	1.03	$70 \cdot 10^{-4}$
Cobalt steel, 36% Co, 4% Cr, 5% W, 0.7% C	1.00	$240 \cdot 10^{-4}$
Alnico 5B, 8% Al, 14% Ni, 24% Co, 3% Cu, 51% Fe	1.27	$600 \cdot 10^{-4}$
Cunife, 60% Cu, 20% Ni, 20% Fe	0.54	$500 \cdot 10^{-4}$
Fine powder, 30% Co, 70% Fe	0.90	$1000 \cdot 10^{-4}$
Ferroxdur, $BaFe_{12}O_{19}$	0.34	$1800 \cdot 10^{-4}$
Bismanol, MnBi	0.43	$3400 \cdot 10^{-4}$

magnetism required to offset the residual magnetism. The coercive force is not equal to the residual magnetism, the relationship being dependent on the material's magnetic hysteresis properties.

Nonferrous alloys have elements other than iron as their primary material, such as aluminum, magnesium, copper, titanium, tin, lead, or zinc. The main differences between ferrous and nonferrous alloys are in magnetic properties, density, color, corrosion resistance, strength, modulus of elasticity, hardness, hardenability, spark resistance, availability, cost, and reaction to manufacturing processes.

Aluminum is probably the most commonly used nonferrous alloy because of its favorable properties and commercial availability. Table 8.4 outlines the numbering classification for commercial aluminum alloys and their heat treatment. Aluminum is generally used for its low density, good electrical and thermal conductivity, antimagnetic properties, and its atmospheric corrosion resistance. Aluminum can be welded, cast, drawn, and machined. Commonly used aluminum alloys are listed in Appendix F-4.

Copper is generally used for its electrical conductivity and its ductility. Copper can be cast, formed, drawn, and welded, but it has low strength and corrodes to a black or green color depending on the corrosion agent. When alloying metals, such as tin, zinc, and lead, are added to copper, the material properties are altered, and the range of applications increased, most notably an increase in abrasion wear resistance. See Appendices F-6 and F-7 for selected properties of brasses and bronzes.

TABLE 8.4

Aluminum identification coding system.

4 Digit Code	Primary Alloy Elements
1---	Unalloyed, > 99% Al
2---	Copper
3---	Manganese
4---	Silicon
5---	Magnesium
6---	Magnesium and Silicon
7---	Zinc
8---	Other elements

Heat Treatment Suffixes

- -F   As fabricated   -O   As annealed
- -H   Strain hardened by cold working
  - -H1h   Hardened only; h = hardness index, 8 = full hard
  - -H2h   Hardened and partially annealed
  - -H3h   Hardened and stabilized
- -T   Heat treated
  - -T2   Annealed, cast alloys
  - -T3   Solution heat treated and cold worked
  - -T4   Solution heat treated and aged naturally
  - -T5   Artificially aged only
  - -T6   Solution heat treated and artificially aged
  - -T7   Solution heat treated and stabilized
  - -T8   Solution heat treated, cold worked, and aged
  - -T9   Solution heat treated, aged, and cold worked
  - -T10   Artificially aged and cold worked

Many less common metals and alloys are used commercially such as nickel and titanium for their high strength, low strength lead for its radioactive shielding, and highly flammable magnesium for its low density. See Appendices F-5 and F-9 for specific material properties of magnesium and titanium.

8.2 Polymers

Plastics, or polymers as they are more properly called, include materials that are not usually crystalline in structure, and can be easily molded or formed into shape because they change viscosity as temperature changes. One main difference between polymers and metals is that polymers lack free electrons in their atomic structure. Most polymers do not change from a liquid to a solid at a constant temperature, and they do not crystallize upon solidification. **Thermoplastic** polymers have the capacity to be repeatedly heated and formed, and cooled and hardened. **Thermosetting** polymers undergo a chemical change when formed and cannot be heated and reformed. Polymers can be formed, extruded, or cast to basic shapes and used as is, or they can be subsequently sawed, sheared, drilled, tapped, and welded to create a more complicated desired shape or assembly.

There are many advantages of polymers as a class of engineering materials. Advantages include the ability to be formed into intricate shapes with tolerances of ± 0.001 inches per inch, the availability of colors without painting, resistance to chemicals, resistance to water and atmospheric corrosion, good surface finish, low density, and good heat and electrical insulating properties. One of the largest uses of polymers is for parts that have to support only their own weight, such as covers, shrouds, and ducts. Containers for liquids, pastes, and foods are a good application for polymers because the contents rarely interact with the polymers. Photoelastic strain analysis is a unique use of polymers, and is made possible because of a change in the polymer's index of refraction as a function of a force-induced strain. Couple the change in refraction with their natural transparency, and polymers become an excellent material to use to model structural components. As the polymer modeled structure is slowly loaded, the regions of high stress and patterns of stress fields become apparent as light and dark bands appear.

The general disadvantages of polymers include their dimensional instability; excessive creep at all temperatures; deterioration when exposed to ultraviolet light, neutrons, and gamma rays; swelling due to liquid absorption; form and shape instability within certain temperature ranges because of polymer's viscoelastic behavior; low strength; and in some cases, flammability. Plastic products do not biodegrade, so disposal after use is a problem, and recycling procedures are another challenge for engineers. The coefficient of expansion for polymers is up to 10 times greater than that for metals, but polymer stiffness is lower than that of metals, so thermal cracking is not a concern as it is for some brittle materials.

Polymers as a class of materials have a wide range of properties, and a specific property of a given polymer also has a wide range of values. This results in errors being made in selecting the proper polymer, as well as designing to an allowable property value. The glass transition temperature, for example, where polymers change from ductile to brittle behavior, ranges from $-20°C$ for low-density thermoplastic polyethylene, to $350°C$ for a thermosetting fiberglass filled polyamide. The modulus of elasticity ranges from $15 \cdot 10^3$ psi (103 MPa) for low-density polyethylene, to $3 \cdot 10^6$ psi (20.7 GPa) for fiberglass filled thermosetting epoxide. Tensile yield strength ranges from 600 psi (4.1 MPa) for low-density polyethylene, to 28,000 psi (193 MPa) for fiberglass filled polyamide. Thermal conductivity ranges from 0.0016 Btu·in/ft²·sec·°F (0.0021 Btu·mm/m²·sec·°C) for thermoplastic polystyrene to 0.0010 Btu·in/ft²·sec·°F (0.0013 Btu·mm/m²·sec·°C) for high-density polyethylene. Electrical resistivity for all polymers ranges from 10^{12} to 10^{18} ohm·cm. The tensile strength of vinyls ranges from 1000 to 9000 psi (6.9 to 62.0 MPa), and the modulus of elasticity for phenolics ranges from $0.5 \cdot 10^6$ to $5 \cdot 10^6$ psi (3.45 to 34.5 GPa). There is not much data available on the fatigue characteristics of polymers. The following paragraphs summarize the most common commercial polymers. See also Tables 8.10, 8.11, 8.12, 8.14, 8.16, 8.17, 8.18, and 8.19 for additional data on polymer materials and their properties.

Polyethylene is manufactured in different density ranges, and its tensile strength increases as density increases. Low density polyethylene is used for trash bags and other throwaway applications. High density polyethylene is used for shrouds, covers, pipes, and chemical storage containers because of its chemical resistance, and it is usually fabricated using injection molding. Ultra high density polyethylene, the most expensive of the three, is hard to fabricate by casting, and is generally used in sheets for applications such as wear strips.

Polypropylene, made from propylene gas, is a slightly crystalline polymer, which gives it higher strength than polyethylene. It is used for structural applications such as bottles and appliance parts, and is also good for making containers to store chemicals because of its chemical resistance.

Polybutylene, made from crude oil, is used for making butyl rubber which is the primary ingredient in some caulking compounds. Polybutylene is also used for gaskets, O rings, and oil seals because of its ability to adhere to a surface or remain flexible under normal temperature conditions.

Polyvinyl chloride (PVC), made from acetylene gas and hydrochloric acid, is used for vinyl upholstery material, as well as wire insulation, cold water pipes, and nonstructural applications. PVC is weak, notch sensitive, and fails in a brittle failure mode.

Polyvinylidene chloride (PVDC), made from trichloroethane, is similar to PVC, but its strength and temperature resistivity are higher. This makes it a better choice for hot water pipes and structural applications such as packaging; Saran Wrap™ is a PVDC.

Polystyrene, made from ethylbenzene, is the cheapest of all plastics. It is rigid, weak, and has poor impact strength. It is used for low strength applications such as sound and thermal insulation, and one time use products such as fast food restaurant spoons and styrofoam cups.

Terpolymer (polystyrene, acrylonitrile, and butadiene) overcomes polystyrene's low strength problem, and is used for high impact molded products such as telephone cases and durable toys. Along with polyethylene and polypropylene it accounts for one-half of the polymer production in the United States.

Polymethyl methacrylate, commonly referred to as an acrylic, is a high strength transparent polymer used for "unbreakable" windows, low cost lenses, machine guards, and false teeth. It is also used for lacquer and paint. Plexiglas™ is a polymethyl methacrylate.

Polyvinyl acetate is a soft thermoplastic polymer whose main use is in latex paints and white glue.

Polyvinyl alcohol is a water soluble, but oil resistant polymer that is used for its good film coating characteristics, including pill coatings.

Polymethyl pentane is a thermoplastic polymer with good temperature resistance, electrical resistance, and transparency, making it good for light fixtures.

Fluorocarbons (PTFE) are a low strength, expensive, hard to fabricate polymer. They do, however, have low friction characteristics, making them a good choice for a dry film lubricant—that's right, Teflon™.

Polyamides can be either thermoplastic or thermosetting and have a very low coefficient of thermal expansion (approximately $28 \cdot 10^{-6}$ in/in/°F), which along with their high temperature properties, make them suitable for elevated temperature structural applications.

Polysulfone is a low water absorption, high temperature resistant thermoplastic polymer. It is used for medical and food handling devices, as well as electronic circuit boards.

Polyphenylene oxide is a thermoplastic polymer similar to polysulfone; it absorbs less water, is tougher, is a fire retardant, but is more heat sensitive than polysulfone.

Polyethersulfone is a high strength, creep resistant, thermoplastic polymer, but its water absorption resistance is low.

Polyphenylene sulfide is a crystalline thermoplastic that is stable at high temperatures and nonflammable.

Polyacetals are high crystalline polymers, some of the best to resist moisture absorption. They are also one of the most important structural engineering polymers and are used for machine components such as bushings and gears, but manufacturing to close tolerances is difficult to control.

Cellulosics, as a class of thermoplastic polymers, are made from the organic molecule cellulose. Because of this organic source, they are highly flammable. However they still have applications such as film products, cellophane, and rayon. They are easily colored, will withstand weathering, and are used for domestic products rather than commercial products.

Polyesters as a category of thermoplastic polymers have a wide range of properties, but they can have strength to weight ratios higher than steel. Fiberglass is a polyester and is used for low strength insulation as well as high strength car bodies, ropes, and fabrics. Polycarbonates are a linear-chained molecular form of polyester. They have good impact strength, temperature resistance, and are used for protective helmets and face guards.

Polyamides are a class of newer high temperature, high strength polymers that do not soften and flow like most polymers. Polyamides are processed in granular form and fabricated to shape by compressing the powder into molds under high pressure, followed by an oven bake at high temperature. The voids can be packed with a lubricant, making the final product excellent for self lubricating bearing applications. This process is the same as powder metallurgy for granular metals. Polyamides are a high strength polymer, used for such applications as bushings and gears; nylon is a polyamide. A disadvantage of polyamides is their water absorption potential and corresponding dimensional change. Kevlar™ is a newer nylon that overcomes some of these limitations. Fabrication is usually by injection molding, but polyamides can also be continuously extruded into thread, and used to make cloth.

Polyurethane has a wide range of properties, permitting an application for which a "rubbery" substance is required, as well as a high hardness application such as a wear pad for moving parts, furniture, varnish, or floor tile. Polyurethane will foam and is used for foam insulation and packaging material. A form of polyurethane developed by Roderic Lakes from the University of Iowa, named 're-entrant foam,' is the only known material to exhibit a negative Poisson's ratio, with values ranging from -0.1 to -0.7.

Melamine formaldehyde is a thermosetting, low cost, tough, abrasion and temperature resistant polymer, used in the food handling processing industry including domestic tableware.

Urea formaldehyde is similar to melamine formaldehyde, but is not as resistant to temperature, and will degrade when exposed to sunlight.

Polyetherimide is a chemical resistant, high temperature thermoplastic polymer that has low flammability, and is used for automotive parts that get hot and are in contact with oils and greases.

Rubber is also categorized as a polymer because of its molecular structure. Rubbers are useful primarily because of their resilient properties. Natural rubber, however, is susceptible to attack by ozone (O_3) and petroleum products, which leads to early deterioration of rubber products and encourages the use of synthetics such as polyisoprene, styrene butadiene, chloropene (Neoprene™), polybutadiene, nitrile, ethylene propylene, chlorosulfonated polyethylene, polysulfide, polyacrylic, fluoraelastomer, and epichlorohydrin.

Polymers are often mixed with fibers or particles (up to 60 percent by weight) to improve particular properties, generally strength, resistance to water absorption, and reaction to chemicals and sunlight. The following three polymers are commonly used to create composite materials, See Section 8.7 for additional discussion on composite materials.

Epoxies have good temperature resistance and a low shrink rate when they solidify. This makes them a good choice for coverings for other materials. They are suitable for composite materials, and are good choices for manufacturing holding devices because they will not scratch the surface of a finished product.

Phenolics are a thermosetting polymer, and when molded, are used for formed parts such as electrical junction boxes and switch plates. When laminated with non-polymer materials, they are useful as low strength gears and bushings, and electrical circuit boards. The addition of fine sawdust as a filler can double the toughness of phenol polymers.

Silicones are high temperature resistant polymers, and are used for oil additives, gaskets, and seals. Silicone is also commonly used in caulking material for domestic applications. Mixed with fiberglass, the mechanical properties improve, making the composite suitable for aircraft applications.

TABLE 8.5

Selected commercial glasses (37, 38 and others).

Type	SiO_2	Na_2O	CaO	B_2O_3	PbO	Al_2O_3	MgO	K_2O
Fused Silica	96	–	–	–	–	–	–	–
Low thermal expansion, high viscosity.								
Pyrex	81	4	–	12	–	2	–	–
Low thermal expansion, low ion exchange.								
Lamp bulbs	73	16	5	–	–	2	4	–
Easy to work.								
Thermometer	73	10	–	10	–	6	–	–
Dimensional stability.								
Container	72	14	10	–	–	2	1	–
High durability, easy to work, resists chemicals.								
Plate	72	14	10	–	–	1	4	–
High durability.								
Window	72	14	8	–	–	1	4	–
Heat absorbing	71	10	9	–	–	4	4	–
Lead tableware	66	6	1	–	16	1	–	10
High index of refraction.								
Lamp stems	55	12	–	–	32	1	–	–
High resistivity.								
Fiberglass	55	1	16	10	–	15	–	–

8.3 Ceramics

Ceramics and glasses are those materials made from a combination of metallic and nonmetallic elements. Their basic properties include the following: they are harder and have lower thermal expansion and thermal conductivity than most metals; they are more wear resistant than metals; they are chemically inert, thermally stable, and good heat and electrical insulators; but they are brittle. Glasses are good thermal insulators with thermal conductivity values ranging from $9 \cdot 10^4$ to $30 \cdot 10^4$ cal·cm/cm^2·sec·°C. Table 8.5 lists common glasses and Appendix F-10 lists their mechanical properties. Table 8.6 ranks some ceramics by hardness. Commercial uses of ceramics include low stress, high abrasive wear applications such as wire drawing dies, thread guides, and metal cutting tools. Ceramics can be used as compressive members, but tensile strength and impact strength are too low for general design applications.

Concrete is one of the most common of structural ceramics. Its advantages include ease of mixing, ease of forming before curing, high density, and relative low cost. Disadvantages include low tensile strength, brittleness, low resistance to salt corrosion, and extreme difficulty in machining to a different shape after hardening. Appendix F-11 lists properties of concrete.

8.4 Wood

Wood and organic materials come from trees and plants. The main advantage of organics is that they can be raised as crops, and therefore are a renewable resource. Disadvantages include that they absorb water, can rot unless specially treated, will burn even if

TABLE 8.6

Relative hardness of selected abrasive materials (22 and others).

Material	Composition	Relative Hardness on Knoop Scale
Diamond	C	1.00
Boron Carbide	B$_4$C	0.44
Silicon Carbide	SiC	0.38
Titanium Carbide	TiC	0.35
Tungsten Carbide	WC	0.26
Corundum	Al$_2$O$_3$	0.25
Topaz	Al$_2$SiO$_4$F$_2$	0.19
Graphite	C	0.15
Quartz	SiO$_2$	0.12
Nickel	Ni	0.080
Silver	Ag	0.0086

TABLE 8.7

Variation in properties of wood (5, 14, 36 and others).

Property	Hardwoods	Softwoods
Density, lb/ft^3	18–50	18–31
Modulus of elasticity, Mpsi	0.9–2.2	0.5–1.9
Compressive strength, psi		
Parallel to grain	2100–9200	1900–8400
Perpendicular to grain	100–900	200–500
Tensile strength, psi		
Parallel to grain	275–1000	200–450
Shear strength, psi		
Perpendicular to grain	700–1500	700–900
Parallel to grain	1400–2300	850–1500

wood products are caused by growth characteristics, growing environment, moisture content, and differences in species. Wood must be inspected and sorted to be sure it meets defined grade specifications. Grading considers such characteristics as knots, end checks, grain slope, moisture content, and method of sawing. Moisture affects strength, which can vary as much as 200 percent when green is compared to kiln dried lumber. The strength of wood decreases as the percent of moisture content increases, reaching a minimum strength at fiber saturation, approximately 30 percent water by weight. Table 8.7 quantifies the range of wood strength properties. Strength values of "dry" wood must be reduced as moisture content increases. The values are for illustration purposes only; see references (5), (14), and (36) for additional design criteria material values for wood.

Common uses of wood include construction lumber, plywood, particle board, Masonite™, and paper products. These materials are easily cut and shaped, and in the case of chips, sawdust, and fibers, can be molded when mixed with the proper bonding agents.

8.5 | Material Properties

Regardless of the material category that is best suited for a product design, specific properties should be considered when making material selections. These properties also apply to the purchase of components for systems design. Materials must be selected so that the components will do the job they are intended to do.

8.5.1 Mechanical Properties

Mechanical properties are those properties that indicate how a material will react to a mechanical force. Following is a summary of the basic and commonly used mechanical properties of materials.

FIGURE 8.1

Elastic and plastic strain.

[Stress-strain curve showing elastic strain region (steep linear portion) and plastic strain region (curved portion after yield), with σ on vertical axis and ε on horizontal axis.]

Normal stress, σ, is an internal force per unit area that a component "feels" from an applied load. The area resisting the applied force for tension and compression forces is perpendicular to the direction of the force. For shear stresses the resisting area is parallel to the applied force. Common units of stress are lbf/in^2, N/m^2 (Pa), or MPa or GPa.

Normal strain, ϵ, is the deformation of the component in the direction of the normal stress. It is calculated by dividing the total deformation by the length before the load was applied, or in the case of true strain the instantaneous length. Common units for strain are in/in or m/m.

Elastic strain is material deformation that disappears when the applied stress is removed. Elastic strain causes the springback encountered during metal-forming operations (see Figure 8.1).

Plastic strain is material deformation that remains when an applied stress is removed. Plastic strain is required to permanently bend metals.

The **modulus of elasticity,** E, is defined as the ratio of the normal stress to the resulting strain, $E = \sigma/\epsilon$. Typical units for E are psi, Mpsi or MPa. Its value is constant in any region of the stress-strain diagram that is linear. See Appendices F-1, F-2, F-3, F-12 and Table 8.20 for selected values of E. In nonlinear regions of the stress-strain curve, the modulus is the slope of the curve at the point in question, and is called the **tangent modulus,** E_t. In general $E_t < E$.

Values of E for tension and compression are nearly equal and only one value is given for any given material structure at a given temperature. The value of E decreases as temperature rises unless a phase change occurs (see Figure 8.2).

Elastic strength, S_y, is related to three concepts: yield strength, yield point, and the proportional limit. These three terms are often used interchangeably with the elastic strength, so caution must be used when tables are consulted. The *yield strength* of a

FIGURE 8.2

The effect of temperature on the value of the modulus of elasticity for selected pure metals.

From Guy, A.G. and J.J. Hren. *Elements of Physical Metallurgy*, 3rd ed. © 1974 Addison-Wesley Publishing Co., Reading, Mass. Figure 2.10, p. 63. Reprinted with permission.

material is the stress that causes a pre-defined percentage of permanent deformation, generally 0.03 percent to 0.2 percent of the original length.

The *yield point* is the location on the stress-strain diagram where increased deformation occurs without an increase in load. It is as though the material resists deformation up to a point and then "gives in." Figure 8.3 illustrates two types of steel, one with a yield point and one without.

The *proportional limit* is the maximum stress that can be applied to a material and still have the material obey Hooke's law, which states that the ratio of stress to strain is a constant, the modulus of elasticity.

Not all materials exhibit all three values. Some nonferrous alloys do not have a constant stress/strain ratio when loaded. Some brittle materials fracture before reaching 0.2% permanent deformation, and some materials do not have a yield point. When a material does exhibit more than one of these characteristics, the values are quite close to each other, so the general term elastic strength is used for the value. *You* must determine if the variation in the three values is significant enough to worry about. The three values can be quite different for tension, S_{yt}, compression, S_{yc}, or shear, S_{ys}. For example, the elastic shear strength of wrought aluminum is about 58 percent of the elastic tensile strength, but the elastic tensile and compressive strengths are nearly equal. Always verify the significance of data given in tables before using materials in designs. See Appendices F-2 through F-10 for selected values of material yield strength.

The **modulus of resilience**, R, is a measure of a material's ability to absorb energy elastically, that is with no permanent deformation. The magnitude of the modulus is dependent upon other material properties and is quantifiable as the area under the stress-strain curve up to the elastic limit, $R = S_{yt}^2/2E$.

FIGURE 8.3

Sample stress-strain curves.

Courtesy Association of Iron and Steel Engineers, Pittsburgh, Penn. Extracted from the 10th edition of "The Making, Shaping and Treating of Steel."

Poisson's ratio, ν, is the ratio of the unit strain perpendicular to the direction of the load to the unit strain parallel to the direction of the load. The general range of values for ν is 0.25 to 0.33. The closer the value of ν is to 0.50, the smaller is the reduction of volume under a compressive load, and the less the increase in volume under a tensile load. Table 8.8 lists selected values for Poisson's ratio.

Shear stress, τ, is also a force per unit area, but the resisting material area, and the resulting shear stress are parallel to the direction of the applied force. When an applied force and the reactive force are not axial they cause a force couple, which leads to a shear stress. A shear stress exists when a load is applied which forces layers of material to slide over one another. Direct shear stress, $\tau = F/A$, is illustrated in Figure 8.4(a). Torsional shear stress, $\tau = Tr/J$ for a round bar, is illustrated in Figure 8.4(b).

Shear strain, Φ, is the tangent of the angle of deformation caused by a shear force see (Figure 8.4(b)). When Φ is "small" its value is approximately equal to tan $\Phi = B'B/L = \Phi r/L$, where Φ is measured in radians.

TABLE 8.8

Selected values of Poisson's ratio @ 20°C (7, 36, and 38).

Material	Poisson's Ratio
Fused quartz	0.17
Concrete	0.19
Tungsten carbide	0.22
Zinc	0.25
Glass	0.25
Tungsten	0.28
Cast iron	0.26–0.31
Magnesium	0.29
Steel	0.30
Nickel	0.31
Titanium	0.36
Aluminum	0.34
Copper	0.35
Molybdenum	0.35
Brass	0.35
Tin	0.36
Silver	0.37
Lead	0.44
Rubber	0.49

The **modulus of rigidity**, G, is the ratio of the shear stress to the shear strain, $G = \tau/\Phi$. The modulus is valid only within the elastic limit and where the stress-strain diagram is linear. The modulus of rigidity is related to the modulus of elasticity, $E = 2G(1 + v)$. See reference (24) for proof of this relationship.

The **ultimate strength**, S_u, of a material is the maximum stress reached on a stress-strain diagram. It is used in design as the maximum stress a component will withstand before complete fracture failure. It is measured in lbf/in^2 or MPa. There are three values of ultimate strength to consider in design: tensile, S_{ut}, compressive, S_{uc}, and shear, S_{us}. These three values are not necessarily equal, and it must be clear which value is used during the design analysis procedures. Examples include gray cast iron, which has $S_{uc} > 3S_{ut}$; SAE 4340 heat treated steel, which has $S_{us} = 0.63S_{ut}$; and cast AZ63-HT magnesium, which has $S_{us} = 0.50S_{ut}$, but $S_{ut} = S_{uc}$.

The **modulus of rupture** for a circular cross section is defined as $S_{su} = (T_u r)/J$ where:

T_u = the maximum torque for failure on a torque-twist diagram,

r = the radius of the test bar used for the torque-twist test,

J = the polar moment of inertia of the test bar.

FIGURE 8.4

Direct and Torsion shear.

(a) Direct shear

(b) Torsion shear

Strain hardening, often called percent cold work (%CW), occurs when a ductile material's hardness is increased with plastic deformation. Many material properties are affected by cold work as shown in Figure 8.5. Percent cold work is quantifiable as follows:

%CW = $(A_o - A_f)100/A_o$, where

A_o = original cross-sectional area, and A_f = final cross-sectional area.

FIGURE 8.5

Strain hardening. Cold work (a) increases the hardness, (b) increases the strength, and (c) decreases the ductility.

From L.H. VanVlack, *A Textbook of Materials Technology*. © 1973 Addison-Wesley Publishing Co., Reading, Mass. Figure 4-4.1, p. 51. Reprinted with permission.

The **hardness** of a material is a measure of its resistance to indentation. Several different measuring scales have been established to cover the range of material hardness encountered. The scales are used depending on the material and its hardness, but some overlap occurs. Table 8.9 lists some of the common hardness scales.

One special relationship occurs between the strength of steels and the Brinell hardness; S_{ut} is approximately equal to 500 times the Brinell hardness number. This applies up to a strength level of approximately 200,000 psi (1380 MPa). For many nonferrous metals S_{ut} is approximately equal to 300 times the Brinell hardness number (35).

Though there are many materials that are both hard and brittle, do not confuse the two terms. Fracture with a minimum of deformation is an indicator of brittleness, whereas hard materials resist indentation and wear. Brittle materials often have much larger strength in compression than in tension, as illustrated in Table 8.10.

TABLE 8.9

Hardness scale comparison (22, 36, and 38).

Rockwell$_C$	Rockwell$_B$	Rockwell$_F$	Brinell	Knoop
80				
70				972
60			614	732
50			484	542
40			372	402
30			283	311
20	98		230	251
10	90		185	201
	80		150	164
	70	97	125	139
	60	91	107	120
	55	88	100	112
	50	85	83	107
	41	80	75	98
	23	70	63	82
	14	65	59	75
		55	50	65

TABLE 8.10

Ultimate tensile and compressive strength of selected brittle materials (7, 8 and 36).

Material	S_{ut}, psi	S_{uc}, psi
Acrylic plastic	10,600	17,000
Gray cast iron		
Class 20	22,000	83,000
Class 60	62,500	188,000
Alumina ceramics	30,000	300,000
28 day concrete	400	5,300
Window glass	10,000	500,000
Aluminum oxide	20,000–30,000	280,000–380,000
Zirconium oxide	4500–12,000	60,000–100,000
Tungsten carbide	130,000	518,000–800,000
Silicon carbide	15,000	150,000

FIGURE 8.6

The Jominy hardness test. (a) test setup; (b) typical data.

From A.G. Guy and J.J. Hren, *Elements of Physical Metallurgy*, 3rd ed. © 1974, Addison-Wesley Publishing Co., Reading, Mass. Figure 10.23, p. 486. Reprinted with permission.

FIGURE 8.7

Hardenability band for 4140H steel.

Courtesy Association of Iron and Steel Engineers, Pittsburgh, Penn. Extracted from the 10th edition of "The Making, Shaping and Treating of Steel."

Hardenability is a measure of the ability of a steel to reach its maximum hardness when cooled rapidly from an elevated temperature. This is determined by measuring the hardness of a standard size sample that has been cooled from elevated temperatures at different rates of cooling. A standard test of hardenability is the Jominy test, which is illustrated in Figure 8.6, where one end is cooled rapidly and the other surfaces are cooled at a reduced rate. Figure 8.7 illustrates the variance in the hardenability of 4140H steel. Figure 8.8 illustrates the effect of alloying elements on the hardenability of steel with a constant carbon content. Figure 8.9 illustrates the effect of increasing the carbon content on steel hardenability.

The **fatigue strength** of a material is a measure of its resistance to cyclic loading when part of the cycle is a tensile load. The cyclic stress can be: **1.** alternating levels of tensile stress, **2.** alternating levels of tensile stress after a preset tensile or compressive load, or **3.** alternating levels of tension and compression stress. The peak levels of the cyclic stress can be constant or varying.

The measure of a material's fatigue strength is the **fatigue limit**, S_e, sometimes called the endurance limit. The limit is usually defined as the maximum value of equal alternating levels of tension and compression stress before a material fails after 10^6 cycles. Figure 8.10 shows typical test data. A component part can fail at a stress level lower than either S_{ut} or S_{yt}. For many steels the fatigue limit is approximately 50 percent of the material's ultimate tensile strength. Appendix F-1 lists S_e values for gray cast iron.

FIGURE 8.8

Minimum hardenability limits for five steel alloys with the same carbon content.

[Figure: Rockwell hardness C scale vs. Distance from quenched end (sixteenths of an inch, top axis 2–32; millimeters, bottom axis 5–50). Curves labeled G43400 (4340), G98400 (9840), G41400 (4140), G51400 (5140), G86400 (8640).]

Courtesy Association of Iron and Steel Engineers, Pittsburgh, Penn. Extracted from the 10th edition of "The Making, Shaping and Treating of Steel."

Part geometry can have a significant effect on the fatigue limit. A sharp notch or small radius at a corner can reduce the limit by 50 percent. A rough surface can also reduce the fatigue limit (see Figure 8.11). Rough surfaces can result from many different sources such as hot rolling, forging, flame cutting, or corrosion.

The best measure of the fatigue limit of a proposed design is obtained if the part is built and subjected to a test that reflects the actual intended use. The test part should be manufactured with the same methods that will be used to make the part in regular production.

Fracture toughness is measured using critical stress intensity, K_c, a measure of the ability of a material to resist crack growth. It has the unusual units of $(lbs/in^2)\sqrt{in}$, or $MPa\sqrt{m}$. See Tables 8.11 and 8.12 for selected critical stress intensity factors.

Ductility, D, is a measure of the amount of deformation that a material can withstand before fracture occurs. A common measure of a material's ductility is calculated using the following model $D = (l_f - l_o)100/l_o$, where

l_o = the length of the part before testing, and l_f = the length of the part at failure.

A calculated value of ductility for a tensile test must be accompanied with a reference to the original length. The majority of the deformation occurs over a small length, and the longer the original length, the smaller will be the calculated value of D.

FIGURE 8.9

Minimum hardenability limits of five steel alloys, changing only the % carbon.

[Graph: Rockwell hardness C scale vs. Distance from quenched end (sixteenths of an inch, top axis 2–32; millimeters, bottom axis 5–50). Curves labeled G86600 (8660), G86500 (8650), G86400 (8640), G86200 (8620), G86300 (8630).]

Courtesy Association of Iron and Steel Engineers, Pittsburgh, Penn. Extracted from the 10th edition of "The Making, Shaping and Treating of Steel."

An original length of 2.00 inches, or 50 mm is a common test length. One rule of thumb for steels is that a sample that deforms less than 5 percent in 2.00 inches is considered brittle, and one that deforms more than 5 percent is considered ductile.

The toughness, or **impact strength**, of a material is a measure of its ability to absorb energy before failure. Toughness is associated with the area under the stress-strain curve and has units of in·lbf/in^3. Ductile materials are generally also tough, and both ductility and toughness increase in metals as temperature increases. Impact also implies that loads are applied during "short" periods of time, making the strain rates "high." Analysis of impact loads is complicated by the fact that material properties change as the strain rate increases; see Figure 8.12. As strain rates increase, material yield strength approaches ultimate strength and the material behaves brittlely. Selected dynamic strengths are listed in Table 8.13. Impact resistance is also affected by material temperature. A critical temperature exists for body-centered cubic structured materials called the transition temperature; see Figure 8.13. It is the temperature below which a normally ductile material behaves brittlely.

Standard tests for measuring material toughness include the Charpy test and the Izod test.

Creep is a phenomenon that generally occurs at an elevated temperature of approximately one-half the melting temperature of a metal. Under this condition stress in

FIGURE 8.10

Typical S-N (stress cycle) diagrams for fatigue tests of various metals.

Key to curves
A 1.20% C steel, oil quench, draw at 860°F
B Chrome-Nickel steel
   C = 0.24%, Cr = 0.87%, Ni = 3.33%
   Oil quench, draw at 1200°F
C 0.53% C steel, oil quench, draw at 1200°F
D Structured steel rod, as rolled
E Duralumin
F Copper, annealed
G Cast iron
o→ Specimen not fractured

From D. K. Allen, *Metallurgy Theory and Practice*. © 1969 American Technical Publishers, Inc.

FIGURE 8.11

Effect of surface conditions on the fatigue strength of steel. Divide N/m^2 by 6.89×10^3 to get psi.

Mirror-polished
Polished
Machined
With sharp circumferential notch
With mill scale
Under ordinary water
Under salt water

Tensile strength of steel, 10^7 N/m^2

From A. G. Guy and J. J. Hren. *Elements of Physical Metallurgy*, 3rd ed. © 1974 Addison-Wesley Publishing Co., Reading, Mass. Figure 4.27, p. 152. Reprinted with permission.

TABLE 8.11

Mechanical properties of selected engineering materials at room temperature.

Material	Density kg/m³	Elastic Modulus, GPa	Strength, MPa	Critical Stress Intensity Factor, MPa√m	Nom. Fract. Strain —	Maximum Strength/ Density, (km/sec)²
Ductile steel	7850–7870	200	350–800	170	0.2–0.5	0.1
Cast iron	6950–7700	55–140	140–650	5–20	0.0–0.2	0.09
Cast Al alloys	2570–2950	65–72	130–300	—	0.01–0.14	0.1
Wrought Al alloys	2620–2820	68–72	55–650	15–100	0.05–0.30	0.2
Polymers	900–2200	0.1–21	5–190	0.5–3	0.0–8.0	0.1
Low-alloy steel	7800–7900	200	1000–2000	15–170	0.05–0.25	0.25
Glasses, ceramics	2200–4000	40–140	10–140	0.2–5	0.0	0.05
Stainless steels	7600–8000	200	300–1400	55–120	0.10–0.30	0.2
Copper alloys	8200–8900	100–117	300–1400	10–100	0.02–0.65	0.17
Titanium alloys	4400–4800	116	400–1600	50–140	0.05–0.30	0.35
Mg alloys	1740–1840	45	80–370	—	0.01–0.25	0.2
Moldable glass fiber resin	1700–2100	11–17	55–440	20–50	0.003–0.015	0.2
Graphite-epoxy	1500	200	1000	45–120	0–0.02	0.65
Tool steels: Hardness to 68 R_c; stable up to 600°C; strength to 2000 MPa				30–100		
Sintered carbides: Hardness at 20°C, 86 to 93 R_A (69 to well above 70 R_c equivalent); to 85 R_A max. at 760°C				3–21		
Superalloys: Strength is approximately 140 MPa at 650°C, 50 MPa at 750°C, 25 MPa at 900°C				—		

From David K. Felbeck and Anthony G. Atkins, *Strength & Fracture of Engineering Solids*, p. 520. © 1984, Prentice-Hall, Inc., Englewood Cliffs, N.J. Reprinted by permission.

metals below the yield strength can cause plastic deformation. Creep occurs over extended periods of time, and like fatigue tests, cannot be speeded up. Data extrapolation is risky at best, so equipment operating at elevated temperatures must be inspected on a regular basis. Some metals, and wood, will show signs of creep at normal temperatures.

Creep deformation is also noted by a change in deformation rate over time. After initial deformation, the rate, $\Delta L/\Delta t$, reduces, and holds fairly constant until a critical deformation is reached. At this point the deformation rate increases rapidly and sudden failure occurs.

TABLE 8.12

Selected values of quasi-static toughness and yield strength.

Material Type	E, GPa	σ_y, MPa	R, kJ/m^2	K_c, MPa$\sqrt{m}$	K_c/σ_y, $\sqrt{m}$
Glasses	70	138	0.01	1	0.01
Ceramics	300	280	0.1	5	0.02
PMMA	2.8	70	0.5	1	0.02
Cast iron	100	300	1	10	0.03
Boron-fiber epoxy composites	170	2000	30	70	0.04
Tough polymers (Polycarbonate)	2.5	63	4	3	0.05
High-strength aluminum-base alloys	70	600	28	44	0.07
High-strength steels (4340)	210	1400	140	170	0.12
High-strength titanium-base alloys	110	1100	170	136	0.12
Maraging steel	210	960	150	180	0.20
Low-carbon steels	210	280	140	170	0.60

From David K. Felbeck and Anthony G. Atkins, *Strength & Fracture of Engineering Solids*, p. 525. © 1984, Prentice-Hall, Inc. Englewood Cliffs, N.J. Reprinted by permission.

A metal with fine grains is more susceptible to creep than the same metal with large grains. This is because high temperatures affect grain boundaries first. Grain boundaries are in a higher stress state due to boundary irregularities. Low temperature strength is just the opposite, where fine grained metals are stronger than large grained metals.

Specific strength is a seldom used but an interesting measure of a material's strength. Specific strength is defined as the ratio of strength to density. All the usual precautions about which strength to use apply. Comparisons can only be made if the same strength criteria are used: yield or ultimate, and tension, compression, or shear. Table 8.14 gives some selected values.

Machinability is a measure of how well a material can be "cut" to shape. Machinability is an elusive material property because so many conditions affect the "cutting" process. Among these conditions are material hardness, grain size, alloying elements, acceptable tool wear, cutting speed, heat treatment, percent cold working, machine tool power requirements, cutting tool configuration, coolant properties, machine tool condition, part tolerances required, and surface finish required. There are conflicts over an absolute rating scale, so the best to hope for is a comparative scale. Table 8.15 illustrates one such rating scale.

FIGURE 8.12

Stress-strain diagrams for Ti-5 Al-2.5 Sn at 75° and 900°F at strain rates indicated.

From D. K. Allen, *Metallurgy Theory and Practice*. © 1969 American Technical Publishers, Inc.

8.5.2 Electric Properties

Electrical resistivity, k, has the units ohm·inch or ohm·centimeter and is related to the resistance of a specific piece of material expressed in the formula $R = kl/A$, where:

l = length of the material in the direction of current flow, inches,

A = the material cross section area perpendicular to the direction of current flow, in^2, and

R = the resistance of the piece of material, ohms.

The ohm·inch units are an unfortunate choice because too often only the length of a wire is considered when calculating total resistance. If the units ohm·inches^2/inch were used they would remind us that both area and length are needed. An analogy exists when

FIGURE 8.13

Typical ductile-brittle transition temperatures. (*p*, plain tension; *r*, reduction of area; *e*, elongation; *c*, Charpy V-notice impact energy).

Y-axis: Percent reduction area / Percent elongation / Charpy impact energy, ft lb (0, 25, 50, 75, 100)
X-axis: Temperature, °K (0 to 800)

Curves labeled: Nb (*p, r*), Sn (*p, r*), Polymethylmethacrylate (*p, e*), Mo (*p, r*), W (*p, r*), Cr (*p, r*), Ta (*p, r*), Mild steel (*c*), Mild steel (*p, r*), Be (*p, e*), AgCl (*p, r*), NaCl (*p, r*), Si (*p, r*).

From A. H. Cottrell, *The Mechanical Properties of Matter*. © 1981 John Wiley & Sons, Inc., N.Y. Reprinted by permission.

deformation is calculated, the length units are not canceled, and the answer is expressed as inches/inch or centimeters/centimeter.

The value of k for metal conductors is $\leq 10^{-5}$, for semiconductors 10^{-2} to 10^4, and for insulators $\geq 10^{12}$. The values depend to some degree on the applied voltage. See Table 8.16 for selected values of electrical resistivity.

Electrical conductivity is the reciprocal of resistivity and has the units mho/inch. Electrical conductivity and resistivity are important in the design of electronic equipment, and power generating and transmitting equipment. Resistivity of metals generally decreases as temperature decreases because the free electrons do not have as high an energy level. Some materials become superconductors at reduced temperatures, their resistance to electron flow approaching zero. Mercury, for example, becomes a superconductor at −452°F, Nb_3Ge at −418°F, and $Ba_xLa_{2-x}CuO_4$ at −396°F. In 1986–1987 research on the superconductivity of materials came to the forefront because of discoveries of new superconductors such as $Y_{0.85}Ba_{0.15}CuO_4$, $YBa_2Cu_3O_7$, and $La_{2-x}Sr_xCuO_4$, which become superconductive at −292°F, −288°F, and −283°F respectively. There are over 6000 material compounds that exhibit superconductivity at low temperatures, but a practical problem yet to be solved is to produce superconductors in strong wire form and capable of carrying high currents.

TABLE 8.13

Static and dynamic properties of selected materials.

Material	Condition	Static yield strength psi	Ult. strength, psi Static	Ult. strength, psi Dynamic
Ingot iron	Annealed	16,000	37,100	57,400
SAE 1015	Annealed	29,250	50,600	63,500
1022	Cold rolled	64,500	84,000	105,000
1022	Annealed	41,000	65,000	82,700
1040	Annealed	43,000	78,050	91,800
1045	Normalized	55,500	97,750	105,700
1045	Anneal. (Low)	48,200	81,500	117,000
1045	Anneal. (Hi)	51,750	94,300	132,200
1045	Spheroidized	66,400	76,600	98,400
1045	Quench & T.	132,250	142,900	169,000
1095	Normalized	74,000	144,600	151,000
1095	Quench & T.	136,000	176,500	170,500
2345	Austemp. R_c 35	121,000	155,700	182,000
2345	Q. & T. R_c 52	210,000	265,000	276,000
4140	Austemp. R_c 31	82,000	139,750	150,700
4140	Q. & T. R_c 49	195,000	244,000	258,000
5150	Austemp. R_c 31	102,500	148,500	165,300
5150	Q. & T. R_c 52	205,500	257,000	279,100
8715	Heat treated	103,700	141,300	170,000
8739	Austemp. R_c 37	137,000	169,000	207,400
8739	Q. & T. R_c 52	205,000	272,000	290,000
Copper	Annealed	4,000	29,900	36,700
Copper	Cold rolled	30,000	45,000	60,000
1100 Al.	$\frac{1}{2}$ Hard	12,000	17,200	22,100
1100 Al.	Annealed	1,700	11,600	15,400
2017 Al.	As received	38,000	59,900	63,800
2024 Al.	As received	47,000	65,150	68,600
2024 Al.	Annealed	14,000	33,950	44,980
Magnesium F	As received	25,500	35,920	51,760
Magnesium J	As received	29,870	43,750	51,360
302 Stainless	As received	44,000	93,300	110,800
Zamak II	Die cast	14,000	34,500	50,200

[13]For a more extensive table, see Clark, D. S., and Wood, D. A., "The Tensile Impact Properties of Some Metals and Alloys," *Trans. Amer. Soc. for Metals*, 42 (1950), p. 45. See also *Trans. Amer. Soc. for Metals*, 46 (1954), p. 34.
From M. F. Spotts, *Mechanical Design Analysis*, p. 365. © 1964 Prentice-Hall, Inc., Englewood Cliffs, N.J. Reprinted by permission.

Materials

TABLE 8.14

Density and mechanical properties of bulk materials and filaments.

Bulk Material	ρ, kg/m³	S_u, MPa	S_u/ρ, (km/sec)²	E, GPa	E/ρ, (km/sec)²
Low-carbon steel	7850	400	0.051	210	27
High-tensile steel	7850	2000	0.25	210	27
Aluminum alloy	2700	550	0.20	74	27
Magnesium alloy	1800	350	0.19	45	25
Nickel	8860	490	0.055	210	24
Copper	8940	240	0.027	120	13
Titanium alloy	4500	1000	0.22	120	27
Wood	600	60	0.1	14	23
Glass	2500	60	0.024	70	28
Filaments					
Glass	2500	1000	0.4	70	28
Iron	7850	14000	1.8	210	27
Nickel	8860	3800	0.43	210	24
Copper	8940	2900	0.32	120	13
Alumina (Al_2O_3)	4000	18700	4.7	390	98
Silicon carbide	3200	24100	7.5	560	175
Boron	2500	3450	1.4	380	152
Graphite	2250	21000	9.3	750	334
Carbon	2250	1600	0.71	360	160
Kevlar 49	1440	2700	1.9	130	90
Nylon	1100	400	0.36	4	3.6
Polypropylene	900	900	1.0	8	9

From David K. Felbeck and Anthony G. Atkins, *Strength & Fracture of Engineering Solids*, p. 528. © 1984, Prentice-Hall, Inc. Englewood Cliffs, N.J. Reprinted by permission.

The **dielectric constant** of a material is the ratio of the charge carried in a capacitor with the sample material between the parallel plates and the charge carried in the same capacitor with a vacuum between the parallel plates. The dielectric constant is important in the design of electrical components where charge capacity is critical; Table 8.17 lists some typical values. The value is a function of both temperature and frequency, having a maximum value at some temperature but reducing as frequency increases.

Magnetic hysteresis is the relationship between induced magnetism and the applied magnetic field for a given material. Materials with no hysteresis are easily magnetized and demagnetized. Higher values of hysteresis means that more magnetism remains in the material after the magnetic field is removed. Hysteresis is an important concern in the design of electric motors, generators, and electromagnets.

Electric permeability of free space is a constant (10^{-7} Wb/A·M) in the Biot-Savert law which is used to calculate magnetic fields around a current carrying wire.

TABLE 8.15

Comparison of machinability for selected materials (15, 21, and others).

Material	Relative Rating, %
Magnesium alloys	500–2000
Aluminum alloys	300–2000
Leaded brass	300–600
Zinc alloys, yellow brass, gray cast iron	200
1113	135
1112	100
1109, 1110, 1117	85
1118, 1120	80
1016, 1022, 1137, 4023, 4119	70
1020, 1141, 4032, 5120	65
1040, 3120	60
3140	55
1320, 2330, 6120, 8650	50
2340, inconel	45
9315, manganese bronze	40
Cast monel (70%Ni–30%Cu)	35
High speed steel, 2515	30
Stainless 18-8	25
White cast iron	0–25
Sintered carbides	0

TABLE 8.16

Selected electrical resistivities @ 20°C, from low to high (22, 36, and 38).

Material	ohm·cm
Silver	$1.6 \cdot 10^{-6}$
Annealed copper	$1.7 \cdot 10^{-6}$
Gold	$2.4 \cdot 10^{-6}$
Aluminum	$2.7 \cdot 10^{-6}$
Brass (70%Cu-30%Zn)	$6.2 \cdot 10^{-6}$
Nickel	$7.8 \cdot 10^{-6}$
Iron	$9.7 \cdot 10^{-6}$
White tin (temp > 13.2°C)	$12.8 \cdot 10^{-6}$
Stainless steel (18-8)	$73 \cdot 10^{-6}$
Nichrome (80%Ni-20%Cr)	$100 \cdot 10^{-6}$
Gray tin (temp < 13.2°C)	$100 \cdot 10^{-6}$
Silicon	$2.0 \cdot 10^{5}$
Electric porcelain and other insulators	$10^{13} - 10^{15}$
Nylon	10^{20}
Polystyrene	10^{24}

TABLE 8.17

Selected dielectric constants (22, 33, 39 and others).

Material	Relative Dielectric Constant @ 10^6 Hz	Dielectric Strength Volts/meter $\times 10^6$
Vacuum	1.00000	∞
Air	1.00059	3
Polytetraflourethylene	2.1	60
Polyethylene	2.3	20
Silicon oil	2.5	15
Polystyrene	2.5	24
Rubber @ 25°C	2.7	—
Lucite @ 23°C	2.8	—
Nylon	3.4	14
Polyvinyl chloride	3.4	—
Paper	3.7	16
Fused quartz	3.78	8
Fused silica	3.8	10
Bakelite	4.9	24
Pyrex glass	4.0	14
Electrical porcelain	6	1.6–7.9
Neoprene rubber	6.7	12
Soda lime glass	7	10
Methanol	33.6	—
Water	80	—
Strontium titanate	233	8

8.5.3 Thermal Properties

The **thermal conductivity** coefficient is a measure of how fast a material conducts heat from a high temperature source to a low temperature sink. Units of measure for thermal conductivity are (Btu·in)/(ft^2·sec·°F), or (joule·cm)/(cm^2·sec·°C). See Appendices F-1 and F-10 for selected values. The reciprocal of thermal conductivity is thermal resistance, and is referred to as the R factor in insulating materials. R values for common building materials are listed in Appendix R. Averaging of R values for heat loss calculations is required whenever there are construction variations. This is analogous to parallel electrical resistors, where current has alternate paths to travel.

EXAMPLE 8.1:
Average insulating value calculation.

A house wall, built with 2 x 4 inch (1.50 x 3.50 actual) construction lumber spaced 16 inches on center, and insulated with 3.5 inches of fiberglass has an average R value as follows; R_{av} = (R of fiberglass)(% of area covered) + (R of 2 x 4)(% of area covered).

Neglect doors and windows, which need to be analyzed separately, and use one 16 inch wall section as typical.

$R_{av} = [(3.50)(3.15)](14.50)/16 + (4.38)(1.50)/16 = 10.4$

Calculations of this kind are important when estimating heating or cooling losses in buildings.

The **thermal expansion** coefficient, γ, is a measure of the change in the dimensions of a part as the temperature changes. The expansion coefficient is calculated using the model

$\gamma = \Delta l / [(l_0)(\Delta T)]$, where

Δl = change in length due to ΔT,

ΔT = the change in temperature, degrees Fahrenheit or Celsius, and

l_0 = the original length in the same units as Δl.

The units for γ are in/(in·°F), or mm/(mm·°C), see Appendices F-1, F-2, and F-10 for selected values. The coefficient does vary with temperature and the variation should be accounted for in critical cases, as shown in Figure 8.14. The error created by not accounting for the variation is often less than the range of variation possible from other limiting assumptions. Thermal expansion phenomena are used when designing a bimetallic temperature sensor, and can be a design factor when parts shrink in cold

FIGURE 8.14

Thermal coefficient versus temperature. The volume expansion coefficient is approximately three times the linear expansion coefficient. The values increase slightly with temperature, and are greater for materials that melt at a lower temperature.

From L.H. VanVlack. *A Textbook of Materials Technology.* © 1973 Addison-Wesley Publishing Co., Reading, Mass. Figure 2-6.1, p. 18. Reprinted with permission.

weather or get larger in hot weather. The dimensional change is equivalent to strain. If the strain is contained by the surrounding structure, the strain can be converted to a stress by using the modulus of elasticity. A cyclic temperature change can induce cyclic stresses and result in fatigue failure.

EXAMPLE 8.2:

Results of temperature variation.

What change in length must be allowed for in the design of a 40-inch (1.02 m) torsion bar made from 1040 steel, when the temperature range is assumed to be −40°F to 150°F (−40 to 66°C)?

From Appendix F-1, $\gamma = 6.3 \cdot 10^{-6}$ in/in/°F, so $\Delta l = (40)(6.3 \cdot 10^{-6}) = 0.000252$ inches (0.0064 mm).

The additional stress on the bar, if the ends are confined, is:

$$\sigma = E(0.000252/40) = (30 \cdot 10^6)(6.3 \cdot 10^{-6}) = 189 \text{ psi } (1.30 \text{ MPa}) \text{ compression.}$$

The **melting temperature** of a metal is the temperature at which transformation from a solid to a liquid occurs. This happens when added heat does not cause a rise in the temperature but does cause a phase transformation. The added heat is called the **heat of fusion**. In a reverse role the heat of fusion is also the amount of energy that must be removed to solidify a liquid phase. This occurs at the same temperature as melting occurs. Melting temperatures are listed in Appendix E.

Pure metals have constant melting temperatures, but the temperature is dependent upon pressure. Alloys can alter the melting temperature of a metal and cause the melting to occur over a range of temperatures. Alloying can also cause metal to melt at a lower temperature than either of the alloying elements. For example, pure iron melts at 2797°F, but iron with 4.3 percent carbon added melts at 2065°F; lead melts at 327°C and tin melts at 232°C, but a 35 percent tin-65 percent lead alloy melts over the range 183°C to 245°C, and a 61.9 percent tin-38.1 percent lead alloy melts at 183°C.

The **glass transition temperature,** T_g, is the temperature at which polymers change from reacting to stress in a ductile manner to reacting in a brittle manner. Technically, it is the temperature at which 100 TPa·sec viscosity is reached. Nontechnically, it occurs at about one-third of the material's boiling temperature. Polymers exhibit the highest damping capacity at their T_g temperature. Some metals exhibit a similar change from ductile to brittle reaction at a transition temperature. This contributes to changes in crack growth behavior.

8.5.4 Other Material Properties

Materials have many properties to consider when being selected for a design application. Some additional properties are listed to remind you that final selection may have to be a compromise between a variety of characteristics.

Crystal structure is the regular atomic array that occurs when a metal solidifies. The more common structures are face-centered cubic (FCC), body-centered cubic (BCC), and hexagonal close-packed (HCP). Many metal properties are related to the crystal structure of a metal such as alloy dissolving potential, number of possible slip planes, and heat treatment potential through phase transformations. Appendix E lists the crystal structure of common metals.

The **density** of a material is a measure of its compactness, measured in lbm/in^3, or kg/m^3. See Appendix E for selected values. Density affects product weight. Low weight is important for the design of fuel efficient automobiles and for this reason many automobile parts are made from plastic and aluminum. High density is important for the design of counterbalance mechanisms, or for gamma ray shielding structures.

The **vapor pressure** of a gas-solid system, a gas-liquid system, or a gas-liquid-solid system is the pressure of the vapor when the system is in equilibrium. This is important in the design of chemical processes and apparatus being used to control the processes.

Viscosity is the internal frictional resistance of a fluid as a result of an applied shear force. A fluid is defined as a substance that deforms continuously under shear forces. A shear force on a fluid may be caused by gravity or other external forces such as a pump. The absolute, or dynamic viscosity is measured in Pa·sec, N·sec/m^2, or lbf·sec/ft^2 (see Table 8.18). Viscosity of liquids generally decreases as temperature increases, whereas the viscosity of gases generally increases as temperature increases. Viscosity is important whenever fluid flow is involved, such as oil lubrication of bearings, water pumps, Freon compressors, or air flow over an airplane wing. Fluidity is the reciprocal of viscosity.

The **porosity** of a material is the ratio of the volume of the internal voids (open pores) to the volume of the true mass of the material without the voids. Porosity is

TABLE 8.18

Viscosity of selected noncrystalline materials, from low to high (21, 38, and others).

Material	Viscosity, N·sec/m^2 or Kg/m·sec
Magnesium @ 680°C	0.0012
Air @ 20°C	0.0018
Aluminum @ 700°C	0.003
Copper @ 1200°C	0.0032
Iron w/3.4% C @ 1300°C	0.0089
Pentane (C$_5$H$_{12}$) @ 20°C	0.025
Water @ 20°C	0.1
Machine oil @ 20°C	0.1–0.6
Phenol (C$_6$H$_5$OH) @ 20°C	1.0
Syrup (60% sugar) @ 20°C	5.6
Window glass @ 700°C	$1.0 \cdot 10^6$
Polymers @ T_g + 20°C	$1.0 \cdot 10^7$
Polymers @ T_g + 10°C	$1.0 \cdot 10^9$
Polymers @ T_g	$1.0 \cdot 10^{12}$
Window glass @ 515°C	$1.0 \cdot 10^{12}$

FIGURE 8.15

Reflected and refracted light waves.

important in the design of a fluid absorbing material such as a sponge, a permeable product such as a filter, or in the design of an oil impregnated sintered bearing.

Permeability is a measure of a material's ability to allow the passage of fluids or gases. Low permeability is needed for the design of a pump diaphragm, whereas high permeability is needed for the design of an air filter or oil filter.

Reflectivity is a measure of the return of light or other waves that have reached a material's surface, either externally or internally onto the surface. Light incident onto a surface is either reflected from the surface or transmitted through the surface. The amount reflected varies with the angle of incidence and wavelength. Waves reflect from a surface at the angle of incidence measured relative to the normal to the surface, as shown in Figure 8.15.

Light reflectivity is important in the design of mirrors, lenses, paints, and other surface coatings. Sound reflectivity is important in the study of acoustics and the design of soundproofing structures. Specular reflection is reflection from a smooth surface compared to diffuse reflection, which occurs from a rough surface. A smooth surface has small surface variations compared to the wavelength of the incident wave.

The **refraction index**, m, is the ratio of the speed of light in a vacuum, c, ($3 \cdot 10^{10}$ cm/sec) to the speed of light in the sample material, v; $m = c/v$. It is also related to the ratio of the sine of the angle of incidence to the sine of the angle of refraction, $m_r/m_i = \sin \theta_i / \sin \theta_r$, (see Figure 8.15). When the incident medium is air, $m_i = 1$. The refraction index is also equal to the square root of the dielectric constant. Refraction is important in the design of optic lenses, laser beam control surfaces, and fiber optic equipment. Table 8.19 lists the index of refraction for some common materials.

Transparency is a measure of a material's ability to scatter incident light waves. High transparency means low scatter. This property is important in the design of mirrors, reflectors, solar collectors, and parabolic dish antennas. Fresnel's equations allow the calculation of the light reflected, I_r, and transmitted, I_t, based on the refraction indices, but only when the angle of incidence is small, $\tan \theta \Rightarrow 0$.

$$I_r = \frac{(m_2 - m_1)^2 I_i}{(m_1 + m_2)^2}, \quad \text{and} \quad I_t = \frac{(4m_1 m_2) I_i}{(m_1 + m_2)^2}$$

TABLE 8.19

Selected indexes of refraction (22, 33, and others).

Material	Index of Refraction
Diamond	2.42
Zircon	1.92
Heavy flint glass	1.66
Zinc crown glass	1.52
Benzene	1.50
Plastics	1.50
Carbon tetrachloride	1.46
Nitric acid	1.40
Ethanol	1.36
Acetone	1.35
Methanol	1.33
Water	1.33
Ice	1.31
Liquid CO_2	1.20
Most gases @ 0°C, 1 atm	1.00+

Light absorption is the ability of a material to absorb incident light. If I is the light intensity at a depth x into the material, then Lambert's Law applies:

$I = I_0 e^{-Bx}$, where

I_0 is the light intensity transmitted into the material, the non-reflected light, and

B is the material's absorptivity measured in cm^{-1} or in^{-1}.

Use caution, because B can be wave length dependent.

The **half-life** of a radioactive material is the amount of time that is required for the material to decay to one-half of its beginning radioactivity level. The radioactivity can come from α, β, or γ rays. The half-life of materials ranges from seconds to years. Plutonium, for example, has a half-life of 24,000 years. Half-life is important in the design of nuclear reactors and storage systems for radioactive waste products.

8.6 Other Selection Criteria

Final material selection may seem like a very complex procedure, and if every selection required a review of every known material, indeed it would be. However, the selection of a material is a sub-problem of the overall problem, and the problem solving method can be used. In all probability, decisions made elsewhere in the overall problem solution have reduced the selection list to a limited list of choices. The following example illustrates how this might happen.

> **EXAMPLE 8.3:**
>
> ## Material selection.
>
> Consider the need for material to be used for a dolly caster support. Four casters must withstand a 400 lbf (1780 N) static load and the forces of moving the load over the floor. These conditions result in tension and compression stresses, so brittle materials such as ceramics and plastics are poor choices.
>
> If a casting is being considered, the use of the higher strength nodular or malleable irons will permit thinner sections than gray iron or many nonferrous alloys. Aluminum is a possibility, but would need thicker sections to overcome its low modulus of elasticity. A caster support that deflects too much will allow the caster to roll unevenly and promote premature wear and failure.
>
> If a stamping design is being considered, 1010 steel could be used because forming without rupture is important. Stamping favors high production quantities because the cost of the stamping dies is high.
>
> If a weldment is being considered, 1020 steel is a logical choice. The cost of welded assemblies is low because component parts can usually be made with standard tooling and weld fixtures are not as expensive as stamping dies. Welding favors low production quantities.
>
> If holes for ball bearings, roller bearings, or self lubricating bearings are required for the wheel axle to ride in, then the caster support must be thick enough to contain the bearing, and close machining tolerances are required. That restriction will favor a casting design because metal can be placed exactly where it is required and machined to low tolerances.

There are also computer software programs available to assist in material selection such as MetSel/2, a software package distributed by The American Society for Metals. This menu-driven program assists in the selection of materials from a list of 500 possibilities based on user selected criteria.

The final selection of a material for a given application may have nothing to do with material properties. Two non-material property restrictions placed on the engineer and the selection of materials are the material's commercial availability, and the manufacturing processes available to convert the purchased material into final product form.

The commercial availability of materials can be determined by contacting material suppliers. For metal availability, suppliers such as Ryerson (an Inland Steel Company) or Central Steel and Wire Company, both with main offices in Chicago, IL are possibilities.

Existing manufacturing capabilities will also have an effect on final material selection. Many manufacturing facilities have evolved to their present state to support "normal" production. For example, a seemingly small change in material from steel to aluminum may cause plant operating problems. A machining chip removal and scrap reclamation system may not be able to handle a mixture of ferrous and nonferrous metals. If changes from ferrous alloys to nonferrous alloys or polymers is desired, a thorough study of the effects on plant facilities is required. Staying within the current limitations reduces the need for additional capital investments. On the negative side, being restricted to current facilities restricts choices and hinders development of new processes. Which company will be the first to venture to new processes? Some companies consider research and development in manufacturing as important as research in new products.

8.7 Composite Materials

When different materials are used together in such a way that their reaction to an applied force are dependent on each other, the result is a composite material. Some examples of the simultaneous use of materials have to do with surface coatings, such as chrome plating of steel to prevent corrosion, but that is a surface application. Ideally, composite materials should be a homogeneous mixture of a filler material (usually fibrous) and a matrix material. If a composite is homogeneous, then material properties can be averaged, as the following analysis illustrates. The method used to average the material properties depends on whether the fibers are aligned with the direction of applied force, whether they are perpendicular to the applied force, or whether they are randomly placed in relationship to the direction of the applied force.

Consider a tensile load W that is applied to a member composed of two different materials, a matrix material a and a fibrous material b aligned with the direction of applied force. Each material will support a portion of the load, equal to the product of their respective areas (A_a and A_b) and their respective stresses (σ_a and σ_b), and $W = \sigma_a A_a + \sigma_b A_b$.

The average stress $\sigma_{av} = W/A_{total} = W/(A_a + A_b)$, so

$W = \sigma_{av}(A_a + A_b)$, and $\sigma_a A_a + \sigma_b A_b = \sigma_{av}(A_a + A_b)$, therefore

$\sigma_{av} = (\sigma_a A_a + \sigma_b A_b)/(A_a + A_b) = \sigma_a P_a + \sigma_b P_b$, where P_a = fraction of total area that is material a, and P_b = fraction of total area that is material b.

If the numerators and denominators of the fractions $A_a/(A_a + A_b)$ and $A_b/(A_a + A_b)$ are multiplied by the material length L then P_a and P_b can also be used as volume ratios. This is valid only if the material mixture is homogeneous. It also assumes both materials have the same strain, otherwise L_a and L_b would not be equal and the ratios would not be part of the whole.

When a composite material is loaded in compression parallel to the aligned fibers, the overall strength depends on the fiber-matrix bond. A weak bond allows the fibers to detach from the matrix. The fibers may buckle, and the advantage of the composite is lost. If the fiber-matrix bond is very strong, matrix shear failure generally occurs. The function of the matrix is to support the fibers so their strength can be utilized. The matrix is generally the cheaper and weaker material. When the composite material is under a tensile load parallel to the aligned fiber direction, failure depends on the tensile strength of the fibers, the shear strength interface bond, or the matrix tensile strength. Whether the load is tensile or compressive, as the direction of the applied load becomes more angular to the fiber direction, the matrix-interface bond is stressed more in shear. This will cause failure before the fibers reach their maximum strength. The stress on the matrix-fiber interface increases as the difference between the values of E increases. Knowledge about the strength of the fiber-matrix bond can be used to determine the volume ratio of matrix to fiber to optimize overall composite strength and cost.

If the load is perpendicular to the aligned fibers, the equal strain assumption is replaced with an equal stress assumption (see Exercise 8.53). To make material properties non-directional, short fibers are mixed so orientation is random, or long fibers are embedded at different angles. Fiber materials include metals, asbestos, ceramics, other polymers, boron, graphitized carbon, glass, paper, and cotton.

An example of a composite material is an aluminum base with Al_2O_3 fibers added.

TABLE 8.20

Density and mechanical properties of selected composite materials.

Material	ρ, kg/m³	S_u, MPa	S_M/ρ, (km/s)²	E, GPa	E/ρ, (km/s)²
Boron–5505 epoxy	1990	1590	0.80	207	104
HTS graphite–5208 epoxy	1550	1480	0.95	172	111
Kevlar 49–resin	1380	1380	1.00	76	55
E glass–1002 epoxy	1800	1100	0.61	39	22
Boron–6061 Al	2600	1490	0.57	214	82
T50 graphite–2011 Al	2580	566	0.22	160	62

From David K. Felbeck/Anthony G. Atkins, *Strength & Fracture of Engineering Solids*, p. 528. © 1984, Prentice-Hall, Inc., Englewood Cliffs, N.J. Reprinted by permission.

Ultimate tensile strength is increased 110,000 MPa by adding 15 percent fibers. The yield strength is doubled, elongation is reduced from 8 percent to 1 percent, and the coefficient of thermal expansion is reduced (6).

Methyl methacrylate reinforced concrete is another example of a composite material structure. Compressive strength is increased 240 percent, tensile strength is increased 275 percent, the modulus of elasticity is increased 77 percent, the modulus of rupture is increased 228 percent, the coefficient of thermal expansion is increased 30 percent, and the resistance to hydrogen-chloride exposure is increased.

Fiberglass added to polymers also creates a composite with improved properties. Other composite materials include Borsic-aluminum, Kevlar-epoxy, Kevlar-polyester, Graphite-polymer, and Glass-polymer.

Table 8.14 lists some commercial filament materials used in composite materials. Table 8.20 lists examples of commercial composites.

Steel reinforcing rod in concrete is a common engineering composite material because concrete is weak in tension, 400 psi (2.76 MPa), while steel rod is strong in tension, $35,000 \leq S_{yt} \leq 70,000$ psi ($241 \leq S_{yt} \leq 482$ MPa). The rod(s) must be placed in such a way that the total structure reacts favorably to the presence of both materials because steel rods improve directional properties.

EXAMPLE 8.4:
Concrete with steel reinforcing rod.

Suppose a 0.75 inch (19 mm) diameter rod with S_{yt} = 60,000 psi (414 MPa) is embedded in a 6 x 6 x 40 inch (0.152 x 0.152 x 1.02 m) long concrete section and loaded with a 14,000 pound (62.275 N) tensile load in the direction of the rod. The composite can be analyzed as follows.

$\sigma_{av} = 14{,}000/6^2 = 389$ psi (2.68 MPa)

At that stress level $\epsilon_{concrete} = \sigma_{av}/E = 389/2 \cdot 10^6 = 0.000194$ in/in (m/m), and $\epsilon_{steel} = 389/30 \cdot 10^6 = 0.000013$ in/in (m/m). *Note:* The value of $E = 2 \cdot 10^6$ psi is from Appendix F-1. Other studies have arrived at values in the range $3.5 - 5 \cdot 10^6$ psi (24.1 – 34.5 GPa), Appendix F-11. If the interface bond does not fail, then the steel must elongate more than the average stress indicates, and the concrete will not elongate as much as the average stress indicates. The elongation difference of 0.000181 in/in (m/m) must be shared between the steel and the concrete.

Let x = the additional elongation the steel must undergo. Then $(0.000181 - x)$ = the reduction in elongation of the concrete. For equilibrium, F, the force compressing the concrete must equal the force pulling the steel in tension.

$F = (2 \cdot 10^6 \text{ psi})(0.000181 - x \text{ in/in})(35.56 \text{ in}^2) = (30 \cdot 10^6 \text{ psi})(x \text{ in/in})(0.44 \text{ in}^2)$, so

$x = 0.000153$ in/in, and $F = 2020$ lbf(8985 N).

The additional stress on the steel = $(30 \cdot 10^6)(0.000153) = 4590$ psi (31.6 MPa), and the total stress on the steel is $389 + 4590 = 4979$ psi (34.32 MPa).

The additional stress on the concrete is $-2020/35.56 = -57$ psi (–0.393 MPa), minus because it is compression. The net stress on the concrete is $389 - 57 = 332$ psi (2.29 MPa).

The shear stress on the rod-concrete interface is 2020 lbf/$(0.75\pi$ in$)(40$ in$) = 21.4$ psi (0.147 MPa).

8.8 Additional Study Topics

1. Sintered carbides
2. White cast iron
3. Ceramic manufacture
4. Metal solidification in a vacuum
5. Metal solidification under zero gravity
6. n-type and p-type semiconductors
7. Making and using nylon
8. Making and using rubber
9. Solders
10. Making and using tempered glass
11. Non-jewelry use of platinum, gold, and silver
12. Stainless steels
13. Fiber optics
14. Damping capacity
15. Refractory materials
16. Superconductors

17. Hardness scales: Mohs, Knoop, Vickers, Rockwell, and Brinell
18. Nonferrous super alloys of nickel, cobalt, molybdenum, tungsten, niobium, tantalum, and zirconium
19. Age hardening and precipitation hardening
20. Phase diagrams
21. Isothermal transformation diagrams
22. Recrystallization temperature
23. Galvanic action and electromotive series
24. High electric conductivity polymers

8.9 References

1. Allen, Dell K. *Metallurgy Theory and Practice.* Homewood, Ill.: American Technical Society, 1971.
2. Altamuro, Vincent. "Design For Automated Manufacturing." *Design Graphics World*, March, 1987, 15–18.
3. Askeland, Donald R. *The Science and Engineering of Materials*. Monterey, Cal.: Brooks/Cole, 1984.
4. Barry, B.T.K. and C. G. Thwaites. *Tin and Its Alloys and Compounds*. New York: John Wiley & Sons, 1983.
5. Bodig, Jozsef and Benjamin A. Jayne. *Mechanics of Wood and Wood Composite.* New York: Van Nostrand Reinhold, 1982.
6. Bowles, R.R., D.L. Mancini, and M.W. Toaz. "Metal Matrix Composites Aid Piston Manufacture." *Manufacturing Engineering,* May, 1987, 61.
7. Boyer, H.E. and T.L. Gall, eds. *Metals Handbook*, Desk Edition. Ohio: American Society For Metals, 1985.
8. Budinski, Kenneth. *Engineering Materials Properties and Selection*, 3rd ed. Englewood Cliffs, N.J.: Prentice Hall, 1989. Includes up-to-date details on polymers and engineering ceramics.
9. *Business & Industry*. "The Surprising Story of Plastics." December, 1989, 17–31.
10. Collins, J.A. *A Failure of Materials in Mechanical Design*. New York: John Wiley & Sons, 1981. Chapter 11.
11. Cottrell, A.H. *The Mechanical Properties of Matter*. New York: John Wiley & Sons, 1964.
12. Stephen H. Crandall, et. al. *An Introduction to the Mechanics of Solids,* 2nd ed. New York: McGraw-Hill, 1972.
13. Cremer, Lothar and H. Muller, eds. *Principles and Applications of Room Acoustics.* New York: Elsevier Science, 1982.
14. Davis, Harmer E., et al. *The Testing of Engineering Materials*, 4th ed. New York: McGraw-Hill, 1982.
15. Doyle, Lawrence E., et al. *Manufacturing Processes and Materials for Engineers,* 3rd ed. Englewood Cliffs, N.J.: Prentice Hall, 1985.

16. Dym, Joseph B. *Product Design With Plastics: A Practical Manual.* New York: Industrial Press, 1983. Lists properties of over 200 different plastics.
17. Elliott, Roy. *Eutectic Solidification Processing: Crystalline and Glassy Alloys.* Boston: Butterworth, 1983.
18. Felbeck, David K. and Anthony C. Atkins. *Strength and Fracture of Engineering Solids.* Englewood Cliffs, N.J.: Prentice Hall, 1984.
19. Ferguson. *Reinforced Concrete Fundamentals.* New York: John Wiley & Sons, 1979.
20. Flinn, Richard A. and Paul K. Trojan. *Engineering Materials and Their Applications*, 2nd ed. Boston: Houghton Mifflin, 1981.
21. Guy, Albert G. and John J. Hren. *Elements of Physical Metallurgy,* 3rd ed. Reading, Mass.: Addison-Wesley, 1974.
22. *Handbook of Chemistry and Physics*, 67th ed. Boca Raton, Fla.: CRC Press, 1985.
23. Herubin, Charles A. and Theodore W. Marotta. *Basic Construction Materials,* 2nd ed. Virginia: Reston, 1981.
24. Higdon, Archie, et al. *Mechanics of Materials*, 4th ed. New York: John Wiley & Sons, 1985.
25. Jacobs, James A. and Tom Kilduff. *Engineering Materials Technology.* Englewood Cliffs, N.J.: Prentice Hall, 1985.
26. Juvinall, Robert C. *Fundamentals of Machine Component Design.* New York: John Wiley & Sons, 1983.
27. Langone, John. *Superconductivity.* Chicago: Contemporary Books Inc., 1989.
28. Lankford, W.T., et al. *The Making, Shaping and Treating of Steel*, 10th ed. Pittsburgh: Association of Iron and Steel Engineers, 1985.
29. McKelvey, John P. "Understanding Superconductivity." *Invention & Technology.* Spring/Summer, 1989, 49-57.
30. Palframan, Diane. "FMS: Too Much, Too Soon." *Manufacturing Engineering.* March, 1987, 34-38.
31. Polakowski, N. H. and E. J. Ripling. *Strength and Structure of Engineering Materials.* Englewood Cliffs, N.J.: Prentice Hall, 1966.
32. Ramachandran, Feldman, and Beaudois. *Concrete Science.* London: Heyden & Sons, 1981.
33. Richerson, David W. *Modern Ceramics Engineering.* New York: Marcel Dekker, 1982.
34. Serway, Raymond A. *Physics for Scientists & Engineers.* New York: Saunders, 1982.
35. Shigley, Joseph E. and Larry D. Mitchell. *Mechanical Engineering Design,* 4th ed. New York: McGraw-Hill, 1983.
36. Thornton, Peter A. and Vito J. Colangelo. *Fundamentals of Engineering Materials.* Englewood Cliffs, N.J.: Prentice Hall, 1985.
37. Tooley, Fay V., ed. *Handbook of Glass Manufacture*, 3rd ed. Vol I. New York: Ashlee, 1984.
38. VanVlack, Lawrence H. *A Textbook of Materials Technology.* Reading, Mass.: Addison Wesley, 1973.
39. VanVlack, Lawrence H. *Elements of Material Science and Engineering,* 6th ed. Reading, Mass.: Addison Wesley, 1989.

40. Williams, A.F. and W. L. Lom. *Liquefied Petroleum Gases, Guide To Properties, Applications and Uses*. England: Ellis Harwood, 1982.

8.10 | Exercises

1. Select a topic from the list of additional study areas and write a research paper and prepare an oral presentation.

2. What basic material properties are used advantageously and account for the material used in each of the following cases?
    a) Rubber washer on a water faucet
    b) Brass fitting on a water faucet
    c) Nichrome wire for an electric heater element
    d) Copper wire for house wiring

3. How does the calculation for ductility change for the following data?

Gage Length	Length at Fracture
2.0 inches	3.12 inches
4.0 inches	5.12 inches
8.0 inches	9.12 inches

4. A material failed under a tensile test with the following results. Is the material more likely ductile or brittle? Why?

Original gage length	Failure gage length
2.00 inches	2.11 inches
6.00 inches	6.35 inches

5. A steel rod at 40°F is 2.00 inches in diameter and 30.00 inches long. It is then heated to 200°F. How much force is required to hold the rod to its original 30.00 inch length?

6. An extension cord is made from #16 copper wire, has 120 volts applied to it, and is being used to run a $\frac{1}{4}$-horsepower motor. Model the system and answer these questions:
    a) How long can the extension cord be before there is a 10 percent voltage loss?
    b) Answer the question if the wire is #14 wire.
    c) Answer the question if the wire is #12 wire.

7. What is the maximum length of a vertically hung, 1.00 inch diameter steel rod before it will yield due to its own weight? When will it fracture due to its own weight?

8. A 0.10 inch diameter aluminum wire 100 feet long is drawn through a 0.0625 diameter die. What is the new length of wire?

9. A guitar player tuned up a guitar in the dressing room before a performance. The performance was outside in the sun. Did the tuned guitar get flat, sharp, or neither as the performance proceeded? Explain.

10. A precast concrete beam is to be installed so its ends are fixed. Should it be installed at temperatures above or below the anticipated use temperature? Why?

11. A piece of lead has a shear stress of 700 psi applied to it. What is the shear strain if the same material exhibits an elastic strain of 0.0004 in/in under a 1000 psi tensile load?

12. What processes seem reasonable for shaping ceramics, considering they are hard and brittle?

13. Explain the advantages of using steel reinforcing rods embedded in concrete.

14. Explain any disadvantages of using steel reinforcing rods embedded in concrete.

15. Graph the modulus of elasticity as a function of melting temperature for several commercial metals. Can you find a graph paper that shows the data as a straight line? If yes, determine a mathematical model relating the two variables.

16. Simplify the equation $S_{su} = T_u r/J$ for a round bar by reducing the number of variables. Discuss how this simplified model would assist you in designing an experiment to study the reliability of a given S_{su} value.

17. The Brinell Hardness number is defined as: BHN = $P/[\pi D/2][D - \sqrt{D^2 - d^2}]$, where P = the load on a hardened ball in kilograms, D = the diameter of the hardened ball in mm, and d = the diameter of the indentation in a test material in mm.
Keeping in mind that this is a three dimensional model, explain the meaning of the model relative to the definition of hardness.

18. Derive a model to determine hardness using a conical point of angle θ based on the same principle as Exercise 8.17.

19. Explain why the modulus of rigidity, G, is approximately equal to one-third the value of E, the modulus of elasticity.

20. Compare the specific strength of steel, aluminum, wood, a polymer, and glass.

21. Discuss the details of an experiment you would run to compare the strength of thin material, comparing side notched samples and non-side notched samples.

22. Explain what will happen to the steel bar in Figure 8.16 as the triangular shaded area is heated from room temperature to 1000°F and then allowed to cool back to room temperature slowly.

23. List an advantage and a disadvantage of using each of the following alloying elements in iron.
    a) Sulfur
    b) Carbon
    c) Manganese
    d) Nickel

Materials

FIGURE 8.16

Diagram for Exercise 8.22.

      e) Chromium
      f) Cobalt
      g) Vanadium
      h) Molybdenum
      i) Silicon
      j) Tungsten

24. List an advantage and a disadvantage of each of the following for use as a base material.
    a) Aluminum
    b) Magnesium
    c) Tungsten
    d) Copper
    e) Titanium
    f) Lead
    g) Zinc
    h) Tin
    i) Gold

25. Which steel will exhibit the largest springback when formed, one with $S_{yt} = 30,000$ psi or one with $S_{yt} = 50,000$ psi? Explain.

26. The same bent part is to be made from steel, then made from aluminum.
    a) If both materials had the same yield strength, which material would require the largest force to make the bend? Why?
    b) Which material would absorb the most energy as it was being formed? Why?
    c) Which material would release the most energy during springback? Why?

27. Explain why a smaller load will cause the same amount of deformation in a metal if the temperature goes up.

28. If aluminum is to be substituted for iron for an electric motor housing, list the critical material properties and compare both materials for the job.

29. Compare bronze and aluminum as materials for use in a fresh water environment. Repeat the comparison for the salt water environment of the ocean.

30. What problems would occur, and what changes would be required, if a company wanted to change from gray iron casting to aluminum casting for its main product line?

31. The material values of two grades of polyethylene are available as tabulated.
    a) Does the difference in density seem significant?
    b) Are the differences in thermal conductivity significant? In what applications would the difference be important?
    c) How much pressure is exerted by polyethylene if the ambient temperature is changed by 100°F? Calculate for both high and low density.

	E psi	Specific Gravity	Thermal Conductivity Btu·in/ft²·sec·°F	Thermal Expansion in/in/°F
Low density	15–40,000	0.92	0.00065	$1 \cdot 10^{-4}$
High density	60–180,000	0.96	0.0010	$70 \cdot 10^{-6}$

32. Low density polyethylene has a heat resistant temperature range for continuous use of 130–175°F while that for high density polyethylene is 180–250°F. Relate this property to the thermal conductivity and density.

33. Derive the model definition for the modulus of resilience, $R = S_{yt}^2/2E$.

34. Explain why the modulus of rupture, which is energy absorbed per unit volume when a material is loaded to failure in tension, is approximately equal to $(S_{yt} + S_{ut})\epsilon_f/2$ for ductile material, where ϵ_f is the strain at fracture.

35. A 0.750 inch diameter bar of 1020 hot rolled steel is 20 inches long between supports. How much twist can the bar withstand before the maximum stress exceeds S_{ut}?

36. A 0.02 percent offset strain is often used to estimate a material's yield strength. Does this seem appropriate for steel alloys considering the range of known yield strengths: $45,000 \leq S_{yt} \leq 230,000$ psi?

37. Verify that all the light incident on a surface, I_i, is equal to the sum of the light reflected and transmitted using Fresnel's equations.

38. Tests on various materials indicate increases in S_{yt} and S_{ut} under impact loading. Would you expect this change to increase or decrease the toughness of a test specimen? Explain.

39. Does the observed fact that S_{yt} and S_{ut} increase as strain rate increases make designs conservative or nonconservative when an impact load is considered?

40. Concrete reinforcing rod has raised ridges on it so the concrete grips to it. Does this gripping promote equal stress or equal strain?

41. Based on material hardenability and the results of rapid quenching, what application would indicate a preference for 4340 steel over 1060 steel? When would 1060 be preferred?

42. If 4063 steel costs $1.50 per pound, and is used to make a 10-inch long part that will carry a 50,000 pound cyclic fatigue load, how much difference in cost between polishing and machining would make one process more advantageous than the other?

43. Would tungsten or mild steel be more susceptible to failure under impact load at:
    a) 0°C?

Materials

    b) 100°C?
    c) 500°C?

44. In Figure 8.2, the line for Fe has a step shown. Explain what causes this step.

45. For a given geometry configuration and loading, compare the following choices for elastic energy absorbing potential:
    a) 1020 HRS
    b) Annealed 304 stainless steel
    c) 6061T-6 aluminum
    d) AZ80A magnesium
    e) Annealed 70% Cu-30% Zn brass

46. Is there any reason to suspect that Poisson's ratio, ν, cannot be greater than 0.50? *Hint:* Consider the material volume change at $\nu = 0.50$.

47. Interpret what happens physically when a given torsion load is placed on two members M_1 and M_2 of the same geometry when $G_1 > G_2$.

48. An unknown material was tested in tension. When the load was removed the sample did not return to its original length. The next day the sample had returned to its original length. Explain some possibilities.

49. A material's thermal shock resistance R is defined as $R = K/E\gamma$, where
    K = the coefficient of thermal conductivity, E = the modulus of elasticity, and γ = the coefficient of thermal expansion.
    Compare the thermal shock resistance of some common materials such as steel, aluminum, brass, glass, and nylon.

50. An estimate of toughness can be made by assuming the stress-strain curve is straight from the yield point to the ultimate strength. Find some stress-strain curves and compare the actual area under the curve to this straight line estimated area.

51. Derive the mathematical model for σ_{av}, for combining three or more materials in a homogeneous composite structure.

52. By using the basic relationship $\sigma = \epsilon E$ and the basic composite equation, show that $E_{av} = P_a E_a + P_b E_b$.

53. In Exercises 8.51 and 8.52 the basic assumption "equal strains" was used. An alternate assumption would be that "equal stresses" exist. Equal stress is more likely when a composite is loaded perpendicular to the direction of the fibers. Show that if equal stress is used as the basic assumption then the results are as follows.

    $1/\sigma_{av} = P_a/\sigma_a + P_b/\sigma_b$, and $1/E_{av} = P_a/E_a + P_b/E_b$

54. Explain what happens physically when composite materials are under equal strain when loaded. Find examples that support your ideas.

55. Explain what happens physically when composite materials are under equal stress when loaded. Find examples that support your ideas.

56. What properties of graphite make it a good material for golf club shafts?

57. Using the data from Example 8.4, verify numerically that $\sigma_{av} = \sigma_a P_a + \sigma_b P_b$.

58. Using the data from Example 8.4, verify numerically that $\epsilon_{av} = \sigma_{av}/E_{av}$.

59. Refer to Example 8.4 and determine the maximum load that can be placed on the sample before the yield strength of either the concrete or steel will be exceeded.

60. Would one steel rod of 0.75 inches in diameter embedded in concrete be better or worse than two 0.50 inch diameter rods? Explain in quantitative terms and suggest an application for each configuration.

61. A concrete post 2 x 2 x 24 inches has a 24 inch long 0.75 inch diameter steel rod cast in its center. The post was subjected to a fire before it could be delivered. Is it likely that the shear stress between the rod and the concrete would exceed 500 psi during the fire? Explain in quantitative terms.

62. Show that for a two-material composite, the maximum tensile load P in pounds that can be placed on the sample before yielding occurs can be determined from either of these equations.

$$P = \frac{S_{yt1}A_t(E_1A_1 + E_2A_2)}{E_1A_1 + E_1A_2} = \frac{S_{yt2}A_t(E_1A_1 + E_2A_2)}{E_2A_2 + E_2A_1}$$

63. Determine the maximum length of each of these materials that can hang freely without fracture failure.

	ρ, Mg/m³	S_{ut}, MPa	E, GPa
a. 1020 steel	7.85	500	210
b. Reinforced tape	1.90	770	27
c. Epoxy	2.34	1800	250
d. Graphite	1.65	20,000	500

9 Manufacturing

"For manufacturing and the manufacturing professional, the question of change and the future is more than merely a strategic issue. . . .It's an issue of critical importance, an issue of survival—not in simplistic nationalistic terms—of global magnitude."
✦ *ROBIN P. BERGSTROM, EDITOR-IN-CHIEF, MANUFACTURING ENGINEERING.*

An engineer designing a product must consider how the component parts of the design can be manufactured economically. A knowledge of manufacturing processes allows the design of functional, reliable, cost-effective products. A design that cannot be manufactured is almost as bad as no design at all. New processes develop as the needs of our technological society demands. Some of these processes are specifically industry oriented, such as the photo-etching process used for manufacturing printed circuit boards; others are used for fabricating more exotic materials, as in the manufacture of ceramic components. The purpose of this chapter is to review common processes used for a majority of products; specialized, hand, and custom processes will not be discussed.

9.1 Process Selection

The processing of purchased material into usable parts depends on many variables and parameters, but the most important are:

1. The available material forms (bars, plates, sheets, coils, or castings),
2. The kind of material (ferrous alloys, nonferrous alloys, wood, polymers, or ceramics), and
3. The manufacturing methods and processes available.

There is no absolute best method to use to manufacture a part just as there is no absolute best design of a product. The choice of which manufacturing method to use depends on availability and the functional tolerances required. Functional tolerances include tolerance of size, position, geometry, and surface finish. For a particular design, one of these requirements may be more important than the others. Relative size and position are critical for assembly; size and geometry affect movement and operating efficiency; surface finish affects component wear and fatigue life. Dimensional tolerances and surface finish requirements placed on a part drawing indicate the desired function of the part in relation to other parts in an assembly.

TABLE 9.1

Reasonable accuracy from industrial processing equipment.

Production Process	ISO Limit Class
	2 3 4 5 6 7 8 9 10 11 12 13 14 15 16
Lapping	
Honing	
Super finish	
Cylindrical grinding	
Centerless grinding	
Plan grinding (fine)	
Plan grinding	
Boring plus reaming	
Boring	
Reaming	
Broaching	
Turning (diamond)	
Turning	
Turning on turret lathe	
Milling	
Cutting	
Shaping	
Drilling	
Extrusion	
Drawing	
Blanking	
Die casting	
Casting	
Flame cutting	
Forging	

From P. Wallach, *Metric Drafting*, Glencoe Publishing Co., Inc. © 1979.

Functional part tolerances in the English system have been standardized by the American National Standards Institute (ANSI) and are summarized in table form under five categories.

1. RC: Running or sliding clearance fits. These fits range from RC1, close sliding, to RC9, loose running.
2. LC: Locational clearance fits. These range from LC1, snug fit, to LC11, freedom of assembly.
3. LT: Transition clearance or interference fits. These range from LT1, minimum interference, to LT6, minimal clearance.
4. LN: Locational interference fits range from LN1 to LN3.
5. FN: Force or shrink fits. These range from FN1, light press fit, to FN5, high stress force fit.

See reference (7) for complete numerical tables.

The tolerance tables should only be used as guidelines because experience with specific applications will result in deviations from table values. For example: A force fit places stress on the component parts and it must be determined if the material selected can withstand the stress. The results will be different as the material and material properties change. Get to know your company's product and manufacturing capabilities to find out what is practical in your facility.

Tables for standardized International Tolerance (IT) values in the metric system also exist. These tables list values in 16 International Organization for Standards (ISO) limit classes; see reference (25) for complete ISO tables. Table 9.1 shows the relationship between some common processes and the most used ISO classes. Class 2, for example, used in the design of gages and other measuring tools, places a 0.003 mm tolerance on a 50 mm basic dimension, while class 16, used in the design of sand castings and flame cut parts, places a 1.90 mm tolerance on a 50 mm basic dimension.

The relationship between the functional tolerances, machining processes tolerances, and product cost is an important one. In general as tolerances go down, manufacturing cost goes up. This occurs if a given process has a tolerance range and the range is not exceeded, or if a tolerance requires a process change. The following example illustrates this concept.

EXAMPLE 9.1:
Tolerance and process selection.

A part is to be made 6.00 ± 0.04 inches long from a $0.25 \times 1.50 \times 120.00$ inch bar. Table 9.3 indicates a band saw to be a satisfactory process. If the tolerance is reduced to ± 0.03, the band saw is still acceptable but it must be maintained more closely. Perhaps the blade will need replacing or resharpening more often, and that will increase overhead cost. If the tolerance is reduced to ± 0.02, then the process must be changed, perhaps to a circular saw that uses a more expensive saw blade and more expensive machine, both of which increase overhead cost. See Section 14.8.2 for additional discussion on overhead cost.

TABLE 9.2

Reasonable tolerance ranges for industrial processes.

Machining tolerances

Range of sizes, in. From	To and Including				Total tolerance, in.						
0.000	0.599	0.00015	0.0002	0.0003	0.0005	0.0008	0.0012	0.002	0.003	0.005	
0.600	0.999	0.00015	0.00025	0.0004	0.0006	0.001	0.0015	0.0025	0.004	0.007	
1.000	1.499	0.0002	0.0003	0.0005	0.0008	0.0012	0.002	0.003	0.005	0.008	
1.500	2.799	0.00025	0.0004	0.0006	0.001	0.0015	0.0025	0.004	0.006	0.010	
2.800	4.499	0.0003	0.0005	0.0008	0.0012	0.002	0.003	0.005	0.008	0.012	
4.500	7.799	0.0004	0.0006	0.001	0.0015	0.0025	0.004	0.006	0.010	0.015	
7.800	13.599	0.0005	0.0008	0.0012	0.002	0.003	0.005	0.008	0.012	0.019	
13.600	20.000	0.0006	0.001	0.0015	0.0025	0.004	0.006	0.010	0.015	0.025	

Operation

- Lapping, honing, and electric discharge
- Grinding, diamond turning and burnishing(1)
- Broaching
- Reaming(2)
- Turning, boring, facing, planing, and shaping
- Milling(3) and lathe cutoff
- Stamping, blanking, and and punching

(1) True position 0.005
(2) True position 0.015. Reaming is a secondary operation. It is used to improve size tolerance and surface finish of drilled holes. Reaming should remove at least 10% of the hole diameter.
(3) True position 0.010

Tables 9.2, 9.3, and 9.4 show similar relationships for some of the process tolerances in the English system. There is no direct conversion from class of fit to process as there is in the IT system.

A design engineer must also be familiar with tolerances of location: position, concentricity, and symmetry; as well as tolerances of form: flatness, straightness, profile, roundness, cylindricity, parallelism, angularity, perpendicularity, and run out. See reference 7, Chapter 22, for a thorough discussion and description of these tolerances.

The surface texture, or surface finish of a part is a functionally critical requirement. Surface finish determines such things as surface-to-surface wear, and resistance to fatigue cracking. Surface finish values follow closely the size tolerance capabilities of processes; as surface finish improves (roughness decreasing), tolerances decrease (parts are more precisely manufactured); see Table 9.5.

TABLE 9.3

Reasonable tolerance ranges for industrial processes.

Processes	Tolerances, inches
Shearing	±(10% of material thickness)*
Arc welding	±(0.04 + 1% of dimension)
Bending dimension	±(0.04 + 10% of material thickness)
Bending angle—cold	±(3° – 5°)
Bending angle—hot	±(1°)
Flame cutting	±(10% of material thickness) From mean profile.
Sawing	
Circular	±0.01
Band	±0.03
Hack	±0.05

*This includes the out of squareness caused by the shearing action. The sheared edge is not flat in the direction of the shear blade motion.

Each manufacturing process has an inherent surface finish range, but this range depends on such things as the age and condition of the machine, the rigidity and condition of the fixture, the selection and condition of the cutting tools, and the skill of the operator.

Each process is also suited to make a given quantity of parts. The quantity is usually based on the cost of the machine and tooling. For example, die casting is a good way to control part tolerances, but if fewer than 30 pieces are needed, the tooling cost would make it an uneconomical choice. Many processes may be available, but processes have

TABLE 9.4

Reasonable tolerances for drilled holes.

	Drilled hole tolerances True position 0.020	
	Tolerance, inches	
Drill size, inches	Plus	Minus
0.0135 (No. 80) through 0.185 (No. 13)	0.003	0.002
0.1875 through 0.246 (Letter D)	0.004	0.002
0.250 (Letter E) through 0.750	0.005	0.002
0.7656 through 1.000	0.007	0.003
1.0156 through 2.000	0.010	0.004
2.0312 through 3.500	0.015	0.005

TABLE 9.5

Surface roughness produced by common production methods.

Process	Roughness average (R_a), micrometers μm (microinches μ in.)
	50 (2000) / 25 (1000) / 12.5 (500) / 6.3 (250) / 3.2 (125) / 1.6 (63) / 0.80 (32) / 0.40 (16) / 0.20 (8) / 0.10 (4) / 0.05 (2) / 0.025 (1) / 0.012 (0.5)
Flame cutting	
Snagging	
Sawing	
Planing, shaping	
Drilling	
Chemical milling	
Electric discharge mach	
Milling	
Broaching	
Reaming	
Electron beam	
Laser	
Electro—chemical	
Boring, turning	
Barrel finishing	
Electrolytic grinding	
Roller burnishing	
Grinding	
Honing	
Electro—polish	
Polishing	
Lapping	
Superfinishing	
Sand casting	
Hot rolling	
Forging	
Perm mold casting	
Investment casting	
Extruding	
Cold rolling, drawing	
Die casting	

Key: ■ Average application   ▨ Less frequent application

The ranges shown above are typical of the processes listed.
Higher or lower values may be obtained under special conditions.

ANSI/ASME B46.1-1985, Published by the American Society of Mechanical Engineers. Reprinted with permission.

size and power limits. Use care not to exceed these limits unless new facilities can be justified. Justification includes consideration for plant space, power requirements, and trained operators.

Plan ahead before starting a new process or expanding an existing one. Properly trained people may not be available, and special training sessions may be required. Providing special training costs money and takes time. Be sure your project schedule includes time for proper training, both for old and new personnel. The following examples illustrate some of the conditions that will limit process selection, (see also Exercise 9.18).

EXAMPLE 9.2:
Process selection limitations.

1. If a machine needed to introduce a new process takes 12 months to receive and install, and parts are required in 6 months, an alternate process must be found or the parts must be purchased. Sometimes an alternate process will be used for a short period of time until a desired process can be obtained and set up to work. The alternate process may be slower, more expensive, or produce less consistent tolerances. Close checks must be made of alternate temporary processes.

2. If your plant does not have chrome plating equipment, chrome plated parts should not be designed unless there is a reliable vendor who can supply the service. Be sure your production schedule includes the additional time and cost for shipping parts to an outside vendor and back to your plant.

3. If a bend die already exists to bend 0.1875 inch thick material to 75°, a new part should be designed to use this existing die if possible. New tooling costs money, ties up capital, requires lead time, and requires storage space when not in use.

4. If the largest press brake available is rated at 1000 tons capacity, and you design a part requiring 1500 tons, additional lead time to obtain a new press will be required. If only one new part will be run on the press, it probably would not be a wise decision to buy the new press. The best approach may be to redesign the part or plan to purchase it.

9.2 Material Forms

The beginning of any manufacturing process analysis requires a knowledge of available material forms and shapes. The form and shape of the purchased material determines the processes that can be used to begin fabrication of the part.

9.2.1 Bar Stock

One of the most common forms of material is bar stock, that includes any material which has a constant cross section. Bar stock comes in a variety of standard lengths such as 8,

10, 12, 14, 16, 18, or 20 feet. Common methods of fabricating materials into constant cross sectional shapes are rolling, drawing, extruding, and casting. Natural materials such as stone or wood are made into bar shapes by cutting material away from the naturally occurring shape.

Bars can be purchased in a large number of different cross sections. A common section form of bar stock is **rectangular**, either solid or tubular.

Common cross section dimensions, in inches, of rectangular metal bars range from 0.250×0.375 to 3.000×8.000.

Thicknesses generally available in inches are:

0.250 to 0.625, in increments of 0.0625

0.750 to 1.250, in increments of 0.125

1.500 to 3.000, in increments of 0.250

Widths generally available in inches are:

0.375 to 1.750, in increments of 0.125

2.000 to 4.500, in increments of 0.250

5.000 to 8.000, in increments of 0.500

All sizes are not available in all alloy combinations. Check the catalogs of steel suppliers for specific sizes available. Converting available inch sizes to metric by direct conversion does not result in sizes that are commercially available. Preferred metric sizes follow a pattern similar to the English system, as shown in Table 9.6.

Common available sizes of rectangular wood material are:

Nominal thickness (actual) in inches

1.00 (0.75), 2.00 (1.50), 4.00 (3.50), 6.00 (5.50)

Nominal widths (actual) in inches

2.00 (1.50), 4.00 (3.50), 6.00 (5.50), 8.00 (7.25), 10.00 (9.25), 12.00 (11.25)

TABLE 9.6

Preferred metric sizes in millimeters.

1	10	55	130	260	700
1.2	12	60	140	280	750
1.6	16	65	150	300	800
2	20	70	160	350	900
2.5	25	75	170	400	1000
3	30	80	180	450	1200
4	35	90	190	500	1500
5	40	100	200	550	2000
6	45	110	220	600	2500
8	50	120	240	650	3000

The difference between nominal and actual dimensions is the result of rough sawn lumber (nominal size) being planed smooth to actual size. *Note*: Actual dimensions have been reduced over the years. This causes problems when matching "new" construction to "old" construction.

Bars also come in standard **round** cross sectional shape. Round metal bars are usually available in these sizes, in inches.

0.1875 to 0.625, in increments of 0.0625

0.750 to 5.000, in increments of 0.125

5.250 to 10.000, in increments of 0.250

Availability depends on alloy content and rolling capacity of the supplier.

Various **special** section shapes of material are available because of a particular use. Special sections include angles, channels, plain, I beams, a variety of moldings, and concrete reinforcing rods.

The actual sizes of available material depends on accepted standards and suppliers' facilities. Companies that supply metal bars must invest money in tooling and equipment to provide different sizes and shapes. Only those shapes and sizes with a high demand or with a high selling price will be available.

There are many production manufacturing processes that can be used on bar stock.

Bar saws can be used to cut bar material to length with cuts perpendicular or at an angle to the length of the bar. With proper fixtures, both simple and compound angle cuts can be made. These cuts can be made on power hack saws, band saws, or circular saws. Sawing is a low cost operation based both on the cost of the machine tool and the low complexity of the sawing operation.

Reciprocating *power hacksaws* are not used for high speed production because they cut only on the forward stroke, but placed in groups under the control of one operator, they are an economical and reliable method to cut parts to length. The forward stroke can be cam controlled to allow cut time to be 60% or more of the total cycle time, but even so, many plants use the power hack saw only in maintenance departments, tool rooms, and repair departments. The cut marks of a hack saw are nearly parallel straight lines, and length tolerances of ±0.06 inch are typical.

Band saws, horizontal or vertical, with automatic hydraulic clamping and bar advancing between cut cycles, are an efficient bar cutoff operation because the blade is cutting 100 percent of the time. Except for compound angle cuts, bar fixturing is done with standard tooling that comes with the machine. The band saw cut marks are also parallel straight lines, and length tolerances from ±0.03 to ±0.06 inches are common.

Band saw blades generally have a smaller blade pitch (distance between successive teeth) than other saws and are better suited for cutting thin-walled cross sections such as angles, tees, moldings, and hollow pipe. Band saw blades are also generally thinner than other types of saw blades so less material is wasted in the saw kerf between parts. Figure 9.1 is a close up view of a horizontal blade band saw. The blade is nearly through the cut and the parts being cut are stacked in front of the blade, while the uncut bar is behind the blade.

A metal cutting *circular saw*, Figure 9.2, is similar to a mill slotting machine, with the bar held stationary in a vise and a cutter blade large enough in diameter to cut through the bar. To cut a 4.00 inch diameter bar, for example, requires a 10 inch diameter blade or larger (4.00 + 4.00 + 2.00 spindle). Cut lines are circular lines equal in radius to the blade radius. Because a circular saw blade is similar to a mill cutter, it is generally thicker than

FIGURE 9.1

Bosses being cut to length on a horizontal band saw.

a band saw blade and produces more waste, but is capable of cutting a flatter surface, and cutting within smaller length tolerances; ±0.01 to ±0.04 inches are common.

Parts that are sawed to length from a bar can be used as is because the cut surface is relatively flat and perpendicular to the bar length, such as a round boss to be welded to a flat plate. The saw operation generally leaves a burr, which should be filed or ground off before the part is used.

Shearing is the process of separating material without requiring any loss of material between parts. Shearing can be done hot, as in the case of a steel mill cutting billets, but it is commonly done with the metal at room temperature. Shearing can be done on almost any kind of material except brittle materials that shatter easily, like glass at room temperature.

Bar stock can be sheared to length with cuts perpendicular to, or at an angle to, the bar length on a *bar shear*. Compound angle shearing is not recommended because the shear blade will not strike perpendicular to the face of the bar and the resulting forces are difficult and expensive to compensate for, and excessive part distortion is likely. For efficient and non-part-damaging shearing, the bar being cut must withstand the force of shearing action without unwanted distortion. Non-rectangular bars can be satisfactorily sheared if the shear blade is designed to support the material, (see Figure 9.3). Tooling costs will be higher for this type of shearing because special blades must be made.

256   Manufacturing

FIGURE 9.2

Circular metal-cutting saw.

FIGURE 9.3

Circular bar shear cutoff.

FIGURE 9.4

Sections that are not practical to shear to length.

Some cross sectional shapes, like the ones shown in Figure 9.4, do not lend themselves to shearing because it is difficult, costly, or impractical to support the bar to avoid distortion during shearing.

Shear cutting is not limited to straight cuts. Curved and contoured blades can be made for shearing bars into lengths with curved ends (see Figure 9.5). The blades for non-straight cuts must be special made for the desired radius, and add tooling cost to the manufacture of the part. Curved shearing may, however, eliminate the need for subsequent machining operations. Shear cutting is a low cost operation because the cutting cycle is rapid with production rates measured in strokes per minute.

Shearing length tolerances of 10 percent of material thickness are common and include the non-squareness of the sheared edge in the direction of the shearing action. The clearance between punch and die results in an edge that is not perpendicular to the face of the material where the shear blade contacts the material or the punch enters the material. The smaller the clearance, the squarer the cut. The sheared end of a bar is not flat because the shearing action cuts about half way though the thickness of the material, and fractures (tears) the second half. For this reason a sheared edge will not match a flat surface, and gaps will be formed. The gaps may be great enough to cause problems with welding, because parts may not stay square with each other without special fixturing.

Bar stock can also be **flame cut** to length at any contour desired because the path of the cutting torch is controlled from an external positive control (see Figure 9.6).

FIGURE 9.5

Layout for shearing non-straight cuts.

Rectangular cross sections are best suited for flame cutting because the torch tip should be a constant distance from the surface of the material during the cut, and the tip opening on the torch is a function of material thickness. Flame cutting is a preferred process on low carbon steel, and is a low cost operation because the machines and tooling are not complex. The bars can be located and clamped with standard reusable tooling and the control template can be a line drawing for an electric eye to follow, a metal template for a stylus follower, or a computer program. The efficiency of flame cutting is enhanced by having several torches cutting at the same time, either at the same time on one bar or at the same time on several bars. Flame cut tolerances range from ±0.06 to ±0.12 inches from mean profile, and depend on the material being cut, the material thickness, and the rigidity and accuracy of the setup.

Laser cutting is similar to flame cutting because the path of cutting is controlled in a similar manner, but the cutting action of a laser is different from that of a flame cutter. A laser is a coherent beam of photons, focused to be parallel, and of the same wavelength. The laser supplies energy to vaporize the material. The basic cutting action of a flame cutter is to heat the metal to the melting temperature of the metal and blow-burn through the thickness using heat and oxygen under pressure. Because of these differences, the speed of laser cutting is higher than that of flame cutting, but is limited to material less than 0.500 inches thick because of power requirements. Cutting tolerances are good, ±0.01 to ±0.03 inches from mean profile being common. Other advantages of laser cutting include:

1. Narrow heat affected zone, 0.007–0.010 inches wide
2. High linear cutting speeds, 40-360 inches/minute
3. Ability to cut many different materials such as metal, plastic, and wood
4. Quiet
5. No gas storage required

Lasers have other industrial applications that can be combined with the profile cutting operation, or done separately. Some common applications include:

1. "Drilling" holes, 0.002–0.050 inches in diameter
2. Heat treating of steels, selective and localized
3. Marking and labeling, replacing etching and stenciling
4. Welding, no filler material needed

Turning operations on lathes can also be used as a first operation on bar material. Bar lathes come as single or multiple spindle machines, and can turn the outside of the part with tools mounted on the cross slide, and the inside of the part with tools mounted on a turret. A part as simple as a round boss could be cut off on a lathe, but this would not be an efficient use of a lathe. A lathe can do other operations such as turning, chamfering, drilling, boring, threading, taping, grooving, and knurling. Because lathes cost more than a saw they should be used for more complicated parts if the work is available. If a saw breaks down and needs repair, bosses could be cut off on the lathe as a stopgap measure. In general a lathe will be used for a first operation when the part is made from a round bar and has some machining required that can be performed prior to cutoff, as shown in Figures 9.28 and 9.30. Many complicated parts with concentric features can be completely machined before cutting off from the bar. On special lathes, using special

FIGURE 9.6

Four torch bar flame cutting operation.

tooling accessories, cross holes and milled flats can also be added to a part before it is cut off from the bar. Tolerances depend on the material being turned, the condition of the machine, the sharpness and appropriateness of the tooling, and operator setup. Common cutoff length tolerances range from ±0.010 to ±0.030 inches. Lathe operations are medium cost; lathes cost more than saws, shears, and flame cutters because of the higher precision required. The machine setup is more complicated than that of saws and flame cutters, and cycle times are measured in minutes per piece.

9.2.2 Plates And Sheets

A second common form of available material is plates and sheets. Plates and sheets have constant thickness and can be purchased in a variety of lengths and widths. Differences between bars, and plates and sheets include:

1. Plates and sheets are wider than bars, generally ranging in width from 2 to 10 feet.
2. Sheets are available in thicknesses not available in bars. Gage thicknesses of sheets include sizes from #6 (0.1943) to #30 (0.0120) inches. Standard metal plate thicknesses are:

    0.1875 to 0.5626 inches, in increments of 0.0625

    0.625 to 1.750 inches, in increments of 0.125

    2.000 to 5.000 inches, in increments of 0.250

3. A catalog of available sheets and plates will list widths up to 8 feet and lengths up to 20 feet.
4. Thickness tolerances for sheets and plates are generally smaller than those for bars because of the requirements of the user and the control of the rolling process

Manufacturing

used to make them. Sheets, in particular, need smaller thickness tolerances because common uses include forming and drawing, where a large tolerance would prevent the use of precision dies.

5. Many high strength alloys are available in bar sizes but not in plates or sheets.
6. There are special configurations that are available as plates such as expanded metal for grating, and floor plate.
7. Wood sheets (plywood) are laminated veneer glued in alternating directions. This gives multidirectional strength qualities and sizes that are not available in standard dimensional lumber. Common plywood sizes in inches are:

Thickness: $\frac{1}{4}, \frac{5}{16}, \frac{3}{8}, \frac{1}{2}, \frac{5}{8}, \frac{3}{4}, \frac{7}{8}, 1, 1\frac{1}{8},$ and $1\frac{1}{4}$.

Width and Length: 24×96, 48×96, and 48×120.

First operations for plates and sheets are similar to those for bars except that the machines used are larger and have higher power ratings.

Sawing is not generally used to cut metal plates and sheets for production because it is slow and relatively inaccurate. Sawing is a common cutting process for wood and plastic sheets.

Shearing of metal plates and sheets on gate shears is a common process. A *gate shear* operates like a bar shear except that blades up to 16 feet are common, but only straight blades are used.

FIGURE 9.7

Six torch plate cutting operation.

Flame cutting on plates, Figure 9.7, is used for cutting parts whose profile is not a series of straight cuts. Parts can be nested to increase production and reduce scrap, and multiple torch machines reduce the production time. Heat distortion is a problem, and thinner parts must be handled carefully. Special clamping and water spray cooling are methods used to avoid excess part distortion.

Laser cutting can be performed on thin sheet material in a manner similar to flame cutting, but because of power requirements, parts are usually cut one at a time. There is less heat to dissipate in laser cutting and speed and accuracy are good.

Stamping and blanking operations are used to make parts from metal sheets. When sheets are purchased in the proper size, the first operation may be the stamping of the part complete in one or more passes of a press. Figure 9.8 shows an example of two

FIGURE 9.8

Hydraulic press forming.

FIGURE 9.9

Parts made on progressive dies.

large hydraulic presses in tandem used to form bends in metal plates. Components such as car hoods, fenders, and appliance cabinets fall into this category. Smaller parts can also be stamped from sheets using progressive dies with multiple positions that allow complex parts to be made. Figure 9.9 illustrates parts made by stamping. The cost of blanking machines and blanking dies is high, so blanking is not economical for low volume requirements.

Nibbling is a process used to make parts from sheets when production quantities are low, and the cost of blanking dies cannot be justified. Nibbling is a process that uses a reciprocating punch, the shape of which is arbitrary. Material is fed past the punch and each stroke removes some material on a predetermined path. Profiles and cutouts are standard operations for nibbling. A sharp scalloped edge left by the punching action of round punches is a disadvantage of nibbling. Grinding or filing of the scallop burr may be required for safety reasons. Rectangular punches used for straight line cuts eliminate some of this problem. Because punching places a compressive stress on the punch side of the metal, close multiple punching can result in material that curves up. Roll straightening is then required.

9.2.3 Coil Stock

Coil stock is an alternate material form for constant thickness material. Coiled material in sheet thicknesses is used to feed stamping, blanking, and other rapid action die processes. This allows longer runs to be made without stopping to feed new material into the press.

Coil stock can also be purchased in plate thickness for use on a **cut-to-length line**. A cut-to-length line permits the roll flattening and square cutting of material from a coil. The main advantage of such a process is that many different length plates can be cut from one width-thickness sized roll. The machine investment is high, but the process avoids stocking many different length plates of the same thickness. It also reduces scrap on any part not made from a standard length plate. Figure 9.10 shows machinery used to process coil stock.

FIGURE 9.10

Processing coil material.

Other materials are also available in coils, such as paper for continuous printing, cellophane tape, and wire, rope, and thread. Large diameter wire, for example, is used in machines to make nails, bolts, and nuts. Figure 9.11 shows how coiled material is fed into a manufacturing operation. Coil stock also feeds the progressive dies used to make the parts shown in Figure 9.9.

9.2.4 Casting

Casting is a common method of starting the manufacture of a part that takes on its first usable shape and chemistry during the casting process. Casting has the advantage of placing material only where it is needed, and irregularly shaped objects can be cast to keep subsequent material removal to a minimum. Machining a casting is required only when functional tolerances are required that are less than can be obtained by the casting process. Some parts are used as cast, with no machining required, such as wheel weights, boat anchors, and handles for cabinets and faucets.

Disadvantages of casting include the relatively slow process cycle of making a pattern, preparing a mold, melting the material, pouring the casting, waiting for material to solidify, and dumping the mold and removing the gates, risers, and flashing. Modern automation techniques, however, have speeded up some of these steps, and in the case of

FIGURE 9.11

Coil material feeding a press operation.

injection molding of nonferrous metals, casting is an efficient, accurate, high speed production operation. Another disadvantage of casting is the cost of tooling including the cost of a part pattern, core patterns, match plates, or metal dies. These are special costs for each part being made. General tooling costs are also required for additional items such as flasks, furnaces, sand, and resin binders to make cores.

There are alternative methods of casting materials because there are many factors influencing the casting process. Some of these factors are:

1. Required surface finish
2. Required size tolerance
3. Internal cavity complexity
4. Minimum wall thickness
5. Melting temperature of the material being cast
6. Production quantity required
7. Available processes
8. Part weight

Sand mold casting is a common casting method usable on all metals commercially cast. The correct mixture of sand and resin binders will withstand molten metal temperatures, but still remove easily from the casting when the metal has cooled and

TABLE 9.7

Relative cost of sand casting various irons.

Type of Iron	Relative cost
Gray Iron	Lowest cost
White Iron	add 20% to gray iron cost
Malleable Iron	add 50% to gray iron cost
Nodular Iron	add 80% to gray iron cost
Cast steel, carbon only	double gray iron cost
Cast steel, other alloys	up to five times gray iron cost

solidified. Internal cavities can be obtained by using cores. The cost of the patterns range from low cost wood for low quantity runs, to higher cost metal for high production runs. Of all the casting processes, sand casting results in the largest part size tolerances; therefore, many sand castings have machining operations performed. Large shapes can be cast in pit molds so size and weight are not a limitation. Variations in relative cost using sand casting exist because of material differences. Table 9.7 shows such a comparison. Costs can be expected to vary depending on the size, shape, quality, vendor, and market conditions, but sand casting is the least expensive casting process. If there is only one available vendor to make castings, the price is apt to be higher than if there are several vendors to choose from.

Die casting is a casting process used on materials with lower melting temperatures such as aluminum, copper, magnesium, zinc, lead, and tin. The molten metal is forced into a steel die under pressure to insure the mold is filled, and to shorten the cycle time. Pressure applied during solidification time of any casting process reduces porosity and improves part fatigue resistance. The molds in die casting are high strength heat resistant steel. This accounts for die cast parts having small tolerances and good surface quality. The metal dies, however, limit the size of parts that can be cast because half the die is moved to open and close the die cavity during the casting cycle. The dies are jacketed for water cooling, which also limits their size. Cores remain intact when removed, which means they must get smaller as they get further into the mold. Die design is complex and costly. Plastics and glasses can also be cast with this process; it is often referred to as injection molding for those materials.

Permanent mold casting also uses metal molds but the molten metal is poured in rather than forced in under pressure. Metals commonly cast using this process are iron, bronze, magnesium, aluminum, zinc, and lead.

Centrifugal casting is done by pouring molten metal into a rotating mold. Only the outside shape of the part is controlled by the mold. The inside size of the part is controlled by the volume of metal poured in. The process is best suited for symmetric hollow parts such as pipes or wheels.

Plaster mold casting uses disposable plaster molds and is preferred for small parts. The smoothness of the plaster results in improved surface finish, and improved tolerance control. Only materials with lower melting temperatures such as aluminum, copper, magnesium, and zinc are cast with this process because of the degradation temperature of plaster.

TABLE 9.8

Surface finish ranges for casting processes.

Ceramic	10–50 μin
Investment	10–85 μin
Plaster	30–50 μin
Die (injection)	40–100 μin
Permanent	100–250 μin
Centrifugal	100–250 μin
Shell	200–800 μin
Sand	250–1000 μin

Investment casting, often called lost wax casting, uses a pattern of wax around which the mold material is formed. The wax is then melted and poured out and the desired metal poured in. Intricate and clustered parts can be efficiently cast this way. Good surface quality and dimensional control are also advantages of this process. Size control is enhanced because the pattern does not require draft allowance for removing the pattern. This process is used for metal or plastic parts.

Shell mold casting uses a mold prepared with sand mixed with plastic resin, which is poured in halves as a thin coating on a master metal pattern. The cured coating halves are then clamped together and liquid metal is poured in. This process results in improved dimensional control and surface quality, but at an increase in cost, and with limits on the size of the part that can be cast.

Ceramic mold casting is similar to shell mold casting except that ceramic material replaces the sand-plastic mold material. Because ceramics are brittle, sand backup is generally used for support. Ceramic molds improve dimensional and surface quality control, but at increased cost and limited part size.

When casting is being considered as the preferred process, consider the range of surface finishes, Table 9.8, the size limit guidelines, Table 9.9, the tolerance range, Table 9.10, and production rates and tooling cost, Table 9.11.

TABLE 9.9

Reasonable weight limits for casting processes.

Ceramic	20 lbs
Permanent	50 lbs
Die (injection)	75 lbs with aluminum, best under 15 lbs
Investment	100 lbs
Shell	100 lbs
Plaster	300 lbs
Centrifugal	300 lbs
Sand	Largest practical, total weight in tons

TABLE 9.10

Normal tolerance ranges for casting processes. All dimensions in inches.

	Part Feature Dimension			
	0–4	4–8	8–16	16–24
Sand	±0.03	±0.04	±0.06	±0.08
Permanent	±0.02	±0.03	±0.04	±0.05
Die and ceramic	±0.002	±0.003	±0.005	±0.010
Shell	±0.003 per inch of dimension			
Investment	±0.005 per inch of dimension			
Centrifugal	±(0.015 + 0.002 per inch of dimension)			
Plaster	±(0.005 + 0.010 per inch of dimension)			

9.3   Secondary Operations

Many parts require more than one manufacturing operation, and many of the processes listed as primary operations can also be used for secondary operations. A flat part, for example, may have several shearing operations performed in sequence and then drilled. A part from a lathe may need a second lathe operation to finish all the required surfaces. There is nothing special about the division between primary and secondary operations.

9.3.1 Forging

Forging is the process of changing the shape of material by applying pressure to a material blank. The pressure can be supplied by a mechanical press, including drop hammer, hydraulic press, or by hand. Metal dies in the machine force the metal into the desired shape, and only machining for close tolerances is required. Parts that are forged resemble parts that have been cast because a parting line (line between die halves) shows. Forging generally begins with bar material and is used to make parts such as wrenches, connecting rods, crankshafts, and gear clusters. Forging can be performed

TABLE 9.11

Comparison of casting processes to make a 3-to 5-pound part.

Process	Production Parts/hour	Relative Tooling cost
Sand casting	10–15	1
Centrifugal casting	30–50	2
Permanent mold casting	40–60	6
Die casting	120–150	15

while the material is hot, above its recrystallization temperature, or cold, below its recrystallization temperature. Hot forging does not raise the mechanical strength properties of metals because hot forging involves very little plastic deformation. Cold working does raise the property values because cold forging does involve plastic deformation.

Characteristics of hot forging include:

1. No change in the hardness or ductility of the material, subsequent machining is no problem.
2. The material is made tougher because the material grains are made smaller, which is good for high stress applications.
3. Grain flow is made directional and parts will have directional strength properties, which helps reduce the effect of geometric stress concentration locations.
4. Less force is required to form the part into the desired shape, so smaller machines can be used.
5. More severe forming can be done without rupturing the material and complex shapes can be formed.
6. Time is required to heat the material to proper forging temperature, and so the process cycle time increases.
7. Material is not cold worked to higher strength.
8. Cooling time is required before subsequent handling or machining operations can be performed (a good place to use robots).
9. Special heat resistant tooling is required and, even so, tooling deterioration occurs more rapidly than with cold forming.
10. Metals, particularly steel, form scale during the process and the parts may need post-forging cleaning before they can be used.

Only ductile materials can be cold forged because part shaping takes place by plastic deformation. Characteristics of cold forging include:

1. The material does not have to be heated, so the process is faster than hot forging.
2. The yield strength of the material is increased.
3. The amount of forging that can be done is limited because excess deformation will rupture the material.
4. More force is required to obtain a desired shape.

There are many methods of forging, ranging from a hammer-swinging blacksmith to a 10,000 ton hydraulic forge.

Rolling is a forging process in which material is forced between rollers to change its form. The rollers can be smooth and the part merely reduced in thickness to create standard bar stock, or the rollers can have a shape and the rolled material will take the shape of the rollers to create I beams, channel sections, angle iron bars, or T shapes. Grain flow in the final form will be oriented in the direction of rolling and so directional strength properties exist.

Drawing is the process of pulling material through a die. The die forces the material into the shape desired. Wire is a common product manufactured by drawing.

Press forging includes all operations in which the material is placed between forming dies. For centuries the blacksmith used hammers and manual skill to form hot metals to shape parts such as wagon wheel rims, horseshoes, and swords. Dies used today can be low cost flat blocks of metal used for flattening distorted parts, or they can be quite complicated and costly, such as a four-position die for forging connecting rods.

Modern production press forging uses dies that have the shape of the final part machined into them. The dies are mounted in a press and the material is placed between the dies. As pressure is applied from the machine, the part takes the shape of the die. In many cases several passes of the press are required to complete the part to shape. Press forging can be done hot or cold.

Tolerances for forging depend on many factors, including the volume of material that is displaced, the temperature of the material, the quality and precision of the dies, the force available from the press used, material composition variation, and operator setup. Acceptable tolerances range from ±0.03 inches for small parts and good tooling up to ±0.25 inches for large parts and average tooling. Precision forging at tolerances of ±0.010 inches can be obtained with tooling in first class condition and using multiple pass operations. For this level of quality expect the cost of tooling and labor to increase, perhaps as much as double.

Extrusion of material is done by forcing the material through dies. This is similar to drawing except that pressure forces material out of the die instead of pulling it from the die. The dies control the shape of the final "part." The "part" is usually a bar, which is later cut to length. Extrusion is generally done hot to reduce the force required and to allow more complex shapes to be made, but cold extrusion can also be done and it improves surface finish and controls tolerance more closely. Chalk ledges for blackboards, and aluminum moldings for windows and doors are examples of parts generally extruded to shape. Unless the parts are subsequently polished, there will be parallel die marks running the length of the extrusion.

Bend allowance is required whenever angle bending of flat metal is done between a **V** punch and a **V** or **U** die. When bends are made in sheet or plate metal it is necessary for a designer to be able to determine how large to make a part blank prior to bending.

FIGURE 9.12

Typical air bending operation.

FIGURE 9.13

Typical matched punch and die bending operation.

The forming of a bend in a part can be done with an air bend die as shown in Figure 9.12, or the part can be made in a die set made just for the part, as shown in Figure 9.13. In either case the material size and the tonnage required to make the bend are dependent on more than just the material thickness and its strength. Variable factors include the condition of the die surfaces, the coefficient of friction between the material and the die, the use of a lubricant, and allowable springback.

It is well known that the inside material in a bend is in compression and the outside material is in tension during bending, and that a neutral axis (neither in compression nor in tension) exists. The neutral axis on a part stays the same length during the forming process, and an estimate of the length of the neutral axis is needed to determine the size of the unformed blank. But it is not known exactly where the neutral axis is, because during elastic deformation, the neutral axis is between the area under tensile stress and the area of compression stress. During plastic deformation the neutral axis can move from the center of the cross section. Similar to many engineering concepts, the following method is a guideline to be used for analysis. Adjustments will be necessary as the variable factors such as material hardness, surface friction, lubrication, and die wear change.

EXAMPLE 9.3:
Prebend length calculation.

How large does the flat rectangular blank need to be to make the part shown in Figure 9.14?

Refer to Figure 9.15 and note that the length of material needed *in the bend* is:

$$L = 2\pi(r + n)(\theta/360), \text{ where:} \tag{9.1}$$

n = the distance from the inside fiber to the neutral axis,

r = the bend radius, and

θ = the angle of bend, degrees.

FIGURE 9.14

Typical 90° bent part.

L, r, and n must all have the same units. Table 9.12 gives the values of n for various ranges of r, and is based on experience; it was not developed from theoretical concepts. Your experience with actual dies, parts, and machines will let you modify and further refine this data.

Based on Equation 9.1 and Table 9.12, the length of the part in Figure 9.14, before bending, is:

$$2[4.00 - (0.50 + 0.75)] + 2\pi[0.75 + (0.34)(0.50)][90/360] = 6.95 \text{ inches},$$

and the blank size is 6.00 × 6.95 inches. If possible the grain flow direction should be parallel to the 6.95 dimension to reduce the possibility of developing cracks during bending. Some reduction in part thickness may occur because of the applied force and the resulting tension and compression stresses.

FIGURE 9.15

Location of neutral axis in a bent part for cold forming. Refer to Table 9.12.

TABLE 9.12

Location of neutral axis based on bend radius. Refer to Figure 9.15.

Range of radius	Value of n
$r \leq t$	$0.33t$
$t < r \leq 2t$	$0.34t$
$2t < r \leq 3t$	$0.38t$
$3t < r \leq 4t$	$0.42t$
$4t < r \leq 5t$	$0.46t$
$5t < r$	$0.50t$

Shearing is a forging process previously discussed, but it has many sub-processes that have been given special names.

Shear cutoff is an operation to make a specific length part from a bar or plate. The cut can be perpendicular or at an angle to the edge of the bar.

Parting is similar to shear cutoff, except that a scrap width between parts exists. The scrap allows the end of the part cut off to be nonparallel to the end of the next part on the bar.

Blanking cuts the complete profile of a part. Blanking is used for irregularly shaped parts and requires special dies. *Fine blanking* is the highest precision shearing available and is used for thin gears and other parts made from sheet metal that require small tolerances. It requires special dies and machines, and can only be justified when part requirements are high, and when subsequent machining steps can be eliminated. Before high cost tooling and machines can be justified, a part must have a fixed design so changes will not be made soon after a die is built. The quantity of parts required must also be high enough to amortize the tooling cost.

Notching changes the profile of a part by removing portions of the material. Notch dies can have any desired shape, square, rectangular, or curved.

Punching removes shapes from the interior of a part. Punches can be round or any other shape depending on the desired profile required.

Slitting and lancing make cuts in the material, but do not remove any material. This process is used in the manufacture of products such as louvered grilles.

Nibbling reciprocates a punch that removes material with each stroke of the machine and is done to avoid the high cost of special punches and dies. The path of the reciprocating punch is controlled to produce a desired external or internal shape.

Shaving is a second pass at a cut shape to remove only a small amount of material. It is used to control tolerances and surface finish.

Drawing and stretching are processes used to form thin material to the shape of a punch. There are also subcategories of these processes.

Stamping is a process for which a high cost punch and die set are designed and built just for a specific part such as those shown in Figure 9.9. It can be justified when the design of a part is stable and larger quantities are required, such as home appliance or automotive parts.

Drawing uses a punch that moves with the machine ram and forces against the material, which is held peripherally with a hold-down pad. As the punch pushes against the material, it deforms the metal through a clearance depression in the lower die. The blank pulls away from the hold-down and conforms to the shape of the punch. This is another case where a ductile material is required. Parts made with this process include shell casings and metal cups for covering exposed bearings.

Stretch forming uses a stationary punch, and the material blank is pulled down over the punch.

There are many variations in the basic drawing and stretch forming operations, but the objective is to form material to a shape that is preset in the form of a punch. Variations include:

1. Using a backup rubber pad rather than a metal die.
2. Using a fluid backup such as oil.
3. Using an explosive charge for the force rather than a press.
4. Using a vacuum to draw the material to the punch. This method is used extensively in plastic molding.
5. Using fluid pressure and a die.
6. Spinning. This process uses a lathe type machine and a method of forcing the metal to the punch as the punch and metal blank are spinning.

Powder metallurgy is a process of making parts by compressing metal powder into a die. Characteristics of the process include the variety of materials that can be used, including the high melting temperature materials like tungsten (6150°F), molten reactive materials like zirconium, and the materials that cannot otherwise be combined, like copper and graphite for motor brushes. Good surface quality and dimensional control are maintained, and subsequent machining is rarely required. Special high pressure machines and dies are required, and this limits the part size. Powder metallurgical parts are porous enough to be packed with a lubricant and can be used as self-lubricating bearings.

Special forging methods are available for specific types of parts and include the following processes: drop forging, swaging, sizing, upset forging, coining, hobbing, roll forging, ironing, riveting, staking, stitching, draw bending, roll bending, compression bending, roll extrusion, and cold heading.

9.3.2 Machining Operations

Many parts are designed so they cannot be manufactured completely by just shearing, forming, or casting. Intricate shapes and some tolerances require machining operations after the part's basic shape is created. The operations may make a hole round, create a low tolerance, or control the surface finish. A cast surface may need to be flat to match another part or seal against a gasket. Diameters on hydraulic or engine components need to be held to close tolerances, and grooves for O rings and seals must be concentric with outside and inside diameters. This section reviews the basic commonly used machining operations required when "first" operations are inadequate.

Lathes have already been discussed as a first operation on bars. Turning on a lathe may be the only major operation required if a part has concentric features, and a deburr and/or wash operation may finish the part.

Lathes can also be used for secondary machining operations if the part blank is held between centers or in a chuck. Occasionally long parts, such as axles or motor shafts, will require an end center support and a movable backup support to counteract tool pressure. The cutting operations possible are the same as for first operations.

A part held between centers cannot be completely machined in one operation if there are internal features to be machined. A part held in a chuck cannot be completely machined on the outside diameter in one operation, but machining can be performed on the exposed end and on the inside diameter.

Lathes come in many common configurations.

1. Single spindle, combination bar, collet, chuck, or face plate; chucker; or bars held between centers with cross slide tool feed.
2. Single spindle, bar collet or chucker, with horizontal turret and multiple cross slide tool feed.
3. Multiple spindle, bar collet or chucker, with vertical turret and multiple cross slide tool feed.
4. Single spindle, chucker or face plate, with vertical turret and multiple cross slide tool feed.
5. Any of the above under numerical or direct computer control.

When parts get too large to be safely handled in a horizontal position, such as large axle housings, parts are machined on vertical lathes. Large diameter tables (4, 6, 8 feet) hold special fixtures for supporting and locating parts. No between-center work is done on vertical lathes. The variety of machining operations done vertically are the same as those done horizontally.

Milling machines are used to remove metal and make surfaces flat, parallel, or perpendicular. One or more surfaces of a part can be machined in one operation if multiple position fixturing is provided. The machine and operation cost for milling is similar to that for lathes. Milling is a versatile machining process because part fixtures and cutters have great variety, and cutters can be made to create any number of shapes.

A *vertical milling machine* has a rotating spindle like a drill press but has a table that can feed a part past the rotating tool left and right, in and out, and up and down. The spindle must have heavy duty bearings to support the multi-directional forces that occur. A vertical mill generally uses an end mill tool bit to face, slot, contour, or notch a part held in a fixture, and the parts are generally small and can be hand loaded.

A *horizontal milling machine* can do the same job as a vertical mill, but it is better suited for larger parts, which may require hoist loading. With an outboard arbor support larger cutters can be used and heavier cuts made. The supported arbor can also accommodate more than one cutter. Two cutters set at some fixed spacing can straddle mill the opposite parallel faces on a part in one operation. Two cutters of different diameters can mill perpendicular surfaces in one operation. Figure 9.16 shows the end of a boom section being milled on a horizontal milling machine. Note the improvised safety shield above the spindle to prevent chips from flying about.

Boring is a process that can be done on several different machines: lathes, drill presses, regular mills, boring mills, and boring machines. Choice is based on the part size and configuration, other operations being performed, surface finish required, and tolerances required.

A *boring machine* is a specially built, high precision machine used primarily to finish the bore diameter of a part to a close tolerance. Boring machines can have a vertical or horizontal spindle, and often have more than one spindle. Additional tools can

FIGURE 9.16

Horizontal milling operation.

be added to face, chamfer, and groove, but these operations are of secondary importance. The machines are not large, and typical parts run on a boring machine would be pump housings, small gear cases, and cylinder sleeves, but machine, tooling, and operation cost is high because of the small tolerances required. Figure 9.17 shows a part setup ready to have a hole bored to size.

As parts get larger, and other operations such as milling, drilling, and tapping are also required, parts can be mounted on an index table on a *horizontal boring and milling machine*. The spindle head holds a rotating tool and is able to feed the spindle in and out (the x-axis), as well as up and down (the y-axis). The part is held on a table that can feed left to right (the z-axis) and can rotate (fourth axis). If the table will adjust to an angle with the horizontal, a fifth axis of versatility is available. Any number of tools can be used during any one operation so a part can be completely machined on four or five sides. Any rotating tool machining operation can be performed. Figure 9.18 shows a part on the table of a horizontal boring and milling machine after machining. The part is mounted on an index table so all four sides can be machined in one setup. A boring tool is shown in the spindle.

If the machine is equipped with a tool changer and is computer controlled, it can be used efficiently to run a series of different parts. A sophisticated machine of this type is called a *machining center* because of its great versatility. Figure 9.19 shows a gear-case being machined on a horizontal machining center. The insert shows the gear-case cover that is machined using the same holding fixture. The machine cost is high and can only be justified if part machining is complex and required tolerances low.

FIGURE 9.17

Part set up for hole boring on a boring machine.

Flexible Manufacturing Systems (FMS) are special arrangements of machines, handling equipment, and checking equipment under the control of a computer to handle and machine a variety of parts. The parts can come to the system in a random order and are identified by spindle probes on the machine that allow the computer to call up appropriate machining sequences from the computer, which in turn loads special tooling from the tool changer. FMS are versatile and can include vertical and/or horizontal machining centers, lathes, drills, and other machines as well as automatic part loading and unloading (often done with robots) and inspection equipment.

A **planer**, in metal working, is a machine that holds a single-tipped tool cutter stationary, and cuts the work piece as it reciprocates past the cutter. With each cycle of the machine the work piece is advanced perpendicularly to the stroke direction of the machine. Planing is a slow process and is not used for high speed production. It is,

FIGURE 9.18

Casting machined on a horizontal boring and milling machine.

however, a good process for machining long straight parts such as press-brake dies and machine tool bed ways, as illustrated in Figure 9.20.

Planing in woodworking is a process used to reduce the size and/or improve the finish of cut wood, using a rotating cutter.

A **shaper** is a machine that uses a reciprocating single tip tool while the part is held stationary, the reverse of a planer. This restricts the stroke length and limits the size of the part that can be machined. Shaping is not a high production process.

A **drill press** is a machine that holds a rotating tool, generally in the vertical position, and has a mechanism for controlling the advance of the rotating tool into a part held in location on a table. In the simplest case the operation is the drilling of one hole through a part in one location. The only tooling required is a simple holding fixture with one drill bushing.

FIGURE 9.19

Gearcase machining on a horizontal machining center.

In a different situation, with a positionable table (generally numerically controlled), a tool changer or a quick change chuck, and a variety of tools, more complicated parts can be machined. The operations possible include drilling through the part, drilling to a fixed depth, reaming a pre-drilled or cored hole, spot-facing, counterboring, countersinking, trepanning, boring, tapping, and burnishing. With a multi-position or indexable fixture, a part can be machined on more than one side in one operation.

There are a variety of drill presses available; each one is individually suited for a special application.

1. Single spindle, stationary table.
2. Several single spindle machines and a common table. This setup allows fixtures to be easily moved from spindle to spindle or multiple fixtures for the same or different parts, one under each spindle. In this way a sequence of operations on the same part can be done efficiently, or multiple parts can be machined by one operator.
3. Single spindle, positionable table, for parts with more than one hole location.
4. Radial arm drill, used to locate the spindle in various positions on a stationary part that is too large or heavy to move around.
5. Single spindle turret drill, table type optional. This type of drill press allows sev-

FIGURE 9.20

Planing a lathe bed-casting.

eral tools to be mounted and used in any sequence. The machine is similar to a turret lathe except that the turret is vertical, the tool rotates, and the part is stationary.

6. Multiple spindle. This type of drill press is used to drill a fixed pattern of holes all at the same time. The spacing between holes is part of the original setup but can be changed from part to part. This is a high production process and used for parts like engine blocks. This type of machine is often included in automatic high production transfer lines.

Broaching is a machining process using a straight tool that has a series of cutting teeth on it. The cutter is pushed or pulled through, or over the surface of a part. The series of cutting teeth cut the desired shape in the part progressively, each tooth removing a small amount of material. For surface broaching, a tool backup support is required. For internal broaching, a pre-broach hole is required, and the broach supports itself against the part as it is pulled or pushed through.

Broach cutters are expensive but are well suited for internal keyways and splines and other hard-to-machine shapes. Production cost is medium if requirements are high enough so cutter cost can be allocated over many parts.

Grinding is a process of paradox. On one hand, for the removal of scale, rust,

Manufacturing

casting gates and risers, or saw cutoff burrs, it is a rough, no tolerance, freehand operation. On the other hand, grinding is a process used to remove small amounts of material from a part while controlling size, with tolerances to ±0.0001, and surface finish to 20 µin. In either case the cutting tool is a rotating grinding wheel made up of fine abrasive particles held together with a binder, available in various shapes and grades.

A *surface grinder* has a table that supports the part under a rotating grinding wheel. The grinding wheel can be rotated on a vertical or horizontal axis. This process is used to produce smooth flat surfaces. The table is usually magnetic so no holding fixture is required for simple parts. For parts without parallel surfaces minimum fixturing is required.

A *cylindrical grinder*, Figure 9.21, is used to make parts round. Cylindrical grinding can be internal, where a part is held on the outside perimeter and rotated while

FIGURE 9.21

Cylindrical grinding operation.

the rotating grinding wheel cuts on the circumference of a hole. Cylindrical grinding can also be external, where the part is held between centers and the rotating grinding wheel passes over the outside diameter as the part is being rotated. In this case the part needs to have center holes drilled on the axis of the ground diameter. Different diameters on the outside surface can be held concentric with this process.

A *centerless grinder* also makes parts round. Centerless grinding is similar to external cylindrical grinding except the parts being ground are fed continually through the machine. A rotating pressure wheel forces the part against a rotating grinding wheel and feeds the part through the machine. The rotating pressure wheel causes the part to rotate as it passes by the grinding wheel. Only parts being ground on the largest outside diameter can be ground this way, and the outside diameter must be continuous enough to support the part as it is fed through the machine.

Grinding, like all processes, gets more expensive as tolerances are reduced and low surface finishes are required. Grinding is not only an important production process, it is also the method used to sharpen the cutting tools used in other manufacturing processes, from the tips of drills to the edges of the blades used in the shearing process.

9.3.3 Other Finishing Processes

Honing is a finishing process used to control size and surface finish on either internal or external dimensions. It is accomplished by using fine particle honing stones held in floating holders, and by oscillating them over a surface or oscillating and rotating them in a hole. Very little material is usually removed (0 to 0.010 inches). Small tolerances (±0.0005 inches) and good surface finishes (8 to 10 μin) are the main objectives. Honing is used on parts such as cylinder bores, bearings, and ring gages.

Lapping is a process very similar to honing except it is generally done on the exterior of a part. The amount of material removed is generally less than 0.004 inch. Surface finishes to 2 μin, and tolerances to ±0.00005 inches are possible. Lapping is a common process for a last step while making gages, master gage blocks, and high precision gears.

Superfinishing is very similar to lapping except the grit of the abrasive is finer. Material removal is less than 0.0004 inches and surface finishes near perfect (0–30 μin) are possible, but size control ranges near ±0.0004 inches.

Low surface finishes can also be accomplished by **polishing, buffing, power brushing, burnishing, or vibratory finishing**. No measurable amount of material is removed: only the surface finish is improved.

9.3.4 Surface Coatings

A clean surface may be the only requirement for a part going into an assembly. Hot water and detergent, or solvent cleaner may be adequate. This process will remove machining chips, machining cutting fluids, and general shop dirt including discarded cigarette butts and coffee cups.

Pickling is an acid bath used to remove surface oxide. Parts must be washed first. Pickling is required when subsequent surface coating, such as plating or painting, requires a "bare" metal surface.

Manufacturing

Organic conversion coatings include phosphate, chromate, oxide, and anodic coatings as primers for subsequent coatings, or as-is use, as in the case of anodized aluminum.

"Paints" include oil base paint, lacquers, varnishes, enamels, and liquid plastic coatings and are used to protect the base material from degradation and to improve appearance.

Hot-dip plating or electroplating, including such methods as chrome plating and galvanizing, involves coating a metal part with a different metal to protect the base part.

Other specialized surface treatments include: shot and sand blasting, metal spraying, chemical vapor deposition, vacuum deposition, plasma deposition, mechanical plating, vitreous coatings, plastisol plastic dipping, polyvinyl chloride coatings, and metal sputtering.

9.3.5 Welding

Welding is a versatile production process that has been used for many years in a variety of applications. The primary use of welding is to create a complex shape by combining simple parts made from bars, plates, and castings. Welding can be used for custom one-of-a-kind parts or can be used for high rate production. The process requires that two or more parts be heated and melted together, or heated and forced together. In either case joined parts function as one, and a crack formed will transmit through the weld. Manual welding is commonly used for low quantity products such as large building structures, bridges, or earth-moving equipment. Automatic welding is often done with robots for high production items such as automobile bodies. Welding is a versatile operation and can be used for joining most carbon steel, some stainless steel (304 and 316), aluminum, some brasses, and nickel. Cast iron and other low tensile strength materials do not weld easily because the heat of welding expands the adjacent material, and places a high tensile stress at the heat zone interface.

Some Specific Welding Processes

- Arc spot welding
- Plasma arc welding
- Laser welding
- Submerged arc welding
- Stud welding
- Electron beam welding
- Thermit welding
- Gas-shielded arc welding
- Friction welding
- Ultrasonic welding
- Explosive welding
- Diffusion bonding
- Tungsten inert
- Gas-shielded arc welding

- Atomic hydrogen arc welding
- Gas welding and brazing
- Resistance welding (projection, seam, upset-butt, flash-butt, and percussion)

The important characteristics of welding include:

1. Shapes can be created that would otherwise require a casting, a forging, or a large amount of machining.
2. Dissimilar materials can be joined, such as 1020 steel to 1080 steel, with gas shielded arc welding.
3. Parts will not vibrate loose after welding as bolted parts can.
4. Surfaces that have been worn down can be built up with welding, and remachined.
5. Welding equipment is fairly compact and relatively portable; gasoline powered engine-generator setups allow welding to be done any place.
6. Heat treatment of the metal takes place because of the heating and cooling that occurs.
7. Heat distortion can occur as well as heat induced residual stress.
8. The parts welded act as one when a crack propagates.
9. Disassembly of parts is slow and usually causes parts to be scrapped or to require rework.
10. Weld splatter can ruin critical surfaces and internal compartments such as fuel tanks and oil reservoirs.
11. Welds are usually not smooth surfaces.
12. Verification of weld integrity requires X-ray inspection.
13. Overhead welding requires special skills.

To produce good welded assemblies requires:

1. Proper design so welds are accessible and part fit-up is consistent.
2. Component parts made to required dimensions and tolerances.
3. Fixturing to locate and hold the component parts in the proper position. Proper fixturing includes the use of rotating positioners to allow welds to be as close to horizontal as possible, and accessible (see Figure 9.25).
4. Proper welding equipment with adequate power and duty cycle specifications.
5. Operators properly trained in the use of available equipment.
6. Proper ventilation for the removal of gases and smoke.

9.3.6 Heat Treatment

Heat treatment is one of the manufacturing processes intended to alter the microstructure of materials. The main purpose of heat treatment is to change the characteristics of a material to obtain properties that otherwise would not exist. These desired properties include hardness, toughness, grain size, strength, or ductility.

Metal can be made soft so cold forging will not fracture the material. With steels, and many nonferrous alloys, this is done by heating to temperatures above the recrystallization temperature and by slow cooling to room temperature. Materials can be made hard so they will resist indentation and wear abrasion. This is done to steel by heating it to a temperature above its recrystallization temperature and cooling it rapidly to room temperature. In the range between soft and hard, materials are tougher and they can withstand impact loads without fracture. Some parts, such as gears, are made hard on the surface to resist wear, and are left ductile on the inside to make them tougher. These "case hardened" surfaces are created with a variety of methods such as carburizing, flame hardening, laser hardening, or induction hardening.

Heat treatments for the control of hardness and toughness are generally done as final operations because machining after hardening is difficult to do. Operations done after hardening are generally limited to final grinding or other abrasive finishing operations such as honing or lapping. These operations must be carefully controlled so the favorable properties created by heat treatment are not lost.

No attempt will be made here to elaborate on all the techniques of heat treatment. Both ferrous and nonferrous alloys can be heat treated, though the methods are different. Other materials such as glass, plastics, and wood are subjected to "heat treatment," but not in the sense of altering the microstructure. Raising the temperature of non-metallic materials, like glass and plastic, makes them soft for easy forming. Wood products are heated to remove water, which increases their strength, or water is added to make them weaker and formable. Steaming of wood is done to raise the grain and repair dents made by careless part handling.

There are problems associated with heat treatment of metals that must be anticipated and controlled to obtain high quality parts. Some of these concerns are:

1. Part distortion can occur due to the applied heat if it relieves residual stresses from other operations.
2. The medium used to cool the material may have a corrosive action on the surface of the part, and affect the surface finish.
3. It is very difficult to determine if a part has been heat treated by looking at it; inspection needs special nondestructive techniques.
4. A part can lose its heat treatment if it is heated from an external source, such as by friction with another part.
5. Microscopic cracks can form when a part is quenched rapidly. These cracks can become sources of fatigue failure cracks.
6. There are limits of hardenability of steel alloys, as shown in Figure 9.22.

9.3.7 Special Machines

The definition of a special machine will differ from person to person and plant to plant. In general a special machine will be a machine or series of machines that produce a specific part or family of parts. A standard process may be specialized for one part or a family of parts, or a combination of standard processes will be combined for one part or a family of parts. It may be that fixtures and tooling, permanently installed, will make a machine "special."

FIGURE 9.22

Limits of hardenability for steel alloys.

Machines can be put together in many combinations to produce parts to an unusual configuration or simply to increase production. One method for doing this is to have special machine heads that contain milling, drilling, boring, or other standard processes clustered around special fixturing. The part in the fixture then can be indexed from position to position so that each special head can perform some machining on the part without removing the part from the fixture. The fixtures can accommodate only one part, or a family of similar parts. The group of special machine heads and the fixture will be reserved for those parts alone.

If production requirements increase, an alternate arrangement is to place the machines in such a way that a fixture can be located in position by each head, and each head will be machining on a different part at the same time. The layout of this type of machine is more complicated because it is important that each station in the sequence have about the same amount of machining time. This will allow for the shortest complete cycle time for the parts being machined. This type of a setup requires that the sequence of stations form a loop. This allows an operator to be at one location to unload a finished part from a fixture and reload it with a new unfinished part. There needs to be as many fixtures as there are machining stations, plus one for unloading and loading. On small parts this can be accomplished with a rotary table. On large parts, like engine blocks, a transfer line with a power conveyor will be required. An alternative method is to have a fixture by each station that will automatically open and close at the end of the machining cycle. The part is then transferred from position to position by automatic handling equipment. Total investment required, part complexity, and production requirements all influence the decision whether to use a special machine or standard ones because special machines are more expensive than standard machines. Each special machine will have engineering time and cost associated with developing tooling, fixturing, and controls for the special application.

FIGURE 9.23

The special machine of Case Study #7.

Figure 9.23 illustrates the complexity of a special machine. The fixture that holds the part is shown in the foreground. The fixture is pulled down the ways by the attached cable. The first station has two opposing milling heads to create parallel faces. Subsequent stations drill, spot face and bore holes. The insert picture shows a closeup of one of the special heads. Note the alternate positions available for drills as the part being machined changes.

Without adequate cost and sales estimates, any analysis for a particular machine sequence line could be completely wrong and result in one of two bad conditions:

1. Machining capacity exists to make more parts than can be sold, meaning too much money has been spent on equipment.
2. Insufficient machining capacity exists to produce all the parts that can be sold. The customers want your product and you can't deliver.

Versatility of machine design can help this situation. If much doubt exists about required production, try to design a machine that can easily be converted to run other parts, and have an alternate method of producing the parts being made.

CASE STUDY #7:
Special Machine

A gray iron casting was being machined into a pump support. The support mounted to an engine, onto which a hydraulic pump was attached. The alignment of the pump shaft to the engine was important. The first vehicle that was designed for this casting did not have enough production to warrant a special process. As a result, the casting was machined in several separate steps, each with a different fixture. Many castings were not machined to specifications, and rework and scrap were common. Over a period of years other vehicles with similar mounting castings were introduced. Each mounting casting performed the same function but was dimensionally different because of a variety of engine and pump sizes. One day an enterprising engineer decided to try to do something about the situation of multiple parts, multiple fixtures, and scrap and rework.

It would have been nice if one part could have been used for all machines, but this was not a workable proposal because of the product size variation. Instead, the "best" solution turned out to be a special machine. The machine was a shuttle machine, a single fixture holding one part shuttled between eight special heads, four opposing each other on each side. The first station milled both sides of the casting so they were parallel. The second station performed drilling from both sides. The third station did spotfacing from one side and dowel pin boring from the other side. The fourth position bored the pump mounting hole. The fixture then shuttled back to the start and was unloaded and reloaded with an unfinished casting. All holding fixtures were built new, and several parts used the same fixture. The result of the project was that parts were being made with a reduction in machine time, rework and scrap were reduced below the factory average, and all mounting castings were run on the one machine with a two shift operation.

One situation did arise that required a change in machining sequence. The face milling was removing material that contained residual stresses, and the casting deformed after it was drilled, bored, and removed from the fixture. The solution was to rough mill the faces, shuttle the fixture back to the start, loosen the clamps, allow the part to relax, re-clamp, and finish mill the faces before any drilling or boring. Figure 9.23 is the special machine designed and built to machine the pump support castings.

9.4 Tooling

Some discussion must be offered about tooling; that is, the dies, jigs, fixtures, cutters, and gaging used to make parts. When craftsmen made their products centuries ago, work was done using hand tools. The quantity of parts made was few, and there was variation from product to product. For example, the first printing was done by carving the words for each page on a separate block of wood. Needless to say, products took a long time to

FIGURE 9.24

Simple drilling operation.

make, and repairs also had to be hand made to fit. Stocking repair parts was not, and could not, be done.

As the skill of workers changed, and the need and desire existed for more products, various ways were devised to allow more consistent parts to be made. One way to increase the consistency of making parts is to have tooling that holds an unfinished part in an orientation that allows each part to be manufactured the same each time. These needs give rise to holding fixtures with built in locators and clamps so that all parts are presented in the same orientation to the cutter. Fixtures must be rigid enough to withstand the weight of the part being held, the reaction clamping forces to hold the part in the fixture, and the forces of the cutting tools. Rigid fixtures cost money, however, so good fixtures can only be justified when production needs are high enough to amortize the cost of the tooling. The amortized tooling cost is part of what is called overhead, or process burden.

Tooling can be as simple as the support fixture used for drilling shown in Figure 9.24, or as sophisticated as the welding fixture shown in Figure 9.25. Figure 9.16 illustrates a fixture used to hold a part while it is being machined.

A machine tool itself is a holding fixture of sorts. The table must be built perpendicular and/or parallel to a rotating spindle or moving ram, or some other machine movement. The precision of table location and spindle advancement all contribute to the level of machining accuracy. The design and building of machine tools and related tooling is as much a part of engineering as is designing the parts that are to be manufactured.

FIGURE 9.25

Welding fixtures.

9.5 Design Considerations

No two engineers usually start a design project the same way. Some will start with the limitations of readily available material and processes and use currently manufactured parts as their guide. This is a conservative approach, and has the advantage of using the tried results and experience of other product engineers. A new product can be designed with less time, but significant changes and/or improvements are less likely to be made.

An alternative approach is to begin a design without considering process or material controls, use the guidelines of ANSI functional tolerances, design a solution, and then find the material and processes needed to make it. This approach will generally take more time, because a new idea needs more test time to prove the idea reliable, new materials and machinery may have to be ordered, and new process capability must be determined. This procedure allows for more innovative solutions and it may cost more in the initial stages, but it allows reanalysis of existing products, perhaps reducing costs or improving reliability.

Still other engineers will combine the two approaches. They will not limit their thinking to existing facilities, current practices, or a completely new technology. They will be innovative; will compromise to maintain adequate cost, time, and reliability; will justify new processes and techniques when practical (Case Study #7), but will stay with conventional methods when expedient.

Assembly procedures and techniques are also important concerns for efficient product design. The following suggestions can improve assembly depending on the actual circumstances encountered. The suggestions listed are to stimulate your thinking; they are not universally applicable.

1. Reduce the number of parts; fewer parts generally mean less complicated assembly.
2. Design a single part to perform more than one function. Use combination parts such as nuts with washers attached.
3. Drill and tap a part instead of using a nut.
4. Make parts self aligning by using tapers and slots.
5. Design component parts to be symmetrical so they can be used either side up.
6. Design the same part for right or left hand use, even if an extra unused hole is added.
7. Avoid press fit design unless absolutely necessary.
8. Avoid drilling holes in assembly. This increases assembly time and makes subsequent repairs difficult.
9. Design so that all parts can be assembled from the same side or provide a rollover assembly fixture.
10. Design parts to be rigid enough to withstand assembly without distortion or fracture.
11. Design welded assemblies to stay mechanically assembled prior to welding and to avoid costly clamping.

As parts are designed to be manufactured, the conditions the manufacturing system must follow are being established. Material and tooling must be available to make the parts as you have designed them. Knowledge about how parts are manufactured helps in designing them so tooling costs can be kept low. The following examples illustrate some of the processes already discussed.

EXAMPLE 9.4:
Manufacturing comparison.

Consider the block shown in Figures 9.26 and 9.27, dimensioned in inches. In Figure 9.26 the two 0.50 holes should be drilled in separate operations because the second hole is located by dimension from the first hole. This will result in two fixtures, one to drill the first hole 0.50 from the edge, and one to drill the second hole 0.88 from the first hole. If part function requires this sequence of operations, then it should be done

FIGURE 9.26

Part for Example 9.4.

and the extra money spent. However, if the holes can be dimensioned as in Figure 9.27, then the two holes can be drilled in one fixture in one operation.

The slot is another place where dimensioning variation can occur. In Figure 9.26 both sides of the slot are located from the same edge. Figure 9.26 indicates the necessity for two operations or two cutters in one operation using only one fixture. That is better than dimensioning one side of the slot from one end and the other side of the slot from the other end, which would definitely require two fixtures and two operations. Figure 9.27, on the other hand, is a case where a 0.62 wide cutter can be used and set up to locate 1.50 from the end. A second problem exists with the slot, that is, the depth. Figure 9.26 requires that the block be fixtured *up* against a stop so the 0.50 depth can be maintained. This may be required, and if it is, then the fixture must be built this way. Figure 9.27, on the other hand, has the slot dimensioned from the bottom of the part, and the fixture only has to hold the part rigid and be set up so the locator on the fixture is 0.75 below the cutter. The clamp can easily come over the top, and the fixture is simpler to design, build, and use; money is saved all the way around. *Note*: Some numerically controlled machines use a spindle touch probe to determine the location of the top of a part. If this capability is available, then dimensioning down from the top of a part is practical. Touch probing a part increases the cycle time to make a part, so don't require its use just for fun.

FIGURE 9.27

Part for Example 9.4.

Many engineers, in a hurry to get a job done, will often design parts that require extra money to be spent on tooling because they fail to consider simple requirements like those discussed. This is another case where communication with a manufacturing engineer or tooling engineer will save money and still give a functional part. Design your product to function as intended, but design parts to accomplish that function keeping the methods and cost of manufacture in mind. Don't compromise a function for low-cost tooling, but, on the other hand, don't require something just for the sake of fancy specification.

Experience is the best teacher when it comes to making parts because of the variety of possible ways they can be manufactured. A common first approach is to ask, "What processes are available?" Parts that could run on a multiple bar lathe may be run on a single spindle lathe only because machine time is available there. There are, however, standard processes that are common to most manufacturing operations. The following examples illustrate the variety of approaches that can be taken with common methods of manufacture.

EXAMPLE 9.5:
Typical manufacturing sequence.

List a possible method of manufacture for the part shown in Figure 9.28, dimensioned in inches.
- 10. Material: 1.00 diameter bar stock.
- 20. Drill 0.437 through, turn 0.81 dia., turn 0.06 × 0.05 groove, and cut off to 1.125 in a bar turret lathe.
- 30. Drill 0.250 dia. cross hole in a single spindle drill.
- 40. Mill 0.06 slot in a horizontal milling machine.
- 50. Deburr and wash.

FIGURE 9.28

Part to turn on a lathe.

Manufacturing

FIGURE 9.29

Cast part to machine.

EXAMPLE 9.6:
Typical manufacturing sequence.

List a possible method of manufacture for the part shown in Figure 9.29, dimensioned in inches.
  10. Material: Sand cast iron, add 0.12 material on all machined surfaces.
  20. Straddle mill 2.00 dia. pads in a horizontal milling machine. Use a two-position fixture to get one finished part for each cycle of the machine.
  30. Drill two 1.00 dia. holes in line through 2 arms, drill 0.50 dia. and spotface to 4.50 dimension. Use an indexable fixture under a radial arm drill.
  40. Deburr and wash.

EXAMPLE 9.7:
Typical manufacturing sequence.

Consider the part shown in Figure 9.30, dimensioned in inches. This is a part that can easily be manufactured in a bar lathe from 2.00 dia. material. If the 2.00 dia. has to be ground to a close tolerance, then larger diameter material would be used, 2.0625 or

FIGURE 9.30

Lathe part, optional dimensioning.

(a)   (b)

2.125. Alternate methods of dimensioning are shown to illustrate how manufacturing setup is controlled by part design.

In Figure 9.30(b), the 2.00 dia. would be the exposed end of the bar, and the overhang from the chuck or collet would be approximately 2.38 to allow for cutoff. The exposed end would probably be faced, and each diameter turned from that reference end at distances of 0.50, 1.50, 1.75, and then cut off at 2.00 length. This may cause problems because the largest mass is farthest from the support of the chuck, and getting the turning tool into the small diameters is difficult.

In Figure 9.30(a), the same amount of overhang would exist, but diameter turning would be done at 0.25, 0.50, and 1.50 from the exposed reference end. There is more room to get turning tools in to the part, and the largest mass is closest to the chuck.

EXAMPLE 9.8:
Typical manufacturing sequence.

Consider the part in Figure 9.31, dimensioned in inches. A likely method of manufacture for this part is:
  10. Material: 2.00 dia. cold rolled steel.
  20. Turn 1.00 dia., 1.00 spherical end, taper, and cut off to 4.00 on a bar lathe.
  30. Straddle mill flats to 0.62 and mill slope to 3.00 on a horizontal milling machine.
  40. Drill starter hole (~0.31 dia.) for slot on single spindle drill.
  50. End mill slot (~0.31 dia.) on a vertical mill.
  60. Broach slot, 1.00 × 0.38.
  70. Deburr and wash.

Operations 40 and 50 could be done in one operation on a vertical or horizontal spindle boring mill.

FIGURE 9.31

Part for Example 9.8.

EXAMPLE 9.9:

Typical manufacturing sequence.

Consider the part shown in Figure 9.32, dimensioned in inches. A method of manufacture for this part is:
    10. Material: Cast iron, sand cast, add 0.12 material on all machined surfaces, cast holes solid.
    20. Mill back side to $\frac{3}{4}$ on a horizontal milling machine with a face cutter.
    30. Drill $1\frac{1}{4}$ dia. and spotface $2\frac{1}{2}$ dia. to 2.75 overall dimension on a single spindle drill.
    40. Drill (6) $\frac{5}{8}$ dia. holes, one at a time in a fixture that rotates under a single spindle drill.
    50. Drill (3) $\frac{1}{2}$ dia. holes, three locations in a fixture under a single spindle drill.
    60. Deburr and wash.

Operation 40 could be done on a multiple spindle drill for high production requirements. Operations 20, 30, and 40 could be done in one operation on a horizontal spindle machining center using an index fixture or by back spotfacing. Operation 20 and 30 could be done on a lathe if it had swing capacity of 5.25 inches or greater.

FIGURE 9.32

Cast part to manufacture.

- 5/8 Drill, 6 holes equally spaced
- 9-1/4 D
- 2-1/2
- 1-1/4 Drill
- 5/8
- 5/8 R
- 1-3/4
- 2-3/4
- 3/4
- 1/2 Drill
- 8 D
- 3/4
- 6-3/4 D

From T.E. French and C.J. Vierck. *Engineering Drawing and Graphic Technology*, 12th ed. © 1978 McGraw-Hill, N.Y.

9.6 Summary

The preceding discussion of manufacturing processes has been brief. As you gain experience in the world of designing and manufacturing products, some of these processes will become second nature to you, others will not be used and will be forgotten. Knowledge of manufacturing processes is the same as other relevant design information; you should know that data exists and be able to recall enough to search out answers to your specific problems. You may even choose to become a manufacturing engineer and become a specialist in how to manufacture parts.

9.7 Additional Study Topics

To give one more emphasis on the variety and range of available manufacturing processes, the following processes and machines are listed for study. They are listed without comment, and are suggested for additional research.

1. Gun driling
2. Vertical shaper, or slotter
3. Copy milling machine
4. Cam milling machine
5. Mill dividing head
6. Climb and down milling

7. Belt grinding
8. Camshaft grinding
9. Thread rolling
10. G-Trac gear generator
11. Photo-etching
12. Thermal deburring
13. Ball sizing
14. Ultrasonic machining
15. Threading machine
16. Tracer lathe
17. Turret cross slide
18. Threading machine
19. Automated assembly
20. Blind broaching
21. Plunge form grinding
22. Flame hardening
23. Tool steels
24. Jolt-squeeze machine
25. Core making
26. Stud welding
27. Seam welding
28. Cutting fluids
29. Vibratory deburring
30. Explosion deburring
31. Automated assembly
32. Chucks and collets
33. Induction hardening
34. Electrodischarge machining
35. Electrochemical machining
36. Abrasive flow machining
37. Multiple spindle milling machine
38. Planer milling machine
39. Titanium coated cutting tools
40. Pattern shrink and draft allowance
41. Nondestructive testing
    a) X-ray inspection
    b) Magnetic particle inspection
    c) Ultrasonic testing
42. Carburizing, cyaniding, and nitriding
43. Machining power requirements
44. Machine drives—see Figure 9.35
    a) mechanical
    b) electrical
    c) hydraulic
45. Allotropic and nonallotropic hardening
46. Interchangeable and noninterchangeable parts in assembly
47. Production assembly lines

9.8 References

1. American Machinist. *Computers in Manufacturing.* New York: McGraw-Hill, 1983.
2. Behringer, Catherine A. "Superplastic Forming." *Manufacturing Engineering.* January, 1987, 41–44.
3. Budinski, Kenneth. *Engineering Materials: Properties & Selection.* Reston, Va.: Reston Publishing, 1979. Chapter 17 on plating, Chapter 18 on welding.
4. DeGarmo, Paul E., I. Temple Black, and Ronald A. Kohser. *Materials & Processes in Manufacturing*, 7th ed. New York: Macmillan, 1988. Contains 31 good manufacturing case study problems.
5. Dorf, Richard C. *Robotics and Automated Manufacturing.* Va.: Reston, 1984.
6. Doyle, Lawrence E. et.al. *Manufacturing Processes and Materials for Engineers*, 3rd ed. Englewood Cliffs, N.J.: Prentice Hall, 1985.

7. Earle, James H. *Engineering Design Graphics*, 5th ed. Reading, Mass.: Addison Wesley, 1987.
8. Farnum, Gregory T. "Reinventing The Lathe." *Manufacturing Engineering*. June, 1986, 36–39.
9. Feigenbaum, Armand.V. *Total Quality Control*. New York: McGraw-Hill, 1983.
10. Gayman, David J. "Composite Fabrication Speeds Up." *Manufacturing Engineering*, July, 1987, 26–31.
11. Groover, Mikell P. *Automation, Production Systems and Computer-Aided-Manufacturing*. Englewood Cliffs, N.J.: Prentice Hall, 1980.
12. Hales, H. Lee. "Producibility and Integration: A Winning Combination." *CAM Technology,* August, 1987, 14–18.
13. Kafrissen, Edward and Mark Stephen. *Industrial Robots and Robotics*. Reston, Va.: Reston Publishing, 1984.
14. Kaye, Alan. *Die Casting Metallurgy*. Boston: Butterworth, 1982.
15. Kochhar, A.K. *Microprocessors and Their Manufacturing Applications*. Baltimore: Edward Arnold, 1983.
16. Lieberman, Norman P. *Process Design for Reliable Operations*. Houston: Gulf, 1983. Author's experience with hardware design in petroleum refining.
17. Machinability Data Center Technical Staff. *Machining Data Handbook,* 3rd ed., 2 vols. Cincinnati, Ohio: Metcut Research Associates, 1980.
18. Oberg, Erik, et al. *Machinery's Handbook*, 22nd ed. New York: Industrial Press, 1984.
19. Phillips, Robert E. "ECG (Electrochemical Grinding): Combining Electrochemical Attack and Abrasion." *Manufacturing Engineering*, December, 1986, 46–48.
20. Rehig, James A. *Introduction to Robotics: A Systems Approach*. Englewood Cliffs, N.J.: Prentice Hall, 1985.
21. Rusinoff. *Manufacturing Processes—Materials & Production*. Chicago: American Technical Society, 1962.
22. Schey, John. *Introduction to Manufacturing Processes*. New York: McGraw-Hill, 1979.
23. Schneberger, G.L. *Adhesives in Manufacturing*. New York: Marcel Dekker, 1983.
24. Stewart, John. *The Welder's Handbook*. Englewood Cliffs, N.J.: Prentice Hall, 1981.
25. "Tech Update: New CO_2 Laser for Cutting and Welding." *Manufacturing Engineering*. July, 1990, 42.
26. Trucks, H.E. *Designing For Economical Production*, 2nd ed. Dearborn, Mich.: Society of Manufacturing Engineers, 1987.
27. Wallach, Paul. *Metric Drafting*. Encino, Cal.: Glencoe, 1979.
28. Weintraub, Chuck. "Design For Manufacturability." *Design Graphics World*. October, 1988, 14–16.
29. Whitney, Daniel E. "Manufacturing by Design." *Harvard Business Review*. July-August, 1988, 83–91.
30. Wick, Charles. "Coatings Improve Tool Life, Increase Productivity." *Manufacturing Engineering,* December, 1986, 26–31.
31. Yankee, Herbert W. *Manufacturing Processes*. Englewood Cliffs, N.J. : Prentice Hall, 1979.

9.9 Exercises

1. Select a topic and prepare a paper and an oral presentation on one of the items listed under Additional Study Topics. Add to your choices the lists of topics on pages 000, 000, 000, and 000.

2. Write a five to ten page paper on any of the processes mentioned in the chapter.

3. One theory of grinding gives a model for estimating the maximum chip thickness as follows (6, 697):

    $t = (2v/Vn)\sqrt{C[(1/D) + (1/d)]}$, where

    t = maximum chip thickness in inches
    v = surface speed of rotating work in inches/min.
    V = surface speed of grinding wheel in inches/min.
    n = grains in wheel per linear inch
    C = depth of cut on centerline between grinding wheel and part
    D = grinding wheel diameter in inches
    d = part diameter before grinding in inches

    a) Will t always be less than C?
    b) Discuss alternative techniques that will allow larger depth of cut. Include % changes in C with % changes in D, v, V and n.
    c) As a grinding wheel wears away D gets smaller. How can this change in D be compensated for?

4. Discuss the differences between push and pull broaching. Compare such things as:
    a) Tool differences
    b) Holding fixtures
    c) Sequence of operator steps
    d) Machine tool differences

5. Think about the specific operation of electroplating and come up with a reason to plate a part other than to protect the base metal from corrosion.

6. Zinc coating (galvanizing) protects the base steel part in at least two ways. What are they?

7. Concrete reinforcing bars are made from scrap carbon steel, and the carbon ranges from 0.40 to 0.70 percent. The bars are hot rolled to shape and come in different configurations and grades.

Grade	S_{ut}, psi	S_{yt}, psi
Structural	55–75,000	35,000 minimum
Intermediate	70–90,000	40,000 minimum
Hard	80,000 minimum	50,000 minimum

a) These do not fit exactly the definition of bars that was given in the chapter. Why not?
   b) Why are they included in bars?
   c) How would you determine which grade a particular bar was if you had a sample?
8. What are the advantages and disadvantages of several individually powered drill spindles mounted sequentially over a common table? When would it be good to have a separate fixture under each spindle compared to one fixture that slides from spindle to spindle?
9. Table 9.5 shows the ranges of surface finishes that are typical of processes that have been mentioned in this chapter. In general, smaller surface finishes increase cost, as do smaller tolerances.
   a) Compare all the processes that will result in as good a surface as die casting and determine if size tolerances given in the chapter support this conjecture.
   b) Discuss reasons for the variation within one process for the range of finishes possible. Pick one process and tell specifically what you would do to get the best finish in the range indicated.
10. One model used for calculating the force F, in pounds, to make a bend in a part of thickness t, in inches, and of bend length w, in inches, is $F = kwt^2/L$, where L is the opening width of the lower die in inches and $k = 0.67S_{yt}$ for elastic bending and $k = 1.33S_{ut}$ for plastic bending (6, 287). Explain why the angle of bend doesn't appear in the model and why w and L are to the first power while t is squared.
11. Discuss how you would go about running an experiment to verify the model in Exercise 9.10. Conduct the experiment if facilities are available.
12. Choose the criteria that generally results in the greatest springback when bending a part.
   a) Larger or smaller bend radii?
   b) Thicker or thinner material?
   c) Larger or smaller bend angles?
   d) Harder or softer materials?
13. Refer to Table 9.11. What would you expect to happen to the production rates as the part weight increases?
14. Refer to Table 9.11. What would you expect to happen to the relative cost as the part weight increases? How about absolute cost?
15. Compare the ratio between the weight of molten metal used and final part weight probable for the following casting processes. List some ways each process might fail to make a good part.
   a) Sand casting
   b) Permanent mold casting
   c) Die casting
16. Why should the weight limits in Table 9.9 be used as relative recommendations rather than absolute?
17. Two guidelines for bending parts made from sheets and plates are:
   1) The lower die opening should be eight times material thickness.

Manufacturing

2) The shortest leg of a bent part should not have a flat portion less than $1\frac{1}{2}$ times material thickness unless special tooling is available.

Are the two guidelines compatible? If necessary consider making a 90° bend in 0.250 inch thick steel.

18. Estimate the relative tooling and labor costs for each of the three methods of manufacture graphed in Figure 9.33. Which method would you use to make 75 pieces? Why?

19. *a*) What is the correlation between casting temperature and die temperature using the following data for die casting?

	Base alloy			
	Yellow Leaded Brass	Zinc	Al	Mg
Die temperature °F	950	425	550	500
Casting temperature °F	2000	760	1220	1400
Die life, # of pieces	10^4	10^6	10^5	10^5
Relative part cost	40	1.8	0.45	0.60

b) Give reasons why relative part cost does not correlate with die life in spite of the fact that die cost is large. Recall that dies for die casting are made from steel and are water cooled.

FIGURE 9.33

Relative costs for three methods of making an aircraft part in various quantities.

Method:
(1) Machined from barstock
(2) Forged and machined
(3) Cold extruded and machined

Cost per piece, dollars vs. Number of pieces

From Doyle/Keyser/Leach/Schrader/Singer, *Manufacturing Processes & Materials For Engineers*, 3rd ed, p. 277. © 1985 Prentice-Hall, Inc., Englewood Cliffs, N.J. Reprinted by permission.

FIGURE 9.34

End view of nested bars.

20. One of the ways that power hacksaws can be made more efficient is to nest bars in the vise of the machine so the blade cuts more than one bar at a time (see Figure 9.34). Discuss advantages and disadvantages of nesting. Consider a 1.00 inch diameter bar, a maximum vise opening of 4.5 inches, a 14 inch blade, a 2 to 7 inch variable stroke, and a vise height of 3.00 inches. Consider such items as loading bars to the machine, unloading parts from the machine, advancing the bars between cuts, and the time difference to cut 100 pieces 3.00 inches long from bars 196.00 inches long.

21. Metal spinning is an alternate method of producing some parts that could be made by drawing. Compare the processes using the following data.

	Drawing	Spinning
Tool cost	$10,000	$1500
Tool lead time	10 weeks	3 weeks
Production rate	200 per hour	18 per hour

22. A forging can be made with hand tools by a skilled operator in 20 minutes using equipment that costs $20,000. A set of forging dies to make the same part costs $5000 and runs on a $100,000 machine. An operator who is paid $\frac{2}{3}$ of the rate of the skilled operator produces a part in 3 minutes. Discuss and compare the two methods.

23. *a*) Compare the cost of making a 2.00 inch diameter boss from 1020 steel on a lathe and on a saw. Use the following data if no other is available and assume the hourly labor rates for each machine are equal. Assume a 10 year machine life and 2000 operating hours per year.

	Lathe	Saw
Machine cost	$80,000	$30,000
Cutoff time in minutes	0.50	1.50

b) If the hourly wage rate of the lathe operator is $10, what hourly wage rate for the saw operator will make the labor plus machine time cost of both processes the same?

c) Does the answer in part (b) seem reasonable? Propose a solution.

304   **Manufacturing**

24. List as many rotating tool operations that can be performed on a horizontal boring-milling machine as you can think of.

25. Discuss the advantages and disadvantages of running a variety of parts, in random sequence, on a horizontal boring machine that has a tool changer. Some items to consider are:

FIGURE 9.35

Various kinds of press drives.

From Doyle/Keyser/Leach/Schrader/Singer, *Manufacturing Processes & Materials For Engineers*, 3rd ed, p. 312. © 1985 Prentice-Hall, Inc., Englewood Cliffs, N.J. Reprinted by permission.

a) The number of tools that the tool changer holds.
b) When do dull tools get changed?
c) How does the machine know which part is on the table of the machine?
d) How are different parts fixtured?
e) Families of parts (similar shape but different sizes).
f) Random parts (not similar; different sizes and different materials).

26. List alternate methods of shuttling a fixture between stations of a special machine such as would be needed for the pump mounting casting in the case report on a special machine.

27. Figure 9.35 illustrates various types of drive mechanisms used for press operations, compressors, and other linear power motion. If the press force is considered constant, list and discuss the advantages and disadvantages of each mechanism.

28. Will a hot forged or cold forged part have a higher resistance to:
    a) Fatigue?
    b) Creep?
    c) Stress corrosion? Why?

29. If over-bending of 5° is required to obtain a part with a 90° bend made from 0.50 thick steel with S_{yt} = 40,000 psi, would you expect the same over-bend to be required if the steel is 0.75 thick? You may have to make some assumptions about the tooling and machine used.

30. Discuss briefly three methods of machining gear teeth on a part. Mention any primary advantage or disadvantage of each method.

31. A nibbler uses a one-inch diameter punch and makes one stroke in two seconds. What linear feed rate in inches/minute is required so that the scallop height is ≤ 0.010 inches?

32. Design the drill fixture for operation 50 in Example 9.9.

33. Design the drill fixture to drill the part in Figure 9.26.

34. Design the mill fixture to mill the slot of the part in Figure 9.26.

35. Figure 9.36 represents a hole pattern in a rectangular shape, dimensions shown in inches. Consider the holes to be 0.50 dia. How would the manufacture of the holes change for each of the following circumstances.
    a) Material: cork, 0.10 thick, use as a gasket.
    b) Material: steel, 0.010 thick, use for a shim.
    c) Material: steel, 0.250 thick, use as filler plate.
    d) Material: steel, 0.750 thick, use as a filler plate.
    e) Material: steel, 1.00 thick, use as a dowel plate.

36. Design the driill fixture for the part in Exercise 9.35, part (d).

37. Figure 9.37 represents a housing cap to be made from bar steel, dimensions shown in inches. The location of the 4 bolt holes relative to the 0.50 diameter is critical, as is the concentricity between the 1.00 and 0.50 diameters. The teardrop shape is a wind fit into a mating casting. The profile is also a wind fit. The 0.38 and the 0.18 thickness relationship is critical. Discuss how to manufacture the part.

FIGURE 9.36

Exercise 9.35.

FIGURE 9.37

Exercise 9.37.

FIGURE 9.38

Exercise 9.39.

38. How will the manufacture of the part in Figure 9.37 change if it is sand cast from iron?
39. Figure 9.38 represents an iron casting to be machined. Some, but not all of the critical dimensions are noted. Discuss other possible critical dimensions and propose a likely method of machining the casting to meet the critical dimensions.
40. Figure 9.39, is a sketch of a part made from cast aluminum, dimensions shown in inches. Two machined surfaces are shown. Complete the dimensions necessary for machining only, and list the required manufacturing steps.
41. Design a mill fixture to mill the part in Exercise 9.40.
42. Figure 9.40 represents a part to be turned on a lathe from 1.00 inch diameter cold rolled steel.
    a) Sketch an orthographic drawing of the part complete with dimensioning so the part can be machined completely in one lathe operation.

FIGURE 9.39

Exercise 9.40.

0.359 drill-thru
0.750 spotface
0.375 ream-thru
2 holes

0.250 drill-thru
0.500 dia × 82°csk
2 holes

0.250 drill-thru
0.375 cbore-0.250 deep
2 holes

0.88
0.38

FIGURE 9.40

Part sketch for Exercise 9.42.

1.00

FIGURE 9.41

Part for Exercise 9.44.

[Drawing of bell crank with dimensions: 3/8, 1 R, 1/2 R, 3/4 R, 2-1/4, 2-1/4, 1/2 R, 7/16 dia 2 holes. Labeled "Bell crank / Cast iron"]

From J.H. Earle. *Engineering Design Graphics*, 5th ed. © 1987 Addison-Wesley Publishing Co. Reading, Mass. Fig. 15.79, p.256. Reprinted with permission.

    b) Suppose the method used in part (a) doesn't work satisfactorily and the turning has to be done in two operations. Discuss alternate methods of performing two operations and the associated dimensioning preferred.

    c) How would the part have to be redesigned if the smallest diameter had to be ground because of tolerance requirements?

43. A weld fixture supports two halves of a structure and holds the spacing between them. The space held by the fixture is critical and allows other parts to be properly assembled. The two halves of the welded assembly vary in such a way that at times they touch each other before they are welded; other times there is a gap that must be welded. In service some of the welded assemblies appear to tear down the weld joint. Strain gage tests indicate moderate levels of load on the joint. Discuss possible causes and solutions.

44.–53. For the parts shown in Figures 9.41 through 9.50 discuss possible methods of manufacture. Assumptions about critical relationships may need to be made. Include notes about any likely functional relationships. Comment on any design shortcomings.

FIGURE 9.42

Part for Exercise 9.45.

3/4 R
1/4 R
3/4 (0.750) Drill
3/4 R
1/4
1/2
2-27/32
7/8 dia, thru
Rod guide
Cast iron

From J.H. Earle. *Engineering Design Graphics*, 5th ed. © 1987 Addison-Wesley Publishing Co. Reading, Mass. Fig. 15.81, p. 256. Reprinted with permission.

FIGURE 9.43

Part for Exercise 9.46.

1-5/16 dia
3 holes
1-11/32 R
1-11/32 R
1
1-1/2
4-1/2
1-11/16
F & R 1/8 R
Lug
Cast iron

From J.H. Earle. *Engineering Design Graphics*, 5th ed. © 1987 Addison-Wesley Publishing Co. Reading, Mass. Fig. 15.83, p. 256. Reprinted with permission.

FIGURE 9.44

Part for Exercise 9.47.

F & R = 1/8 R

Pedal crank
Cast iron

From J.H. Earle. *Engineering Design Graphics*, 5th ed. © 1987 Addison-Wesley Publishing Co. Reading, Mass. Fig. 15.86, p. 257. Reprinted with permission.

FIGURE 9.45

Part for Exercise 9.48.

Shaft set
Magnesium M11910
SI

F & R 4 R

From J.H. Earle. *Engineering Design Graphics*, 5th ed. © 1987 Addison-Wesley Publishing Co. Reading, Mass. Fig. 15.88, p. 257. Reprinted with permission.

Exercises

FIGURE 9.46

Part for Exercise 9.49.

Spacer
CRS
SI

From J.H. Earle. *Engineering Design Graphics*, 5th ed. © 1987 Addison-Wesley Publishing Co. Reading, Mass. Fig. 15.94, p. 258. Reprinted with permission.

Manufacturing

FIGURE 9.47

Part for Exercise 9.50.

Bearing cap
Cast iron
SI

From J.H. Earle. *Engineering Design Graphics*, 5th ed. © 1987 Addison-Wesley Publishing Co. Reading, Mass. Fig. 15.92, p. 258. Reprinted with permission.

FIGURE 9.48

Part for Exercise 9.51.

From J.H. Earle. *Engineering Design Graphics*, 5th ed. © 1987 Addison-Wesley Publishing Co. Reading, Mass. Fig. 15.106, p. 260. Reprinted with permission.

FIGURE 9.49

Part for Exercise 9.52.

Support
Matl: aluminum alloy
Fillets & rounds 0.12 R

From J.H. Earle. *Engineering Design Graphics*, 5th ed. © 1987 Addison-Wesley Publishing Co. Reading, Mass. Fig. 24.46, p. 453. Reprinted with permission.

FIGURE 9.50

Part for Exercise 9.53.

Spindle head
1045 Steel
Fillets & Rounds 3 R
SI

Rear view

From J.H. Earle. *Engineering Design Graphics*, 5th ed. © 1987 Addison-Wesley Publishing Co. Reading, Mass. Fig. 24.48, p. 454. Reprinted with permission.

10 Probability and Statistics

"We use statistics the way a drunk uses a lamppost. For support, not illumination."
◆ MIKE KEENAN, PHILADELPHIA FLYERS COACH.

The purpose of this chapter is to review some commonly used general probability and statistical concepts and models. It is not an attempt to develop statistical theory or to make you an expert statistician. See references (10), (12), and (13) for theory development. Expertise in the use of statistics comes after extensive use of the concepts. It is difficult to gain an intuitive grasp of statistical methods until many problems and situations have been encountered, and various procedures used. Statistical analysis is required and frequently used by engineers to make inferences from data, but many pitfalls await the unwary.

Probability and statistics helps answer questions that come up from time to time during design development, and an engineer should know how to deal with them. Among the situations and questions that can arise are:

1. Did a change in a manufacturing process make a difference in the output product?
2. Is product A "better" than product B?
3. Is it likely that two samples came from the same batch?
4. What is the meaning of material property data in tables?
5. How can table data be of value in the design of a product?
6. Are two machines producing the same quality parts?
7. Is experimental data normally distributed?

10.1 Probability

Because the word probability is used frequently in this chapter and in many engineering communications, a review of some probability terms and uses is appropriate.

Probability is used to indicate the likelihood of an event occurring or not occurring, and is quantified with a numerical value from 0 to 1, or from 0 to 100 percent. A probability of 0 (0 percent) means an event will never happen, a probability of 1 (100 percent) means an event is sure to happen. A probability value close to 0 indicates a low chance of an event occurring, and a value close to 1 indicates a good chance that the event will occur. An event that is 95 percent probable is more likely to occur than one that is only 35 percent probable. A 95 percent probability does *not* mean that 95 times out of 100 the event will occur and 5 times it will not occur. A probability value is a relative indicator, nothing more. Five non-events may not occur for 213 trials, or they might occur in 20 trials, and it doesn't matter if samples are chosen sequentially or in some random pattern. A term closely associated with probability is reliability.

Reliability is the probability of a product or system operating and reacting as designed. Engineers want a product to be highly reliable so that everything goes as planned. An event that is 0.999 probable is 99.9 percent reliable; the chance for failure is 1 in 1000. The terms probability and reliability are often used interchangeably.

The probability of an event E occurring, out of all possible outcomes, is defined as the ratio of the number of times the desired event actually occurs to the number of times the event was attempted.

$$P(E) = \frac{\text{occurrences of } E}{\text{total number of tries}} \tag{10.1}$$

The sum of all probabilities in an experiment or observation is one, so the probability of an event *not* occurring $= 1 - P(E)$.

EXAMPLE 10.1:
Single event probability.

Consider a six-sided fair die. The probability that any given number will come up on any given roll of the die is 1/6. This is true since only one side can come up, and there are six different sides. A 5 has as much chance of coming up as a 2, or a 4, or any other of the six numbers.

P(rolling a 3) = 1/6 = 0.167.

P(rolling a 1.8) = 0.

10.1.1 Multiple Event Probability

When success is achieved if more than one event happens, (a win if a 3 *or* a 4 is rolled), the probability of success will increase because the number of possible outcomes has not changed, but the number of succeeding events has increased. How can the probability of two or more independent events be greater than the probability of each event when the probability of each event is less than one? The combined probability of **or** independent mutually exclusive events occurring is the **sum** of the individual probabilities.

If success is only achieved when two or more sequential events occur, the probability decreases because the number of successful events has diminished and the total number of events has not changed. How can the net probability decrease if each probability is less than one? The combined probability of **and** *independent* events occurring is the **product** of the individual probabilities.

If two events are *not independent* (not mutually exclusive), then the joint probability is expressed as $P(A \text{ or } B) = P(A) + P(B) - P(A \text{ and } B)$ where $P(A \text{ and } B)$ are all events that satisfy both criteria at the same time. Do not confuse $P(A \text{ and } B)$, which are simultaneous events, with $P(A)$ and $P(B)$, which are sequential events.

EXAMPLE 10.2:

Multiple event probability.

a) What is the probability of rolling a 3 **or** a 4 on one roll of a fair die?
$P(\text{rolling a 3 or 4 on one roll}) = 1/6 + 1/6 = 1/3 = 0.333$.

b) What is the probability of rolling a 1, 2, 3, 4, 5, **or** 6 on one roll of a fair die?
$P(\text{rolling a 1 or 2 or 3 or 4 or 5 or 6}) = 1/6 + 1/6 + 1/6 + 1/6 + 1/6 + 1/6 = 1$.

c) What is the probability or rolling a 3 followed by rolling a 4 on two rolls of a fair die?
$P(\text{rolling 3 on one roll})$ **and** $P(4 \text{ on the next roll}) = (1/6)(1/6) = 1/36 = 0.028$.

d) What is the probability of getting an odd number **or** a prime number on one roll of a die?
$P(\text{odd } \textbf{or} \text{ prime}) = P(\text{odd}) + P(\text{prime}) - P(\text{odd } \textbf{and} \text{ prime})$
$P(\text{odd or prime}) = 3/6 + 3/6 - 2/6 = 4/6 = 0.667$.

10.1.2 Histogram

A **histogram** is a graphical method of displaying data. It shows by way of frequency of occurrence the more probable events that can occur. For example, Figure 10.1 shows a histogram of the results of 36 rolls of a pair of perfect dice, which were first recorded in an outcome matrix, Table 10.1. The horizontal axis of the histogram shows the possible outcomes, and the vertical axis shows the frequency of occurrence of each outcome. The histogram can assist in the probability analysis, as the next example illustrates.

EXAMPLE 10.3:

Single event probability.

Consider two dice rolled at the same time. To determine the probability of any particular sum being rolled, all the possibilities must be known. There are 36 different combinations that can occur, but some sums are more likely to occur than others, as shown in Table 10.1 and Figure 10.1.

For a two-dice roll the following probabilities exist.

$P(\text{rolling a sum of 2}) = 1/36 = 0.028$ (quite unlikely)

$P(\text{rolling a sum of 7}) = 6/36 = 0.167$ (more likely)

FIGURE 10.1

Histogram of 36 rolls of a pair of perfect dice.

P(rolling a sum of 2 **or** a sum of 11) = 1/36 + 2/36 = 0.083

P(rolling a sum of 3 on roll 1) **and** P(rolling a sum of 2 on roll 2) = (3/36)(1/36) = 3/1296 = 0.002 (very unlikely)

10.1.3 Counting Methods

Experimental probabilities are easy to determine because the only requirement is to tally the number of trials and the number of favorable events that occur. Theoretical probabilities are more difficult to determine because some method is needed to count the number of ways that events can happen (the denominator of the probability ratio), and how many ways the event can happen in the way desired (the numerator in the probability ratio). Several models exist for counting events to aid in the theoretical probability calculation.

TABLE 10.1

Two-dice roll outcome array.

		\multicolumn{6}{c	}{Die #1}
		1	2	3	4	5	6
	1	2	3	4	5	6	7
	2	3	4	5	6	7	8
Die #2	3	4	5	6	7	8	9
	4	5	6	7	8	9	10
	5	6	7	8	9	10	11
	6	7	8	9	10	11	12

The first method used to count events is the **basic principle of counting**. Consider a sequence of n events that are required to complete a given function. The total number of different ways the function can be completed is the product of the w_i different ways of performing each individual task: Πw_i, where $i = 1, 2, 3, \ldots, n$. (10.2)

EXAMPLE 10.4:
Product rule of counting events.

An engineer is studying the effect of wind on a model of a windmill. Three variables are to be controlled: four different angles of the blade; two-blade, three-blade, and four-blade designs; and six different wind speeds. The total number of different experimental runs will be $(4)(3)(6) = 72$. No wonder complete experiments can take so much time to run.

A second method of counting is done using **permutations**. A permutation, symbolized as $_nP_r$, is an *ordered* selection of r objects from a source of n different objects; the grouping ABC being different than BCA. The formula for calculating permutations is:

$$_nP_r = n! / (n - r)!$$ (10.3)

EXAMPLE 10.5:
Permutation count.

An engineer is going to test six different grades of tires on a test car. If it is assumed that a tire may wear differently depending on whether it is on the left front, right front, left rear, or right rear, and interaction of tire performance might be a concern, then order of selection becomes important. Under this assumption the number of different test runs is $_6P_4 = 6!/2! = 360$ different test runs.

A third method of counting is to use **combinations**. A combination is not concerned with order, but just selects r objects from a source of n different objects. For the same values of n and r, combinations give a smaller count than do permutations because order doesn't matter; grouping ABC is the same as BCA, CAB, or ACB. The symbol for combinations is $_nC_r$, and is calculated using the formula:

$$_nC_r = n! / [r!(n - r)!]$$ (10.4)

EXAMPLE 10.6:
Combination count.

If the engineer in Example 10.5 thought that the location of the tire on the car was not important, how many different test runs are possible?

$$_6C_4 = 6!/(4!2!) = 15 \text{ different test runs.}$$

Probability

Another tool that can be used to keep track of the number of ways something can happen is to draw a **tree diagram**. This is particularly helpful for sequential events and provides a visual representation. Most engineers feel more comfortable with visual problem analysis.

EXAMPLE 10.7:

Tree diagram probability.

Three coins are placed in a container. One coin has two heads (2H), one coin has two tails (2T), and one coin has a tail and a head (TH). What is the probability of drawing a coin, flipping it twice, and getting two heads?

Drawing any one of the coins is assumed equally likely, so $P(2H) = P(2T) = P(TH) = 1/3$. If the 2H coin is drawn and flipped twice, only heads can occur. If the 2T coin is drawn and flipped twice, only tails can occur. If the TH coin is drawn, four different two flip combinations can occur (TT, HH, HT, or TH). Figure 10.2 shows the tree diagram for this problem. The sum of the probabilities at any node must equal one, and if the tree is properly done, the sum of the probabilities of the final branches will equal one. Using the tree, the probability of any of the six different sequences can be determined. *Note*: More than one branch of the tree can end with the same result. If this occurs, then the probability of that event occurring will be the sum of the branch probabilities. P(Drawing a coin, flipping twice, and getting two heads) = $1/3 + 1/12 = 5/12 = 0.417$.

FIGURE 10.2

Tree diagram for Examples 10.7 and 10.8.

Select one coin	Flip #1	Flip #2	Sequence probability
$P(2H) = 1/3$	$P(H) = 1$	$P(H) = 1$	$(1/3)(1)(1) = 1/3$
$P(2T) = 1/3$	$P(T) = 1$	$P(T) = 1$	$(1/3)(1)(1) = 1/3$
$P(TH) = 1/3$	$P(T) = 1/2$	$P(T) = 1/2$	$(1/3)(1/2)(1/2) = 1/12$
		$P(H) = 1/2$	$(1/3)(1/2)(1/2) = 1/12$
	$P(H) = 1/2$	$P(T) = 1/2$	$(1/3)(1/2)(1/2) = 1/12$
		$P(H) = 1/2$	$(1/3)(1/2)(1/2) = 1/12$

EXAMPLE 10.8:
Tree diagram probability.

For the same circumstances as Example 10.7, what is the probability if a coin is drawn, flipped twice, and two tails come up, that the coin selected is two tailed?

P(event happening with a two tailed coin) = 1/3,

P(event happening in any possible way) = 1/3 + 1/12 = 5/12. Use Equation 10.1, and P(event happening with a two tailed coin, if it happens at all) = (1/3)/(5/12) = 4/5 = 0.80. A likely event.

EXAMPLE 10.9:
Tree diagram probability.

A company uses a product supplied by two different vendors. Vendor A supplies twice as many parts as does vendor B. Vendor A has historically supplied parts that are 90 percent good, and vendor B has supplied parts that are 95 percent good. If a part is selected from the assembly line and it is bad, what is the probability that it came from vendor A? Figure 10.3 shows a tree diagram for this example.

P(bad part from A out of all parts) = 0.067

P(bad part out of all parts) = 0.067 + 0.017 = 0.084

P(bad part from A, if the part is bad) = 0.067/0.084 = 0.798. More likely than not.

FIGURE 10.3

Tree diagram for Example 10.9.

Select a part	Kind of part	Sequence probability
$P(A) = 2/3$	$P(\text{good}) = 0.90$	$(2/3)(0.90) = 0.600$
	$P(\text{bad}) = 0.10$	$(2/3)(0.10) = 0.067$
$P(B) = 1/3$	$P(\text{good}) = 0.95$	$(1/3)(0.95) = 0.317$
	$P(\text{bad}) = 0.05$	$(1/3)(0.05) = 0.017$

10.1.4 Series Reliability

A special design concern exists when the completion of an event in sequence with other events depends on the completion of a previous event in the sequence. The reliability, R_t, of a system of n components in series, each with a reliability, R_i, is calculated as the product of the individual reliabilities as follows: $R_t = \Pi R_i$, where $i = 1, 2,\ldots, n$. (10.5)

EXAMPLE 10.10:

Series event probability.

Consider a four machine battery where each machine operates independently of the other machines, and the part being made progresses sequentially from machine to machine. If the probability of each machine's being in a running mode is 90 percent, what is the probability that the four machine battery will be producing parts? If the assumption is made that all machines must be running, then the answer is the product of the individual probabilities, $(0.90)^4 = 0.656$. Recall that this is an **and** situation. Notice how quickly the overall reliability of the machine battery goes down.

Even if there were only two machines involved, the overall reliability (probability of both machines running) is 81 percent. Series reliability is one of the disadvantages of sequential machine batteries. From an engineering standpoint, the job that you might be assigned is to come up with ways to increase the reliability of each machine. To begin the analysis, data would be collected to help define the problem. Among the data you might collect are the causes for each machine to be down. Some reasons you might find, and suggestions for remedy, are as follows:

1. Operator absent. If the operator is absent, and the process is a critical one, a substitute operator should be brought in. The substitute might require additional training, and this must be taken into account. If the process is not a critical one, the absence and reduced uptime might not be so important.

2. Sharp perishable tools not available. If sharp drills, reamers, and other tools are not available, perhaps additional sets of tools should be kept ready for tool changes. Additional tools may have to be purchased. It also might be that the tool grinding room has a bottleneck and there are problems there that need solving.

3. Fixtures out of adjustment and being repaired. A fixture out of adjustment and needing repair is not an unusual condition. All tools wear as they are used. Perhaps tool repair could be done on a shift when this machine battery is not being run. If the process and the parts coming off the battery are critical, a second set of fixtures would be justified.

4. Normal maintenance downtime. Normal maintenance downtime should be scheduled for times when the machine is not scheduled to run: during off shifts, weekends, or plant shutdown and vacation times. Do not skip maintenance; the downtime problems will only be greater at a later date. The bearings may need replacing, or the bed ways may need regrinding, rather than just lubricating.

5. No material to be machined. Material flow is critical to keep any machine running, particularly batteries of machines. Be especially concerned if the material is a special casting or a hard-to-buy steel. The less common the material required, the higher the potential for delay.

6. No room left to store completed parts. Material handling must get raw material to the machine and must also remove finished parts safely from the area so there is no damage to the finished part and there is room to operate.

7. Machine malfunction, hydraulic or electric. Basic machine malfunctions will occur. They can be reduced by preventive maintenance, such as daily checks on fluid levels. Inspection of parts being produced on the machine is also a good monitoring device. Unscheduled downtime is more likely to occur as the sophistication of the machining increases. Numerical control, high precision part requirements, and complicated machining sequences all add to the likelihood that something will go wrong.

Ordinarily the probability of failure does not depend on the failure occurring at any regular interval of time. If, however, the failure rates of various components are constant for some incremental unit of time, then the reliability of a series system of these components is calculated as follows.

$$R_{system} \text{ (for time } t) = \exp[(-\Sigma\lambda_i)t], \text{ where} \tag{10.6}$$

λ_i = failures per unit time for the i_{th} component, and $i = 1, 2, \ldots, n$.

EXAMPLE 10.11:

Constant rate series reliability.

Component A of a two component series system has a known failure rate of 0.02 failures/1000 hours (0.00002) while component B of the system has a failure rate of 0.01 failures/700 hours (0.0000143). The 500 hour reliability of the two component system = exp[−(0.00002 + 0.0000143)500] = 0.983, while the 1000 hour reliability = exp[−(0.00002 + 0.0000143)1000] = 0.966.

10.1.5 Parallel Reliability

One method of improving the performance of multi-station processes is to have parallel lines of activity. For the same individual reliability, parallel lines of activity increase overall reliability. The reliability of a system, R_t, of n components in parallel, each with a reliability of R_i is calculated as follows.

$$R_t = 1 - \Pi\,[1 - R_i], \text{ where } i = 1, 2, \ldots, n. \tag{10.7}$$

EXAMPLE 10.12:

Parallel event reliability.

The overall reliability of a two unit, parallel system, each with a reliability of 90 percent, is 1 − (0.10)(0.10) = 99 percent.

If the failure rates of various components are constant for some unit of time, then the reliability of a parallel system of these components is calculated as follows.

R_{system} (for time t) = $1 - \Pi[1 - \exp(-\lambda_i t)]$, where (10.8)

λ_i = failures per unit time for the i_{th} component, and $i = 1, 2, \ldots, n$.

EXAMPLE 10.13:
Constant rate parallel reliability.

Using the same two components from Example 10.12 in parallel will result in a more reliable system. The 500 hour reliability of the two component parallel system = $1 - [1 - \exp(-0.00002)(500)][1 - \exp(-0.0000143)(500)] = 0.99993$. This analysis assumes that either component can perform the required function independent of the other component.

10.2 Normal Distribution

One of the most common statistical procedures used is to assume that data follows a normal distribution, and then make calculations on the data based on that assumption. This assumption is often erroneous because many real situations do not follow a normal distribution. Data that follows a normal distribution function is defined as $f(x) = \exp[-(x-\mu)^2/2\sigma^2] / [\sigma\sqrt{2\pi}]$, where σ is the standard deviation of the of the x data values and μ is the arithmetic mean of the x data values and the data plots as a bell shaped curve, as shown in Appendix G.

If data is normally distributed, then the calculated mean divides the data into two symmetric distributions. Normal data is also assumed to be continuous rather than discrete. In fact, both of these restrictions are usually violated. The first reason for violating the assumptions is that a physical object cannot be zero or negative size, and so the lower limit of a measurement is bounded while the upper limit is not. The second reason is the accuracy of measuring devices. If the smallest division measurable is 0.001 inch, then all data is discrete at the 0.001 level. This is not the same as the discrete measuring of such events as flaws per foot or cars per minute, where integer values occur. If a set of data is normally distributed, then the mean, **x**, and standard deviation, S, can be calculated based on a sample of the data, and 68.27 percent of all data will be within $\pm S$ of the mean, 95.45 percent of all data will be within $\pm 2S$ of the mean, 99.73 percent of the data will be within $\pm 3S$ of the mean, 99.994 percent of the data will be within $\pm 4S$ of the mean, and 99.99994 percent of the data will be within $\pm 5S$ of the mean.

10.2.1 Measures of Central Tendency

Measures of central tendency help quantify the distribution of the data.

The **mean**, **x**, is defined as the sum of the individual data values, $x_1, x_2, \ldots$, divided by n, the number of values. The mean can be thought of as the centroid or balance point of all the data.

$\mathbf{x} = (\Sigma x_i) / n$, where $i = 1, 2, \ldots, n$. (10.9)

The **median** is the data value in the middle of the list of values if they are arranged in order of magnitude. If an even number of values is listed, and hence no middle value, the mean of the two middle values is used. On the histogram it is the data value that divides the histogram into two equal areas.

The **mode** is the data value that occurs most often in a list of data. If a histogram is plotted, either singularly or by class groups, the mode is the value at the highest frequency occurrence on the graph.

10.2.2 Measures of Dispersion

Measures of dispersion help quantify the manner in which data is distributed about the mean of the data. The most common measure of dispersion is the **variance**, σ^2, and is defined as:

$$\sigma^2 = [\Sigma(x_i - \mu)^2]/n, \text{ where} \tag{10.10}$$

μ = mean of the universe,

n = the number of data values, and

x_i = individual data values, $i = 1, 2, \ldots, n$.

The **standard deviation**, σ, is commonly used rather than the variance and is calculated as the square root of the variance. If the mean of the universe, μ, is not known, then an estimate of σ^2 is made by calculating S^2, the sample variance.

$$S^2 = [\Sigma(x_i - \mathbf{x})^2]/(n - 1), \text{ or the alternate form} \tag{10.11}$$
$$S^2 = [\Sigma x_{i^2} - n\mathbf{x}^2]/(n - 1). \tag{10.12}$$

The **sample standard deviation**, S, is commonly used rather than the variance and is calculated as the square root of the sample variance.

The **range** is another measure of dispersion that simply notes the smallest and largest values from the list of data. Range generally increases with sample size, so use caution when using range values for comparison between groups of data with different sample sizes. Range does not have much significance if the sample size is less than 10.

EXAMPLE 10.14:

Sample data treated as normally distributed.

Consider the data taken from an experiment on the life of 16 incandescent light bulbs. The bulbs were on continuously until burnout. The hours until burnout for each bulb were 517, 493, 492, 678, 829, 443, 530, 549, 578, 338, 685, 674, 475, 552, 571, and 517.

$\mathbf{x} = 557$ hours, and $S = 115$ hours.

The range is $829 - 338 = 491$, the median is $(549 + 530)/2 = 540$, and the mode is 517. The difference in the three values, 557, 540, and 517, indicates the data is not exactly normally distributed. (See additional discussion on the normality of the data in Example 10.25.)

In this example a histogram requires some condensing to make it meaningful. If the horizontal axis has every possible outcome marked, 491 vertical lines are

TABLE 10.2

Cell tally for data from Example 10.14.

Cell range	Cell midvalue	Number of data
300.5–349.5	325	1
349.5–400.5	375	0
400.5–449.5	425	1
449.5–500.5	475	3
500.5–549.5	525	4
549.5–600.5	575	3
600.5–649.5	625	0
649.5–700.5	675	3
700.5–749.5	725	0
749.5–800.5	775	0
800.5–849.5	825	1

required and the graph would be too large for efficient use. Simplifying requires the creation of cells of values. Cells should be equal in size and selected so it is clear if a data value is in the cell or not. The values should also be selected so the midrange value is a possible data value. Generally the number of cells is maintained between 10 and 15. A heuristic formula to use for an estimate of the number of appropriate cells is to let the number of cells = 1 + 3.3[ln (number of data values)]. If different sets of data are to be compared, the same cell size must be used for all sets of data. Table 10.2 shows one possible tally by cells for the data and Figure 10.4 shows the associated histogram.

FIGURE 10.4

Histogram for Example 10.14.

10.3 Z Test

The Z test is used to assign a probability value to the likelihood that a set of data is representative of a known normal distribution. Appendix G summarizes the statistical values to use for this test. For derivation of the Z test model, and other tests reviewed in subsequent sections, see references (10), (12), and (14).

The question answered with the Z test is whether a data sample is likely to have come from a known or assumed universal group. The question often arises when a company receives parts or material from a supplier and the shipment is being inspected. A universe has a known or accepted standard from which mean, μ, and standard deviation, σ, values are available. To perform the Z test, take a sample of size n, measure a critical attribute on each sample, calculate the mean of the measurements, calculate the z value using Equation 10.13, and then determine the associated probability using Appendix G.

$$z = \frac{|x - \mu|}{\sigma/\sqrt{n}} \quad (10.13)$$

The sample size taken depends on the number of objects available and the consequence of having an incorrect conclusion. A value of $n \geq 30$ is recommended for reliable comparison. If the size of the sample is small, and the size of the universe is known, then calculate z with Equation 10.14 where N is the size of the universe, and n is the sample size.

$$z = \frac{|x - \mu|\sqrt{N - 1}}{(\sigma/\sqrt{n})\sqrt{N - n}} \quad (10.14)$$

The value calculated for z is located in a table of cumulative normal distribution values, Appendix G. The table gives the probability of having this value, and will be an indicator of whether the sample is from the universe.

EXAMPLE 10.15:

Z test.

Consider that the sample data in Example 10.14 was taken from a universe of 1000 with a known mean, $\mu = 750$, and standard deviation, $\sigma = 50$, then

$z = [|557 - 750|/(50/\sqrt{16})]\sqrt{999/984} = 15.56$.

The highest z value in Appendix G is 3.59, which is associated with a probability of 0.99983. This means that if the calculated value of z is 3.59, then only 17 times out of 100,000 would the data be representative of the universe. The symbol α is used to indicate this probability, in this case $\alpha = (1 - 0.99983) = 0.00017$. Because $15.56 > 3.59$, there is less chance than 17 out of 100,000 that the data is representative of the universe. Additional discussion of α error is contained in the next section.

The Z test is also used to determine if an attribute of a redesigned part has changed. The Z test can be extended to compare two test samples against each other to determine if there is any difference in the sample means. Models #4, #6, or #8 in Table 10.3 can be used depending on what is known about the universe. If there is no universe established, then model #8 is appropriate.

10.4 Decision Error

Engineers must draw a conclusion about sample data after a statistical calculation is made. Four possibilities exist after each decision is made.

1. The conclusion is made that the data is from the universe and it is from the universe, an acceptable result.
2. The conclusion is made that the data is from the universe and it is not from the universe, a β type error. The β error is often called a "status quo" error, for no change in the present status is detected or recognized.
3. The conclusion is made that the data is not from the universe and it is not from the universe, an acceptable result.
4. The conclusion is made that the data is not from the universe and it is from the universe, an α type error. This is a rejection-without-cause error, for a change in the present status is recognized when a change should not have been recognized.

Another way to interpret the α and β errors is in the comparison of means. This Z test and the associated error answers the question, "Has the average value been changed due to action taken?" If the conclusion is reached that there is a change, two conditions are possible:

1. There was a change and the correct decision was made.
2. There was no change and to recognize a change is a Type I or α error.

If a conclusion is reached that there is no change, two conditions are possible:

1. There was no change and the correct decision was made.
2. There was a change and the decision not to recognize the change is a Type II or β error.

There is a relationship between α and β errors but it is not a simple one. For a given sample size, if the probability of an α error is decreased then the probability of a β error is increased. The only way to reduce both types of errors is to increase n, the sample size.

EXAMPLE 10.16:
α error analysis.

Consider the experimental data $n = 30$ and $x = 195$, and the universe reference data $\mu = 200$, and $\sigma = 10$.

FIGURE 10.5

Relationship between α error and β error.

H_1, test hypotheses a change

H_0, null hypotheses status quo no change

α error

β error

Reject H_0 ⟶ ⟵ Accept H_0

Is the sample from the universe? Calculate z relative to the universe and determine α.

$z = |195 - 200|/(10/\sqrt{30}) = 2.74$.

From Appendix G, $z(2.74) = 0.9969$ which implies α = 0.0031 for a one-sided test. The sample is probably not from the universe and this conclusion will be wrong only 31 times out of 10,000.

To discuss the relationship between α and β, a value of α must be chosen. If a value of α is taken as 0.05, then a decision reliability of 0.95 is considered adequate, or there is a 5 percent chance of making an α type error. This may not be satisfactory for some situations. Note that the statistical Appendices H, I, and J have various α values shown. Statistical tables rarely show values for α greater than 0.10. Who wants to be wrong more than 10 percent of the time? In most cases a 0.05 or 0.01 error is used.

Figure 10.5 shows the graphical interpretation of the α and β error. Recall that when n increases the width of the standard curve generally reduces, and this lowers both the α and β errors. The α and β decision error analysis can be associated with any of the statistical tests made, as subsequent examples will illustrate.

10.5   *t* Test

The *t* test is used to answer the same question as the Z test, is the sample from an accepted universe, but it is used when the standard deviation of the universe is not known, only the mean is. The model is changed slightly from the Z test model as shown in Equation 10.15.

$$t = \frac{|x - \mu|}{S\sqrt{n}}, \text{ where} \qquad (10.15)$$

S is the sample standard deviation defined by Equations 10.11 or 10.12. Note again that the sample size n influences the calculation. Calculate a value of t, use the degrees of freedom equal to $(n - 1)$, and find the associated α error in Appendix H to estimate the probability that the sample is from the universe. Always note if the comparison being made could be greater or less than the mean of the universe, a two-tail test, or if the comparison can only be greater or only less than the mean, a one-tail test.

EXAMPLE 10.17:

t test.

Consider the data from Example 10.14, and assume the standard deviation of the universe is not known. Then $t = |557 - 750| / (115/\sqrt{16}) = 6.71$ with 15 degrees of freedom, one less than the number of data samples. In this case the sample mean could be greater or less than 750 so a two-tail test is made.

A t value of 4.073 indicates a probability of 0.001 that the sample is from the universe. The t value of 6.71 indicates less chance than that; that is, the α error is less than 0.001. This test agrees with the z test performed in Example 10.15 and confirms that the data is not from the assumed universe.

EXAMPLE 10.18:

Process control check.

A machine that controls the weight of a bagged product is being checked. Eleven samples were weighed at 49.2, 50.1, 49.8, 49.7, 50.1, 50.5, 49.6, 49.9, 50.4, 50.2, and 49.7 pounds. Is the machine maintaining a 50 pound mean?

$\bar{x} = 49.9$ lb and $S = 0.38$ lb

Because the standard deviation of the universe is not known, the t test is used:

$t = |49.9 - 50| / (0.38/\sqrt{11}) = 0.873$

Interpolating from Appendix H, α is approximately equal to $0.20 + (1.372 - 0.873)(0.20 - 0.10)/(1.812 - 1.372) = 0.31$. The 0.31 value indicates that if the decision is made to reject the 50 pound mean hypothesis, it will be a wrong decision approximately 31 times out of 100. Conclusion: The machine is maintaining a 50 pound mean. The original hypothesis is continued if it cannot be rejected by analysis. That does not mean that the original hypothesis is correct, only that it cannot be rejected at this time with the data on hand. Failing to reject, and accepting are not the same thing. Failing to reject simply means that at this time there is no data that supports rejection. Additional data may be required.

10.6 Effect Of Sample Size

A test is more reliable, and engineers feel more confident, if the number of samples tested increases. The limit is to test 100 percent of a product, and in some cases this is done. However, testing 100 percent of a product is expensive, so representative samples are generally taken. The size of the sample is important for good decision making as illustrated in the next example.

> **EXAMPLE 10.19:**
>
> ### Sample size.
>
> Consider a sample of 7 from a universe whose mean is thought to be 0.300. The sample mean and standard deviation are 0.353 and 0.070. Is the sample from a universe whose mean is greater than 0.300?
>
> $$t = |0.353 - 0.300| / (0.070/\sqrt{7}) = 2.00.$$
>
> For a one tailed test, α is approximately equal to $0.025 + (2.447 - 2.00)(0.025)/(2.447 - 1.943) = 0.047$. If the conclusion is made that the sample is not from the universe, it will be a wrong conclusion only 47 times out of 1000.
> Suppose that the sample whose mean and standard deviation are 0.353 and 0.070 is in fact from a universe whose mean is 0.400. What sample size, n, would be required to conclude, within an α error of 0.01, that the sample was from this new universe? For this to be true, it is necessary for $t_{\text{calculated}}(n) > t_{\text{table}}(n-1)$. For this sample $t_{\text{calculated}}(n) = |0.400 - 0.353| / (0.070/\sqrt{n})$. Trying different values of n until $t_{\text{calculated}}(n) > t_{\text{table}}(n-1)$ leads to $t_{\text{calculated}}(16) = 2.686 > t_{\text{table}}(16) = 2.583$, so at least 16 samples are required.

10.7 F Test

The F test exists as an additional aid to help determine if a sample comes from a known universe, or if two samples come from the same universe, when only the sample deviations are known. The model to use is shown in Equation 10.16.

$$F = (S_1/S_2)^2 \tag{10.16}$$

The largest calculated standard deviation is taken as S_1 because the table values of F are all greater than one. The table in Appendix I lists the F test values associated with various α errors. The sample size of both S_1 and S_2 must be known, as the table requires both their degrees of freedom to be used, usually one less than the sample size. The next example illustrates the F test procedure.

> **EXAMPLE 10.20:**
>
> ### F test.
>
> Consider a test of 15 light bulbs whose times until burnout were 569, 456, 637, 522, 578, 798, 650, 732, 552, 728, 655, 619, 456, 441, and 690 hours.

$\bar{x} = 605$ hours, $S = 108$ hours,

Range = 798 − 441 = 357, median = 619, and mode = 456. All the values are different from the sample in Example 10.14. Are the bulbs from the same universe?

In this case $F = (115/108)^2 = 1.13$; the degrees of freedom of the numerator is 15 and that of the denominator, 14. From Appendix I, the F value for $\alpha = 0.10$ is 2.01, so $\alpha > 0.10$ because $1.13 < 2.01$, hence it is safe to conclude that the samples are from the same universe. If the calculated value of F is greater than 2.01, the conclusion that the samples were from different universal sets would be wrong less than 10 percent of the time. If the calculated F value is greater than 3.66 the conclusion that the samples were from different universes would be wrong less than 1 percent of the time (99 percent confident).

Because the F test only considers the comparison of sample deviations, an alternate test can be used to compare two samples if it is thought that they are from different universes. The test considers the difference between sample hypothesis means with the relationship $\mu_1 - \mu_2 = \delta$; see Table 10.3, models #4, #5, #6, #7, and #8. The smaller the value of δ, the more likely the samples are from the same universe. The model to use depends on what is known: #4 requires σ_1 and σ_2 to be known; #5 does not require σ_1 and σ_2 to be known but does assume they are equal and allows the sum of sample sizes to be less than 30; #6 is the same as #5 but requires the sum of the sample sizes to be greater than 30; #7 and #8 do not require σ_1 and σ_2 to be known or to be equal, sample size dictates which to use. The next example illustrates this technique.

EXAMPLE 10.21:

Sample mean difference.

Use the data from Examples 10.14 and 10.20 and determine if the data are from the same universe by letting $\delta = 0$ in model #7 from Table 10.3.

$k = [108^2/15]/[(115^2/16) + (108^2/15)] = 777.6/1604 = 0.485$

°F = $1/[0.485^2/15) + (0.515^2/14)] = 29$

$t = (605 - 557)/1604 = 0.0299$.

In Appendix H, for a two tailed test, the smallest t value is 1.311 and is associated with an α error of 0.20. Because 0.0299 is much less than 1.311, α is much larger than 0.20, and this confirms the hypothesis that the data are from the same universe.

10.8 χ^2 Test

The Chi squared, χ^2, test is a method of comparing a sample against a standard reference value, a comparison between observed data and expected results. It is another tool to use to compare data to determine if a redesign , or other change, has resulted in significant difference in a product. The model for the Chi squared test is:

$$\chi^2 = \Sigma[(A - S)^2/S], \text{ where} \quad (10.17)$$

A = number of actual occurrences of a particular event, and

S = the standard to compare against.

The summation is used because all possible occurrences must be considered. The value calculated is then compared with a table of values for χ^2, Appendix J, for a selected significance level (α error level). The question answered is whether or not the events are occurring as expected when compared to a past, or known occurrence level. The reference value(s) must be known to perform the test as illustrated in the next example.

EXAMPLE 10.22:
Chi squared test.

A die is rolled 100 times with the following results. Is the die a fair die? Fair implies the observed frequency of any number = 100/6 = 16.7.

Number	1	2	3	4	5	6
Frequency	10	18	13	22	20	17

$\chi^2 = (10 - 16.7)^2/16.7 + (18 - 16.7)^2/16.7 + (13 - 16.7)^2/16.7 + (22 - 16.7)^2/16.7 + (20 - 16.7)^2/16.7 + (17 - 16.7)^2/16.7 = 5.95$. Find 5.95 in Appendix J for 5 degrees of freedom. Interpolating between 4.35 and 9.24 results in an α value approximately equal to 0.30, so the die is fair. To declare the die unfair would be a wrong decision 30 times out of 100. The related β error can be calculated by reversing the argument and assuming the rolled results are the standard and comparing them to the theoretical fair results:

$\chi^2 = (10 - 16.7)^2/10 + (18 - 16.7)^2/18 + (13 - 16.7)^2/13 + (22 - 16.7)^2/22 + (20 - 16.7)^2/20 + (17 - 16.7)^2/17 = 7.46,$

and α is approximately equal to 0.20. This α error is the β error associated with the first comparison, so if the die is not declared unfair it is an incorrect decision 20 times out of 100. Note that the smaller error occurs with the possibility of making no change (die is fair) compared to making a change (die is unfair).

EXAMPLE 10.23:
Chi squared test.

Two different models of a pump were being considered for use by a company. A 1000 hour experiment was run to establish the failure rate. Neither pump had been tested before so the premise is made that the failure rates are the same, $F_A = F_B$, and the average failure rate is assumed to be $(F_A + F_B)/2$. By using these assumptions each group of pumps is one variable, and the number of pumps is not considered a variable. Data collected from the proposed life of the pumps showed that 5 pump A's failed and 9 pump B's failed.

$\chi^2 = \{[F_A - (F_A + F_B)/2]^2/(F_A + F_B)/2\} + \{[F_B - (F_A + F_B)/2]^2/(F_A + F_B)/2\}$
$= (5 - 7)^2/7 + (9 - 7)^2/7 = 1.14$

For one degree of freedom (2 groups less one) and $\alpha = 0.20$, $\chi^2 = 1.642$. Because $1.14 < 1.642$, then $\alpha > 0.20$, and it is concluded that the pumps do have the same failure rate. Even though the number of pumps is not a concern, it is assumed that the same number of both types of pumps are being tested at all times. This means a pump that failed is repaired and placed back on test.

TABLE 10.3

Standard statistic tests.

Null hypothesis	Assumptions	Calculate	Alternate hypothesis	Critical value to reject null hypothesis
1. $\mu = \mu_0$	σ known, $n > 30$ preferred	$z = \dfrac{x - \mu_0}{\sigma/\sqrt{n}}$	$\mu \ne \mu_0$	$z < -z_{\alpha/2}$, or $z > z_{\alpha/2}$
			$\mu < \mu_0$	$z < -z_\alpha$
			$\mu > \mu_0$	$z > z_\alpha$
2. $\mu = \mu_0$	σ not known, $n < 30$ OK	$t = \dfrac{x - \mu_0}{S/\sqrt{n}}$	$\mu \ne \mu_0$	$t < -t_{\alpha/2}$, or $t > t_{\alpha/2}$
	μ known, $°f = n - 1$		$\mu < \mu_0$	$t < -t_\alpha$
			$\mu > \mu_0$	$t > t_\alpha$
3. $\sigma = \sigma_0$	μ not known, $°f = n - 1$	$\chi^2 = \dfrac{(n-1)\sigma^2}{\sigma_0^2}$	$\sigma \ne \sigma_0$	$\chi^2 < \chi^2_{\alpha/2}$, or $\chi^2 > \chi^2_{1-\alpha/2}$
			$\sigma < \sigma_0$	$\chi^2 < \chi^2_\alpha$
			$\sigma > \sigma_0$	$\chi^2 > \chi^2_{1-\alpha}$
4. $\mu_1 - \mu_2 = \delta$	σ_1 and σ_2 known, Any n	$z = \dfrac{x_1 - x_2 - \delta}{\sigma_{x_1 - x_2}}$	$\mu_1 - \mu_2 \ne \delta$	$z < -z_{1-\alpha/2}$, or $z > z_{1-\alpha/2}$
		$\sigma_{x_1-x_2} = [(\sigma_1^2/n_1) + (\sigma_2^2/n_2)]^{1/2}$	$\mu_1 - \mu_2 < \delta$	$z < z_{1-\alpha}$
			$\mu_1 - \mu_2 > \delta$	$z > z_{1-\alpha}$
5. $\mu_1 - \mu_2 = \delta$	σ not known, $\sigma_1 = \sigma_2$, $n_1 + n_2 < 30$, $°f = n_1 + n_2 - 2$	$t = \dfrac{x_1 - x_2 - \delta}{S_{x_1-x_2}}$	$\mu_1 - \mu_2 \ne \delta$	$t < -t_{\alpha/2}$, or $t > t_{\alpha/2}$
		$S_{x_1-x_2} = [k(1/n_1 + 1/n_2)]^{1/2}$		
		$k = \dfrac{(n_1-1)S_1^2 + (n_2-1)S_2^2}{°f}$		
			$\mu_1 - \mu_2 > \delta$	$t > t_\alpha$
			$\mu_1 - \mu_2 < \delta$	$t < t_\alpha$

μ = null hypothesis mean
μ_0 = alternate hypothesis mean
μ_1, μ_2 = comparative hypothesis means
σ_1, σ_2 = comparative hypothesis deviations
n = sample size
S = sample deviation
σ = null hypothesis deviation
σ_0 = alternate hypothesis deviation
x = sample mean
$°f$ = degrees of freedom
Sample and alternate hypotheses are often the same.

TABLE 10.3 (cont'd.)

Standard statistic tests.

Null hypothesis	Assumptions	Calculate	Alternate hypothesis	Critical value to reject null hypothesis
6. $\mu_1 - \mu_2 = \delta$	σ not known $\sigma_1 = \sigma_2$ $n_1 + n_2 > 30$	$z = (x_1 - x_2)/[(S_1^2 + S_2^2)/(n_1 + n_2)]^{1/2}$		
			$\mu_1 - \mu_2 \neq \delta$	$z < -z_{1-\alpha/2}$, or $z > z_{1-\alpha/2}$
			$\mu_1 - \mu_2 < \delta$	$z < z_{1-\alpha}$
			$\mu_1 - \mu_2 > \delta$	$z > z_{1-\alpha}$
7. $\mu_1 - \mu_2 = \delta$	σ not known $n < 30$ OK $\sigma_1 \neq \sigma_2$	$°f = 1/[(k^2/°f_1) + (1-k)^2/°f_2]$ $k = [S_1^2/n_1]/[S_1^2/n_1 + S_2^2/n_2]$ $t = [x_1 - x_2 - \delta]/[S_1^2/n_1 + S_2^2/n_2]^{1/2}$		
			$\mu_1 - \mu_2 \neq \delta$	$t < -t_{\alpha/2}$, or $t > t_{\alpha/2}$
			$\mu_1 - \mu_2 < \delta$	$t < t_\alpha$
			$\mu_1 - \mu_2 > \delta$	$t > t_\alpha$
8. $\mu_1 - \mu_2 = \delta$	σ and μ not known $\sigma_1 \neq \sigma_2$ n_1 and $n_2 > 30$	$z = [x_1 - x_2 - \delta]/[\sigma_1^2/(n_1 - 1) + \sigma_2^2/(n_2 - 1)]^{1/2}$		
			$\mu_1 - \mu_2 \neq \delta$	$z < -z_{1/\alpha-2}$, or $z > z_{1-\alpha/2}$
			$\mu_1 - \mu_2 < \delta$	$z < z_{1-\alpha}$
			$\mu_1 - \mu_2 > \delta$	$z > z_{1-\alpha}$
9. $\sigma_1 = \sigma_2$	Data is normal and independent $S_1 > S_2$ $F = (S_1/S_2)^2$		$\sigma_1 \neq \sigma_2$	$F > F_{1-\alpha/2}$
			$\sigma_1 > \sigma_2$	$F > F_{1-\alpha}$

μ = null hypothesis mean
μ_0 = alternate hypothesis mean
μ_1, μ_2 = comparative hypothesis means
σ_1, σ_2 = comparative hypothesis deviations
n = sample size
S = sample deviation
σ = null hypothesis deviation
σ_0 = alternate hypothesis deviation
x = sample mean
$°f$ = degrees of freedom
Sample and alternate hypotheses are often the same.

In general, statistical tests are made using a basic hypothesis that there is no change, and then analyzing data in an attempt to contradict that premise. See Table 10.3 for a summary of tests that can be conducted using various statistical hypotheses.

10.9 Normal Distribution Test

A test can be made to see how close data is to being normal. This test will help support or reject the hypothesis that the data values are normally distributed. The method to be discussed involves graphing on normal probability paper, as the next example illustrates.

EXAMPLE 10.24:
Normal distribution test.

The data from Example 10.14 will be used to illustrate this method.

Some preliminary calculations are required as shown in Table 10.4. Cells are set so all the data can be placed in 10–20 cells because the graph paper is divided into 20 major divisions. The frequency of data occurring in each cell is tallied. Cumulative frequency for any cell is determined by adding the cell frequency, the preceding cell frequency, and the preceding cumulative frequency. If this is done properly, the cumulative frequency after the last cell value should equal twice the sample size. Cumulative frequency percent is calculated by dividing the cumulative frequency by twice the sample size.

The results of the preliminary calculations are then graphed on normal probability graph paper. Label the linear vertical axis with the mean cell values and center the range of cell values on the axis so the scale values can be extended equally in both directions. Plot the cumulative percent value for each cell value as shown in Figure 10.6.

The closer the plotted points are to a straight line, the closer the data is to being normally distributed. By drawing a "best" straight line passing through the data, the mean and standard deviation can be estimated from the graph. In this case **x** = 555 and $\sigma = 555 - 435 = 120$. By extrapolating to the 3σ line, an estimate can be made on probable maximum bulb life for 99.77 percent bulb life range. For this example the value is 920 hours.

If the data as presented does not prove to be normally distributed, it may be that if the data is transformed it will become normally distributed. Typical transformations

TABLE 10.4

Cell tally for Example 10.24.

Cell Range	Frequency	Cumulative frequency	Cumulative frequency, %
300.5–349.5	1	1	3
349.5–400.5	0	2	6
400.5–449.5	1	3	9
449.5–500.5	3	7	22
500.5–549.5	4	14	44
549.5–600.5	3	21	66
600.5–649.5	0	24	75
649.5–700.5	3	27	84
700.5–749.5	0	30	94
749.5–800.5	0	30	94
800.5–849.5	1	31	97
		32	

FIGURE 10.6

Normal probability plot for Example 10.24

include taking the log, a power, a root, or the reciprocal of the data; all of the models used for mean and standard deviation must also change as a result. For example, the sample variance on log normal paper is:

$$S^2 = \frac{\Sigma(\log x_i - \log \mathbf{x})^2}{(n-1)} \tag{10.18}$$

10.10 | Bionomial Distribution

Many industrial process and system inspections only check to determine if something works or doesn't work; a part is either good or bad. No attempt is made at determining a degree of good or bad. The distribution used for this situation is the **binomial distribution**, where only two data distinctions are made, success or failure. The only graph that is generally used in this case is a plot of the number of sample defects, the bad parts, as a function of the interval of check. The probability of an event succeeding exactly r times out of n trials is given by Equation 10.19.

$P_r = {_nC_r}p^r q^{n-r}$, where (10.19)

p = the probability of success (good),

q = the probability of failure (bad) = $1 - p$,

n = the number of trials, and

${_nC_r}$ is defined by Equation 10.4.

The binomial distribution has the mean number of successes in n trials = np, and the standard deviation of the number of successes = $npq = np(1-p)$. The next examples will illustrate the use of the basic model.

EXAMPLE 10.25:
Binomial distribution.

Consider a process that historically has been making 95 percent good parts and 5 percent bad parts. If a sample of four parts is taken, what is the probability of 0, 1, 2, 3, or 4 good parts being found?

$P_0 = 1 \times 0.95^0 \times 0.05^4 = 0.000006$

$P_1 = 4 \times 0.95^1 \times 0.05^3 = 0.000475$

$P_2 = 6 \times 0.95^2 \times 0.05^2 = 0.013538$

$P_3 = 4 \times 0.95^3 \times 0.05^1 = 0.171475$

$P_4 = 1 \times 0.95^4 \times 0.05^0 = 0.814506$

The sum of the probabilities is 1 because there are no other events that can happen other than those listed. Recall that the values of ${_nC_r}$ can be gotten from the terms of Pascal's triangle.

EXAMPLE 10.26:
Binomial distribution.

An engineer in charge of a particular system has as a directive to report to management if the average error rate of a given process in the system is above 2 percent. The procedure calls for reviewing 10 transactions a day. The results of a five-day period sample are shown in Figure 10.7. Should the engineer make a favorable or unfavorable report?

If the system is under control then $p = 0.98$, $q = 0.02$, $P_r = {_{10}C_r}(0.98^r)(0.02^{10-r})$, and

$P_0 = 1(0.98^0)(0.02^{10}) = 1 \cdot 10^{-17}$ $P_1 = 10(0.98^1)(0.02^9) = 5 \cdot 10^{-15}$

$P_2 = 45(0.98^2)(0.02^8) = 1 \cdot 10^{-12}$ $P_3 = 120(0.98^3)(0.02^7) = 1 \cdot 10^{-10}$

$P_4 = 210(0.98^4)(0.02^6) = 1 \cdot 10^{-8}$ $P_5 = 252(0.98^5)(0.02^5) = 7 \cdot 10^{-7}$

$P_6 = 210(0.98^6)(0.02^4) = 3 \cdot 10^{-5}$ $P_7 = 120(0.98^7)(0.02^3) = 8 \cdot 10^{-4}$

$P_8 = 45(0.98^8)(0.02^2) = 0.0153$ $P_9 = 10(0.98^9)(0.02^1) = 0.167$

$P_{10} = 1(0.98^{10})(0.02^0) = 0.817$

FIGURE 10.7

Data for Example 10.26.

[Plot: Number of errors vs Day of the week; points at (1,1), (2,1), (3,0), (4,1), (5,0)]

The probability of finding one error is only 16.7 percent, but this week there was one error on three out of five days, 60 percent of the time. No errors should occur 81.7 percent of the time, but in this sample no errors occurred only 40 percent of the time. A formal comparison with expected results can be made using the χ^2 test.

$$\chi^2 = [0.817(5) - 2]^2/4.085 + [0.167(5) - 3]^2/0.835 + 0.0153(5) + 8 \cdot 10^{-4}(5)$$
$$+ 3 \cdot 10^{-5}(5) + 7 \cdot 10^{-7}(5) + 1 \cdot 10^{-8}(5) + 1 \cdot 10^{-10}(5) + 1 \cdot 10^{-12}(5) + 5 \cdot 10^{-15}(5)$$
$$+ 1 \cdot 10^{-17}(5) = 6.76, => \alpha \sim 0.75$$

for 10 degrees of freedom, an alpha value of 0.75 is not nearly small enough to discredit the process. Note that the last seven terms, each with zero occurrences, add very little to the sum.

A comparison can also be made using the entire sample of 50 and the success goal of 98 percent.

$$\chi^2 = [0.98(50) - 47]^2/49 + [0.02(50) - 3]^2/1 = 4.082, => \alpha \sim 0.05 \quad \text{for} \quad 1$$
degree of freedom

Now it is clear that there is a problem, and a report should be made. The reason that the first χ^2 test failed to show a problem was because of the small sample size, $n = 5$.

10.11 Poisson Distribution

Closely associated to the binomial distribution is the **Poisson distribution**, where occurrences are counted during a fixed interval, generally an interval of time. The main concern is whether an event happens or does not happen, but nonoccurrence is not a physical event. Examples include customers arriving at a checkout counter, the arrival of truck loads of material at the receiving dock, or traffic patterns on a stretch of highway. Occurrences are counted per unit time and an average occurrence determined, typically occurrences/minute or occurrences/hour. Once a historical occurrence has been determined then a probability of different occurrences can be calculated. This is done with the Poisson distribution formula:

$$P_r = \frac{e^{-\lambda t}(\lambda t)^r}{r!}, \text{ where} \tag{10.20}$$

r = the number of occurrences per unit to be considered,

e = the base of natural logarithms,

λ = the historical occurrence rate per unit, and

t = the number of units under study.

A Poisson distribution has a mean = λ, and a standard deviation = $\sqrt{\lambda}$.

EXAMPLE 10.27:
Poisson distribution.

If the average traffic flow over a period of time has been four cars per minute, what is the probability of observing 1, 2, 3, 4, 5, 6,...cars per minute?

λ = 4 cars per minute, t = 1 minute, and r = 0, 1, 2,...

$P_0 = e^{-4}4^0/0! = 0.018$ $P_1 = e^{-4}4^1/1! = 0.073$ $P_2 = e^{-4}4^2/2! = 0.146$

$P_3 = e^{-4}4^3/3! = 0.195$ $P_4 = e^{-4}4^4/4! = 0.195$ $P_5 = e^{-4}4^5/5! = 0.156$

$P_6 = e^{-4}4^6/6! = 0.104$ $P_7 = e^{-4}4^7/7! = 0.060$ $P_8 = e^{-4}4^8/8! = 0.030$

$P_9 = e^{-4}4^9/9! = 0.013$

It is 53 percent (0.104/0.195 = 0.53) as likely to observe six cars per minute as it is four cars per minute. It is more likely to observe five cars in a minute than it is no cars per minute.

EXAMPLE 10.28:
Poisson distribution.

Use the data from Example 10.27 and compare the probablity of observing six or more cars per minute to observing four or fewer cars per minute. This requires a comparison between $P_0 + P_1 + P_2 + P_3 + P_4$ and $P_6 + P_7 + \cdots + P_\infty$. That is, compare [$\Sigma P_i$ for i = 0, 1, 2, 3, and 4] = 0.627 to [$(1 - \Sigma P_i)$ for i = 0, 1, 2, 3, 4, and 5] = 0.217. It is nearly three times as likely, (0.627/0.217) = 2.9, to observe four or less cars per minute than it is six or more cars per minute.

The Poisson distribution is an important analysis tool for the design of systems where the number of units to be handled per unit time is critical. Examples include phone calls/minute, cars/minute, customers/hour, or computer data/second. The distribution is also useful for monitoring a process such as sheet steel manufacture, or paper or cloth manufacture, where quality control is based on the number of flaws per unit length. A goal may be to have no more than one flaw in some length interval. The next example illustrates this application.

EXAMPLE 10.29:

Process control check.

A paper mill strives for a maximum of one flaw per 25 feet of paper rolled and has historically met this goal. Ten random samples taken during one day found the following number of flaws per 50 feet of paper: 0, 2, 0, 2, 0, 2, 0, 3, 1, and 2. Is there a control problem?

$\lambda = 1$ flaw per 25 feet = 0.04 flaws per foot, $t = 50$ feet, $\lambda t = 2$

$P_0 = e^{-2} 2^0/0! = 0.13534$, $P_1 = e^{-2} 2^1/1! = 0.27067$,
$P_2 = e^{-2} 2^2/2! = 0.27067$, $P_3 = e^{-2} 2^3/3! = 0.18045$.
$P_4 = e^{-2} 2^4/4! = 0.09022$, $P_5 = e^{-2} 2^5/5! = 0.03609$, and
$P_6 = e^{-2} 2^6/6! = 0.01203$

Number of flaws	Theoretical probability	Actual probability
0	0.14	0.40
1	0.27	0.10
2	0.27	0.40
3	0.18	0.10

$\chi^2 = [0.14(10) - 5]^2/1.4 + [0.27(10) - 4]^2/2.7 + [0.27(10) - 1]^2/2.7 + [0.18(10) - 0]^2/1.8 = 11.95$,

and from Appendix J, χ^2 ($n = 3$) => $\alpha \sim 0.01$, and the process is probably in trouble.

When the actual occurrence of flaws is higher than expected based on history, then a process check is justified, and suggests a more frequent inspection be made until the problems are found, corrected, and the process is brought back under control. Another possibility is that the inspection process needs reviewing. Perhaps the person making the inspection needs retraining, or the equipment used for the check needs recalibration. Always consider the design of the whole checking procedure before criticizing any subpart of it.

10.12 Weibull Distribution

The Weibull frequency distribution is an analysis tool that is often used for design failure analysis to predict the life of a component or assembly, or to determine a product failure rate at a particular moment of time. It is a distribution that is not symmetric, because the lower limit is zero life and there is a possibility of a sample having an unusually long life. The basic Weibull model is:

$F(X) = 1 - \exp\{-[(X - x_0)/(\theta - x_0)]^m\}$, where   (10.21)

x_0 = lower bound of data, $0 \le x_0 \le x_1$, and

θ and m are parameters depending on the data to be molded.

If $x_0 = 0$, then the form of the distribution simplifies to: $F(X) = 1 - \exp(-X/\theta)^m$.

For the component failure application, where $X = t$ = time or cycle count, and $\lambda = (1/\theta)^m$:

$$P(t) = 1 - \exp(-\lambda t^m), \text{ where} \tag{10.22}$$

λ and m are the parameters dependent on experimental data.

For any given product test data, the larger the value of m, the more uniform the product, that is, the more predictable is the product life. This is a relative indicator usable for comparison only, there is no absolute value to strive for. If component failure data follows a Weibull distribution, the failure rate is not constant and is found by taking the first derivative of Equation 10.22, $dP(t)/dt = (m\lambda t^{m-1})\exp(-\lambda t^m)$.

The Weibull distribution is a straight line on log-log graph paper as the following discussion shows. Start with Equation 10.22 and manipulate using the laws of logarithms.

$1 - P(t) = \exp(-\lambda t^m)$

$\ln[1 - P(t)] = -\lambda t^m$

$\ln\{-\ln[1 - P(t)]\} = \ln(\lambda) + m \ln(t)$, where m is the slope of the line, and $\lambda = -\ln[1 - P(t)]$ when $t = 1$.

If $m = 3.57$, the distribution of the data will be nearly normal. If $m = 1$ the product is failing at a constant rate; if $m > 1$ the product is failing at an increasing failure rate; and if $m < 1$ the product is failing at a decreasing failure rate. The following example illustrates the Weibull analysis technique.

EXAMPLE 10.30:

Weibull analysis.

Twenty assemblies picked at random from a production line were tested until they all failed. The results of the test are summarized in Table 10.5 and graphed in Figure 10.8. The graph shows $\{-\ln[1 - P(t)]\}$ plotted as a function of cycles to failure. From the graph, $\lambda = 0.012$, and $m = 1.96$, so $P(t) = 1 - \exp(-0.012t^{1.96})$.

This model can be used to predict the fraction of the sample that failed after t number of cycles within the experimental range of $1.90 \cdot 10^5$ to $20.63 \cdot 10^5$ cycles. Recall the hazards of extrapolation beyond the experimental range.

In the event the Weibull graph cannot be approximated with a straight line, that is, the graph is concave or convex, then a value of x_0 must be estimated and the procedure repeated using Equation 10.21. An estimate of x_0 can be found by plotting the data on rectangular paper and extending the best curve to obtain $(x_0, 0)$. There is no assurance, of course, that any set of data will fit the model.

The model for $P(t)$, Equation 10.22, was arrived at using completely mathematical methods and with no limiting assumptions. Any data that plots as a straight line on log-log paper will fit the model. The data does not have to be failure data. See Chapter 12 for additional discussion on data curve fitting.

TABLE 10.5

Work table for Weibull distribution, Example 10.30.

Order of failure	Cycles to failure·10^5	Fraction of failures, $p(t)$	$-\ln[1 - p(t)]$
1	1.90	1/20	0.05
2	3.34	2/20	0.11
3	3.65	3/20	0.16
4	4.20	4/20	0.22
5	4.72	5/20	0.29
6	5.89	6/20	0.36
7	6.10	7/20	0.43
8	6.62	8/20	0.51
9	7.92	9/20	0.60
10	8.40	10/20	0.69
11	8.50	11/20	0.80
12	9.00	12/20	0.92
13	9.60	13/20	1.05
14	11.02	14/20	1.20
15	11.95	15/20	1.39
16	12.40	16/20	1.61
17	13.03	17/20	1.89
18	13.42	18/20	2.30
19	18.07	19/20	3.00
20	20.63	20/20	—

10.13　Process Capability

A manufacturing process is expected to make parts within an acceptable tolerance range because the proper functioning of a product design depends on it. Checking manufacturing processes on a regular basis is normal to verify that all is well. One method of making this check is to inspect parts from the process at regular intervals. A second method of monitoring a manufacturing process involves conducting a process capability study. The process capability study is made after a machine has been properly set up to run a part, which includes having the necessary tools on the job properly sharpened and adjusted, the machine tool running with the correct speeds and feeds, and the material within the correct specifications, including alloy composition and hardness.

Parts coming from the process are measured and the sizes recorded. If more than one dimension is being machined on the process, each dimension is measured and tallied separately. The mean, **x**, and the standard deviation, S, of each set of dimensions are calculated. If the process is 99.73 percent capable of machining the dimension (270 bad parts in 100,000), then the **x** ± 3S limits should be smaller than the limits of the tolerance

FIGURE 10.8

Weibull plot for Example 10.30.

on the part drawing. This is shown graphically in Figure 10.9. Because of recent industrial emphasis on quality, some companies are requiring that the $x \pm 4S$ process limit be within the part print tolerances, which results in a 99.994 percent reliability (6 bad parts in 100,000). Tables 9.1, 9.2, 9.3, and 9.4 are the results of process capability studies.

In a very critical situation, the allowable variation region in Figure 10.9 can be zero. If that occurs, then close attention must be given to the process to insure that proper tooling and operating conditions are always used, and that the process is centered in the

FIGURE 10.9

Relationship between process control limits and drawing dimension limits.

capability range. If the allowable variation region is very large, the process being used may be inappropriate and more expensive than required to make the part. Recall the discussion of using a bar lathe to manufacture bosses normally done on a saw.

The upper and lower process control limits, UCL and LCL, are arbitrary guidelines, but usually taken at $3S/2$. As long as part dimensions lie between these limits, no corrective action is required. If the dimensions fall outside the PCL lines but inside the part drawing limits, adjustments can be made to the process before bad parts are made.

EXAMPLE 10.31:
Process capability check.

A grinding process is making a part with a print specification diameter of 0.680 ± 0.002. The following diameter measurements were taken from 20 parts: 0.681, 0.682, 0.681, 0.682, 0.681, 0.680, 0.680, 0.683, 0.680, 0.681, 0.679, 0.678, 0.679, 0.678, 0.679, 0.680, 0.680, 0.677, 0.680, and 0.687.

$\mathbf{x} = 0.6804$, $S = 0.0021$, $\mathbf{x} - S = 0.6783$, and $\mathbf{x} + S = 0.6825$.

Although only 3 of 20 samples are bad (85 percent good), the process is only about 68 percent reliable because $(\mathbf{x} \pm S)$ is approximately equal to the part print tolerance specifications. Either the process is out of control, or it is not capable of making the part to print specifications.

The sample size of 20 in Example 10.31 is not large enough to establish an absolute process control limit. If the purpose is to establish an absolute process control, a sample size greater than 50 is appropriate, and the analysis is usually done using subgroups. The subgroup process control study procedure used in the following example is the result of Walter Shewhart's work in statistics that he did in the 1920s while working for Bell Labs.

EXAMPLE 10.32:
Process control-capability study.

Consider a turning operation on a lathe, where parts are measured and sizes recorded sequentially. Sequential recording is essential if time dependent variations are going to be discovered. The size of the subgroup can range from 2 to 25, but 5 to 10 is normal. Table 10.6 shows 75 measurements, taken sequentially, and recorded in 15 subgroups of 5 values in each group. The mean and range of each subgroup is calculated (shown in the table), and then the mean of the entire group of 75, and the mean of the range values is calculated; $x = 0.18155$, $R = 0.00043$.

An estimate of the standard deviation of the universe is calculated as $\sigma_{est} = R/F_1$, where F_1 is a function of the subgroup size, and listed in Table 10.7, $\sigma_{est} = 0.00043/2.326 = 0.000185$. Note that the value of σ_{est} is less than the mean of the sample of 75, which is 0.000193. Note also that the F values used here are not related to the F test discussed in Section 10.7.

The UCL = $x + F_2 R$ = 0.18155 + (0.577)(0.00043) = 0.1818, and the LCL = $x - F_2 R$ = 0.18155 − (0.577)(0.00043) = 0.1813. For exactly 3 σ_{est} control, the part print dimension should be 0.18155 ± 3(0.000185) = 0.1821/0.1810, and for exactly 4 σ_{est}

TABLE 10.6

Data for Example 10.32.

Subgroup number	\multicolumn{5}{c}{Subgroup data element number}	Subgroup total	Subgroup mean	Subgroup range
	1	2	3	4	5			
1	0.1817	0.1814	0.1815	0.1817	0.1817	0.9080	0.18160	0.0003
2	0.1817	0.1817	0.1816	0.1813	0.1818	0.9081	0.18162	0.0005
3	0.1815	0.1815	0.1817	0.1817	0.1817	0.9081	0.18162	0.0002
4	0.1816	0.1818	0.1814	0.1815	0.1815	0.9078	0.18156	0.0004
5	0.1818	0.1816	0.1815	0.1817	0.1818	0.9084	0.18168	0.0003
6	0.1813	0.1816	0.1816	0.1815	0.1813	0.9073	0.18146	0.0003
7	0.1818	0.1812	0.1815	0.1814	0.1815	0.9074	0.18148	0.0006
8	0.1819	0.1819	0.1817	0.1815	0.1812	0.9082	0.18164	0.0007
9	0.1817	0.1816	0.1816	0.1816	0.1813	0.9078	0.18156	0.0004
10	0.1818	0.1813	0.1818	0.1817	0.1815	0.9081	0.18162	0.0005
11	0.1814	0.1817	0.1817	0.1819	0.1816	0.9083	0.18166	0.0005
12	0.1815	0.1814	0.1813	0.1816	0.1813	0.9071	0.18142	0.0003
13	0.1813	0.1813	0.1817	0.1815	0.1818	0.9076	0.18152	0.0005
14	0.1814	0.1815	0.1815	0.1812	0.1815	0.9071	0.18142	0.0003
15	0.1812	0.1811	0.1818	0.1817	0.1815	0.9073	0.18146	0.0007

TABLE 10.7

Factors for process data analysis using subgroup *range* calculation.

N, Subgroup size	F_1	F_2	F_3	F_4
3	1.693	1.023	2.574	0
4	2.059	0.729	2.282	0
5	2.326	0.577	2.114	0
6	2.534	0.483	2.004	0
7	2.704	0.419	1.924	0.076
8	2.847	0.373	1.864	0.136
9	2.970	0.337	1.816	0.184
10	3.078	0.308	1.777	0.223

General expression	Range	Accuracy over range
$F_1 = 9.27\, N^{0.0879} - 8.314$	$5 \leq N \leq 25$	$\pm 2\%$
$F_2 = 3.19\, N^{-1.13} + 0.071$	$4 \leq N \leq 25$	$\pm 3\%$
$F_3 = 3.17\, N^{-0.867} + 1.353$	$3 \leq N \leq 25$	$\pm 1\%$
$F_4 = 0.555\, N^{0.237} - 0.717$	$11 \leq N \leq 25$	$\pm 3\%$

control, $0.18155 \pm 4(0.000185) = 0.1823/0.1808$. If these values are greater than part print specifications, then the process is not capable of making the part.

If $\mathbf{x} \pm 3\,\sigma_{est}$ is less than the part print specification, then the process does not have to be centered to be in control. The allowable variation region equals the part print tolerance range minus six standard deviations.

Additional analysis of the data will give clues to process stability (not part processing capability). A stable process will produce parts within a small range of tolerances, a precise operation. A process may be capable of manufacturing parts within an acceptable range of tolerances but it may not be a precise operation, (see Exercise 10.16 (c)). The following is a list of signals that suggest process instability exists.

1. If any subgroup has its mean greater than the UCL, or less than the LCL.
   a) If the cause of the out-of-range condition can be identified as an isolated case, then the out-of-range subgroup can be ignored and the calculations repeated using the remaining data.
   b) If the cause of the out-of-range condition cannot be identified as an isolated case, but can be identified and corrected, then the correction should be made and the study repeated.
   c) If the cause of the out-of-range condition cannot be identified, then additional data must be collected and the study started over at the subgroup following the out-of-range condition. If an out-of-range mean does not occur again, the first one can be assumed to be a nonrepeatable abnormality.

FIGURE 10.10

Graphical analysis, Example 10.32.

(a) Subgroup average analysis

(b) Subgoup range analysis

2. If seven or more consecutive subgroup means are greater than or less than the overall mean. This condition indicates something may be out of adjustment or worn, or a measuring device may be out of adjustment. The source should be located, repaired, and the study started over. If the parts have been saved in sequence, and the problem is with the measuring device, then the parts already run can still be measured and used in the study.

3. If seven or more consecutive subgroup means are increasing or decreasing in value. This condition indicates a drift in the process. This drift can be caused by process variation, measuring device variation, or material variation. The condition should be corrected and the study started over. If the parts have been saved in sequence, and the problem is with the measuring device, then the parts already run can still be measured and used in the study.

The subgroup range values must also be analyzed. High and low values within a subgroup can average out, and the mean analysis may show no problem, but the range may be out of control. The upper control limit for the subgroup range value is $UCL_R = F_3\mathbf{R}$; for the data in this example, $UCL_R = 2.114(0.0043) = 0.0009$. The lower control limit is $LCL_R = F_4\mathbf{R}$; for the data in this example, $LCL_R = 0$. Use the same three criteria for the analysis of the subgroup range values that were used for the analysis of the subgroup mean values.

Plot the data on a graph for additional aid in the analysis as trends are easier to locate on graphs. Figure 10.10 shows the analysis of this example in graph form.

As sample sizes get larger, the standard deviation of the subgroup becomes more efficient than range as an indicator of process variability. The procedure using standard deviation follows the same pattern as the one outlined in Example 10.32:

1. Tabulate data in sequence using subgroups.
2. Calculate the mean and standard deviation of each subgroup.
3. Calculate the mean, $\mathbf{x}$, and standard deviation, S, of the total sample.
4. Calculate $\sigma_{est} = S/G_1$. Refer to Table 10.8 for values of G_1.
5. Calculate the $UCL = \mathbf{x} + G_2 S$.
6. Calculate the $LCL = \mathbf{x} - G_2 S$.
7. Compare $\mathbf{x} \pm 3\sigma_{est}$ or $\pm 4\sigma_{est}$ to part print specifications.
8. Check subgroup mean values using the three instability criteria.
9. Calculate $UCL_S = G_3 S$ and $LCL_S = G_4 S$.
10. Check subgroup standard deviation values using the three instability criteria.

10.14 | Table Data Analysis

The numerical values of material properties found in tables and handbooks should be used with caution. Are the values shown average values? Are they minimum values? Are they theoretical values? In most cases table data has come from experiments and tests on samples of the material. Experimental data always has scatter because the material is

TABLE 10.8

Factors for process data analysis using subgroup *standard deviation* calculation.

N, subgroup size	G_1	G_2	G_3	G_4
5	0.9400	1.427	2.089	0
6	0.9515	1.287	1.970	0.030
7	0.9594	1.182	1.882	0.118
8	0.9650	1.099	1.815	0.185
9	0.9693	1.032	1.761	0.239
10	0.9727	0.975	1.716	0.284
11	0.9754	0.927	1.679	0.321
12	0.9776	0.886	1.646	0.354

General expression	Range	Accuracy over range
$G_1 = 1.8927\, N^{0.0162} - 0.9979$	$5 \leq N \leq 25$	$\pm 1\%$
$G_2 = 3.621\, N^{-0.647} + 0.158$	$5 \leq N \leq 25$	$\pm 1\%$
$G_3 = 3.628\, N^{-0.879} + 1.229$	$5 \leq N \leq 25$	$\pm 2\%$
$G_4 = 0.684\, N^{0.247} - 0.923$	$10 \leq N \leq 25$	$\pm 5\%$

inconsistent, so absolute property values should not be expected. If a table shows average material property values, then the standard deviation is also needed so ranges of values and probabilities of performance can be determined, but unfortunately most data tables do not give this information. Some material suppliers will list minimum strength values for material properties. This is still not precise. Is the value a 99.5 percent reliability? A 99.9 percent reliability? How then does an engineer determine an acceptable value to use? Tests can be run on the material available, such as a tensile or impact test, or the actual part manufactured to the design specifications can be tested. The second choice is always the best if time permits, but that doesn't answer the questions about table values.

One procedure to use is to assume the table values are average values and use a standard deviation of some percentage of that value; 8 percent is often used. The next example illustrates the effect of this variation. See also Section 11.1.1.

EXAMPLE 10.33:

Material data variation.

A value for yield strength for a specific material was found in a table to be 30,000 psi (207 MPa). If 30,000 was assumed to be an average value, then half of the material samples would fail below 30,000 psi and half would fail above 30,000 psi. This would not be an acceptable level to use in a design. If the standard deviation is assumed to be 8 percent of the table value, then a 99.86 percent ($x - 3\sigma$) reliability strength level would be $30{,}000 - 3(0.08)(30{,}000) = 22{,}800$ psi (157 MPa). Use this approach only if no other data is available and a first iteration value is needed.

Other cases occur when a company uses the same grade material over and over in the design of their products. In the course of extended use, in-house design strength values will be developed. This is good because it represents actual use of the material. It can be bad if a value obtained for one application is used for a unrelated application. Either over-design and excessive cost, or under-design and failure might be the result. Always be sure of the source and meaning of any material strength data used, whether it be in-house or reference table values. The following example illustrates one potential problem.

EXAMPLE 10.34:
In-house material data.

One company used 1020 HR steel for many of their products. The strength value of 16,000 psi (110 MPa) was the accepted in-house strength limit. An engineer, new to the company, questioned the source of the value, and after much investigation discovered it was a fatigue limit value for a product no longer in production. No one remembered if the fatigue loading was axial, torsional, constant amplitude, varying amplitude, for a welded structure, or some other combination loading. Consensus was that it was a conservative value, but no one could be quantitative in their answer. How much money was being unnecessarily spent because of over-design?

10.15 Additional Study Topics

1. The use of a Weibull location parameter.
2. Sample size for a given α and β error.
3. Gamma distribution.
4. Exponential distribution.
5. Hypergeometric distribution.
6. Permutation and combination theory.
7. Chauvent's criteria for data rejection.
8. Nair's method for data rejection.
9. Root mean square tolerancing.
10. Pareto 80/20 distribution analysis.
11. In-process and post-process inspections.
12. Air gaging.
13. Factorial experiments.
14. Fractional experiments.
15. Markov chains of sequential probability.
16. Grubbs method of rejecting suspect data observations.
17. Dixon method of rejecting suspect data observations.

18. Graeco-Latin squares for experimental sequencing.
19. Success-Run theorem: $R_c = (1 - C)^{1/(n+1)}$, where R_C = reliability at confidence level C and n = sample size.
20. Newman-Keuls range test.
21. Stem and leaf display of data.
22. Operating characteristics curves.
23. Geometric mean, used for averaging rates of growth: $x_g = (\Pi x_i)^{1/n}$, where $i = 1, 2, \ldots$.

10.16 References

1. Barker, Thomas B. *Quality by Experimental Design*. New York: Marcel Dekker, 1985. Includes applicable BASIC computer programs.
2. Bethea, Robert, Benjamin Duran, and Thomas Boullion. *Statistical Methods for Engineers and Scientists*. New York: Marcel Dekker, 1975. Emphasis on Chemistry.
3. Beyer, W. H., ed. *Handbook of Tables for Probability and Statistics*, 2nd ed. Boca Raton, Fla.: The Chemical Rubber Co., 1968.
4. Blank, Leland. *Statistical Procedures for Engineering, Management, and Science*. New York: McGraw-Hill, 1980.
5. Budnick, Frank S. *Applied Mathematics for Business, Economics, and the Social Sciences*. New York: McGraw-Hill, 1979. Chapter 16.
6. Ford Motor Company Staff. *Continuing Process Control and Process Capability Improvement*. Dearborn, Mich.: Ford Motor Company, 1983.
7. Gallin, Daniel. *Finite Mathematics*. Glenview, Ill.: Scott Foresman, 1984. Chapters 7, 8, 9, 10, and 11.
8. Grant, Eugene L. and Richard S. Leavenworth. *Statistical Quality Control*, 5th ed. New York: McGraw-Hill, 1980.
9. Holman, J.P. *Experimental Methods for Engineers*. New York: McGraw-Hill, 1971.
10. Hoog, Robert V. and Elliot A. Tanis. *Probability and Statistical Inference*, 2nd ed. New York: Macmillan, 1983.
11. Hooke, Robert. *How to Tell the Liars from the Statisticians*. New York: Marcel Dekker, 1983.
12. Kachigan, Sam Kash. *Statistical Analysis*. New York: Radius Press, 1986. A very readable book.
13. Kennedy, John B. and Adam M. Neville. *Basic Statistical Methods for Engineers and Scientists*. New York: Harper and Row, 1976.
14. Lipson, Charles and Narenda J. Sheth. *Statistical Design and Analysis of Engineering Experiments*. New York: McGraw-Hill, 1973.
15. Little, Robert E. *Probability and Statistics For Engineers*. Beaverton, Ore.: Matrix Publishers, 1978.
16. McCormick, Norman J. *Reliability and Risk Analysis: Methods and Nuclear Power Applications*. New York: Academic Press, 1981.

17. McDonald, T. Marell. *Mathematical Methods for Social and Management Sciences.* Boston: Houghton Mifflin, 1974. Chapters 5 and 6.
18. Messina, William S. *Statistical Quality Control for Manufacturing Managers.* New York: John Wiley, and Sons, Inc., 1987.
19. Montgomery, Douglas C. "Experimental Design and Product and Process Development." *Manufacturing Engineering.* September, 1988, 57–63.
20. Piascik, Chester. *College Mathematics with Applications.* Columbus, Ohio: Charles E. Merrill, 1984. Chapter 7.
21. Resnick, William. *Process Analysis and Design for Chemical Engineers.* New York: McGraw-Hill, 1981.
22. Schenck, Hilbert Jr. *Theories of Engineering Experimentation.* New York: McGraw-Hill, 1961.
23. Schmid, Calvin F. *Statistical Graphics: Design Principles and Practices.* New York: Wiley-Interscience, 1983.
24. Stout, Ken. *Quality Control in Automation.* Englewood Cliffs, N.J.: Prentice Hall, 1985.
25. Taguchi, Genichi. *Introduction to Quality Engineering.* New York: Asian Productivity Organization, 1986.

10.17 | Exercises

1. The F test in Example 10.20 showed that data from Examples 10.14 and 10.20 are probably from the same universe. Combine the data, prepare a histogram, and calculate x and S.
2. One game involving two dice has the player winning the game if a 7 or an 11 is rolled on the first roll of the dice and losing if a 2 or 12 is rolled on the first roll.
   a) What is the probability of losing the game on the first roll?
   b) What is the probability of not winning on the first roll?
   c) Not winning is not the same as losing. The non-winning number rolled on the first roll becomes the player's number, and the player will win if the same number comes up on the second or subsequent rolls. Which non-winning number(s) would you prefer to roll on the first roll if you were the player, and what is your probability of winning with that number?
   d) What number, other than a 7 or 11, would the operator of the game want you to roll on the first roll? Why? What is your probability of winning with that number?
3. A sample of a particular universe was taken and $x = \mu$. Determine the ratio σ/S for this sample if the sample size is 30. In general how does sample size affect the σ/S ratio?
4. Each machine in a battery of machines is never completely independent from the other machines in the battery. Discuss modes of "failure" that would affect more than one machine at the same time.
5. Discuss solutions to the conditions raised in Exercise 10.4.

6. A sequential three-machine battery is handled by one operator and a jib boom crane. Each machine has its own fixture, and the part being machined weighs 200 pounds. Discuss possible solutions to each of the conditions (a) through (g). Consider advantages and disadvantages of each of your solutions.
   a) The jib boom crane breaks down.
   b) Machine #2 has a fixture problem.
   c) Machine #1 fails hydraulically.
   d) Machine #3 runs out of sharp milling cutters.
   e) Machine #2 starts producing parts out of print tolerances.
   f) The motor in machine #1 burns out.
   g) The motor in machine #3 burns out.

7. Each of the systems in Figure 10.11 performs the same function; parallel paths are redundant design. If the reliability of part A is 92 percent and costs $1000, and the reliability of part B is 85 percent and costs $700, which of the systems shown in Figure 10.11 would you consider the "best?" Why?

8. A process machine is checked hourly, and records indicate possible defect rates of 0, 1, 2, and 3 failures per hour with respective probabilities of 1/3, 1/3, 1/6, and 1/6.
   a) Estimate the total number of defects in an eight-hour shift.
   b) If 12 defects were found during an eight-hour shift, would there be a problem? Explain.
   c) If there were 30 defects in an eight-hour shift, would there be a problem? Explain.

9. A process is assumed to be in control if the output product averages no more than two defects in a 100 foot roll.
   a) What is the likelihood of selecting a 100 foot roll that has no defects?
   b) Three or more defects?

FIGURE 10.11

Configurations for Exercise 10.7

10. What level of process reliability would you assume if in a batch of 50 items checked there are five defects? What additional information and details would you like to have before being confident of your answer?

11. The number of available computer terminals in a room with 20 terminals was noted over a five-day period and the following results noted:

Number of terminals available	0	1	2	3	4	5	6	7 or more
Observed frequency	30	35	42	20	15	5	3	none

   a) If checks were taken on a regular basis, how often were checks made?
   b) If a student goes to the terminal room three times a day, what is the likelihood that all the terminals will be busy on each visit?

12. A new material was selected to use for making a spring. A sample of 100 machines was assembled. Fifty machines had the spring made with the old material, and fifty were assembled with the spring made from the new material. The 100 machines were tested, and the following results noted.

	Machines with longer life	Machines without longer life
New spring	30	20
Old spring	19	31

Is the new spring significantly better? If yes, at what level of significance?

13. Two different machines turn the same diameter part. On a given day some samples were measured and the diameter measurements in inches recorded.

   Machine 1: 0.117, 0.130, 0.122, 0.139, 0.132
   Machine 2: 0.120, 0.127, 0.133, 0.115, 0.137, 0.126

   Are the two machines producing the "same" size part?

14. A product "second" is often sold at a reduced price. Discuss some considerations that might affect your decision to sell a product as a second rather than repairing it and selling it at full price.

15. An automatic size gage makes a correct decision on determining a good or bad part 90 percent of the time. If 85 percent of the parts being checked are actually good, what is the probability that a part that is checked as being good is actually good?

16. Discuss the process conditions that would result in each of the cases shown in Figure 10.12. Explain what has probably happened and what you would do about the situation.

17. Three components of a series design have reliabilities of 0.90, 0.85, and 0.80.
   a) Demonstrate that the overall system reliability can be improved the most by duplicating in parallel the component that is least reliable.
   b) Prove mathematically that the premise in part (a) is always true.

18. If one coin in a pile of 12 coins is counterfeit, that is, heavier or lighter than the rest, what is the probability of having a balanced scale if two coins are drawn from the pile and one placed on each side of a balance scale? Repeat the problem for 4, 6, 8, 10 and 12 coins drawn from the pile and divided equally on both sides of the scale.

FIGURE 10.12

Charts for Exercise 10.16.

19. Determine if the combined data from Examples 10.14 and 10.20 is "normal" by graphing the data on normal probability paper.

20. Choose one of the topics from the list of additional study topics and prepare a paper and an oral presentation.

21. The mean of a set of n data values is defined as the value that makes the sum of the squares of the deviation from each data x_i from this mean a minimum. The expression can be written as:
Sum = $\Sigma(x - x_i)^2$, where $i = 1, 2, \ldots, n$.
Derive Equation 10.9 from this expression and prove that it is a minimum.

22. Use the Weibull distribution equation in the form $P(t) = 1 - \exp[-(t/\theta)^m]$, and let $t = \theta$. Then $P(t) = 0.63$, and 63 percent of the samples lie on one side of θ, and 37 percent lie on the other side. Prove this special case, and find this critical value for the data in Example 10.30.

23. The following data is the time in hours to failure of a sample of 20 electrical components. Determine a Weibull distribution equation and estimate what percent of the components will last more than 10 days.
Hours to failure: 106.2, 159.6, 194.3, 215.9, 227.6, 271.6, 274.3, 297.7, 299.0, 345.5, 357.6, 373.4, 398.0, 430.5, 447.9, 485.3, 512.4, 548.9, 564.5, 608.9

24. Analyze the combined data from Examples 10.14 and 10.20 using the Weibull distribution.

25. What distribution form does the Weibull distribution approximate when $m = 1$?

26. What is the significance as the value of the Weibull parameter m gets larger?

27. Is the data in Exercise 10.23 normally distributed? Use two test methods.

FIGURE 10.13

Cut blocks for Exercise 10.28.

(a) Grain direction

(b) Grain direction

28. Blocks of wood $2 \times 2 \times 2$ inches are cut from lumber 2×2 inches $\times$ various lengths. The blocks are fed into an automatic machine that splits the blocks on a diagonal to make prism-shaped glue blocks. The desired result has the grain direction shown in Figure 10.13(a). The automatic machine, however, does not orient the blocks the same way each time, and some blocks come out with the grain direction shown in Figure 10.13(b). The second grain orientation is satisfactory as long as the wood is hardwood, but not if it is softwood. Hardwood and softwood are used in equal amounts but are mixed as the machine is running. The finished blocks are dumped into a common bin.
   a) What is the probability of selecting no bad blocks in a set of six blocks?
   b) What is the probability of selecting a set of six blocks from the bin and having more than one bad block in it?
   c) What is the probability of selecting a set of six blocks and having at least one bad block?
   d) What happens to the probability of getting six good blocks if a set of seven blocks is chosen? Select seven and throw one away.

29. Run the following experiments using paper straws and a paper cutter. Keep track of time to cut ten straws to size and the measured length of the straws after cutting. Conduct a process capability analysis using the procedure of Section 10.13.
   a) Cut the straws into three equal-length pieces with only a mark on the paper cutter.
   b) Cut the straws into three equal-length pieces with a positive stop on the paper cutter.
   c) Compare the data and discuss the results.

30. If you have access to a band saw or table saw, run the following experiment. Use a 2×2 (1.50×1.50 inches) piece of wood of fixed length (2 foot minimum).
   a) Cut the length of board into 1.50 inch lengths with only a line mark on the saw table. Record the time to cut and the accuracy of the pieces cut.
   b) Repeat part (a), but use a positive stop to control the cut length.
   c) Compare the data and discuss the results.

31. Using the data from the blocks cut in Exercise 10.30 part (b), determine the probability of stacking six blocks and reaching a height < 9.00 inches, > 9.00 inches, and exactly 9.00 inches.

32. Two different tachometers were used to verify an experiment; the assumption made was that the actual RPM = 1000. Analyze and discuss the results.

   Tach #1 readings: 950, 952, 948, 951, 950
   Tach #2 readings: 910, 1050, 990, 1030, 1080, 890, 1050, 1060

33. A dial gage is used to check a production component. An inspector checks the same part many times to determine the reliability of the gage. The readings and frequency are listed. Later the same gage is used to check a different part, and a 1.33 reading is obtained. Discuss the situation. How big do you think the second part is? With what confidence?

Reading	1.31	1.32	1.33	1.34	1.35	1.36	1.37	1.38	1.39	1.40	1.41
Frequency	2	6	22	38	57	44	19	10	0	0	2

34. The relationship between ship speed, V, and engine horsepower for a particular ship is: $hp = a + bV^3$, where a and b are parameters.
   a) What kind of a graph will show the relationship as a straight line?
   b) Determine the values of a and b if 29 horsepower are required at 5 knots, and 230 are required at 12 knots.
   c) Determine the horsepower required for a speed of 10 knots.

35. What conclusions can be drawn from the following data, taken from samples of the same material? The numbers are values of S_{ut} in ksi.

Temperature:	Room	Room + x	Room + 2x	Room + 3x
	66	74	55	52
	65	71	56	49
	72	60	55	55
	69	65	49	53
	70	66	53	51

36. A four machine $300,000 fixed part battery is being serviced by a $30,000 robot. The reliability of each machine is 90 percent and that of the robot is 95 percent.
   a) What is the net reliability of the battery?
   b) The battery has work scheduled to keep it busy two shifts (16 hours) per day, so maintenance work can be done on the third shift. Does the battery reliability need to be increased to be able to meet schedule? If no, how much reserve is there? If yes, how would you suggest improving the situation?
   c) Will having a spare robot available help much for overall reliability?

37. Discuss the design of an experiment to determine if the popping of a batch of popcorn follows a normal distribution.

38. A manufacturer of brake shoes wants to prorate warranty replacement cost based on lining wear, adding the cost of warranty replacement to the price of the product. Tests for shoe wear at the warranty mileage, for a sample of ten shoes, show wear of 0.08, 0.09, 0.06, 0.07, 0.10, 0.05, 0.08, 0.07, 0.10, 0.12 inches. Determine a level of wear for 100 percent warranty replacement cost.

39. An engineer is working on a new pump design, and one specification goal is a pumping rate of 9.0 gal/min. Twelve pumps were tested and the following pumping rates measured: 7.8, 8.4, 9.0, 9.2, 9.5, 9.8, 10.2, 10.3, 10.7, 11.0, 11.6, 12.2. How is the design progressing?

40. Compare the following test data on the life of two competitive products. The values are thousands of cycles to failure. Are the product performances the same?

    Product A: 16, 30, 40, 52, 70, 85, 185.
    Product B: 25, 45, 65, 75, 97, 120, 140, 160, 210, 250.

41. For the process in Example 10.26 to be under control, that is, an error rate ≤ 2 percent, how many days of the week would you expect no errors to be found?

42. An ad for a new machine states, "Well designed, well built, $\pm 4\sigma$ production." Discuss the meaning and/or implication of this statement.

43. A grinding process for bearings makes parts with a mean dimension of 1.490 inches diameter with a standard deviation of 0.005. A bearing larger than 1.500 will cause the related assembly to fail. This failure causes a warranty cost of $50. How much money can be spent on improving the manufacturing quality if 10,000 bearings are produced each year?

44. Three parts, $A, B,$ and C, are used in an assembly. Historically the percent of good parts of A has been 95 percent, for B 98 percent, and C 92 percent.
   a) Determine the probability of having an assembly with no bad parts.
   b) One bad part.
   c) Two bad parts.
   d) Three bad parts.

45. A part is processed through four sequential operations. The operations have a record of making good parts at rates of 80 percent, 85 percent, 95 percent, and 90 percent, in that order.
   a) What is the probability of getting a bad part after the four operations if bad parts are removed after each operation?
   b) What is the probability of getting a bad part after four operations if bad parts are not removed after each operation?

46. Show that the equation for χ^2 from Example 10.23 can be simplified to:

$$\chi^2 = (F_A - F_B)^2/(F_A + F_B).$$

47. Two dice are rolled 72 times with the following results. The data is (dice sum, frequency): (2,4), (3,6), (4,4), (5,9), (6,8), (7,12), (8,9), (9,6), (10,7), (11,3), and (12,4). Is it likely that the dice are "fair"?

48. From the data in Example 10.23, does it follow that if a standard failure rate of six pumps was established, pump A is OK and pump B is not? Explain.

49. An electrical component has a constant failure rate of 1/1000 hours.
    a) What is the 750-hour reliability of a two-component series circuit?
    b) What is the 750-hour reliability of a two-component parallel circuit?

50. Refer to Examples 10.11 and 11.13 and demonstrate that the 1000 hour parallel reliability of the two components is higher than the 1000 hour series reliability.

TABLE 10.9

Data for Exercise 10.59 and 10.60.

Sample #	Size	Sample #	Size	Sample #	Size
1	0.763	26	0.764	51	0.764
2	0.764	27	0.763	52	0.764
3	0.764	28	0.763	53	0.764
4	0.763	29	0.764	54	0.765
5	0.764	30	0.765	55	0.764
6	0.764	31	0.763	56	0.764
7	0.763	32	0.764	57	0.763
8	0.764	33	0.764	58	0.763
9	0.764	34	0.764	59	0.764
10	0.763	35	0.764	60	0.762
11	0.764	36	0.764	61	0.764
12	0.764	37	0.765	62	0.764
13	0.764	38	0.765	63	0.763
14	0.764	39	0.763	64	0.763
15	0.765	40	0.764	65	0.764
16	0.764	41	0.764	66	0.764
17	0.763	42	0.764	67	0.764
18	0.765	43	0.764	68	0.764
19	0.764	44	0.764	69	0.763
20	0.764	45	0.765	70	0.763
21	0.764	46	0.764	71	0.764
22	0.764	47	0.764	72	0.763
23	0.764	48	0.763	73	0.763
24	0.764	49	0.765	74	0.764
25	0.764	50	0.764	75	0.764

51. Two test engineers took concrete samples for a compression test study. The samples had been poured eight days before and were thought to be from the same batch of cement. Is it likely that they are from the same batch?

    Engineer A's sample: 4350, 3850, 4240, 3110, 3890, 3850, 3260, 3780 psi.
    Engineer B's sample: 4240, 3750, 4100, 4130, 4660, 4310, 3950, 4210, 4490, 4060 psi.

52. It is desired to run a test on the performance of a product to determine if a design change has altered the average output. The present average is 21.0 with a deviation of 2.0. How many samples must be run with an average of 21.5 and a deviation of 2.0 to determine if the design change has raised the average? *Hint*: Assume a value for α and let n be a variable.

53. How do the results in Exercise 10.52 change if the universe mean is 10.0 with a deviation of 2.0, and a change to a mean of 12.0 is desired?

54. Suppose the desired change in Exercise 10.52 was to keep the same mean but change the deviation to 1.0. Reanalyze the problem.

55. Suppose the desired change in Exercise 10.53 was to keep the same mean but change the deviation to 1.0. Reanalyze the problem.

56. Combine the desired changes from Exercises 10.52 and 10.54 and reanalyze.

57. Show how Equation 10.11 can be transformed into Equation 10.12.

58. Suppose the part in Example 10.32 was dimensioned 0.1815 ± 0.0005. What percent of all parts made are likely to be out of part print tolerance?

59. The data in Table 10.9 came from consecutive parts from a grinding operation. What part print dimension will allow this process to be used with a ± 4σ confidence? Use a subgroup size of five.

60. Use the first 72 values in Table 10.9 and a subgroup size of six, and determine if this makes any difference in the results of the process capability study.

11 Failure Analysis

"I believe that the concept of failure...is central to understanding engineering, for engineering design has as its first and foremost objective the obviation of failure."
♦ *HENRY PETROSKI*

Chapter one summarized a number of engineering failures, and briefly mentioned likely causes. Many of those causes are related to the load conditions imposed on various materials. In this chapter many of the causes of material failure will be discussed in greater detail. Sometimes, however, the first problem is to agree on what conditions must exist before a part or system is considered a failure. In general a failure occurs when, under use or implementation, a part or system no longer performs the function it was designed to do.

The term product failure can have different meanings depending upon the circumstances. One person will say a car has failed when it won't start. A race car driver might say a car has failed if it won't go over 100 miles per hour. Someone else will consider a car a failure when the body rusts out. Because failure means different things to different people, engineers must consider a wide range of possible conditions and failure modes when designing a product.

A wooden pencil fails when the lead breaks or becomes dull. This is a temporary failure and can be fixed by sharpening. A total failure occurs when the pencil gets too short to sharpen or hold on to. A fix on a total failure is to buy a new pencil or borrow one from a friend. Another solution is to invest in a lead holder. Temporary failure for a lead holder is running out of lead. This can be fixed by inserting more lead. Total failure occurs when the lead advance mechanism fails. The purpose of both pencils is the same, but the modes of failure are different.

Some failure modes to consider when designing physical products are included in the following list. Failure can be from too much or not enough of each of the criteria. For example, lack of ductility can cause metal to rupture when it is being formed in a die, while too much ductility can cause failure if elongation causes a part to bend out of shape.

1. Ductility
2. Brittleness
3. Fatigue
4. Creep
5. Tension
6. Compression
7. Corrosion or rotting
8. Color change
9. Volume change
10. Phase change
11. Shear
12. Bending
13. Torsion
14. Buckling
15. Impact
16. Crack growth
17. Transparency

There are many reasons why failures occur, the primary ones being:

1. Wrong interpretation of loads applied to system.
2. Unforeseen use by the customer.
3. Faulty manufacture, parts not made to specifications, system not set up as planned.
4. Faulty material, raw material or purchased components not to specification.
5. Incorrect material or component choice.
6. Incorrect specification requested.
7. Incorrect assembly or use procedures.

Component parts used in assembly or individually must be designed to perform their function in a reliable and safe manner. They must be economically manufacturable so that the consumer will get a product at a reasonable cost.

A reliable part can be counted on to do the job it was designed to do, when it is supposed to, and in the desired manner. A shear pin may not be overloaded for many cycles of use, but when it is overloaded, it must fail or it has not performed reliably. A fire extinguisher may not be used for months, but when a fire breaks out, the extinguisher's reliability must not be in doubt. A light switch must reliably make and break an electric circuit thousands of times during its life.

A part and its assembly must perform in a safe manner, not only for the user of the product but for the protection of innocent bystanders. A lawn mower must cut the grass but not throw grass or other debris at the operator. A parking lot gate arm must open on signal, allow a car to pass safely under, and not come down on top of the car, even if the car stalls under the arm. If the hub holding the blade of a lawn mower fails, the loose blade creates an unsafe condition. If the welded joint on a bridge fails, results can be catastrophic. If the seal on the door of a spacecraft leaks, the results can be fatal to the occupants and perhaps to the entire program.

A part and its assembly must be economical to supply to the consumer. An automobile designed to prevent the occupants from being harmed regardless of the type of collision or accident might be possible, but the cost would exclude purchase by the average consumer. Cost-benefit relationships must be analyzed. Engineers must also consider system designs, whose failures might include:

1. Too much time needed to operate
2. Too many people required to operate
3. Security easily violated, unauthorized use

4. Proper data recording forms impossible to purchase
5. Output doesn't meet postal regulations
6. Output doesn't meet IRS regulations
7. Records are lost or inaccessible
8. Audit of records is difficult or impossible
9. Instructions are inadequate
10. Procedures are user hostile, prone to input errors, with no error traps or user escape routes.

11.1 Design Interferences

The design of a shear pin is governed by knowing the load at which failure is desired. This desired failure load must be less than the strength of the parts being protected by the pin. The pin's design is also governed by the shear strength of the material from which the pin is made and the cross-sectional area that resists the shear. These values are not controllable to exact values and are subject to variations.

11.1.1 Material Strength Variations

The shear strength of a material is a variable. A material could have a mean shear strength of 50,000 psi (345 MPa) and a standard deviation on that value of 4000 psi (27.6 MPa). That means that the actual strength of a particular pin could range from a low value of 38,000 psi (262 MPa) to a high value of 62,000 psi (427 MPa) within 99.7 percent ($\pm 3\sigma$) confidence limits. For a pin whose diameter = 0.1875 inch (4.76 mm), the range of fail load would be 1050 lbf (4671 N) to 1710 lbf (7606 N): (38,000 psi)(π)(0.1875 in)^2/4 = 1050 lbf (4671 N), and (62,000 psi)(π)(0.1875 in)^2/4 = 1710 lbf (7606 N). Can the design accommodate this range of fail values?

11.1.2 Manufacturing Tolerances

The tolerance associated with the manufacturing processes used to make the pin can also have an effect on the range of fail values. Consider a diameter tolerance on the manufacture of the pin of ± 0.006 inch (0.1524 mm) with a standard deviation of 0.002 inch (0.0508 mm). The result is a 3σ range of areas from 0.0259 in^2 (16.71 mm^2) to 0.0294 in^2 (18.77 mm^2): π(0.1815)^2/4 = 0.0259 in^2, and π(0.1935)^2/4 = 0.0294 in^2. Combine the variation in area with a strength of 50,000 psi (345 MPa), and the failure load range is 1295 to 1470 lbf (5760 to 6539 N). While this variation is less than the range caused by the material strength range, it can be significant in a design that cannot tolerate much variance.

Combine the range of diameters with the range of strengths, the actual load at failure could be as low as (0.0259 in^2)(38,000 psi) = 984 lbf (4377 N) or as high as (0.0294 in^2)(62,000 psi) = 1823 lbf (8109 N). That represents an 85 percent variation, (1823 − 984)/984 = 0.85, a significant amount.

11.1.3 Loading Variations

If the load applied to the assembly is an abrupt shock load, the load effect is magnified. A load abruptly placed on an assembly will result in at least a doubling of the applied load effect. Is the component designed to fail at an impact load of 50 lbf (222 N) or a static load of 100 lbf (445 N)? Will the component fail at a lower than maximum load if the load is applied over and over again? Fatigue failures can occur at less than half of the one-cycle, static-test ultimate strength.

11.1.4 Other Variations

Will the component operate at a constant temperature or over some temperature range? The strength of a material is affected by temperature changes. Impact strength and ductility increase as temperature increases, but shear strength decreases.

Will the component be subjected to a corrosive environment and deteriorate, resulting in failure at less than the designed load value? Galvanic action between two metals, and general atmospheric corrosion can reduce critical cross sectional areas or create a nucleus for crack growth.

This chapter reviews various failure modes that must be considered in the design of any physical product. It is important to consider all the variables that might have an effect on the design of a reliable, safe, and economical product. Guidelines will be presented that are the result of years of experimental and real product test work. When a particular guideline is applied to a specific design, the underlying assumptions and conditions of the parameters must be understood. If not, bad applications of sound principles will result. The guidelines presented here will only be a starting point. All designs must be tested under the conditions that will actually be encountered in use. There is no substitute for testing a product under real-time use, but there is a need for thorough pre-build analysis. Designs must be built and tested under real conditions, but the better the first design is (the paper, pencil, computer design), the quicker a product can be brought to the marketplace, and the lower will be the cost. No guideline should ever be accepted as gospel. Every design situation has its own peculiarities, and guidelines, experience, common sense, and tests must be properly balanced for optimum results.

11.2   Measure of Reliability

A part or assembly designed to a level of reliability of 98 percent means that on the average 2 parts out of 100 will fail. A 98 percent probability of performing satisfactorily is the same as a 2 percent probability of failing. This does not mean that 2 parts of a random sample of 100 parts are going to fail. As many as 500 parts may be tested before any fail. What it does mean is that all parts cannot be guaranteed to perform satisfactorily. Do not confuse reliability with value. If a product has value, it not only performs reliably, but economically and without undesirable side effects. A car may start every morning, reliably, but if it is noisy, runs with annoying vibration, and gets 2 miles per gallon of fuel, it is not a good value.

11.2.1 Series Component Reliability

The reliability of an assembly, R_t, of n series components, where the failure of any one component can cause the assembly to fail, is determined as the product of the component reliabilities, R_i as shown in Equation 11.1, (previously discussed in Section 10.1.4). Example 11.1 shows how the model is used.

$$R_t = \Pi R_i, \text{ where } i = 1, 2, 3, \ldots, n. \tag{11.1}$$

EXAMPLE 11.1:

Series reliability.

Consider two parts, each with a reliability of 98 percent, used in an assembly where failure of either part will make the assembly fail, such as two bearings on a rotating shaft. The contribution to reliability due to the two bearings is the product of their individual reliabilities, $(0.98)(0.98) = 0.96$. If the shaft that the bearings are mounted on has a reliability of 0.95, and the housing that contains the bearing has a reliability of 0.94, then the four part assembly has a reliability of $(0.98)(0.98)(0.95)(0.94) = 0.86$. This means that on the average, 14 assemblies out of 100 will fail. Most companies could not stay in business with this level of product reliability because warranty claim expense would be excessive. The effect of the individual reliabilities of components on the overall design reliability must be considered.

11.2.2 Parallel Component Reliability

If an assembly is critical, such as when failure of a system will cause certain death, redundancy in design is often required. Two or more systems are run in parallel so that if either system fails, the remaining one takes over. This improves total effective reliability because failure of the system now requires the failure of both components.

The reliability of an assembly, R_t, of n parallel components, where failure of all of the components must occur before assembly failure occurs, is determined by considering the product of the failure rates of each component $(1 - R_i)$ as shown in Equation 11.2, (previously discussed in Section 10.1.5). Example 11.2 shows how the model is used.

$$R_t = 1 - \Pi(1 - R_i), \text{ where } i = 1, 2, 3, \ldots, n. \tag{11.2}$$

EXAMPLE 11.2:

Parallel reliability.

Consider two safety latches holding a door closed where either latch can hold the door closed. If each latch has a reliability of 0.95, then the reliability (probability) of the door staying latched $= 1 - (1 - 0.95)(1 - 0.95) = 0.998$.

This type of analysis explains the design of such items as multiple cable winch systems, backup pump systems, and parallel computer systems.

A seemingly parallel design will not follow Equation 11.2 if the failure of the system is in fact controlled by one component as illustrated by the following example.

> **EXAMPLE 11.3:**
> **Series reliability.**
>
> An automobile has four tires operating at the same time, but failure of any one tire will cause the auto to fail to perform as designed. If the reliability of each tire is 0.99, then the reliability of the four tire set is $0.99^4 = 0.96$.

Never let parallel backup systems stand idle. Their reliability can deteriorate because of nonuse, and fail just when they are needed most. Backup computer systems, for example, are kept on and in communication with the system through the primary computer. This is sometimes referred to as "hot" backup. Among the things that can occur through nonuse of backup systems include the following.

1. Grease and oil seals can dry out and leak.
2. Rubber components can deteriorate due to attack by ozone.
3. Fuels can lose their volatility and not ignite when required.
4. Water can condense, creating rust and other surface deterioration that cause parts in contact to corrode and possibly become welded together. Electrical contacts may corrode and resistance become too high to control the circuit properly.
5. Vibration from external sources can cause parts to come loose. Parts bolted together may fall off; computer chips may become unplugged.
6. Chemical changes can take place. Substances intended to be fluid can coagulate, those intended to be mixture or compounds may separate.

A nuclear power plant, for example, was cited for a safety violation when a second level backup generator had not been run for six months. Each of three cooling towers have a backup generator in case of a power failure. The generator that had not been run for six months was a single second level backup in case any one of the three first level backup generators failed when placed into service.

11.2.3 Determining Reliability

The best method of determining reliability of a product is to test the part or assembly in its intended environment. An engineer can use all the formulas, theories, and past practices available and still fail to conclude with a reliable design. The reasons for this include:

1. Material specifications variability
   a) Alloy variations
   b) Contamination variations
   c) Crystal structure variations
2. Uncontrollable environmental conditions
   a) Temperature
   b) Humidity
   c) Gases in the atmosphere
   d) Vibration from external sources

       e) Consumer misuse
       f) Product maintenance
3. Variations in manufacture
       a) Refining process material composition
       b) Size control
       c) Surface finish
       d) Heat treatment
       e) Part handling

The component parts and assemblies tested for reliability should come from normal material sources, be manufactured using the methods that will be used in final production, and be tested as the consumer will use the product. Parts that are special made for test do not allow conclusions to be made about product design standards.

A light bulb built by hand and tested under laboratory conditions of constant voltage, constant room temperature, and free from vibration, will not give the results necessary to predict average consumer use. A good lab test will try to account for as many of the customer use variations as possible. A field test under less than ideal conditions will generally show up more problems than will a lab setup.

This does not mean that lab testing is not worthwhile, because it is. Single design features can be isolated for test and help rule out unreliable ideas before money is spent for complete test models. Components can be tested without having a complete product available. Ideas for changes and new techniques can be tried under close monitoring. Combinations of ideas can be tried to see if something is being overlooked, and test monitoring and data collection can be better controlled. The following example illustrates a reliability test performed on incandescent light bulbs.

EXAMPLE 11.4:

Reliability test.

Consider the test of 15 light bulbs with a rated life of 550 hours. The bulbs were left on continually, and the time in hours to burnout for each bulb was 550, 580, 585, 520, 515, 515, 580, 560, 550, 545, 535, 530, 580, 590, and 520.

Burnout is not the only type of failure that could be considered, a specified level of illumination might also be a criterion for failure. Notice again that failure criterion is not a constant.

A quick calculation indicates that indeed the light bulbs have an average life of 550 hours, with a standard deviation of 27 hours. If it is assumed that the scatter of the data is normally distributed, then the following conclusions can be made.

1. The light bulb is 50 percent reliable at a 550 hour life. Half of the bulbs will not last 550 hours, half will last longer than 550 hours.

2. The reliability of the average bulb having a rated life of some other value than 550 can be determined from standard normal distribution tables.

Once test data is available, then other questions about reliability can be answered, as the next example illustrates.

> **EXAMPLE 11.5:**
>
> ## Alternate reliability level.
>
> What is the reliability at a rated life of 500 hours for the bulb sample in Example 11.4? A life of 500 hours is −1.85 standard deviations from the mean; (500 − 550)/27 = −1.85. Appendix G shows −1.85 corresponding to a 96.78 percent reliability level. Would an advertised life of 500 hours for the bulbs be more accurate information for the consumer?

Reliability values and ratings are critical to good product design. If the reliability goal for a product is 99.99 percent (one failure in 10,000) and that level of reliability cannot be achieved with a series component design, then parallel design redundancy may be required. The concept of limited life design must also be considered. Under limited life design philosophy a component is replaced at less than its rated life so that the total product continues to function without failure. This is commonly done with products with limited life such as oil filters, spark plugs, antifreeze, furnace filters, winch cables, light bulbs, and brake linings. Everything will fail sometime, but a good engineer will be guided in product design by a knowledge and awareness of expected reliability.

11.3 Safety Factors

Regardless of the care an engineer takes, and the details checked, there are always unanswered questions that need careful consideration.

1. Have proper load levels been assumed?
2. Are the material strength values reliable?
3. Are the models relating loads to material stress valid?
4. Will the parts be manufactured to the design specifications?
5. Will the material used meet minimum standards?
6. Will the consumer use the product as intended?
7. Will the consumer maintain the product properly?
8. How will the product deteriorate after it is placed in service?
9. Will real use load the material beyond the yield point?

To compensate for these and other unknowns, most designs include a safety factor. A safety factor is a hedge on the unknown and the unpredictable. Safety factors can be used in different ways, but in all cases the intent is to make the product safer when used by the customer. If people will die or incur a large property loss as a result of a product failure, then safety factors must be high enough to avoid failure.

A safety factor can be used to increase the size of an anticipated load, reduce the life of a replaceable part, increase the size of a component, lower the strength value of a material, reduce the spacing of support structures, or increase the number of components. Safety factors are also used when the cooling area for heat exchangers is increased, when a fuel tank is made larger, when material is double stitched, when additional computer memory storage is provided, and when equipment is designed to control voltage peaks. Many product codes require the use of safety factors. Elevator wire cable codes, for example, call for a safety factor of 8 for mine shaft elevators, and 12 for high speed passenger elevators (32).

Safety factors as low as 1.1 can be employed if the production and use of a product is carefully controlled and monitored. Low safety factors are used when excessive component weight is a problem, such as in the aircraft industry. The next example illustrates the use of a safety factor.

EXAMPLE 11.6:
Calculation of a safety factor.

A bolt is to be used to withstand a 5000 lbf (22.2 kN) tensile load, and is to be made from a material which has $S_{yt}(min) = 20{,}000$ psi (138 MPa). What safety factor is in effect against yielding if the bolt used is 0.625 inches (15.88 mm) in diameter?

$\sigma = 5000$ lbf $/ (\pi(0.625^2)/4)$ in$^2 = 16{,}300$ psi (112 MPa), and SF $= 20{,}000/16{,}300 = 1.23$

11.4 Probabilistic Design

An alternative to the safety factor design method is the probabilistic design method. Probabilistic design requires that the mean and standard deviation of all variables be known or estimated. By combining the mean and standard deviation values in all calculations, the probability of failure can be determined, or if a reliability requirement is given, the necessary variable or tolerance levels can be determined.

The first concern to address is to determine how means and standard deviations can be combined in calculations. If Z is a function of several variables, $Z = F(x_1, x_2, \ldots)$, the standard deviation of Z, S_Z, is defined as: $S_Z = \sqrt{\Sigma[(\partial Z/\partial x_i)(S_{x_i})]^2}$ (1). (11.3)

The next three examples illustrate the use of Equation 11.3.

EXAMPLE 11.7:
Standard deviation of a product function.

Consider a function Z, which is the product of two terms, $Z = XY$, where X and Y have standard deviations S_X and S_Y. Apply Equation 11.3 to Z.

$\partial Z/\partial X = Y$, and $\partial Z/\partial Y = X$, so $S_Z = \sqrt{(YS_X)^2 + (XS_Y)^2}$ (11.4)

EXAMPLE 11.8:
Standard deviation of a sum function.

Consider a function Z, which is the sum of two terms $Z = X + Y$, where X and Y have standard deviations S_X and S_Y. Apply Equation 11.3 to Z.

$$\partial Z/\partial X = 1, \text{ and } \partial Z/\partial Y = 1, \text{ so } S_Z = \sqrt{S_X^2 + S_Y^2} \qquad (11.5)$$

The sum model, Equation 11.5, is the basis for root mean square tolerancing of assemblies, where the size of each part and each individual tolerance are accounted for in an assembly dimension. If n components are assembled and their dimensions, $D_i \pm T_i$ accumulate, that is, the assembled dimension $D_A = \Sigma D_i$, then the tolerance on the assembled dimension is $\sqrt{\Sigma(T_i)^2}$.

EXAMPLE 11.9:
Probabilistic design.

Use the data from the shear pin calculation from Section 11.1 with the model $F = \sigma A = \sigma \pi D^2/4$, and determine the reliability and range of F.

$\partial F/\partial \sigma = \pi D^2/4$, and $\partial F/\partial D = \sigma \pi D/2$, and

$S_F = \sqrt{[(\pi D^2/4)(S_\sigma)]^2 + [(\sigma \pi D/2)(S_D)]^2} = (\pi D/2)\sqrt{(D^2/4)S_\sigma^2 + \sigma^2 S_D^2}$

$S_F = \pi(0.1875/2)\sqrt{(0.1875^2/4)(4000)^2 + (50{,}000^2)(0.002^2)} = 63$ lbf (280 N)

F(mean) $= 50{,}000\pi(0.1875)^2/4 = 1381$ lbf (9.52 MPa), and based on Appendix G,

0.14% of the pins will not shear at $1381 + 3(63) = 1570$ lbf (6984 N)
2.28% of the pins will not shear at $1381 + 2(63) = 1507$ lbf (6703 N)
15.87% of the pins will not shear at $1381 + (63) = 1444$ lbf (6423 N)
84.13% of the pins will not shear at $1381 - (63) = 1318$ lbf (5863 N)
97.73% of the pins will not shear at $1381 - 2(63) = 1255$ lbf (5582 N)
99.86% of the pins will not shear at $1381 - 3(63) = 1192$ lbf (5302 N)

Notice that this method gives a narrower range of probable failure values than the calculations in Section 11.1 which were based just on minimum and maximum values.

The probabilistic method can be extended to include any number of functionally related variables, but the calculations do get lengthy. An engineer can also perform a theoretical sensitivity analysis using the probabilistic approach. Taguishi, in his methods of design analysis, stresses the importance of establishing design specifications and tolerances on the specifications to reduce the sensitivity of product performance to variation in the design parameters. A product whose performance is insensitive to component variation is said to be robust.

CASE STUDY #8:
Probabilistic Design Analysis

Consider a component that functions as a simply supported beam, made from a square cross section $x \pm 0.006$ inches (± 0.1524 mm), with a distance between supports 18.00 ± 0.09 inches (457.200 ± 2.286 mm), and made from a material with $S_{yt} = 30,000$ psi (207 MPa) with a standard deviation of 8 percent. Assume the load the beam will carry is 50 ± 15 lbf/in. (8.76 ± 2.63 N/mm) of span, and that the ± 0.006 in. (± 0.1524 mm), ± 0.09 in. (± 2.286 mm), and ± 15 psi (± 0.103 MPa) tolerances are all ± 3 standard deviations in size. What minimum size should the square cross section be for a 99 percent reliability, no-yield design?

$\sigma_{max} = Mc/I = 3wL^2/4x^3$, where $w = 50 \pm 15$, $L = 18.00 \pm 0.09$, and x is the side of the square cross section to be determined, so $\sigma(\text{mean}) = 3(50)(18^2)/4x^3 = 12,150/x^3$. Use Equation 11.3 and

$\partial\sigma/\partial w = 3L^2/4x^3$, $\partial\sigma/\partial L = 3wL/2x^3$, and $\partial s/\partial x = -9wL^2/4x^4$

$S_\sigma = \sqrt{[(3L^2/4x^3)(S_w)]^2 + [(3wL/2x^3)(S_L)]^2 + [(-9wL^2/4x^4)(S_x)]^2}$

$S_\sigma =$

$\sqrt{[(3 \cdot 18^2/4x^3)(5)]^2 + [(3 \cdot 50 \cdot 18/2x^3)(0.03)]^2 + [(-9 \cdot 50(18)^2/4x^4)(0.002)]^2}$

$S_\sigma = \sqrt{67.5^2/x^6 + 40.5^2/x^6 + 72.9^2/x^8}$ (11.6)

To insure that the design is 99 percent reliable the difference between the mean material strength and the mean stress load, $30,000 - 12,150/x^3$, must have a mean and standard deviation distribution that lies 99 percent greater than zero (non-negative difference, strength > stress.) Let P = the mean value of $S_{yt} - \sigma$ and S_P be the deviation of that mean. For any x_i in a normal distribution, $z = (x_i - \mu)/\sigma$, and from Appendix G, $z = -2.33$ for 99 percent reliability. The nonnegative difference means that $x_i \geqslant 0$ so $-2.33 = (0 - P)/S_P$, where $P = 30,000 - 12,150/x^3$.

Use Equation 11.5 and $S_P = \sqrt{2400^2 + 67.5^2/x^6 + 40.5^2/x^6 + 72.9^2/x^8}$, so $-2.33 = -(30,000 - 12,150/x^3) \div \sqrt{2400^2 + 67.5^2/x^6 + 40.5^2/x^6 + 72.9^2/x^8}$, and $5.429 \quad [2400^2 \quad + \quad 67.5^2/x^6 \quad + \quad 40.5^2/x^6 \quad + \quad 72.9^2/x^8] \quad = 30,000^2 - 2(30,000)(12,150)/x^3 + 12,150^2/x^6$, which simplifies to
$-8.687x^8 + 7.290x^5 - 1.476x^2 + 0.0002885 = 0$, and (11.7)
$x = 0.793$ inches (20.14 mm). The next largest standard size material is 0.8125 inches or 25 mm.

Check the validity of the results: $\sigma(x = 0.793) = 12,150/0.793^3 = 24,364$ psi, and $P = 30,000 - 24,364 = 5636$ psi (39 MPa).

$\sigma(x = 0.8125) = 12,150/0.8125^3 = 22,652$ psi (156 MPa), 7% less than the theoretical value of 24,364 psi (168 MPa), but the beam will weigh $[(0.8125/0.793)^2 - 1] = 5\%$ more.

To determine how sensitive the design is to variations beyond those assumed, the tolerances can be increased and the size of the square section recalculated.

Consider what will happen if S_w increases from 5 to 10.

Equation 11.6 becomes $S_\sigma = \sqrt{135^2/x^6 + 40.5^2/x^6 + 72.9^2/x^8}$.

Equation 11.7 becomes $-8.687x^8 + 7.290x^5 - 1.475x^2 + 0.0002885 = 0$, and

$x = 0.793$, so the design is insensitive to changes in w. See Exercises 11.10 and 11.11 for additional sensitivity analysis.

11.5 Material Strength Theories

The yield strength, ultimate tensile strength, ultimate compression strength, ultimate shear strength, fatigue endurance limit, and notch fracture strength of materials are important design criteria. When designing to avoid excess deformation or fracture failure, and to use material properties advantageously, two important questions arise. What is the actual load on the component? What is the actual strength of the material being used?

The load on a component is difficult to determine because the use of the part by the consumer is hard to control. It is not difficult to design a component of a bridge to withstand a given load. Bridges have been built for a long time, and the accumulated knowledge of bridge building is enormous. Bridge failure is generally caused by material deterioration and/or the application of loads above the design limit. The actual failure caused by overload may occur long after the overload happens. Consider a bridge designed to carry a maximum load of 10,000 lbf (44.5 kN) per axle. If a truck carrying 15,000 lbf (66.7 kN) per axle drives over the bridge, the bridge may not collapse because a safety factor was undoubtedly used. However, a microscopic crack may be created, if not from the first overload perhaps from the second or third, particularly in colder weather when material strength may be reduced. From this crack, failure may propagate because of fatigue even though no other loads over 10,000 lbf (44.5 kN) per axle occur. Years later, when the bridge collapses, the cry will be heard, "The engineers made a mistake!" when in fact it was customer abuse, long since forgotten, or even unknown, that caused the failure. It is not practical to design a product to be 100 percent tolerant of all the things that can happen to it. The cost of a product to withstand all possible uses and abuses would be uneconomically high; again the cost of a particular benefit must be considered.

The best that can be done is to match the strength of a material to the known or assumed loads, design the components, test them under real use, redesign based on test results, manufacture the product to the specifications established, issue operating and maintenance instructions to the user, and keep the channels of communications to the customer open so continual monitoring can be done.

11.5.1 Maximum Normal Stress Theory

Engineers consider many theories of material failure when designing products to match the strength of material to the intended design loads. One of the most basic theories is the **maximum normal stress theory**. This theory predicts that component failure will occur when the largest principle stress exceeds the compressive or tensile strength of the material. Either yield or fracture failure may be considered.

Figure 11.1 shows an element of material acted upon by three normal stresses. Assume a load orientation such that $\sigma_1 \geq \sigma_2 \geq \sigma_3$. Yield failure will occur if $\sigma_1 > S_{yt}$ or $\sigma_3 < S_{yc}$, and fracture failure will occur if $\sigma_1 > S_{ut}$ or $\sigma_3 < S_{uc}$. (11.8)

Note: Compressive stresses and strengths will be designated by negative values in examples and problems.

The maximum normal stress theory correlates well with actual tests of ductile isotropic materials as long as loading is all tensile or all compressive. It does not correlate well on a combined tensile-compressive load test, or on a torsion load test. Experiments show failure in torsion at values as low as 60 percent of expected strength values.

Mohr's circle helps explain this reduction in torsional strength as shown in Figure 11.2. Figure 11.2(a) represents a one dimensional tensile test, $\sigma_2 = \sigma_3 = 0$. A one dimensional tensile stress creates a shear stress, τ_3, on a plane inclined 45° to the direction of loading. This is the plane of slip in ductile materials. Note that face centered lattice structures are more prone to slip than other lattice structures because they have more slip planes.

Now consider what happens if σ_3 becomes compressive and σ_1 remains constant as shown in Figure 11.2(b). The maximum shear stress is now higher than before, even though the applied tensile and compressive loads do not individually exceed the strength of the material. In ductile materials, when the tensile and compressive strengths are nearly equal, the compressive and tensile axial loads do not directly cause failure, but they do cause the shear stress to exceed the shear strength of the material, and shear

FIGURE 11.1

Normal stresses.

FIGURE 11.2

Mohr's circle stress analysis.

failure occurs. When σ_1 and σ_3 are increased equally from zero, as shown in Figure 11.2(c), one in tension and one in compression, the shear strength of the material will be reached when the tensile and compressive loads are one-half of the uniaxial failure load. This result led engineers to the maximum shear stress theory.

11.5.2 Maximum Shear Stress Theory

The **maximum shear stress theory** states that a material will fail when any combination of loads causes the shear stress to exceed the shear strength of the material. As long as $\sigma_1 \geq \sigma_2 \geq \sigma_3$, Mohr's circle geometry gives the value for maximum shear stress $\tau = (\sigma_1 - \sigma_3)/2$, (recall that $\sigma_3 < 0$ for compression). If $\sigma_3 = \sigma_2 = 0$, then $\tau = \sigma_1/2$, but this is the uniaxial tensile test, so τ_{max} for non-yielding is $S_{yt}/2$ and τ_{max} for non-fracture failure is $S_{ut}/2$. It is not the magnitude of the stresses σ_3 and σ_1 that causes failure, only the difference between them.

Tensile yield failure will not occur if $(\sigma_1 - \sigma_3) < S_{yt}$ (11.9)

Tensile fracture failure will not occur if $(\sigma_1 - \sigma_3) < S_{ut}$ (11.10)

Compressive yield failure will not occur if $(\sigma_1 - \sigma_3) > S_{yc}$ (11.11)

Compressive fracture failure will not occur if $(\sigma_1 - \sigma_3) > S_{uc}$ (11.12)

Take this analysis one step further and consider a physical member subject to a large but constant force from all sides, like a solid object on the ocean floor. Mohr's circle becomes a point and no shear stress exists. Experiments have verified that solid, ductile, isotropic materials can withstand large external applied forces without failure. This would not necessarily be true of objects with voids however, as the high pressure would likely crush the surfaces inward. Non-isotropic materials do not have equal strengths in all directions and would fail in the direction of lowest strength.

11.5.3 Distortion Energy Theory

The phenomenon of non-failure under large externally applied forces led engineers to the **distortion energy theory**. This theory states that energy causing volume change does not lead to failure. The volume change energy is subtracted from the total energy applied, and only the remaining energy can cause material yield or fracture failure.

Recall that the stress-strain curve for many engineering materials is a straight line up to the yield point, and that the area under the curve is the energy absorbed or being transmitted to the object. Let the energy applied to the object be designated as work energy (WE). Then up to the yield point, $WE = \sigma\epsilon/2$. The total work energy, WE_T, for an element under a triaxial load will be:

$$WE_T = \sigma_1\epsilon_1/2 + \sigma_2\epsilon_2/2 + \sigma_3\epsilon_3/2. \tag{11.13}$$

Recall that the triaxial components of strain can be expressed as:

$$\epsilon_1 = \sigma_1/E - \nu\sigma_2/E - \nu\sigma_3/E,\ \epsilon_2 = \sigma_2/E - \nu\sigma_1/E - \nu\sigma_3/E,\ \text{and}$$
$$\epsilon_3 = \sigma_3/E - \nu\sigma_1/E - \nu\sigma_2/E,$$

where ν is Poisson's ratio. Substitute the ϵ values in Equation 11.13, and

$$WE_T = [\sigma_1^2 + \sigma_2^2 + \sigma_3^2 - (2\nu\sigma_1\sigma_2 + 2\nu\sigma_2\sigma_3 + 2\nu\sigma_1\sigma_3)]/2E. \tag{11.14}$$

If $\sigma_1 = \sigma_2 = \sigma_3 = \sigma_A$, then the theory says that all the energy from the applied loads goes into volume change and none contributes to yielding or fracture so $WE_T = WE_{\Delta V}$, and Equation 11.14 becomes:

$$WE_T = WE_{\Delta V} = (3\sigma_A^2 - 6\nu\sigma_A^2)/2E = (3\sigma_A^2)(1 - 2\nu)/2E. \tag{11.15}$$

Since $\sigma_A = \sigma_1 = \sigma_2 = \sigma_3$, then $\sigma_A = (\sigma_1 + \sigma_2 + \sigma_3)/3$, and Equation 11.15 becomes:

$$WE_{\Delta V} = 3(1 - 2\nu)(\sigma_1^2 + \sigma_2^2 + \sigma_3^2 + 2\sigma_1\sigma_2 + 2\sigma_2\sigma_3 + 2\sigma_1\sigma_3)/18E. \tag{11.16}$$

The energy that may lead to yielding or fracture is the difference between the total energy applied and the energy that goes toward volume change: $WE_Y = WE_T - WE_{\Delta V}$, Equation 11.14 minus Equation 11.16.

$$WE_Y = \{(1 + \nu)/3E\}\{[(\sigma_1 - \sigma_2)^2 + (\sigma_1 - \sigma_3)^2 + (\sigma_2 - \sigma_3)^2]/2\}. \tag{11.17}$$

If a uniaxial test is performed so that $\sigma_3 = \sigma_2 = 0$, then

$$[(\sigma_1 - \sigma_2)^2 + (\sigma_1 - \sigma_3)^2 + (\sigma_2 - \sigma_3)^2]/2 = \sigma_3^2\ \text{and}$$
$$\sigma_3 = \sqrt{[(\sigma_1 - \sigma_2)^2 + (\sigma_1 - \sigma_3)^2 + (\sigma_2 - \sigma_3)^2]/2}.$$

If the load is increased to yield, then $\sigma_1 = S_{yt}$. If σ_1 is interpreted as the limiting value of σ_e, an arbitrary state of stress, then yield failure will result whenever

$$\sigma_e = \sqrt{[(\sigma_1 - \sigma_2)^2 + (\sigma_1 - \sigma_3)^2 + (\sigma_2 - \sigma_3)^2]/2} > S_{yt},\ \text{or when}\ \sigma_e > |S_{yc}| \tag{11.18}$$

For the more common engineering problems of biaxial stress, $\sigma_2 = 0$ and

$$\sigma_e = \sqrt{\sigma_1^2 - \sigma_1\sigma_3 + \sigma_3^2}.$$

For pure torsion in a biaxial stress state, the shear stress $\tau = -\sigma_1 = \sigma_3$ and $\sigma_e = \sqrt{3\tau^2} = \tau\sqrt{3}$, where $\sigma_e < S_{yt}$ for no yield.

Therefore the maximum shear stress before yielding is $\tau < S_{yt}/\sqrt{3} < 0.58\ S_{yt}$. $\tag{11.19}$

This is a higher allowable shear stress than the maximum shear stress theory permits and is closer to the 60 percent level observed in many experiments.

FIGURE 11.3

Comparison of failure theories.

11.5.4 Summary Of Ductile Failure Theories For Biaxial Loading

The maximum normal stress theory can be safely used when the applied loads are both tensile or both compressive. This results in a conservative component design. If the situation reduces to one of *uniaxial* loading, then the design is not conservative but agrees exactly with the other two theories.

The maximum shear stress theory is usable under all biaxial loading situations. It is also on the conservative side unless the load reduces to a uniaxial one, in which case it agrees exactly with the other theories.

The distortion energy theory relates best to experimental data and has a mathematical model as its basis. This basis is only valid for applications where the material stress levels are in the elastic region because of the theory development assumption.

Figure 11.3 compares the three theories, but shows only the comparison for loads up to the yield strength because the distortion energy theory has no model basis past that point. Notice also that the diagram assumes equal tensile and compressive yield strength values. For unequal values the diagram becomes non-symmetric. Remember also that predictions of failure made with any theory should be verified with a real build and test. There may be a load or material condition unaccounted for that might cause unpredictable results.

11.5.5 Modified Theory For Brittle Materials

Brittle materials usually fail at low tension stress levels but generally have high compressive strength and are used because of that property. Loading of brittle materials in tension is dangerous because the ultimate tensile strength is nearly equal to the yield strength (low plastic deformation), and fracture failure can occur without a distortion warning. The failure modes for brittle materials can be analyzed as was done for ductile materials, but the first quadrant of the load analysis diagram must be changed. The tensile strength is lower than compression strength for brittle materials, and the acceptable stress region in the first quadrant gets smaller. The third quadrant will be large compared to the first, and for some brittle materials the first quadrant is often considered to have no acceptable design region at all.

Low brittle-metal tensile strength, and data collected from experiments on brittle materials have shown the need for another theory of failure. The result is the modified maximum shear stress theory, often referred to as the Coulomb-Mohr theory (see Figure 11.4). To use this theory both the tensile *and* compressive strength of the material must be known.

First quadrant failure testing is the same as for ductile materials except the tensile yield strength and the ultimate tensile strength are considered equal. For safe biaxial loading, $\sigma_x < S_{ut}$ and $\sigma_y < S_{ut}$.

The test for non-failure under biaxial compressive loading is also the same, $\sigma_x > S_{uc}$ and $\sigma_y > S_{uc}$; recall that compression stresses are minus values.

Variation arises in the second and fourth quadrants. These two cases are the same except for identifying which stress is compression and which is tension. The test for non-failure under combined loads of tension and compression in the fourth quadrant is:

$$\sigma_y/S_{uc} + \sigma_x/S_{ut} < 1. \tag{11.20}$$

Equation 11.20 is the intercept form of a straight line, and other forms of the equation, usable if one load is known and the limit on the other load is desired, are:

$$\sigma_y > -(S_{uc}/S_{ut})\sigma_x + S_{uc}, \quad \text{or} \quad \sigma_x < -(S_{ut}/S_{uc})\sigma_y + S_{ut}. \tag{11.21}$$

FIGURE 11.4

Failure theory for brittle materials.

The Coulomb-Mohr theory does not agree with experimental data under combined tensile-compressive loading in the fourth quadrant, and because of this the modified Mohr theory was developed. This theory is shown in Figure 11.4 superimposed on the Coulomb-Mohr theory. In the critical fourth quadrant region, the limits of nonyielding, depending on which stress is known, are:

$$\sigma_y < -[1 + (S_{uc}/S_{ut})]\sigma_x + S_{uc}, \text{ or} \tag{11.22}$$

$$\sigma_x < [S_{uc} - \sigma_y]/[1 + (S_{uc}/S_{ut})]. \tag{11.23}$$

Keep in mind that σ_y is compressive, σ_x is tensile, and $\sigma_z = 0$ for biaxial loading.

11.6 Impact Loading

The first variation from static loading to be considered is impact loading. Impact loading occurs when a force is applied to a structure in a relatively short period of time. Relatively short is not very scientific and certainly not quantifiable, and there is a gray area when trying to define a precise time limit for impact loading. Some examples of situations where impact loading should be considered are:

1. A drop forge press
2. A car running into a stationary object
3. A batter getting hit with a pitched ball
4. A hydraulic system when a valve is suddenly opened or closed
5. A tennis ball hitting the strings of a racket
6. The effect of a moving truck on a bridge
7. "Popping" the clutch on a dragster
8. An electric impulse when a circuit switch is opened or closed
9. A pneumatic impact wrench on a bolt head
10. The keys of a typewriter striking the paper

11.6.1 Linear Impact

An impact load caused by a moving mass striking a stationary or slower moving object results in a higher deflection and stress level on a structural member than if the same mass is at rest on the member. This is because the energy of motion must be dissipated during impact. If the impacting object does not deform (very large assumption), the result on the impacted object can be:

1. An increase in localized plastic deformation resulting in cracks or high deformation at the point of impact (a hammer striking a piece of wood),
2. A higher deformation at the point of support or someplace between the support and the point of impact (resulting in a bent support), or
3. A dissipation of energy by heat, sound, vibration or light. Any energy dissipated in this way will make subsequent calculations conservative, as all the energy will

be assumed to go into deformation. This conservatism allows for later simplifications that might tend toward nonconservative design.

It takes time for an impact load to be absorbed by a stationary object, and it is not unusual for a physical member to fail at a location other than at the highest static load stress location. This occurs when the velocity of impact is *very* high. The inertia effect of the member absorbing the impact load contributes to the variance in failure location. The stress caused by the impact load does not have time to dissipate through the whole member due to the short interval of time. For example, consider a bullet from a gun impacting a wood board. The impact may occur in such a short period of time that the area of the board away from the point of impact isn't affected by the bullet striking the board.

The method used to analyze an impact load is to convert the impact load to an equivalent static load that already has a mathematical model solution. The analysis is started by considering a mass falling due to gravity, as in Figure 11.5. What equivalent static load, F_e, will deform the structure in the same way as the falling weight, W?

One boundary condition of this analysis is to limit it to the elastic region of deformation. This means that the potential energy of the falling mass, $W(h + \delta_I)$, when absorbed by the structure, must be equal or less than $\sigma\epsilon/2$, the area under the stress-strain curve. To calculate the total energy absorbed, the unit energy must be multiplied by the volume of the material affected, volume = AL (A = cross section area, L = length). A second limitation on the analysis is that the effect on the stationary object is not a function of time (see Section 11.6.5 for the time dependent theory solution). The mathematics of the non-time dependent theory is as follows.

FIGURE 11.5

Dynamic and static deformation.

a. Condition before weight falls.
b. Condition on impact
c. Equivalent static load to cause same deformation as impact.
d. Condition after impact, at equilibrium.

$W(h + \delta_I) = (1/2)(\sigma\epsilon)(AL)$, but $\sigma = F_e/A$ and $\epsilon = \delta_I/L$, so
$$W(h + \delta_I) = (F_e/A)(\delta_I/L)(AL)/2 = (F_e)(\delta_I)/2, \text{ where} \qquad (11.24)$$
h = Height from which weight W is dropped, inches,
W = Weight of falling object, pounds,
δ_I = Dynamic deflection caused by W falling through h, in inches, and
F_e = Equivalent static force needed to cause δ_I, in pounds.

Let δ_{st} = static deflection caused by W at rest, in inches. Because loads and deflection are linearly proportional within the elastic limit, then $F_e/\delta_I = W/\delta_{st}$, and
$$F_e = W\delta_I/\delta_{st}. \qquad (11.25)$$
So Equation 11.24 becomes $Wh + W\delta_I = (W\delta_I^2)/2\delta_{st}$, which simplifies to $\delta_I^2 - 2\delta_I\delta_{st} - 2h\delta_{st} = 0$. Solve for δ_I: $\delta_I = \delta_{st} \pm \sqrt{\delta_{st}^2 + 2h\delta_{st}}$. Substitute into Equation 11.25, and

$$F_e = W(1 + \sqrt{1 + 2h/\delta_{st}}). \qquad (11.26)$$

The quantity $1 + \sqrt{1 + 2h/\delta_{st}}$ is called the **impact factor** because it converts the effect of a falling weight to an equivalent static force.

If the impact velocity of the weight W is known instead of the height from which it was dropped, then Equation 11.26 can be altered. Recall the free-fall velocity model, where the kinetic energy of motion, $mV^2/2$, is set equal to the potential energy before motion, mgh, so $V^2 = 2gh$. Solve for h, substitute into Equation 11.26, and $F_e = W(1 + \sqrt{1 + V^2/g\delta_{st}})$.

When the magnitude δ_{st} is small compared to the magnitude of h or V, $2h/\delta_{st}$ and $V^2/g\delta_{st}$ are large compared to 1, and the following simplification can be made:

$$\delta_I = \sqrt{2h\delta_{st}} \text{ and } F_e = W\sqrt{V^2/g\delta_{st}}. \qquad (11.27)$$

Caution must be exercised when using this simplifying assumption. The variation in the value of the impact factor can be significant, as shown in Table 11.1.

The gravity constant g occurs in Equation 11.27 because of the conversion from h to V, the velocity on impact. Velocity at impact can be horizontal, vertical, or any other direction, and the models can be used for impact from any direction as long as δ_{st} is calculated as if W acts downward. Recall that:

TABLE 11.1

Comparison of impact factors.

$V^2/g\delta_{st}$ or $2h/\delta_{st}$	Impact factor	Impact factor with simplification	% difference
0	2.00	0	∞
10	4.32	3.16	27
50	8.14	7.07	13
100	11.04	10.00	10
500	23.38	22.36	4

1. Kinetic energy of a moving mass is $KE = mV^2/2 = WV^2/2g$,
2. The stress on a member subjected to an axial load is $\sigma = F_e/A$, so $\sigma^2 = F_e^2/A^2$, and
3. The modulus of elasticity is $E = WL/A\delta_{st}$.

Start with $\sigma^2 = F_e^2/A^2$, and substitute F_e from Equation 11.27, then

$\sigma^2 = W^2V^2/A^2g\delta_{st} = (WV^2/2g)(WL)/(A\delta_{st})(2/AL) = (KE)(2E)/AL = (KE)(2E)/$Volume, and

$KE = (\sigma^2/2E)(\text{Volume})$.

This result indicates that the kinetic energy absorbed by a physical member is a function of the volume of the material subjected to the force. Doubling the volume doubles the energy absorbing potential. Be cautious when using the results of this analysis because of the assumptions made.

11.6.2 Alternate Forms

A spring rate k can also be used in the impact analysis if that data is available. The rate k, lbs/in, is determined from the applied load W and the associated deflection δ_{st}, $k = W/\delta_{st}$, and Equation 11.27 becomes $F_e = kV\sqrt{\delta_{st}/g}$. The use of the spring rate also allows the natural period of oscillation, t, of the component mass, m, to be determined.

$$t = 2\pi\sqrt{m/k}, \text{ with } t \text{ in seconds, and } m \text{ in lb·sec}^2/\text{in}. \tag{11.28}$$

The natural frequency is a criterion used to determine if an applied load should be considered an impact load. Recall the opening paragraph of Section 11.6 and the comment about a relatively short period of time. If the time taken to load a member is less than $t/2$, impact effect must be considered. Measuring the time of load application is no easy task; strain gages and an oscilloscope may be helpful.

11.6.3 Torsion Impact

There is a direct analogy between linear impact and torsion impact. The mathematical models are the same, but the meanings of the variables change.

θ, radial deflection (radians) replaces δ, linear deflection (inches).

T_e, equivalent static torque (in·lbf) replaces F_e, equivalent static load (lbf).

I, moment of inertia (lbf·sec^2·in) replaces m, mass (lbf·sec^2/in).

ω, impact velocity (radians/sec) replaces V, velocity (in/sec).

K, spring rate (in·lbf/radian) replaces k, spring rate (lbf/in).

KE, kinetic energy stays the same (in·lbs).

The equivalent units in the metric system are:

θ, Radians	δ, meters
T_e, Newton·meter	F_e, Newton
I, Newton·meter/radian	m, Kilograms
ω, radian/second	V, Meter/second
K, Newton·meter/radian	k, Newton/meter
KE, Newton·meter	

11.6.4 Other Factors

Experiments have shown that the yield strength and ultimate strength of some materials increase as the rate of loading increases, and this is an advantage when a member is subjected to an impact load. Experimental design test analysis should be used for your specific material.

Stress concentration geometry also has an effect on impact strength and failure. Impact loads are applied so suddenly that even a ductile material will not have time to yield locally, and it will behave as a brittle material, even at room temperature. Whether this is offset by the increase in yield and ultimate strength depends on the material and the geometry of the part. Run tests on your design to be sure.

11.6.5 Stress Wave Propagation

When an object is subjected to an impact load, the total system does not come to equilibrium immediately; it oscillates to equilibrium via wave propagation and mass inertia damping. Consider an object at rest supported rigidly at one end but free on the other end. If the free end is struck axially with a compression load, at the instant of impact only a thin layer on the surface of the stationary object knows that anything has happened. It takes time for the elements on the surface to transfer the energy of impact to the next layer, and that layer to transfer the energy to the next layer, and so on. The situation is similar to electricity "flow" where electrons migrate very slowly, 10^{-4} m/sec, but electricity "travels" approximately $3 \cdot 10^8$ m/sec. It is the impulse reaction that passes at the high speed. And so it is with a mechanical impact load. Stress wave theory concludes that a stress wave propagates from the impact surface at a speed that is a function of the material's modulus of elasticity and density (4,507).

$$c = \sqrt{E/\rho}, \text{ where} \tag{11.29}$$

c = the speed of the stress wave in the object being struck

E = the modulus of elasticity of the material being struck, and

ρ = the density of the material being struck.

Be sure to check units when using this model; for example, if E is in lb/in², and ρ is in lb/ft³, and c is desired in ft/sec, then the model becomes $c = (12 \text{ in/ft}) \sqrt{(E)(32.17 \text{ ft/sec}^2)/\rho}$.

To complete the electricity analogy, the velocity of the particles in the material being struck must be determined. The particles in the first layer struck by the moving mass will have nearly the same velocity as the moving mass, assuming the moving mass deforms a small amount compared to the object being struck. If the case of a falling mass is considered, then the free fall potential energy/kinetic energy analysis used in Section 11.6.1 can be used and $v = \sqrt{2gh}$, where v = velocity in ft/sec, g = 32.17 ft/sec², and h = height from which the mass is dropped in feet. The speed of the particles in the struck object is restricted by physical contact and surface deformation.

If the stress wave theory is extended, it can also be shown that the stress, σ, on the elements of the struck object moving at speed v is

$$\sigma = v\sqrt{E\rho}, (4,513), \text{ where again} \tag{11.30}$$

E = modulus of elasticity of the struck object, and

ρ = the density of the struck object.

When the compressive stress wave reaches the fixed end of the object, it is reflected back along the object as a compressive wave, resulting in a doubling of the compressive stress. If the free end is actually free when the reflected wave reaches it, then the second reflection back along the object is a tension wave and cancels out one of the compression waves. When the tension wave reaches the fixed end it is reflected back as a tension wave canceling out the second compression wave, resulting in a stress neutral object. The four wave cycle repeats until it is damped out by the molecular friction action or until the object fractures, which will occur during the first four wave cycle in which the least stress wave damping has taken place.

A more common occurrence than the four wave cycle occurs when an object is struck and the striking object remains in contact with the object, creating a second wall for wave reflection. This condition occurs because the wave speed is so much greater than the particle speed. The result is that the stress waves continue to build rather than cancel out, in compression or in tension depending on the original force application. Mass inertia damping prevents the stress accumulation from becoming infinite, but the stress does build to a maximum level that is higher than the initial stress wave. An empirical approximation to the maximum stress level reached at the *fixed end* of the object is (4,533):

$$\sigma_{max} = \sigma_0 [\sqrt{W/W_b + 2/3} + 1], \text{ where} \tag{11.31}$$

σ_0 = stress at the struck end of the object at time = zero, calculated with Equation 11.30

W = weight of the striking object, and

W_b = weight of the struck object.

As the first stress wave moves from the struck end of the object, or when the wave is reflected during subsequent cycles, the stress in the wave front is higher than the stress level behind the wave front. For this reason the stress at any other point in the object will be less than σ_{max}. The time dependent solution theory requires that some impact velocity exists, otherwise the initial velocity $\sigma_0 = 0$ and $\sigma_{max} = 0$. For the non-time dependent theory results shown in Equation 11.26, however, the stress of impact with zero velocity is twice that of the static stress.

The next example shows how the stress wave, time dependent impact theory compares with the non-time dependent impact analysis of Section 11.6.1.

EXAMPLE 11.10:

Stress wave analysis.

A 50 pound (22.7 kg) object falls on the end of a one inch (25.4 mm) square steel rod six inches (152 mm) long from a height of five inches (127 mm).

The velocity of the falling mass at impact = $\sqrt{2gh}$ = $\sqrt{2(32.17 \text{ ft/sec}^2)(5/12 \text{ ft})}$ = 5.18 ft/sec (1.58 m/s). The velocity of the stress wave in the struck object = $\sqrt{E/\rho}$ = (12 in./ft) $\sqrt{(30 \cdot 10^6 \text{ lb/in}^2)(32.17 \text{ ft/sec}^2) \div 490 \text{ lb/ft}^3}$ = 16,840 ft/sec (5134 m/s), nearly 3000 times faster than the velocity of impact.

The non-time dependent impact stress is determined using Equation 11.26 by first calculating the static deflection: $\delta_{st} = 6$ in. $\cdot \epsilon = 6\sigma/E = (6 \text{ in.})(50 \text{ lb})/(1 \text{ in}^2)(30 \cdot 10^6$

lb/in²) = 0.00001 in. (0.000254 mm), and then calculating the force; $F = 50$ lb[1 + $\sqrt{1 + (2)(5)/0.00001}$] = 50,050 lbf (22.5 kN), and σ = 50,050 lbf/1 in² = 50,050 psi (348 MPa).

The time dependent impact stress is determined using Equation 11.30. At the instant of impact

$\sigma_0 = v\sqrt{E\rho}$ = (5.18 ft/sec)(1/12 in./ft) $\sqrt{(30 \cdot 10^6 \text{ lb/in}^2)(490 \text{ lb/ft}^3) \div (32.17 \text{ ft/sec}^2)}$
= 9230 psi (63.6 MPa).

The first reaction is that this stress seems too low, being about one-fifth of the non-time dependent stress level; however, 9230 is the first wave stress level and the maximum must be determined using Equation 11.31.

$\sigma_{max} = \sigma_0[\sqrt{W/W_b + 2/3} + 1]$ =
9230{$\sqrt{(50 \text{ lb}/(1 \text{ in.})(1 \text{ in.})(6 \text{ in.})(490 \text{ lb/ft}^3)/(12 \text{ in./ft})^3 + 2/3}$ + 1} = 59,830 psi (412.4 MPa).

Both methods determine stresses larger than actual because they do not consider losses due to heat, light, sound, or striking object deformation.

11.7 Fatigue

If a structure is subjected to a series of loading and unloading cycles, failure of components may occur at stress levels below those calculated as safe by static or impact analysis. Failure of this kind is classified as fatigue failure. The loads may vary for each load-unload cycle, but part of each cycle must be a tensile load.

11.7.1 Characteristics

The general characteristics of fatigue failure are:

1. Failure is caused by repeated plastic deformation. Recall that brittle materials do not exhibit much plastic deformation and therefore will fail at low cycle counts.
2. The critical deformation begins at a microscopic level and can occur at safe macroscopic stress levels.
3. The location of the beginning of microscopic failure is usually at a macroscopic geometric stress concentration point or a microscopic material variation or flaw location.

Macroscopic geometry variations include such features as holes, keyways, threads, welds, splines, or abrupt changes in cross section, all of which can create stress concentration. Manufacturing methods can also create stress concentration sites that may or may not be visible. Such things as machining tool marks; variations in surface finishes; heat checks from heat treatment, flame hardening, carburizing, torch cutting, or welding; or operator identification stamps can be locations of stress concentrations.

Microscopic material variations in grain boundaries, grain size, grain direction, internal crystal voids and impurities, and the effects of alloying elements can also serve as crack initiation locations. Under tensile stress these critical areas undergo plastic strain even though at a macroscopic level the stress level is below S_{yt}.

4. Crack initiation at the stress concentration is followed by crack growth until the reduced material cross section is small enough to raise the average macroscopic stress to the ultimate strength of the material. The condition is generally reached abruptly, and the final fracture appears as a brittle material fracture.

A large amount of empirical data has been collected relating to fatigue failure. As with all design generalizations made from experimental data, the specific part you are designing should be tested under real use to verify the reliability of preliminary analysis calculations.

11.7.2 Standard Test

The standard fatigue test consists of applying constant levels of compression-tension cycles of stress. This type of loading is done using a rotating shaft loaded under pure bending, and each surface element of the shaft undergoes a cycle of tension-compression stress during one rotation. The maximum stress level at which the specimen does not fracture after a given number of cycles is defined to be the fatigue strength of the material at that cycle level.

The generalized results of fatigue tests for steel alloys show three regions of failure, as shown in Figure 11.6. The first region exists for low cycle failure, 1 to 10^3 cycles, where fatigue strength ranges from S_{ut} to $0.90 S_{ut}$. In region two, 10^3 to 10^6 cycles occur before failure, and fatigue strength ranges from $0.90 S_{ut}$ to $0.50 S_{ut}$. These results are consistent for steels with S_{ut} up to 200,000 psi (13.78 GPa), but above 200,000 psi (13.78 GPa) the fatigue strength at 10^6 cycles is relatively constant at 100,000 psi (6.89 GPa). Region three occurs above 10^6 cycles, where the specimen does not fail regardless of the number of cycles applied. This level of stress where failure will not occur is called the fatigue limit, S_e. The standard rotating beam test consistently results in a "small" range of stress for a given life, but a "large" range of life for a given level of stress.

FIGURE 11.6

Idealized fatigue life curve for steel.

Charles Mischke has run tests and shown that for low carbon steels S_e is approximately equal to $0.44S_{ut}$ with a standard deviation of $0.06S_{ut}$, for low-alloy steels S_e is approximately equal to $0.48S_{ut}$ with a standard deviation of $0.06S_{ut}$, and for special alloy steels S_e is approximately equal to $0.52S_{ut}$ with a standard deviation of $0.07S_{ut}$ (25). Cast irons exhibit characteristics similar to steel except that S_e is approximately equal to $0.40S_{ut}$ instead of $0.50S_{ut}$.

Many nonferrous metals do not exhibit a fatigue limit, and it appears from experiments that they fail after some number of cycles regardless of the stress level. For this reason the fatigue limit of nonferrous metals is usually given as the stress level for a life of 10^8 cycles. For various aluminum alloys the 10^8 fatigue limit is approximately $0.40S_{ut}$ when $S_{ut} < 50,000$ psi (344.7 MPa), and stays fairly constant at 20,000 psi (137.8 MPa) for higher strength alloys. For magnesium the 10^8 fatigue limit is approximately $0.35S_{ut}$. Because of the lack of a constant value for a fatigue limit in nonferrous alloys, structures made from them are usually visually inspected on a regular basis for fatigue cracks. Airplanes, for example, must undergo regular inspections for skin cracking so repairs can be made before complete failure occurs.

11.7.3 Non-symmetric Loading—Goodman Diagram

When the levels of tension and compression cyclic loads are not equal, a condition closer to real-time use of structural products than equal cyclic loads, modifications to the fatigue life values must be made. Some parts undergo cyclic tension variations only, or unequal tension-compression stress loading. These conditions can be analyzed using a Goodman diagram. Figure 11.7 shows a typical S-N curve (stress plotted as a function of the number of cycles to failure), and Figure 11.8 is the associated Goodman diagram.

FIGURE 11.7

Idealized fatigue life curve for steel.

$S_3 = 0.90\ S_{ut} = 1.80\ S_e$
$S_4 = 0.77\ S_{ut} = 1.54\ S_e$
$S_5 = 0.63\ S_{ut} = 1.26\ S_e$
$S_e = 0.50\ S_{ut}$

FIGURE 11.8

Typical Goodman diagram for steel.

The Goodman diagram is constructed using data from the *S-N* curve which comes from real test data, or estimated values using the average values mentioned and standard material properties. Point S_3, in Figure 11.8, can be above S_{yt} depending on the actual material values for S_{ut} and S_{yt}. The condition is most apt to occur when the material is in a "dead soft" condition. More commonly point S_3 will be lower than S_{yt} because the stress limits for various cycle levels are influenced by factors other than the material strength.

The slope of line *L* from the origin is $m = \sigma_{ac}/\sigma_{mc}$, where σ_{mc} is the actual mean stress during the loading cycle, and σ_{ac} is the actual stress amplitude during loading cycle.

$$\sigma_{mc} = (\sigma_{max} + \sigma_{min})/2, \text{ and } \sigma_{ac} = \sigma_{max} - \sigma_{mc}. \tag{11.32}$$

The stresses σ_{ac} and σ_{mc} come from stress calculations or from stress levels found during strain gage test analysis. The equation of line *L* is $\sigma_{ac} = m\sigma_{mc}$. This equation can be solved simultaneously with the equations of any of the lines $S_e - S_{ut}$, $S_5 - S_{ut}$, $S_4 - S_{ut}$, or $S_3 - S_{ut}$. The solution values of σ_a and σ_m are then compared with the values of σ_{ac} and σ_{mc} which come from calculation or test to determine if the part is safe from fatigue failure. The next example illustrates unequal stress amplitude analysis.

EXAMPLE 11.11:
Fatigue life analysis.

Given a cyclic loading on a part of σ_{max} = 25,000 psi (172 MPa) tension, and σ_{min} = 15,000 psi (103 MPa) tension, and a material with S_e = 30,000 psi (207 MPa), and S_{ut} = 50,000 psi (345 MPa). Is the component safe from fatigue failure?

$\sigma_{mc} = (25{,}000 + 15{,}000)/2 = 20{,}000$ psi (138 MPa), and

$\sigma_{ac} = 25{,}000 - 20{,}000 = 5000$ psi (34.5 MPa), so

$\sigma_a = (5000/20{,}000)\sigma_m = 0.25\sigma_m$, and the equation of line $S_e - S_{ut}$ is

$\sigma_a = (-30{,}000/50{,}000)\sigma_m + 30{,}000 = -0.6\sigma_m + 30{,}000$.

Solve the equations simultaneously and $\sigma_a = 8830$ psi (60.9 MPa) and $\sigma_m = 35{,}290$ psi (243 MPa). Because $8830 > 5000$ and $35{,}920 > 20{,}000$ the component will have a life greater than 10^6. Graphically this means that the plotted point (20,000, 5000) is below the $S_e - S_{ut}$ line. If the plotted point is above the $S_e - S_{ut}$ line but below the $S_5 - S_{ut}$ line, the component will have a life greater than 10^5 cycles but less than 10^6 cycles.

Any combination of σ_{ac} and σ_{mc} that plots below the S_{yc}-A-S_e-B-S_{yt} line means the component will not yield and will have a life greater than 10^6 cycles. In the region A-S_e-B-F-A there will be no yielding, but fatigue life will be lower than 10^6 cycles. The cycles-to-failure breaks occur at 10^6, 10^5, 10^4, and 10^3 cycles. The values of S_e, S_{yt}, and S_{ut} are not precise enough to estimate fatigue life closer than powers of 10. In the regions C-S_{uc}-S_{yc}-A-C and B-S_{ut}-S_{yt}-B yielding will occur, but 10^6 minimum cycle life will exist. In the region B-F-S_{ut}-B yielding will occur on the first cycle and the fatigue life will be less than 10^6 cycles. Moving out the line L is equivalent to an increase in σ_{ac} and σ_{mc}. Changing the slope of line L requires a change in the ratio σ_{ac}/σ_{mc}.

One final observation about the Goodman diagram. The stress limit lines originating from S_3, S_4, S_5, and S_e, and projected into the compression region are horizontal because an all compressive σ_a-σ_m cycle will not cause fatigue failure. Fatigue failure occurs only when part of the stress cycle is tension. Tension is required to start and/or open up a crack. The limit to this restriction of course is the compressive strength of the material, line S_{uc}-E.

11.7.4 Factors Affecting Fatigue Strength

When analyzing fatigue life, factors to consider include:

1. Applied load
2. Part geometry
3. Material properties
4. Manufacturing methods
5. History of prior loadings (fatigue damage is cumulative).

Of these five items, the applied load is probably the one over which the engineer has the least control. The applied load is the result of customer use and can only be estimated. Part geometry and material selection are under the control of the engineer. Part geometry deals with such features as cross-sectional area, moment of inertia, diameter, length, width, symmetry, and whether the part is hollow or solid. Part geometry may be restricted because of limited space, manufacturing methods, or standard material selections, but within that space the engineer is free to be creative and use whatever shape seems appropriate. The choice of material is important because material governs design criteria such as S_{yt}, S_{ut}, and S_e.

The choice of geometry and material also determines the methods of manufacture that can be used. The choice of manufacturing methods in turn determines surface finishes, which also affect fatigue life. A low surface finish requirement on a steel part will usually require grinding, so a grinder is required. A stepped shaft requires a cylindrical grinder, a constant diameter shaft can be ground on a centerless grinder, while a flat surface requires a surface grinder. Before making decisions determine which manufacturing methods are available. The optimum selection, considering material, geometry, and manufacturing, is not always easy to achieve and requires compromise between all criteria considered.

Any condition that creates a variation from the standard rotating beam specimen must be accounted for. The standard test specimen is 0.30 inch (7.62 mm) in diameter, highly polished, and is loaded as a rotating beam. The following sections discuss adjustments to make during analysis as conditions vary from this standard.

11.7.4.1–Loading Variations The standard rotating beam fatigue test places stress on each point on the circumference of the beam. Each point on the circumference of the beam is placed in a compression-tension cycle in which the tension and compression stress are the same. The material away from the circumference and toward the axis is stressed less than the external fibers; recall $\sigma = Mc/I$. Crack propagation and fatigue failure will originate at or near the surface.

A *stationary* circular beam under a fatigue test, with an alternating up and down load, has only two points subjected to the highest stress. The weakest point of the beam material may not be at the top or bottom where the highest stress level will occur. Therefore, the stationary beam may likely exhibit a fatigue strength higher than the rotating beam. Experiments verify this hypothesis, and fatigue strengths 10% higher than in the rotating beam test are common.

In an *axial* cyclic load test, the entire cross section of the specimen is subjected to the same stress level. Fatigue failure will begin at the weakest point in the section rather than at the weakest point on the surface. Fatigue strengths under these conditions will be less than those determined by a rotating beam test. Fatigue strength at 10^3 cycles is lowered to $0.75 S_{ut}$ for steels, and at 10^6 cycles to $0.45 S_{ut}$. If the load is not concentric, then further reductions of fatigue strength are required because of the bending stress caused by the eccentricity of the load. This will be considered in the eccentricity analysis for column failure.

Fatigue tests performed with *cyclic torsion* loading behave similar to the rotating beam specimen because it is the surface elements that reach the highest stress levels. Recall, however, that the ultimate shear strength of most materials is not equal to the ultimate tensile strength. Tests show that the ultimate shear strength, S_{us}, is approximately equal to $1.30 S_{ut}$ for cast iron, $0.75 S_{ut}$ for steel, $0.60 S_{ut}$ for aluminum, and $0.70 S_{ut}$ for other ductile materials (27). Caution: Use these estimates only when test data are not available, and then be sure to test components thoroughly after they are built.

Recall from the discussion of the distortion energy theory that a lower apparent strength occurs under combined tension-compression loading in ductile material compared to a combined tension or combined compression loading. A similar phenomenon occurs under reverse torsion loading, which combines tension and compression stresses. The fatigue strength of a reverse torsion stressed ductile material is estimated at 58% of the rotating beam specimen. The 10^3 cycle fatigue strength is estimated at 90% of torsional shear ultimate strength, not tensile ultimate strength.

TABLE 11.2

Affect of stress cycle speed on S_e.

Cycle speed, cycles/sec.	Effect on S_e
0–3	Slight reduction from standard
30	Standard test speed
50–100	Slight increase from standard
100–1000	Approximately a 10% increase from standard
> 3000	Decrease from peak level

11.7.4.2–Cycle Speed The speed of the reversed cycle loading also has an effect on the fatigue limit as summarized in Table 11.2.

11.7.4.3–Stress Concentration Factors A component that has any variation in the cross-sectional area exposed to a load will have a region of stress concentration. The change can be any cross section change, such as a stepped diameter on a shaft, a cross hole, threads, grooves, or splines. The stress lines in the part must deviate from their normal flow to get past the irregularity and become concentrated at the point of geometry change. The geometry change causes nonlinear stress distribution, and the effects of various geometry changes have been well documented. Tables of these data are shown in Appendix L. The tables give a value, K_t, which is a multiplier for applied stress. The reciprocal of these values could be multipliers for S_e. Stress concentration factors should be used whenever brittle fracture can be anticipated: a brittle material loaded under tension, a component loaded under impact, or a component subjected to fatigue loading. Because the stress concentration is critical for brittle type fracture, K_t is used only on stress levels within the elastic limit. Stress concentration factors are properly used as a multiplier on stress. Because the factors are different for tension and torsion, the factors should be applied to individual stresses before combining them using Mohr's circle or other stress combination methods.

Stress concentration locations can also cause residual compressive stresses when a load is removed. This occurs when the stress concentration is high enough to cause local plastic deformation. Ductile materials under static loading are not ordinarily affected by stress concentration because of localized yielding in the area of high stress.

11.7.4.4–Statistical Scatter Of Data If the data tables for S_e, or other material strength properties, are average values, then a part designed to the table value would only be 50 percent reliable. Better data would include the mean value and the standard deviation so that any reliability level could be determined. If it is assumed that table data are mean values and no deviation data are available, then an assumption about the variation is required. One approach is to assume a standard deviation of 8 percent of the value listed and adjust the value of S_e for a desired reliability. The 8 percent estimate is common for data of this kind. Remember that this is just a method to use if experimental data on your material is not available and is a starting point only. Table 11.3 lists S_e multiplication factors to use if S_e data is given as an average value and an 8 percent standard deviation is assumed.

TABLE 11.3

Reliability level factors for fatigue strength @ 8% standard deviation.		
Reliability level		**Multiply S_e by**
50.00%	x	1.000
84.13%	(x − S)	0.920
90.32%	(x − 1.3S)	0.896
95.54%	(x − 1.7S)	0.864
97.73%	(x − 2S)	0.840
99.86%	(x − 3S)	0.760
99.997%	(x − 4S)	0.680
99.99997%	(x − 5S)	0.600

x = sample mean, S = sample standard deviation.

11.7.4.5–Adjustments For Size Changes As a member increases in size (compared to the 0.30 inch diameter of the standard rotating beam test specimen) S_e must be reduced. Reasons for this include the following. A larger cross section has a greater chance of containing a flaw. The flaw could be a grain boundary irregularity, an entrapped impurity, a void, or inconsistent hardness or strength properties. A larger area also increases the surface exposed to high stresses, further increasing the probability of a flaw being subjected to the highest stress. A third reason that larger areas result in lower S_e values is the stress gradient in subsurface grains. The larger part size results in a smaller stress gradient (high stress levels are reduced more slowly), and so more material is exposed to higher stresses. This increases the chance that a flaw will be exposed to a stress high enough to initiate cracking. A low stress gradient and a subsurface flaw can combine to cause a crack to initiate below the surface.

If no test data on your part or material are available, then adjust the value of S_e by the factors listed in Table 11.4. An alternate approach is to calculate the multiplier by using Mischke's empirical relationship, $0.872d^{-0.1133}$, where d is the diameter in inches.

TABLE 11.4

Shaft diameter or component maximum dimension effect on fatigue strength (7).	
Size, inches	**Multiply S_e by**
0–0.40	1.0
0.41–2.00	0.9
2.10–3.70	0.8
3.71–5.20	0.7
5.21–6.70	0.6
6.71–8.20	0.5
8.21–9.70	0.4

FIGURE 11.9

Surface finish factors for steel.

From J. E. Shigley and L. D. Mitchell, *Mechanical Engineering Design* © 1983, McGraw-Hill Publishing Co, New York.

11.7.4.6–Surface Finish The standard rotating beam test specimen has a highly polished surface. Most parts do not receive this special treatment, and parts with surface finishes less than perfect will exhibit a reduced fatigue strength. The multiplication factors for surface finish differences for steel are shown in Figure 11.9 and Figure 8.11. The factor is also a multiplier of S_e.

Surface finish multipliers have also been quantified by Mischke in the form [x(mean), σ(standard deviation)]S_{ut}^n, with S_{ut} in ksi. For ground surfaces $x = 1.34$, $\sigma = 0.18$, and $n = -0.0848$; for machined surfaces $x = 2.70$, $\sigma = 0.16$, and $n = -0.265$; for hot rolled surfaces $x = 14.4$, $\sigma = 1.58$, and $n = -0.718$; and for forged surfaces $x = 39.9$, $\sigma = 3.19$, and $n = -0.995$ (25).

EXAMPLE 11.12:

Fatigue life surface finish multiplier.

The surface quality multiplier for a hot rolled steel bar with $S_{ut} = 75,000$ psi is: $(14.4)75^{-0.718} = 0.65$ with a deviation of $(1.58)75^{-0.718} = 0.07$.

EXAMPLE 11.13:
Fatigue life analysis.

Estimate a 97.7 percent reliable fatigue limit for a 3.00 inch (76.2 mm) constant diameter shaft, turned on a lathe to size, loaded in reversed torsion, and made from a low carbon steel with a mean strength value of S_{ut} = 90,000 psi (620 MPa).

Estimated unadjusted shear S_e = (0.50)(90,000) = 45,000 psi (310 MPa).
Adjustment factor for torsion loading = 0.58, see page *394*.
Adjustment factor for stress concentration = 1.0 (no stress concentration).
Adjustment factor for lathe turning = 0.75 from Figure 11.9 or 0.82 using Mischke's method.
Adjustment factor for reliability level = 0.840 from Table 11.3.
Adjustment factor for 3.00 inch diameter = 0.8 from Table 11.4.

Therefore the adjusted fatigue limit S_e' = (45,000)(0.58)(1.0)(0.75)(0.840)(0.8) = 13,150 psi (90.6 MPa). An acceptable torque limit on the shaft can also be calculated: 13,150 = (Torque)(radius)/J, T = (13,150)π(3.00³)/16 = 69,760 in·lbf = 5810 ft·lbf (7878 N·m).

EXAMPLE 11.14:
Fatigue life analysis.

A support is loaded as shown in Figure 11.10 where load P is an equal tension-compression load cycle. If the steel used for the support is hot rolled and has a table value for S_{ut} = 60,000 psi (413.5 MPa) and a 99.86 percent reliability is required, determine the maximum value for P for 10^6 minimum life.

Estimated unadjusted S_e = (0.45*)(60,000) = 27,000 psi (186 MPa)
Adjustment factor for reliability level = 0.760 from Table 11.3.
Adjustment factor for maximum dimension = 0.9 from Table 11.4.
*Adjustment for axial loading made by using 0.45 instead of 0.50, see page *394*.
Adjustment factor for surface conditions = 0.7 from Figure 11.9.
Stress concentration factor for cross hole = 2.70 from Appendix L.
S_e' = (27,000)(0.760)(0.9)(0.7)(1/2.70) = 4790 psi (3 MPa), and
P_{max} = (4790)[($\pi(1)^2/4$ − (0.125)(1.00)] = 3160 lbf (14 kN).

If this value of P is not acceptable for the intended function, design changes are needed; possibilities include:

1. Use a larger cross section area. Space constraints and material cost will limit this choice. Increasing the maximum dimension also reduces the fatigue limit because of the size factor.

2. Use a lower reliability. The choice of this possibility depends on the consequences of a component failure. The more dangerous the result of failure, the less likely this is an acceptable choice.

FIGURE 11.10

Schematic for Example 11.14.

3. Use a smaller maximum dimension to reduce size effect. While this choice would increase S_e by 10 percent, it could reduce the area by more than 10 percent and the net result might be worse.

4. Change the loading to cyclic bending. This would require a major redesign, but S_e could be increased by 20 percent.

5. Reduce the stress concentration. Determine the function of the cross hole and try to find an alternate design.

6. Select a material with a higher S_e. Cost and material availability would be the main concerns with this choice.

7. Accept less than 10^6 cycles of part life and require replacement on a regular basis.

None of the possibilities is particularly easy to implement and the final choice is likely to be a compromise between several alternatives.

11.7.5 Other Fatigue Concerns

There are many design requirements made that have an effect on fatigue life. Any deviation from the standard test specimen must be considered as a variable.

Metallic plating generally reduces fatigue life because coatings have inherent irregularities. Cracks starting in the plating material transfer to the base material, and thicker coatings are worse than thin ones. The larger volume of irregular metal increases the probability that a flaw will be exposed to a high stress. In the case where plating is required for corrosion protection, the reduction in fatigue life may be more important than allowing corrosion failure. Zinc plating, cadmium plating, and aluminum anodizing reduce fatigue life less than does nickel plating and chromium plating.

Many component parts are subjected to **non-constant amplitude fatigue stress loads**: car springs react to many different road conditions, engines do not always run at the same speed, and the magnitude of the wind force on bridges and buildings is not constant. Because fatigue stresses are rarely constant and effects are cumulative, the Palmgren-Miner Theory was developed. It states that fatigue failure will occur when $\Sigma(n_k/N_k) \geq 1$, where k represents stress levels $\geq S_e$, n_k is the number of cycles applied to the part at that stress level, and N_k is the expected number of cycles to failure at that stress level. Ordinarily, $k = 1, 2, \ldots, 6$ and $N_k = 10^1, 10^2, 10^3, \ldots, 10^6$. To use this theory an assumption about actual service cycles or field test data is required. Field test data is ordinarily obtained from strain gage testing.

EXAMPLE 11.15:

Palmgren-Miner theory.

If a standard rotating beam sample is made from a steel with $S_{ut} = 90,000$ psi (620 MPa), and has been exposed to 81,000 psi (558 MPa) for 150 cycles, 69,300 psi (478 MPa) for 2000 cycles, and 56,700 psi (391 MPa) for 25,000 cycles, how many more cycles at 56,700 psi (391 MPa) can the specimen be expected to withstand before failure?

Using Figure 11.7 as a fatigue life base, the specimen has already been exposed to 150/1000 of its life at 150 cycles of 81,000 psi (558 MPa) loading, 2000/10,000 of its life at 200 cycles of 69,300 psi (478 MPa) loading, and 25,000/100,000 of its life at 25,000 cycles of 56,700 psi (391 MPa) loading. Therefore it has $1 - (150/1000 + 2000/10,000 + 25,000/100,000) = 0.4$ of its life left. If L represents the remaining cycles of life at 56,700 psi (391 MPa) loading, then $L/100,000 = 0.4$ and $L = 40,000$ cycles.

11.7.6 Fatigue Testing

The best fatigue analysis comes from applying real service loads to actual parts or assemblies. While the test is being performed, visual checks for cracks can be made and can be enhanced by using common nondestructive tests such as the use of fluorescent dye or ultrasonic testing. Fatigue tests can be interrupted for part inspection because parts do not "rest up" from fatigue testing; total damage is cumulative. Strain gages allow stress level data to be collected, and load levels must be monitored if there is randomness in the test. Save peak load testing for the end of a test after a required number of cycles are reached. Overloading of a structure can produce residual compressive stress at a crack end and thus inhibit crack growth. Continue product testing on critical parts after regular production has started. Material and manufacturing variations may occur, and a potential failure may be prevented with timely action.

Many methods exist for determining fatigue limits, but the standard rotating beam test seems to be adequate for generalized data. Many samples should be tested because of the scatter associated with experimental data. If there is only one test machine available tests must be done sequentially. This may cause variations in fatigue life because of changes in temperature, oxidation of test samples caused by exposure to humid air, and age hardening if nonferrous alloys are being tested.

If more than one test machine is available parallel testing can be done. This allows the n extreme value sample test to be used. In this test n samples are tested simultaneously at the same stress level. When the first sample fails all the machines in the group are stopped. The probability of surviving the applied stress load for the number of cycles until first failure is:

$$P(\text{survival}) = (0.5)^{1/n} \tag{11.33}$$

The test loses validity when any attempt is made to determine probabilities for failure beyond the first failure. Carried to its extreme the maximum probability is 0.50 which implies that no matter how large a sample was started, they all will fail before the mean failure life.

EXAMPLE 11.16:
Multiple specimen, simultaneous running fatigue test.

If five parts are tested at $\sigma_m = 0$ and $\sigma_a = 25,000$ psi and the first sample fails after 10^4 cycles, the probability of all five parts surviving 25,000 psi for 10^4 cycles is $(0.5)^{1/5} = 0.87$.

11.7.7 Part Design

The best design criterion to employ to eliminate fatigue failure is to avoid surface tensile stresses, but unfortunately, that isn't usually possible. The second criterion to use is to induce surface compressive stresses in component parts prior to use. The pre-use compressive stresses must be overcome before the surface can be exposed to a tensile stress, so tensile stress levels are reduced. Among the ways to create pre-use compressive stresses on the surface of a part are:

1. Cold forging. Rolled threads, for example, are 50 percent better than cut threads in avoiding fatigue failure, particularly for steels with $S_{ut} \leq 80,000$ psi (551 MPa).
2. Shotpeen the surface.
3. Surface harden using carburizing, nitriding, flame hardening, or induction hardening.
4. Allow bent parts to spring back; don't over-bend during subsequent welding or assembly.
5. Pre-stress parts in compression before use.

Pre-stressed concrete, for example, has steel reinforcing rods embedded in it. The rods can be placed under a tensile load, and the concrete poured and cured around the rods. When the load is removed from the rods the concrete is placed in compression. The real loads have to overcome the compressive pre-load before subjecting the concrete to a fatigue tensile stress.

11.7.8 Welding Fatigue

Welding has its own series of problems when it comes to fatigue, and they are difficult to quantify, so testing of components is again required. Welding reduces fatigue limits because of:

1. Material homogeneity gradients
2. Cracks caused by cooling shrinkage
3. Incomplete penetration

4. Weak fusion
5. Weld undercut
6. Weld overlap, and
7. Slag inclusion or porosity.

These problems can be kept to a minimum by careful design and manufacturing control including:

1. Remove the bead from butt welds
2. Grind fillet weld toes
3. Use butt joints rather than lap joints
4. Avoid starting or stopping a weld in a high stress area
5. Grind out bad welds and reweld
6. Insure proper penetration
7. Avoid stiffness changes from one part to the next, and
8. Avoid intermittent welds.

11.8 Column Buckling

A physical member, with material properties S_{uc} and S_{yc}, subjected to an axial compression load can be considered to have failed under several different performance criteria.

1. If the stress $\sigma = F/A < S_{uc}$ and the member fractures,
2. If the stress $\sigma = F/A < S_{yc}$ and the member does not return to its original length when the load is removed, or
3. If the member buckles under an applied compressive stress that causes a condition of unstable equilibrium.

This state of unstable equilibrium is a function of load and geometry and can occur even though the macroscopic stress does not exceed S_{uc} or S_{yc} of the material used. Column buckling can also occur in unsuspecting locations such as the compressive side of a loaded beam.

11.8.1 Euler Model

Failure by buckling requires a more complex analysis than failure criteria #1 and #2. One method used for column failure analysis is the Euler model:

$$P_c = \frac{\pi^2 EI}{l_e^2}, \text{ where} \tag{11.34}$$

P_c = Critical load at which unstable equilibrium occurs, lbs

E = Modulus of elasticity of the column material, psi

I = Moment of inertia of the cross-sectional area of the column perpendicular to the direction of the load, and whose neutral axis is perpendicular to the direction of buckling failure, in^4, and

l_e = Effective length of the column, a function of actual length, l, and geometric end conditions, inches (see Figure 11.11).

This model is based on certain assumptions:

1. The column is straight, no eccentricity exists.
2. The load is axial and symmetrically applied.
3. The material is homogeneous.
4. The stress is within the elastic limit.

When the load P_c is reached a situation of unstable equilibrium occurs, and any variations from the ideal assumptions 1, 2, or 3 cause sudden buckling and structural failure.

A second form of the Euler model is possible if the substitution for I is made: $I = Ar^2$, where A = the cross-sectional area, in², and r = the radius of gyration, inches, then Equation 11.34 becomes:

$$P_c = \frac{\pi^2 EA}{(l_e/r)^2} \tag{11.35}$$

FIGURE 11.11

Values of l_e for column analysis, $l_e = Kl$, where l is the actual column length.

	(a)	(b)	(c)	(d)	(e)	(f)
Buckled shape of column is shown by dashed line						
Theoretical K value	0.5	0.7	1.0	1.0	2.0	2.0
Recommended design value when ideal conditions are approximated	0.65	0.80	1.2	1.0	2.10	2.0
End condition code		Rotation fixed and translation fixed				
		Rotation free and translation fixed				
		Rotation fixed and translation free				
		Rotation free and translation free				

Courtesy "American Institute of Steel Construction."

The ratio l_e/r is called the slenderness ratio of the column. Studies have shown that the Euler model gives reliable results for aluminum and steel if $l_e/r \geq 140$, and for wood if $l_e/r \geq 80$. For l_e/r values between 140 and 10 the Johnson model, discussed in the next section, is more applicable. If $l_e/r < 10$, column effect is generally neglected, buckling is not a problem, and $P_c = S_{yc}A$.

The geometric end conditions of the column also influence P_c by changing the value of l_e as shown in Figure 11.11. A high l_e value reduces P_c. The reason for using values higher than theoretical is that theoretical end conditions cannot be manufactured. Manufacturing tolerances and material elasticity prevent a perfectly rigid end support from being built.

11.8.2 Johnson Model

A modification to Euler's model was proposed by J. B. Johnson to bring theory closer in line with experimental results. The modification is appropriate for certain l_e/r ratios.

For ratios of $l_e/r < \sqrt{(2\pi^2 E)/S_{yc}}$, Johnson's model for P_c is: (11.36)

$$P_c = A\{S_{yc} - [[S_{yc}(l_e/r)]^2/4\pi^2 E]\}. \tag{11.37}$$

Johnson's model results in a smaller P_c for a given cross section, or it requires a larger area for a given P_c than does the Euler model.

11.8.3 Eccentricity And Secant Model

A column is never perfectly straight from one end to the other. The straightness variation is termed eccentricity. Eccentricity should be kept small, for as the next model shows, eccentricity lowers P_c. A visible physical eccentricity or a nonvisible material eccentricity can exist. Physical eccentricity can be measured while material eccentricity cannot. Most materials have microscopic variations and are not homogeneous in composition and strength. In either case the effect of eccentricity is to reduce the value of P_c, and the secant model is used to account for eccentricity.

$$P_c = \frac{AS_{yc}}{1 + (ec/r^2)(\mathrm{secant}[(l_e/r)\sqrt{P_c/4AE}])}, \text{ where} \tag{11.38}$$

c = the distance from the neutral axis to the extreme fiber of the cross section A, inches, and e = eccentricity, the largest variation from straight between ends of the column, inches (see Figure 11.12).

When using Equation 11.38 it is important to remember that e, c, and r must all be measured relative to the same axis. The larger the value of ec/r^2, the smaller the resulting value of P_c. If the axis of largest e and largest c does not coincide with the axis of smallest r, then several axis orientations must be checked to find the largest value of ec/r^2. The secant model holds for any l_e/r ratio.

Various minimum values of e have been suggested for preliminary design analysis because material microstructure eccentricity is practically impossible to quantify. One guideline is to assume that $e = l_e/400$. A second guideline is to assume $ec/r^2 = 0.25$, which is commonly used for structural steel in commercially available sizes because it correlates well with the results of experiments run to measure P_c. Material suppliers often list maximum allowable out-of-straightness in their catalogs, but out-of-straightness tolerances cannot account for nonhomogeneous material properties; time again to run tests on the material to be used.

FIGURE 11.12

Variables used in column failure analysis.

In the secant model P_c appears on both sides of the equation and cannot be separated from the equation as a single term because of the secant function, and solving for P_c is best done using computer iteration methods.

If column analysis is required at stress levels above the elastic limit, or when a material is used that does not have a linear stress-strain curve, then E is replaced in all the previous models with E_t, the tangent modulus. The tangent modulus is the slope of the stress-strain curve at the stress level under analysis. Its value requires the knowledge of the stress-strain curve or an assumption about it. Two assumptions are commonly made:

1. The slope of the stress-strain curve is linear up to the stress value equivalent to the 0.2% offset yield point, and
2. The stress-strain curve is linear from the yield point to the ultimate strength.

Never use these assumptions for design analysis without planning to run tests.

The analysis of columns involves different approaches depending on the data given and the assumptions made, as the following example illustrates.

EXAMPLE 11.17:
Column failure analysis

Consider a post used in the support of a building with its base embedded in concrete, the top free to translate but not rotate, made from a pipe with the following dimensions and allowable material properties. What maximum load can the post be expected to support without failure?

$ID = 3.75$ inches (95 mm) $OD = 4.00$ inches (102 mm)
$l = 96$ inches (2438 mm) $E = 30 \cdot 10^6$ psi (207 GPa)
$S_{yt} = S_{yc} = 48{,}000$ psi (331 MPa) $S_{ut} = S_{uc} = 65{,}000$ psi (448 MPa)

$A = (\pi/4)(4.00^2 - 3.75^2) = 1.52$ in^2 (980 mm^2)
$r = \sqrt{(4.00^2 + 3.75^2)/16} = 1.37$ inches (34.8 mm), refer to Appendix P.
$l_e/r = (1.2)(96)/1.37 = 84.1$

Maximum load before fracture failure = $(65{,}000)(1.52) = 98{,}800$ lbf (439 kN).
Maximum load before yielding failure = $(48{,}000)(1.52) = 73{,}000$ lbf (325 kN).
Maximum load before buckling (first use Equation 11.36 to determine which model to use):

$$\sqrt{(2\pi^2 E)/S_{yc}} = \sqrt{2\pi^2(30 \cdot 10^6)/48 \cdot 10^3} = 111 > 84.1.$$

Therefore Johnson's modified model must be used, and $P_c = 1.52\{48 \cdot 10^3 - [(48 \cdot 10^3)(84.1)]^2 / [4\pi^2(30 \cdot 10^6)]\} = 52{,}000$ lbf (231 kN).

This value of P_c is less than the load that would cause yielding, so it is the maximum load to use to avoid any type of column failure, but eccentricity may change this. If a value of eccentricity of $l_e/400$ is assumed, then $e = (1.2)(96)/400 = 0.288$ inches (7.31 mm) and:

$$P_c = \frac{(1.52)(48{,}000)}{1 + [(0.288)(2.00)/1.37^2][\sec(84.1\sqrt{P_c/(4)(1.52)(30 \cdot 10^6)})]}$$

An appropriate approach for an iteration solution is either to start at 52,000 and increment down, or to start at zero and increment up. In either case a computer program will be helpful, and P_c is approximately equal to 38,500 lbf (171 kN).

If the critical compressive load for buckling is calculated without allowing for eccentricity, an alternate estimate of maximum load at failure can be made using the parallel failure mode model which states that the probable failure load = $1/[(1/P_c) + (1/F)]$, where P_c is calculated using either the Euler or Johnson model, and F is either S_{uc} or S_{yc} depending on the type of failure that is unacceptable. For the conditions in Example 11.17, the probable fracture failure load would be $1/[(1/52{,}000) + (1/65{,}000)] = 28{,}900$ lbf, a more conservative answer than obtained using the eccentricity estimate. *Note*: The strength values were assumed to be allowable limits. If they were assumed to be mean values, then they would need statistical adjusting using Table 11.3.

EXAMPLE 11.18:
Column failure analysis.

Consider a solid round shaft that will be axially loaded with a compressive load of 40,000 lbs (178 kN). The shaft is 20 inches (0.508 m) in length and pinned at both ends, $l_e = 1$. If the material to be used has allowable strength limits $S_{yc} = 47{,}000$ psi (324 MPa) and $S_{uc} = 68{,}000$ psi (469 MPa), and $E = 30 \cdot 10^6$ psi (207 GPa), determine the minimum diameter necessary to avoid column buckling failure.

To avoid compressive fracture, $A = 40{,}000/68{,}000 = 0.588$ in$^2 = \pi D^2/4$, and $D = 0.86$ inches (21.84 mm). The next largest standard material size is $7/8$ inch diameter. Always specify a size that can be purchased as standard.

To avoid compressive yielding, $A = 40,000/47,000 = 0.851$ in$^2 = \pi D^2/4$, and $D = 1.04$ inches (26.42 mm). The next largest standard size, depending on supplier, is probably $1^1/_{16}$ inches in diameter.

To avoid column buckling, first calculate D using the Euler model (Equation 11.35) and then verify the l_e/r ratio: $40,000 = \pi^2(30 \cdot 10^6)(\pi D^2/4) \div [20/(D/2)]^2$ and $D = 0.724$ inches. This diameter will allow yielding and fracture, and had this calculation been done without checking for yielding and fracture, a serious design error would have occurred.

Before concluding that column failure is not critical, a check on the l_e/r ratio should be made: $l_e/r = 20/(0.724/2) = 55$, but $\sqrt{2\pi^2 E/S_{yc}} = 112 > 55$, so the Johnson model (Equation 11.37) should be used, and $40,000 = \{\pi D^2/4\}\{47,000 - [47,000^2/(4\pi^2 \cdot 30 \cdot 10^6)][20/(D/2)]^2\}$; $D = 1.07$ inches (27.18 mm); purchase $1^1/_8$ inch diameter. Note that the Johnson model gives a component size that avoids yielding.

Without data to the contrary, some eccentricity should be allowed for, whether it be geometric or material variations. If an eccentricity of $20/400 = 0.05$ is assumed, then:

$$40,000 = \frac{(\pi D^2)(47,000)/4}{1 + \{(0.05)(D/2)/(D/2)^2\}\{\sec[(20/(D/2))\sqrt{40,000/(\pi D^2)(30 \cdot 10^6)}]\}}$$

Another computer iteration problem that results in $D = 1.185$ inches (30.10 mm). The next largest size available to purchase is probably $1^3/_{16}$ inch diameter.

11.9 Fracture Mechanics

The need for new failure analysis criteria was given impetus when 1289 of 4694 Liberty Ships failed structurally during the 1940s, when vessel tanks and steam turbines failed during the 1950s, and when solid rocket motor tanks failed during the 1960s. Brittle type failures were promoted by low temperatures in welded structures made from low strength steel. Researchers learned that cracks existed in the metal prior to fabrication and were nuclei for extended crack growth, which resulted in sudden failure.

11.9.1 Design Criteria

The design criteria in these cases is the magnitude of stress induced in the member at the location of the crack, and the material's resistance to crack growth. The cracks can be present because of flaws in the base material, or they can be induced by high microscopic stress due to material or geometry irregularities. Fracture toughness is the measure of the material property for this design criterion. Fracture toughness, K_c, is the critical value of stress intensity, that is, the threshold value of stress that causes unstable crack propagation. Unstable crack growth is characterized by cracks growing at rates as high as hundreds of feet per second, while stable crack growth occurs at rates of inches per minute or lower. The value of K_c is higher when the service temperature is above the material's transition temperature, and also increases as tensile strength and yield strength decreases. The value of K, which is the calculated stress intensity for a member being designed, depends on specimen thickness, physical constraints, the loading rate, and the

FIGURE 11.13

Modes of crack propagation.

Mode I   Mode II   Mode III

state of stress at the crack tip. Care must be exercised during fabrication not to lower material strength or its transition temperature with manufacturing processes such as welding or heat treatment. Also detrimental to the fracture failure resistance of a design are residual tensile stresses induced by cold-working, machining, and assembly mismatch, and/or cracks in the material. A sharp bend parallel to the grain flow lines of sheet steel is an example of how material cracks might be created during manufacture.

The stress intensity of a tensile loaded member is expressed as $K = c\sigma\sqrt{a\pi}$ or $K = c\tau\sqrt{a\pi}$ for shear loading, and has units of psi$\sqrt{\text{in.}}$ (MPa$\sqrt{\text{m}}$). The value of c depends on the type of loading and the geometry away from the crack area, and a is related to the crack length. For "small" cracks in "large" components, $c = 1.12$ is often used as a beginning approximation. See Appendix M, Figure 11.22, and Exercises 11.58, 11.59, and 11.60 for additional information to determine c. The values of σ and τ are the calculated stresses away from, and not including the area loss due to the crack. Failure by unstable crack growth to fracture is predicted when $K > K_c$.

There are three modes of crack propagation to consider, as shown in Figure 11.13. Each mode has a different effect on crack propagation, as noted in Appendix M. The general symbol K_c is better expressed as K_{Ic}, K_{IIc}, or K_{IIIc}, depending on the crack loading mode. There are standard tests for determining the values of K_c for the different modes of loading. The plane strain fracture toughness of mode I loading is K_{Ic}. The K_{Ic} value is the lower limiting value of K_c as crack tip stress conditions change. The changes are caused partially when the constraint of the member is maximum, but mostly as material thickness increases to a critical value. The value of K_{Ic} also gets smaller as temperature goes down and as strain rate increases.

11.9.2 Plane Stress—Plane Strain

The Anderson model of fracture analysis distinguishes between plane stress failure and plane strain failure using linear elastic fracture mechanics as illustrated in Figure 11.14.

Plane stress analysis is appropriate for thin members, where the fractured surface is inclined to the material surface (shear fracture). Plane strain analysis is appropriate for

FIGURE 11.14

Plane stress and plane strain values of K_c.

[Figure: Graph showing K_c vs Thickness. Y-axis shows $K_c$max at top and K_{Ic} at bottom. Plane stress region is flat at $K_c$max up to t_1, then decreases (idealized data plot) to plane strain region which is flat at K_{Ic} beyond t_2.]

thicker members, where the fracture surface is square with the material surface (brittle fracture). Plane strain analysis is valid if member thickness is greater than t_2 where $t_2 = 2.5(K_{Ic}/S_{yt})^2$, if crack length $a > 2.5(K_{Ic}/S_{yt})^2$, and material width $> 5(K_{Ic}/S_{yt})^2$. Plane stress analysis applies if thickness is less than t_1 where $t_1 = (K_{Ic}/S_{yt})^2/3$. Thicknesses between t_1 and t_2 are in the transition region where combined failure (shear and tension) will occur.

Plane stress failure is noted by slow crack growth and often referred to as leak-before-failure fracture. Plane stress is associated with high energy failure while plane strain is associated with low energy failure. Plane stress values of K_c can be 2 to 10 times the value of K_{Ic}. The value of K_{cmax} is not directly obtainable from K_{Ic} and must be determined from tests. Standard tests for determining K_c and K_{Ic} are available; see reference (9).

As the calculated value of K approaches K_c in value, an adjustment for the crack tip plastic deformation zone is appropriate. For plane stress, the radius of the zone is $r_y = K^2/(2\pi S_{yt}^2)$ and for plane strain $r_y = K^2/(6\pi S_{yt}^2)$. Plane strain behavior may be expected when $2r_y < 0.1t$. Plane stress behavior may be expected when $2r_y \geq t$. The value of r_y is used to modify the value of K as follows, $K = c\sigma\sqrt{\pi(a + r_y)}$.

EXAMPLE 11.19:
Fracture mechanics analysis.

A support plate 6.00 inches (152.40 mm) long, 2.00 inches (50.80 mm) wide, and 0.75 inches (19.05 mm) thick is made from a steel with allowable material values: S_{yt} = 220,000 psi (1516 MPa), and K_{Ic} = 50,000 psi$\sqrt{\text{in.}}$ (55 MPa $\sqrt{\text{m}}$). The plate is subjected to a 60,000 lbf (267 kN) tensile load perpendicular to the 2.00 × 0.75 inch (50.80 × 19.05 mm) cross section, so its macroscopic stress level is 40,000 psi (276 MPa). A 0.12 inch (3.05 mm) long crack is found at the edge of the plate parallel to the 2.00 inch (50.80 mm) dimension. Is the crack a dangerous one?

From Appendix M the value of $F(a/b)$ is approximately equal to 1.16, so substitute 1.16 for c in the formula $K = \sigma c \sqrt{\pi a}$.

$K = (40{,}000)(1.16)\sqrt{(\pi(0.12)} = 28{,}500$ psi$\sqrt{\text{in.}}$ (31 MPa $\sqrt{\text{m}}$). The part is safe because $50{,}000 > 28{,}500$ ($55 > 31$).

A check can be made for plane stress or plane strain conditions to verify that the proper model was used.

$t_2 = 2.5(50{,}000/220{,}000)^2 = 0.129 < 0.75 \Rightarrow$ plane strain

$5(50{,}000/220{,}000)^2 = 0.258 < 2.00$ width $\Rightarrow$ plane strain

$0.129 > 0.12 \not\Rightarrow$ plane strain

If plane strain exists then $r_y = (28{,}500/220{,}000)^2/6\pi = 0.0009$ inch, and

$K = (40{,}000)(1.16)\sqrt{\pi(0.12 + 0.0009)} = 28{,}600$, not a significant difference.

If plane stress is assumed then $r_y = (28{,}500/220{,}000)^2/2\pi = 0.0027$ inch, and

$K = (40{,}000)(1.16)\sqrt{\pi(0.1227)} = 28{,}800$ psi$\sqrt{\text{in.}}$, not a significant difference.

The linear elastic fracture mechanics (LEFM) analysis presented here applies well to high strength metals for either plane stress or plane strain. Brittle materials such as glass, ceramics, rocks, and ice also fit the model. As materials become more ductile, there is need to advance to elastic-plastic fracture mechanics (EPFM) models. If short cracks exist, $r_y \gg 0.03a$, LEFM loses validity and advanced EPFM models are required. In Example 11.19, $0.03a = 0.0036$, which is greater than the calculated value of r_y, so LEFM analysis is acceptable.

11.9.3 Crack Growth

From a design analysis standpoint, another question needs to be answered because cyclic tension loads on a structure can cause an existing crack to grow. How long will it take a crack to grow from noncritical to critical size? One model to use for this analysis was developed by Paris. It is:

$da/dN = B(\Delta K)^n$, where (11.39)

da/dN is the rate of crack growth in inches per cycle,

ΔK is the tensile stress intensity range in psi, and

B and n are constants determined by experimentation.

Values of n generally range between 2 and 7. If the log of da/dN is plotted versus the log of ΔK, then n is the slope of the line and B is the value of da/dN when $\Delta K = 1$. To solve Equation 11.39 for N_f, the number of cycles to failure, substitute $\Delta K = c\Delta\sigma\sqrt{\pi a}$ and integrate both sides within the limits $a_0 \leq a \leq a_f$ and $0 \leq N \leq N_f$. The assumption made is that c and $\Delta\sigma$ are not functions of the crack length a. Occasionally c is a function of a, and then numerical methods of integration are required.

Then $N_f = \dfrac{(a_f)^{(2-n)/2} - (a_0)^{(2-n)/2}}{[(2-n)/2][B(c\Delta\sigma)^n \pi^{n/2}]}$, where $a_f = (K_{Ic}/c\sigma_{max})^2/\pi$.

This method of estimating crack growth to failure only applies to data that is linear on a log-log graph, and recent studies have shown the need for other model assumptions in some applications. The method shown here is, however, a good first approximation. The use of the calculated number of cycles is an estimate on the life of the component and helps determine the length of time between visual inspections. The visual inspections are used to detect cracks at less than critical size and allow repair to be done, either by welding a crack or replacing a component.

11.10 Creep

Creep failure is the result of plastic deformation at stress levels below the yield strength of a material. Unchecked, this deformation can continue until fracture failure occurs. The fractured surface appears as a brittle fracture. This effect is more important to consider at elevated temperatures, especially those above 35 percent of the material's melting temperature measured on an absolute scale. Creep failure occurs at grain boundaries, so large grains deter creep failure.

Creep failure occurs in two or three phases. In phase one an initial load-environment deformation, which may contain both elastic and plastic deformation, occurs in a relatively short period of time. Phase two is marked by a uniformly increasing deformation (constant deformation rate) over extended periods of time. Phase two may end with fracture without entering a third phase. If phase three does occur, additional deformation appears at a more rapid rate and also results in component failure. Because phase two is a steady regular deformation process, experimental data from tests can usually be quantified and used to predict component service life much like predicting fatigue life using the Palmgren-Miner theory.

The elapsed time for any of the phases is governed by the stress level and the temperature. Creep tests take more time than other material property tests because time is one of the variables being measured, and at low stress levels and low temperatures failure may not occur for 10,000 hours or more. For this reason, data on creep failure is more limited than material properties such as yield strength or tensile strength. Data that is available is nicely represented as straight-line functions on log-log graph paper (refer to Section 12.6). The next example illustrates the use of creep data.

EXAMPLE 11.20:
Creep failure analysis.

A particular steel, with 2.25% chromium and 1% molybdenum, has creep failure characteristics as follows, where T = time to rupture in hours, and S = stress level in psi.

At 1000°F: $T = (2.03 \cdot 10^{-5} S)^{-8.13}$
At 1100°F: $T = (2.24 \cdot 10^{-5} S)^{-5.75}$
At 1200°F: $T = (3.08 \cdot 10^{-5} S)^{-4.63}$

If S = 10,000 psi, then part life is 426,600 hours at 1000°F, 5446 hours at 1100°F, and 233 hours at 1200°F. If a part made from this steel is subjected to 1100°F for 2723 hours, half its creep resistant life is used, 2723/5446 = 1/2. The part can then last 116.5 hours at 1200°F or 213,300 hours at 1100°F before complete failure.

11.11 Corrosion

Corrosion failure is the result of material surface deterioration beyond an acceptable level. This could be as simple as a dirty and bad looking surface caused by dust or paint fading. At the other extreme, corrosion could be so bad that a part could be completely corroded away, such as extreme galvanic action, the rusting of iron, or the rotting of wood. In between these extremes are cases of corrosion pitting that may cause grease seals to fail, may create locations for fatigue cracks to initiate, or may reduce the cross section of a component causing deformation or fracture from the increasing stress level. Various categories of corrosion exist:

1. Galvanic corrosion
2. Inter-granular corrosion
3. Cavitation
4. Crevice corrosion
5. Leaching
6. Biological corrosion
7. Pitting
8. Erosion
9. Stress corrosion

Stress corrosion is a common occurrence and affects nonmetals as well as metals. A tensile stress on a part in a critical environment can cause failure. For example, polystyrene in tension in the presence of acetone is a bad combination. Stress corrosion occurs when a part is subjected to a "high enough" tensile stress in the presence of a corrosive environment. "High enough" is in the range of 50–75% of S_{yt}. The corrosive activity is highest at the location of highest tensile stress, causing cracks to form perpendicular to the direction of tensile stress. On the microscopic level the atomic bonds in the region of high stress will break with less additional energy than in areas of low stress, and corrosion supplies this energy. Surface corrosion also reduces fatigue life by providing locations for cracks to form on the surface. Corrosion problems between dissimilar metals can be solved by using gaskets, couplings, or sleeves made of plastic or other electrically inert material to separate the dissimilar metals from each other. Corrosion of metal under water or buried in the ground can be reduced by using magnesium or zinc sacrificial anodes. For additional study on bad material-environment combinations, see reference 11, Appendix 15.

11.12 Fretting

Fretting failure is the result of two components in contact under a large normal force and aggravated by any oscillating or sliding motion. Particles of the materials in contact are worn away, stay on the contact surfaces, and increase the abrasive action. The debris from fretting on iron is a rust color, while on aluminum and magnesium it is black. Failure can occur from fatigue cracks, or by loss of ability to perform if a functional tolerance or surface quality is lost. Other variations of fretting failure are welding and tearing of the surfaces, excess friction and subsurface shearing, surface delamination, micro-crack initiation, and particle abrasion. The pressure between surfaces can be caused by bolted, riveted or keyed joints; wound wire rope on pulleys; or loaded bearings and leaf springs.

11.13 | Hydrogen Embrittlement

Hydrogen can diffuse into the lattice structure of a metal, accumulate as a gas, and cause pressures as high as 180,000 psi (1241 MPa) at a microscopic level. Add a tensile stress to this condition and an abrupt brittle fracture can occur. The hydrogen can originate from pickling operations (the part acts as a cathode in an aqueous solution), electroplating, stray electric currents, as well as service environments. Steels with $S_{ut} >$ 145,000 psi (999 MPa) are highly susceptible to hydrogen embrittlement. Steels with $S_{ut} <$ 100,000 psi (689 MPa) are relatively safe from hydrogen embrittlement. Steels having S_{ut} between 100,000 and 145,000 psi (689 and 999 MPa) are moderately susceptible. The hydrogen pressure alone is not enough to cause failure, so the excess hydrogen can be baked out before the component is placed into service.

11.14 | Vibration

Vibration is cyclic variation of physical quantities such as pressure, stress, voltage, or current. Ordinarily vibration is of concern when it causes the cyclic change in the position of a physical object. A vibration results in a physical displacement, a strain, and can be converted into a force by using the modulus of elasticity. Vibration induced into a structure or component can cause as much harm as an applied force. When induced vibrations last over an extended period of time, the vibration converted into a force is similar to a fatigue loading. Vibrations are particularly bad when the frequency of an applied vibration matches the natural frequency of a structure. The vibration oscillation will last longer, and the magnitude of the vibration can increase depending on the other aspects of the design.

Buildings normally vibrate at 0.1 to 5 hertz and will sustain damage at vibrations of 0.1 mm @ 10 hertz or 0.01 mm @ 100 hertz. Industrial machinery vibrates from 10 to 100 hertz, and the human body is susceptible to major resonance of vibrations in the range of 1 to 15 hertz.

Vibrations can be desirable, as in the case of a violin string, or undesirable, as in the case of an earthquake shock wave. Unwanted vibrations can originate from many sources. One source is machine tool vibration caused by irregular cutting action. Another source is the wind acting on structures. Wind force is never constant, and the varying wind force, coupled with a material's elasticity, causes vibration. There is less chance for vibration from rotary balanced motion than from reciprocating linear motion. The acceleration needed for the change in linear motion causes a force that can easily be transmitted to other structure members and cause a vibration. Atomic vibrations are an example of a naturally occurring undamped vibration.

Recall that the natural frequency of vibration of a mass is $f_n = \sqrt{k/m}$ radians/second, or in the more common units of hertz (cycles per second) $f_n = [\sqrt{k/m}]/2\pi$, where $k =$ spring constant, lbf/inch or Newton/meter, and $m =$ mass, lbf·sec^2/inch or kilograms. The time in seconds for one complete oscillation is $1/f_n$.

Recall that the mass, m, of the weight, W, can be expressed as $m = W/g$, and then $f_n = [\sqrt{gk/W}]2\pi$. Recall also that $W/k = \delta$, and $g = 32.17$ ft/sec^2 = 386.04 in./sec^2, so $f_n = [\sqrt{386.04/\delta}]/2\pi$.

A structure is generally designed to have a high natural frequency because outside disturbances generally have low frequencies. An outside frequency, f, or its harmonics, kf, where $k = 1, 2,\ldots$, can cause vibration in a structure. Springs, for example, can be affected by frequencies as high as the 13th harmonic. The larger the difference between the applied frequency and the natural frequency of the structure, the less chance there is for induced vibration. If a known source of vibration has a high frequency, the absorbing structure should be designed at a low natural frequency.

To increase the natural frequency of a structure requires either that the mass be decreased, or the spring rate (the material's stiffness) be increased. To reduce the mass requires design adjustments such as:

1. Less dense material for the same volume
2. Smaller cross sections, and
3. Fewer number of parts.

To increase the stiffness requires design adjustments such as:

1. Making structural members shorter
2. Adding support members
3. Using material with a high E
4. Designing tight joints between members
5. Designing large surface contact areas, and
6. Arranging material to increase moments of inertia.

If the direction of an anticipated vibration force is known, then a structure can be designed with increased rigidity in a particular direction in a manner analogous to directional column buckling. Buildings in earthquake zones are designed this way. If vibration reduction is not possible with the methods suggested, then vibration damping may be required to reduce the transmission of vibrations. Damping can be accomplished by using resilient materials such as cork, rubber, carpet, felt, or fiberglass between structural members. More complicated methods can also be used, such as adding springs, hydraulic cushions, air cushions, or friction pads. Common applications of vibration control are engine motor mounts, suspension shock absorbers, spring mounted paint mixers, and building pillar cushions. For additional information on vibration isolation see references (5), (21), (23), (28), and (33).

EXAMPLE 11.21:
Vibration analysis.

A cantilevered 2 × 12 inch (51 × 305 mm) board is being cut off with a circular saw as shown in Figure 11.15. The saw blade is 6 inches (152 mm) in diameter, has 36 teeth, and rotates at 1200 revolutions per minute. Is there a chance for unwanted vibrations?

FIGURE 11.15

Schematic for Example 11.21.

If the board is held tight against the support then only the mass overhanging is critical. Assume fir construction lumber, 32 lb/ft³ (513 kg/m³), nominally sized at 1.50 × 11.25 inches (0.038 × 0.286 m).

The deflection at the end is $\delta = wl^4/8EI$ (Appendix O).

$$\delta = \frac{(32 \text{ lb/ft}^3)(1.5 \text{ in})(11.25 \text{ in})(49 \text{ in})^4}{(1728 \text{ in}^3/\text{ft}^3)8(1.7 \cdot 10^6 \text{ lb/in}^2)[(11.25)(1.5)^3/12 \text{ in}^4]} = 0.0419 \text{ in. } (1.064 \text{ mm})$$

$f_n = [\sqrt{386.4/0.0419}]/2\pi = 15.3$ cycles/sec

f_s(saw teeth on wood) = (1200 rev/min)(36 teeth/rev)/(60 sec/min) = 720 cycles/sec >> 15.3.

There should be no induced vibration.

Note: Using a hand saw at one stroke cycle in three seconds is more apt to induce vibration, as anyone who has used a hand saw can vouch for. The fifth harmonic induces vibration. This is the same problem that occurs when cutting tools on machines vibrate, they are too long and lack mass to counteract the induced vibrations.

11.15 Summary

The next case study illustrates how different engineering failure analysis concepts are joined together in a complete design study. It also illustrates how a computer can be used to enhance analysis.

CASE STUDY #9:
Hydraulic Cylinder

The following variables are associated with the analysis of the design of a hydraulic cylinder (refer to Figure 11.16).

FIGURE 11.16

Cylinder schematic for Case Study #9.

D_c = cylinder barrel inside diameter, inches

t = cylinder barrel wall thickness, inches

D_r = rod diameter, inches

A_c = cross sectional area of the cylinder, in²

A_r = cross sectional area of the rod, in²

L = cylinder stroke length, inches

P_e = fluid pressure for rod extension, psi (P_e causes the rod to be in compression)

P_r = fluid pressure for rod retraction, psi; (P_r causes the rod to be in tension)

F_e = extension compressive force on the rod, lbf

F_r = retraction tensile force on the rod, lbf

σ_e = extension compressive stress in the rod, psi

σ_r = retraction tensile stress in the rod, psi

S_{yt} = yield strength of the rod material, psi

S_{ut} = ultimate strength of the rod material, psi

S_e' = adjusted fatigue strength of the rod material, psi

e = eccentricity of the rod, inches

I_r = area moment of inertia of the rod, in⁴

r_r = radius of gyration of the rod cross section, inches

P_c = critical load for rod column analysis, lbf

T_e = time of extension, minutes

T_r = time of retraction, minutes

C_p = capacity of pump, gallons per minute

All cylinder designs do not start with the same limiting conditions. Sometimes the primary constraint will be stroke length, at other times extension force.

Summary

Other constraints may be available pressure, or a time limit on extension and retraction. Another time the goal will be maximum extension force using available materials; and don't forget function for minimum cost.

In some designs no pressure is required for cylinder retraction because a system load acts to retract the cylinder. Control of the oil flow from the cylinder acts as a brake on the retracting mechanism.

Let's see how the variables are related and why a computer may make analysis more efficient. One goal will be to relate the variables to each other so that the fewest number of values will be required for analysis.

To avoid material yielding in the rod, a calculation of stress and a comparison with material properties are required for extension and retraction. Searching material data files for materials with acceptable strength levels is a natural application for a computer.

Generally the minimum values required for F_e and F_r are known based on other design analysis for the function of the cylinder. The values of P_e and P_r are usually equal and are based on available pumps, generally 2250, 2500, or 3000 psi. The value for D_c is selected from available material, either sizes already available in-house, or sizes available from suppliers. These values should be in the form of a table so choices can be easily made. It may also be that the analysis will not begin with D_c, but will end with a calculated size for D_c starting with other restrictions. If that happens then a standard size must be selected and reverse analysis done to insure adequate design function. Never end with a size that must be specially made, because that increases the cost.

Assume that the known criteria, the first data input, are F_e, L, P_e, P_r, and C_p, then $\pi D_c^2/4 \geq F_e/P_e$, and $D_c \geq 2\sqrt{F_e/\pi P_e}$. A file of available sizes will be accessed and the minimum diameter material selected. The minimum barrel wall thickness is determined using the basic hoop stress formula $\sigma = P_e(D_c/2)/t$; $t = (P_e D_c)/(2\sigma)$, where σ = maximum allowable stress in the available barrel material.

The rod diameter is calculated based on the maximum stress applied to the rod.

$$\sigma_e = F_e/A_r = P_e A_c/A_r = P_e(\pi D_c^2/4) \div (\pi D_r^2/4) = P_e(D_c/D_r)^2 > S_{yc},$$ (remember that compression forces are assumed negative), and $D_r \geq \sqrt{P_e D_c^2 S_{yc}}$.

A data file will be accessed that contains available rod diameters and the material values of E, S_{yc}, S_{yt}, and S_{ut}. Several calculations may be required to select the minimum diameter. If material cost is not proportional to rod diameter, then calculations must iterate for minimum cost and still meet strength parameters.

A check must be made to be sure that the retraction stress does not cause the rod material to yield. Calculate σ_r and compare to S_{yt}.

$$F_r = P_r(A_c - A_r) = P_r[(\pi D_c^2/4) - (\pi D_r^2/4)]$$
$$\sigma_r = F_r/A_r = F_r(\pi D_c^2/4) = P_r(D_c^2 - D_r^2)/D_r^2 \leq S_{yt}$$

If this test fails then calculate $D_r \geq \sqrt{P_r D_c^2/(S_{yt} + P_r)}$, and find a different size rod.

To avoid column buckling of the rod a calculation of P_c is made followed by a comparison with F_r. The computer is of great value here because of the iteration required to determine P_c using Equation 11.38. Remember to use a value for eccentricity based on material specifications, accepted design standards, or codes. If the column analysis comparison fails, then reverse the analysis and determine the minimum diameter that avoids column buckling. Any change in D_r requires a check to be sure F_r is adequate.

$$I_r = \pi D_r^4/64 = A_r r_r^2 = (\pi D_r^2/4) r_r^2$$

$$r_r = \sqrt{D_r^2/16} = D_r/4$$

$$P_c = \frac{(\pi D_r^2/4) S_{yt}}{1 + [8e/D_r][\sec((0.8L/r_r)\sqrt{P_c/4A_r E})]}$$

Solve for P_c using iteration and compare to F_e: $P_c \geqslant F_e = P_e \pi D_c^2/4$.

If this test fails the next larger rod diameter must be selected and the check repeated.

To avoid fatigue failure of the rod more calculations, more data file searches, and comparisons are made; $\sigma_{mean} = (\sigma_r + \sigma_e)/2 = (P_r + P_e)(D_c^2/2D_r^2) - P_r/2$. If $P_e = -P_r$, then $\sigma_{mean} = -P_r/2 = P_e/2$.

$\sigma_{amp} = \sigma_r - \sigma_{mean} = (\sigma_r - \sigma_e)/2 = (P_r - P_e)(D_c^2/2D_r^2) - P_r/2$. If $P_e = -P_r$, then $\sigma_{amp} = (P_r D_c^2/D_r^2) - P_r/2$.

S_e' of rod = $(0.50)(S_{ut})$(correction factors for reliability, diameter, axial loading, surface finish, stress concentration, and surface plating).

The factors for reliability can be stored in a data file and accessed by using a reliability value as input, reference Table 11.3, or the factor can be a menu-selected value accessed by showing a table on the screen and providing for user selection. The factor for diameter can be calculated based on rod diameter, $0.872 D_r^{-0.1133}$. The axial loading factor = 0.9. The surface finish factor is a function of the method of manufacture and the material ultimate strength, see Figure 11.9. If the rod surface is highly polished then this factor equals one. Use a factor for plating from 0.50 to 0.90 (conservative to less conservative).

Once S_e' is determined then fatigue life prediction can be made. For fatigue life > 10^6 cycles, then $\sigma_a \leqslant (S_e')\sigma_{mean}/S_{ut} + S_e'$.
For 10^6 > fatigue life > 10^5 cycles, then $\sigma_a \leqslant (1.26 S_e')\sigma_{mean}/S_{ut} + 1.26 S_e'$.
For 10^5 > fatigue life > 10^4 cycles, then $\sigma_a \leqslant (1.54 S_e')\sigma_{mean}/S_{ut} + 1.54 S_e'$.
For 10^4 > fatigue life > 10^3 cycles, then $\sigma_a \leqslant (1.80 S_e')\sigma_{mean}/S_{ut} + 1.80 S_e'$.

If fatigue life > 10^6 cycles is required and the test fails, then a stronger material must be selected or σ_a must be reduced. A decision must be made: live with the reduced life, decrease P_r, which reduces F_r and σ_{mean}, increase D_r, which increases A_r and decreases σ_{mean}; or decrease D_c, which decreases A_c and σ_{mean}.

The time of extension and retraction can also be considered:

$$T_e = A_c L/231 C_p, \; T_r = (A_c - A_r) L/231 C_p.$$

If T_e is too large another decision must be made; live with the reduced time, reduce D_c, reduce L, or increase C_p. To decrease T_r decrease A_c, increase A_r, decrease L, or increase C_p.

A program written to solve the cylinder problem must be flexible and iterative because each analysis will have different restrictions, limitations, and compromises. The program, and relative material data files, is a large project, but if many cylinders have to be designed, the time spent on the program is worth the effort.

11.16 Additional Study Topics

1. Larson-Miller theory in creep analysis
2. Manson-Haferd theory in creep analysis
3. Marco-Stankey theory of failure
4. Henry theory in fatigue
5. Gatts theory in fatigue
6. Carten-Dolen theory in fatigue
7. Marin theory in fatigue
8. Mansion theory in fatigue
9. Probit test methods in fatigue
10. Step test method in fatigue
11. Prot test method in fatigue
12. Staircase test method in fatigue
13. Strain gages
14. Photoelasticity
15. Brittle lacquer stress coat
16. Photo-stress
17. Crack growth threshold values
18. Pre-service overload effects
19. Elastic-plastic fracture mechanics
20. Standard linear-elastic test procedures
21. Haigh diagram analysis
22. Column failure in horizontal beams
23. Thermal relaxation failure
24. Thermal shock failure
25. Impact spalling
26. Pin connected joints
27. Rivet connections
28. Bolted connections
29. Use of set screws
30. Spline joints
31. Springs
32. Clutches
33. Curved beams
34. Fatigue arrest design
35. Stress gradients in the region of geometric changes

36. Baseline notch factor, $K_b = 1 + (1/(K_t(1 + \sqrt{a/r})))$
37. Plastic stress concentration, $K_p = 1 + E_t/(E \cdot K_t)$
38. Myklestad-Prohl method of finding nodes of natural frequency of composite structures
39. Rayleigh method of spring-mass natural frequency calculations
40. Mean stress effects on the Paris crack growth rate model

11.17 References

1. Baird, D. C. *Experimentation: An Introduction to Measurement Theory and Experimental Design*. Englewood Cliffs, N.J.: Prentice Hall, 1962.
2. Burr, Arthur C. *Mechanical Analysis and Design*. New York: Elsevier Science, 1981.
3. Billington, David P. *Thin Shell Concrete Structures*, 2nd ed. New York: McGraw-Hill, 1982.
4. Collins, J. A. *Failure of Materials in Mechanical Design*. New York: John Wiley & Sons, 1981.
5. Crawley, Stanley W. and Robert M. Dillon. *Steel Buildings Analysis and Design*. 3rd ed. New York: John Wiley & Sons, 1984.
6. Deutschman, Aaron D. et. al. *Machine Design: Theory and Practice*. New York: Macmillan, 1975.
7. Dieter, George E. *Engineering Design*. New York: McGraw-Hill, 1983.
8. Dudley, Darle W. *Handbook of Practical Gear Design*. New York: McGraw-Hill, 1984.
9. Ewalds, H.L. and R. J. H. Wanhill. *Fracture Mechanics*. Baltimore: Edward Arnold, 1984.
10. Farkas, Jozsef. *Optimum Design of Metal Structures*. New York: John Wiley & Sons, 1984.
11. Felbeck, David K. and Anthony C. Atkins. *Strength and Fracture of Engineering Solids*. Englewood Cliffs, N.J.: Prentice Hall, 1984.
12. Fisher, John. *Fatigue and Fracture in Steel Bridges: Case Studies*. New York: Wiley Interscience, 1984.
13. Fuchs, H. O. and R. I. Stephens. *Metal Fatigue in Engineering*. New York: John Wiley & Sons, 1980.
14. Harker, Ralph J. *Elastic Energy Methods of Design Analysis*. New York: Elsevier Science, 1986.
15. Harvey, John F. *Theory and Design of Modern Pressure Vessels*, 2nd ed. New York: Van Nostrand Reinhold, 1974.
16. Higdon, Archie, et. al. *Mechanics of Materials*, 4th ed. New York: John Wiley & Sons, 1985.
17. Hindhede, Affe, et. al. *Machine Design Fundamentals*. New York: John Wiley & Sons, 1983.
18. Jones, James V. *Engineering Design*. Blue Ridge, Penn.: TAB Books Inc., 1988. Design as it relates to U.S. government standards.

19. Juvinall, Robert C. *Fundamentals of Machine Component Design*. New York: John Wiley & Sons, 1983.
20. Macaulay, M. A. *Introduction to Impact Engineering*. New York: Chapman and Hall, 1987.
21. Macinante, Joseph A. *Seismic Mountings for Vibration Isolation*. New York: Wiley Interscience, 1984. Practical isolation mechanisms.
22. Magner, Ray. "Designing For Shock Resistance." *Civil Engineering*. October, 1986, 63-65.
23. Meirovitch, Leonard. *Elements of Vibration Analysis*. New York: McGraw-Hill, 1975.
24. Merritt, Frederick S. *Building Design Construction Handbook,* 4th ed. New York: McGraw-Hill, 1981.
25. Mischke, C. R. *"A Probabilistic Model of Size Effect in Fatigue Strength of Rounds in Bending and Torsion."* Transactions of The American Society of Mechanical Engineers, Vol. 102, No. 1. Jan. 1980, 32-37.
26. Parker, Harry and H. D. Kauf. *Simplified Design of Structural Wood,* 3rd ed. New York: John Wiley & Sons, 1979.
27. Peterson, R. E. *Stress Concentration Factors*. New York: John Wiley & Sons, 1974.
28. Pretlove, A.J. *Basic Mechanical Vibrations*. London: Butterworth, 1985. Includes some BASIC computer programs.
29. Roark, Raymond J. and Warren C. Young. *Formulas for Stress and Strain,* 5th ed. New York: McGraw-Hill, 1976.
30. Rolfe, Stanley T. and John M. Barsom. *Fracture and Fatigue Control in Structures*. Englewood Cliffs, N.J.: Prentice Hall, 1977.
31. Seeley, F.B. and J.O. Smith. *Advanced Mechanics of Materials,* 2nd ed. New York: John Wiley & Sons, 1982.
32. Shigley, Joseph E. and Larry D. Mitchell. *Mechanical Engineering Design,* 4th ed. New York: McGraw-Hill, 1983.
33. Tse, Francis S. et. al. *Mechanical Vibrations: Theory and Applications*, 2nd ed. Boston: Allyn and Bacon, 1978.
34. Ugural, A.C. and S.K. Fenster. *Advanced Strength and Applied Elasticity*. New York: Elsevier Science, 1981.
35. Walter, Podlny. *Construction and Design of Prestressed Concrete Segmental Bridges*. New York: Wiley Interscience, 1982.
36. Wilde, D. J. *Globally Optimum Design*. New York: John Wiley & Sons, 1978.

11.18 Exercises

1. Try to find parts that have failed due to the failure modes listed on page 367.
2. Consider each of the seven items listed under "failures occur for many reasons" on page 367. Discuss who is responsible for each item and what could be done to avoid or prevent the problem.

3. A particular engine has six cylinders. If one cylinder fails, the engine can still operate, but inefficiently. If two cylinders fail, the engine stops running. Assume the reliability of each cylinder is 0.96.
   a) What is the probability of only cylinder #3 failing?
   b) What is the probability of some cylinder failing?
   c) What is the probability of only cylinders #3 and #4 failing?
   d) What is the probability of any two cylinders failing?

4. A particular electronic control has a reliability of 95 percent. How many controls must be placed in parallel to achieve a system reliability of 99 percent? 99.9 percent?

5. Determine the most economical system to achieve a system reliability of 99 percent using some combination of the following components. Each component has the same function and can be used in series or parallel mode.

Vendor	Reliability	Cost per unit
A	0.98	$1000
B	0.96	$700
C	0.93	$100

6. Refer to Example 11.4 and determine an advertised life of the light bulb for a 99 percent reliability. For a 99.9 percent reliability.

7. Four machine tools used in a sequential manufacturing process are loaded and unloaded by a robot because contamination from the process is harmful to humans. The cost of the machine tools are $100,000, $150,000, $75,000, and $25,000. The robot costs $30,000 and has a reliability of 95 percent. Assume each machine tool has a historical run time efficiency of 92 percent. Discuss the situation if management wants the whole process to have an uptime of 70 percent. Propose operating changes. Do the same for an uptime of 80 percent.

8. Use Equation 11.3 and find the standard deviation of the quotient of two variables, $Z = X/Y$, where the standard deviations S_X and S_Y are known.

9. Use Equation 11.3 and find the standard deviation of a power function, $Z = X^n$, where the standard deviation S_X is known.

10. Refer to Case Study #8 and determine the system sensitivity for each of the following changes in standard deviation.
    a) Change S_X from 0.002 to 0.004
    b) Change S_L from 0.03 to 0.06

11. Refer to Case Study #8 and determine the system sensitivity to a change in reliability from 99 percent to 99.9 percent.

12. Explain why the maximum normal stress theory might be usable for very brittle materials.

13. Start with Equation 11.18 and show that $\sigma_e = \sqrt{\sigma_x^2 + \sigma_y^2 - \sigma_x\sigma_y + 3\tau_{xy}^2}$ is an equally valid expression for S_e for a specific two dimensional loading problem.

14. Continuation of Exercise 11.13. What condition results in $\sigma_e = \sqrt{\sigma_x^2 + 3\tau_{xy}^2}$?

Refer to Figure 11.17 for Exercises 11.15–11.19.

FIGURE 11.17

Exercises 11.15–19.

Diameter = 2.50"

6'

W = 1000 lbs.

FIGURE 11.18

Exercises 11.20–24.

6'

4'

Turned diameter = 2.00"
0.250" × 1.250" attached bar

F

15. What is the reliability of the shaft if the material has S_{ut} = 80,000 psi and S_{yt} = 50,000 psi? Consider both yield and fracture failure.

16. What is the maximum load W allowed for a yield safe design reliability of 99.9 percent? What free end deflection would be expected at that load?

17. Discuss possible designs for the joint where the shaft attaches to the vertical plate.

18. Discuss the design of attaching methods for the cable to the shaft and the cable to the block.

19. Does anything done in Exercises 11.17 and 11.18 affect any of the analysis done in Exercises 11.15 and 11.16?

Refer to Figure 11.18 for Exercises 11.20–11.24.

20. Determine the reliability of the shaft if $F = \pm 500$ lbf, S_{ut} = 90,000 psi and S_{yt} = 55,000 psi.

21. What is the maximum load F allowed if a shaft reliability of 97.7 percent is desired?

22. What is the maximum load F allowed for a shaft reliability of 99.86 percent?

23. Discuss the design of the joint where the rod is attached to the vertical wall. Suggest alternative methods of attaching.

24. Discuss various methods of attaching the bar to the shaft.

25. Is the data in Example 11.4 normally distributed?

Use these conditions for Exercises 11.26–11.29. A 1100 pound ball is dropped in the middle of a nominally sized 2 × 4 inch, 60-inch long simply supported beam from a height of 10 inches above the beam. The beam material is white pine with $E = 10^6$ psi and an allowable stress limit of 6000 psi.

FIGURE 11.19

Exercise 11.30.

26. Will this condition stress the beam above the allowable limit?
27. If a spring is placed under each end of the support of the beam, what spring rate (lbs/in.) will be acceptable so the beam is not stressed above the allowable limit?
28. What weight ball can be dropped from a height of 12 inches and not exceed the allowable stress limit if no springs are present?
29. What height can a 100 pound ball be dropped from and not load the beam above the allowable stress limit? Assume no springs are present.
30. Refer to Figure 11.19. The post is made from concrete and holds a static load as shown, S_{uc} = 5000 psi, S_{ut} = 750 psi, and $E = 2 \cdot 10^6$ psi. Will the post fail under a collision with a 4000 pound car traveling at 40 miles per hour? The bumper of the car is 12 inches off the ground, the hood top is 30 inches from the ground, and the car is a front engine car.
31. The firing of a rocket may not be an impact load if the thrust is applied slowly enough. If the interval of firing time is considered short enough to be an impact load, what equivalent load must be considered for the members near the rocket engine compared to the force of the engine?
32. Theory states that the impact energy absorption is expressible as $KE = \sigma^2$(component volume)$/2E$. This often does not give reliable results because of:
    a) Local strain effects in the area of applied load,
    b) Mass inertia of the members themselves, and
    c) The effect of the type of member support. Design experiments to investigate these effects.
33. Design an experiment to determine the velocity of impact for which an entire member does not have time to react to the applied load, and fails at some point in the center. Consider a cantilevered beam such as shown in Figure 11.20. Add variables to the function as required.
34. List the assumptions made or implied in arriving at $KE = \sigma^2$(volume)$/2E$.
35. Refer to Figure 11.21.
    a) Compare the impact resisting capacities of the two circular rods made from the same material. Neglect end effects and inertia unknowns. Assume F is a symmetrically applied impact load.

FIGURE 11.20

Exercise 11.33.

W = weight
V = velocity at impact
L = length of beam
I = moment of inertia of beam
X = distance to point of failure

Find $X = f(W, V, L, I, ?)$

Failure Analysis

FIGURE 11.21

Exercise 11.35.

b) If the impact load is high enough, each rod will fail. Where would you expect each rod to fail? Is this the same place they will fail if the load is increased slowly until failure? Explain.

c) List some possible design changes you would try to improve the load carrying capacity of either rod. Consider both static and impact loads.

36. Using the spring constant k, show that $F_e = \sqrt{2(KE)k}$, and for the case of an impact load applied perpendicular to the cross section, that $k = AE/L$.

37. How does the multiplier for reliability change if the standard deviation for values of S_e is assumed to be 10% instead of 8%. Refer to Table 11.3.

38. Discuss the details of an experiment to study creep failure. Consider the possible effects of such variables as stress level, temperature, time, grain size, and material yield strength.

39. If you thought that creep failure was possible in a design and the consequences would be fatal, what methods of monitoring the structure would you consider to avoid the creep failure?

40. Find examples of parts that have failed because of corrosion or stress corrosion. Discuss methods of avoiding the conditions found, or ideas to design around the problem.

41. Find samples of parts that have failed due to fretting action. Discuss methods of avoiding the condition that caused the failure, or ideas to design around the problem.

42. Design a device to use to test humans and their discomfort due to physical vibrations.

43. The natural frequency of a spring, in cycles per second, can be determined from the model:

$f = (13,900)d/(ND^2)$, where

d = spring wire diameter, inches,

N = number of active coils, and

$D + 2d$ = the spring outside diameter, inches.

A valve spring can be affected by an applied vibration and all its harmonics up to the 13th order. Find a valve spring, take measurements, and determine the maximum RPM of the engine allowable before the spring is "in trouble." If you can't find a spring use $d = 5$ mm, $D = 47$ mm, and $N = 5$.

44. *a*) Refer to Equation 11.35 and determine the l_e/r ratio that would reduce $P_c = S_{yc}A$ by 10 percent for each of the following cases:
    1) $S_{yc} = 65,000$ psi and $E = 30 \cdot 10^6$ psi.
    2) $S_{yc} = 31,000$ psi and $E = 10.4 \cdot 10^6$ psi.
    b) What percent reduction occurs using the $l_e/r < 10$ rule?

45. A supplier lists a straightness tolerance for bars of material as 0.125 inch in any 5 feet of bar length. Reanalyze Example 11.17 using this tolerance.

46. What is the value of l_e/r that makes P_c for the Euler model and the Johnson model equal?

47. A particular aluminum alloy has an E_t value that varies from $10.5 \cdot 10^6$ to $1.0 \cdot 10^6$ psi for $S_{yt} = 57,000$ psi to $S_{ut} = 68,000$ psi. What is the effect on P_c in this interval of stress?

48. A construction 2×4 (actual size = 1.50×3.50 inches) is 8 feet long and has a 0.10 inch eccentricity in the 3.50 inch direction. Will the post fail as a column? Use design criteria from Exercise 11.70.

49. Many hydraulic cylinders are designed as shown in Figure 11.16. The hydraulic pressure P_e is supplied by a pump and forces the piston to the right for extension of the cylinder, and to the left for retraction. Valves and proper porting allow pressure on the side opposite the moving force to be near zero. The diameter of the rod affects the retraction force because it takes away from the pressure area. Let the piston diameter = 6.00 inches; the maximum pressure = 3000 psi; and rod material with $S_{yc} = S_{yt} = 60,000$ psi, $S_{uc} = S_{ut} = 90,000$ psi, and $E = 30 \cdot 10^6$ psi.
    a) Determine the rod diameter and the difference between the extension force F_e and the retraction force F_r based on yielding failure on the rod.
    b) Determine the rod diameter and the difference between the extension and retraction force based on buckling failure.

50. Will fatigue analysis change the answers in Exercise 11.49? Design for 99.97% reliability and assume the rod is highly polished and chrome plated.

51. Many structural shapes are not symmetrical about a central axis so their moment of inertia and hence their resistance to buckling is directional. Consider a 13.75 lb/foot, 8-inch structural channel used as a post. How much difference is there, between the strongest and weakest direction, in the load the post can hold without buckling failure?

52. Would two channels from Exercise 11.51 placed back to back make the post equally likely to buckle in the two primary directions? Explain.

53. Discuss a solution to obtaining equal column buckling resistance in the two primary directions using the post from Exercise 11.51.

54. If the only material available for a 10 foot post is a $2 \times 2 \times 1/4$ inch thick angle iron,

how would your design change as the anticipated load increased? Determine the loads at which you would change designs. Assume $l_e = (0.8)l$.

55. Compare the crack growth behavior of these three materials.
    a) Ferritic-pearlitic steel: $da/dN = 3.6 \cdot 10^{-10}(\Delta K)^{3.0}$
    b) Martensitic steel: $da/dN = 6.6 \cdot 10^{-9}(\Delta K)^{2.25}$
    c) Austenitic stainless steel: $da/dN = 3.0 \cdot 10^{-10}(\Delta K)^{3.25}$
    Answer such questions as:
    a) Which material will fail in the fewest number of cycles at a given load?
    b) Which material allows cracks to grow the fastest?

56. A fail safe design incorporates multiple load paths or crack stoppers. If component failure could be fatal, this is a necessary design criterion. Identify products or designs that have this feature built into them.

57. Damage tolerant design is permissible if some degree of failure is not fatal. Identify products or designs that can be or are designed this way.

58. For the center crack analysis, Figure 11.22(a), alternate values of c have been proposed. Write a computer program to compare the value of c obtained with these formulas and a heuristic value of 1.12 used for "small" cracks. Use $0.0 \leq a/2b \leq 0.30$ with an increment of 0.01. Comment on the results.
    a) $c = 1 + 0.256(a/2b) - 1.152(a/2b)^2 + 12.20(a/2b)^3$
    b) $c = \sqrt{\sec(\pi a/2b)}$ (Fedderson)
    c) $c = 1/\sqrt{1 - (a/b)^2}$ (Dixon)
    d) $c = \sqrt{(2b/\pi a)\tan(\pi a/2b)}$ (Irwin)

59. For the double edge crack, Figure 11.22(b), a formula for c has been proposed: $c = [1.122 - 0.561(a/2b) - 0.205(a/2b)^2 + 0.471(a/2b)^3 - 0.190(a/2b)^4]/\sqrt{1 - a/2b}$. Write a computer program to compare the formula results with the value of 1.12 commonly used for "small" cracks.

FIGURE 11.22

Crack/plate width geometry.

(a)   (b)   (c)

60. For the single edge crack, Figure 11.22(c), a formula for c has been proposed as long as $a/b \leq 0.6$: $c = 1.12 - 0.231(a/b) + 10.55(a/b)^2 - 21.72(a/b)^3 + 30.95(a/b)^4$. Write a computer program to compare the formula results with the value of 1.12 commonly used for "small" cracks.

61. Four, 0.50×2.00 inch, 24.00 inch long bars are used to support a static hanging load of 350,000 pounds. The material selected for the application has $S_{yt} = 125,000$ psi, $S_{ut} = 145,000$ psi, and $K_{Ic} = 90,000$ psi$\sqrt{\text{in}}$.
   a) Is a 0.25 inch center crack length perpendicular to the direction of load critical?
   b) How about an edge crack of 0.25 inch?

62. A design engineer has decided that a 1.50 inch diameter cold rolled 4140 steel bar will be safe from fatigue for a cyclic axial load from 80,000 lbs tension to 30,000 lbs compression. Do you agree or disagree? Why?

63. Is the following data consistent for a reversed bending fatigue test on a material with $S_{ut} = 218,000$ psi and $S_{yt} = 200,000$ psi?

Sample #	σ_{max}	Cycles to failure
1	$17.0 \cdot 10^4$	$2 \cdot 10^4$
2	$15.1 \cdot 10^4$	$5 \cdot 10^4$
3	$14.1 \cdot 10^4$	$1 \cdot 10^5$
4	$12.7 \cdot 10^4$	$2 \cdot 10^5$
5	$12.5 \cdot 10^4$	$5 \cdot 10^5$
6	$12.3 \cdot 10^4$	$1 \cdot 10^6$
7	$12.1 \cdot 10^4$	no failure

64. Discuss a design situation where temperature induced stress could cause fatigue problems.

65. Which keyway in Figure 11.23 has the highest fatigue life? Why? Assume equal length and equal depth.

FIGURE 11.23

Exercise 11.65.

66. Sketch a typical S-N curve for a steel with $S_{ut} = 110{,}000$ psi. Include $10^0 - 10^7$ cycles in the domain.

67. What is the temperature at which creep should be a concern for the following?
    a) 1020 steel
    b) aluminum

68. Estimate the fatigue limit of a steel that has a Brinell hardness of 300.

69. Under what circumstances are the size and weight of a simply supported beam as critical for failure analysis as the load placed on the beam? Is the condition different for a cantilevered beam?

70. Analyze the following design guidelines taken from the National Design Specification for Wood Construction.
    a) A 2×10 inch joist made from wood with $E = 1.5 \cdot 10^6$ psi has a maximum allowable span of 16 feet when placed on 16-inch centers and has a maximum bending stress of 1200 psi.
    b) Same as a. except a 2×8 inch joist and a maximum allowable span of 12 feet-7 inches.
    c) A short length column of diameter d and length l is a column where $l/d \leq 11$ and P_c' = the applied compressive stress.
    d) A medium length column is a column where $11 < l/d \leq k$, $P_c = F_c(1 - (l/dk)^4/3)$, and $k = 0.671\sqrt{E/F_c}$.
    e) A long column is a column where $l/d > k$ and $P_c = 0.3E/(l/d)^2$.

71. Discuss how you would go about determining the eccentricity of a buckled column after it failed if you assumed the load.

72. Discuss how you would go about determining the load that made a column buckle if you assumed eccentricity.

73. Why is a highly polished part more susceptible to fatigue failure after surface damage than is an as-cast or forged part after surface damage?

74. The stresses of $\sigma_x = 6000$ psi and $\tau_{xy} = 4000$ psi are determined at a point on a component made from 4340 heat treated steel. Analyze the situation.

75. A particular component of an assembly is subjected to cyclic tensile loads of unequal magnitude: 25,000 psi 50 percent of the time, 30,000 psi 30 percent of the time, and 35,000 psi 20 percent of the time. If the steel used has $S_{yt} = 40{,}000$ psi and $S_{ut} = 50{,}000$ psi, at what cycle count would you predict fatigue failure?

76. A 6000 pound tow truck is trying to pull a 3000 pound car out of the mud with a 0.50 inch diameter, 10 foot long steel cable, $E = 29 \cdot 10^6$ psi, $S_{yt} = 40{,}000$ psi, and $S_{ut} = 65{,}000$ psi.
    a) What happens if the truck takes up all the cable slack before trying to pull the car out of the mud, but the car doesn't move?
    b) What happens if the truck is moving at 5 mph as the slack is taken up?

77. Some steels will absorb the maximum impact load at a contact velocity in the neighborhood of 100 ft/sec as illustrated for three steels in Figure 11.24. Discuss possible reasons for this.

78. An element of an aluminum structure is subjected to the stresses as shown in

FIGURE 11.24

Exercise 11.77.

Figure 11.25. If the material has $S_{yt} = 25{,}000$ psi and $S_{ut} = 46{,}000$ psi, would you expect failure if the stresses were static?

79. Would you expect failure by fatigue for the element in Figure 11.25 if each stress level was a cyclic 0 to 5000 psi tensile stress?

80. Discuss a design analysis plan to follow if a pipe is going to carry fluid at 5000 psi and be subjected to a 60,000 in·lbf torque.

FIGURE 11.25

Exercise 11.78.

81. Consider a situation where the component from Example 11.20 is subjected to a constant 10,000 psi stress, but the temperature changes every 10 hours, repeating the temperature cycle from 1000°F to 1100°F to 1200°F every 30 hours. Estimate how long the component will last before failure and the temperature that exists at the time of failure.

82. A cutter on a milling machine is leaving chatter marks on the machined part. List and discuss the advantages and disadvantages of various methods to reduce the possibility of chatter.

83. Will ammonia attack annealed or cold worked brass the most? Explain.

84. Does the following tensile test data seem reasonable?

Load, lbf	Diameter, inches	Load, lbf	Diameter, inches
0	0.3645	5800	0.3160
3500	0.3520	5700	0.3020
4300	0.3500	5600	0.2900
4800	0.3470	5300	0.2700
5175	0.3440	5050	0.2600
5400	0.3400	4950	0.2550
5650	0.3340	4650	0.2350
5775	0.3240	Fracture	

85. Cars often run "smoother" at a particular speed. Explain why this is likely based on vibration considerations.

86. Why are heavy machine processing plants generally built on a single level?

87. For a given material, what ratio of σ_a/σ_m will tend toward higher fatigue life, a high ratio or a low ratio? Why?

88. Analyze the data from Example 11.4 and determine if it agrees with the n test, (Equation 11.33), by calculating the corresponding life probabilities.

89. What are some causes of systems that are assumed undamped to come to rest as though they are damped, such as a cantilevered diving board?

90. A 20 kg test apparatus is being used on an experiment. It is mounted on four springs to isolate it from the experiment. The springs are rated at 3 kN/m. A vibration coming from the experiment is affecting the apparatus.
    a) What vibration frequency is probably coming through the system?
    b) Would a three spring suspension help?
    c) Suggest solutions to the vibration problem.

91. For each of the following environmental conditions list possible effects on physical products and probable modes of failure.
    a) High temperature
    b) Low temperature
    c) High relative humidity
    d) Low relative humidity
    e) High contact pressure

- f) Low contact pressure
- g) Solar radiation
- h) Sand and dust
- i) Salt water
- j) Wind
- k) Rain
- l) Rapid temperature change
- m) Zero gravity
- n) Ozone
- o) Acceleration
- p) Vibration
- q) Magnetic fields

92. Sixty components were divided into 6 groups of 10 each for test purposes. Each group of 10 was run until the first component failed. The hours to first failure of each group are: 120, 65, 155, 300, 200, and 250. What conclusions can be drawn?

93. What is the reliability of the bulbs represented in Example 11.4 at a 750 hour rating?

94. Each of the circular cross section posts shown in Figure 11.26 are subjected to the same cyclic load P. P is applied as an alternating load on the top of the post. Force F is an axial load, tension on one post, compression on the other. Determine a functional relationship between F and P so that in post #1 there is no tension stress during the cycle, and in post #2 there is no compression stress during the cycle.

95. For any given level of F and P in Figure 11.26, which post would be most fatigue resistant? Why?

FIGURE 11.26

Exercises 11.94 and 11.95.

FIGURE 11.27

Exercises 11.96 and 11.97.

96. Which cantilever beam in Figure 11.27 will best resist the fatigue loading P? Why?

97. Which cantilever beam in Figure 11.27 will best resist the fatigue load P if the beams are rotating in a ball bearing mount at the wall? Why?

98. A proposed diving board is shown in Figure 11.28. Force F can be considered to be any load, from a 70 lb child walking to the end of the board and diving off, to a 250 lb adult springing on the board to a height of 3 feet above the board. Three materials are being considered for the board. Analyze and make a recommendation.
    a) Aluminum: $S_{yt} = 27{,}000$ psi, $S_{ut} = 29{,}000$ psi, $S_e = 10{,}000$ psi, $E = 10^7$ psi.
    b) Stainless steel: $S_{yt} = 55{,}000$ psi, $S_{ut} = 155{,}000$ psi, $E = 30 \cdot 10^6$ psi.
    c) Douglas fir: $S_{yt} = 8100$ psi, $S_{yc} = 6400$ psi, $S_{uc} = 7400$ psi, $E = 1.9 \cdot 10^3$ psi.

99. Explain how the addition of a torsion load to a column affects its buckling resistance.

100. Before engineers knew so much about reliability factors, stress concentration factors, fatigue effects, and other reasons why structures fail prematurely, they used large safety factors. For each of the following, calculate the safety factor that would be required if engineers had no knowledge of any of the life reducing factors.
    a) Example 11.9
    b) Exercise 11.17
    c) Case Study #8

101. Redo Exercise 11.10 for an aluminum rod.

102. Redo Exercise 11.10 for a hardwood rod.

103. Analyze each of the following design guidelines. Are they generally useful or specifically applied?
    a) Avoid notches, sharp corners, and perforations.
    b) Spread part material out from its axis.
    c) Keep maximum forces between stationary parts rather than moving parts.
    d) Control fluid circuits at low pressure and low flow rates.
    e) Avoid sudden changes in structural cross section or fluid flow sections.

FIGURE 11.28

Exercise 11.98.

f) Design probable fluid leak paths toward the center of rotation instead of away from the center of rotation.
g) Electric switching should be done at low voltage and low current levels.
h) Vacuum tubes are better for high frequency and high power applications than are transistors.
i) Small sizes in general can operate at higher speeds.
j) A large ship wake means speed and efficiency loss.
k) Design members to be in compression when fatigue life is a concern.

12 Engineering Models

"Modeling is the means we use to ignore what we cannot understand and to consider what we do understand."
♦ ROBERT F. STEIDEL AND JERALD M. HENDERSON

Models are an important and useful tool for the engineer and are used to analyze products, systems, and technical phenomena so the design process results in a safe, reliable product. Models include any visualization analysis tool that assists the clarification of solution ideas. The objective is to gain an understanding of the problem and of potential solutions so the good ideas can be improved, the poor ones rejected, and questions answered.

12.1 Types of Models

Models may look like a real product but not necessarily operate like one, such as a full size model of an automobile to show styling. Models may operate like a real product but not be the size of one, such as an airplane model in a wind tunnel. Models may look completely different from the real product but give precise data and information about the product, such as an electric analog simulation of a mechanical device, or the volume formula for determining how much oil will fit in a storage tank. The model, $V = \pi r^2 h$, doesn't look at all like a cylindrical storage tank, but it accurately predicts the volume based on a given radius and height; units, of course, must be consistent. A model may also be completely different from the product under test but behave the same. A mechanical system and an electrical system can be analyzed using the same differential equations; only the meanings of the variables change (see Table 12.1).

TABLE 12.1

System model analogies.

Mechanical	Thermal	Electrical
Force F, lbf	Temperature θ, °F	Voltage E, volt
Displacement x, ft	Heat H, Btu	Charge q, coulomb
Velocity $v = dt/dt$, ft/sec	Heat flow $Q = dH/dt$, Btu/sec	Current $i = dq/dt$, ampere
Acceleration $a = d^2x/dt^2$, ft/sec^2	Heat flow change $dQ/dt = d^2H/dt^2$, Btu/sec^2	Current change $di/dt = d^2q/dt^2$, amp/sec
Viscous friction D, lbf·sec/ft	Thermal resistance R, °F·sec/Btu	Resistance R, ohm
Inertia M, lbf·sec^2/ft	—	Inductance L, Henry
Elasticity K, lbf/ft	Thermal capacity C, Btu/°F	Capacitance C, farad
$M\dfrac{d^2x}{dt^2} + D\dfrac{dx}{dt} + Kx = F$	$R\dfrac{dH}{dt} + \dfrac{H}{C} = \theta$	$L\dfrac{d^2q}{dt^2} + R\dfrac{dq}{dt} + \dfrac{q}{C} = E$

An **iconic model** looks like the thing it is representing. It can be the same size, smaller, or larger. The model may show all of the features of the object being modeled or just some. The features chosen to show are the ones to be checked or emphasized for a presentation or test study. Figure 12.1 is an example of an iconic model.

An **analog model** behaves like the thing it represents in that it obeys the same laws of action. In some cases it may also look like the thing it is representing, like a model ship being tested for reaction to water turbulence. At other times it may not look like the thing at all, such as an electrical system that follows the same equations as a mechanical system as illustrated in Table 12.1. The equations shown in Table 12.1 are themselves models.

FIGURE 12.1

Three dimensional non-functional model.

A **symbolic model** represents in an abstract way the thing it stands for. Symbolic models include mathematical formulas and graphic drawings, situations where one symbol arbitrarily stands for something else. Symbolic models are intended to condense principles to a form that can be manipulated and studied on paper or with a computer. The level of abstraction can be very high compared to the iconic and analog models. This is because the symbols used can be associated with many different physical problems. Following are four examples of symbolic models.

EXAMPLE 12.1:
Multi-use symbolic model.

Consider the symbolic model $A = bh/2$ and what the model can be used for.
    One use is to calculate the area of a triangle, where b is the base and h is the altitude upon that base, where the units of each quantity must be the same. Another possibility is to calculate the net price of b number of items, selling at unit price h, which are on the half-price sale table. A third possibility exists if A is known instead of b and h. Then $bh = 2A$ represents the equation of a degenerate hyperbola (two straight lines).

Symbolic models are frequently used by engineers to assist in performing their job of design. A symbolic model can be applied whenever it matches a physical problem. Engineers are familiar with, and use, many symbolic models. Some symbolic models are exact models, while others are only approximations of the real thing.

EXAMPLE 12.2:
Exact symbolic model.

A common exact symbolic model is $A = \pi r^2$, used for finding the area of a circle when the radius is known. No one doubts the validity of the model, and it is used by engineers and others without hesitation, but π itself is inexact to any number of decimal places because it is an irrational number.

EXAMPLE 12.3:
Exact symbolic model.

The exact symbolic model for calculating the base of natural logarithms is an infinite series:

$$e = 1 + 1/1! + 1/2! + 1/3! + \cdots + 1/(n-1)! + \cdots$$

Any degree of precision can be obtained by evaluating successive terms of the series, but it is only an approximation (inexact) when a finite number of terms are used.

EXAMPLE 12.4:
Exact symbolic model.

The symbolic model $S_e = 0.50(S_{ut})$ is an exact statement used to estimate the fatigue strength of a steel, but it does not necessarily give exact results. It is an model to use in place of more accurate data that would take time and test to determine. Recall all the variations that can occur and affect the process of determining fatigue limits. Is the model less usable? No. Can the model be used any time for any situation? No. *You* will determine if the model is usable in a given situation. The ability to determine when a model is usable is one of the abilities of a good engineer.

The following case study is an example of how existing mathematical models can be used to help analyze and quantify raw data. This case is an example of an exact model used in an inexact situation.

CASE STUDY #10:
Tree Age Analysis.

To illustrate the use of exact models in inexact applications, consider the problem of determining the age of trees. To begin the analysis, an assumption about the growth pattern of trees is necessary. Trees grow faster in wet weather than dry, and in general have a fast growth period in the spring followed by a slow growth period the rest of the year. This alternating growth pattern causes a variation in wood density that makes rings visible when a tree is cut. Counting growth rings is the accepted model used to estimate the age of a tree. This method is just an estimate, however, because the effects of unusual climate conditions and/or other environmental conditions can cause unusual ring patterns. Variations in ring patterns include incomplete rings and multiple rings during a one year period.

Counting growth rings is a straightforward model to follow and is easy to apply, but a problem exists. How is the age of a tree that is still standing to be determined without cutting the tree down? Some enterprising person, perhaps an engineer, solved the problem by designing the Swedish Increment Borer. The Borer is a hand instrument that drills a hole about 0.1875 inches (4.76 mm) in diameter in a tree and removes a core sample. This core sample is then studied, and the rings counted. Carefully done the procedure causes no harm to the tree, and the core can be replaced and waxed over to prevent any infection This is a nice technique and uses the same modeling scheme, counting the rings. It has some difficulties, however. The boring tool must be aimed at the center of the tree to pick up all the growth rings, and it is not an easy task to drill by hand into a wet tree. There is also some potential for dropping the core and losing the ring count. Multiple samples from the same tree are required to verify the count and to average out irregularities in ring growth. On a cut tree multiple counts can be done with little additional effort. With a borer, additional time and effort are required to obtain multiple counts.

A second method to estimate the age of a tree also uses the growth ring principle and assumes that the rings are circles. In fact, some tree ring patterns are nearly perfect circles. If the circumference of the largest growth ring circle and the spacing between rings are known, an estimate of a tree's age can be made using the following analysis. Let D = the diameter of the largest ring, and C = the circumference of the largest ring, then $C = \pi D$. If d = the distance between rings, the age of the tree can be modeled as:

$$\text{Age} = D/2d \text{ or Age} = C/2\pi d \qquad (12.1)$$

Either model can be used, depending on whether D or C is known. Using Age = $C/2\pi d$ will give more consistent results because when the circumference is measured, diameter variations are averaged out. If Age = $D/2d$ is used, a single measure of diameter may be an extreme value and give less accurate results. These models also assume a constant ring spacing and this is probably never true, ring spacing being dependent on annual growing conditions. An average value for d can be obtained from cut trees or boring samples. A large number of samples for values of d should be used to average out growth variations. The model, Age = $C/2\pi d$, is an exact one, but the physical situation being modeled is an inexact one. The engineer must always be prepared to make decisions about how a model will be used.

To use the model Age = $C/2\pi d$, two pieces of data are required, the circumference of the tree and the ring spacing. The circumference of a tree can easily be measured, but be sure to make a circumference reduction for bark thickness recalling that each half inch of bark thickness will change the circumference by π inches. The ring spacing, d, can be estimated from other trees in the area of the same type that have been cut, or with core samples. It is not surprising to find trees from different regions that have ring distances twice the spacing of trees of the same type from other regions. The ring spacing is a signature of the type of tree and how it was grown, a reflection of the species and the local growing conditions.

To use these tree-age models requires data collection. Many engineering design models fall into the same situation. Table 12.2 lists sample data collected from trees that have already been cut. The data is graphed in Figure 12.2 and a ring spacing grid overlaid on the graph. The straight line overlay assumes a 0.250 inch bark thickness. This figure illustrates the wide variation in ring spacing.

Using Age = $C/2\pi d$ will only be as valid as the accuracy of d, so a decision must be made on its value. This is another example of the necessity of making a decision and accepting the results. Always record the assumptions and decisions made. Careful data collection is also important, such as restricting measurements to one species of tree.

How old is a tree with a circumference of 40 inches? Using $d = 0.06$ inches, and Equation 12.1, the estimate of age is 106. If $d = 0.10$ inches, the age would be only 64 years.

Suppose a piece of wood has been processed through a mill and all that is visible is a portion of the ring pattern, and the center of the tree is not on the piece of wood. How can the age of the tree that the piece of wood came from be determined? An extension of the previous model is appropriate, where equally spaced circular rings was assumed.

The rings can be counted to get a minimum age, but without the center of the tree shown to obtain a total count; further analysis is needed. Recall from geometry that three points not on a straight line determine a circle. The approach used here is to locate three points on one ring and extend this to a full circle to obtain the circumference. One mathematical model that can be used for a circle is:

$$X^2 + AX + Y^2 + BY + C = 0. \tag{12.2}$$

Whose radius is: $r = \sqrt{-C + (A/2)^2 + (B/2)^2}$. (12.3)

This calculated r and an estimated value for d can then be used in the model Age = r/d. Substitute three pairs of (x, y) coordinates into Equation 12.2 and solve simultaneously for A, B, and C. Observe that the data points are only estimates of true values, but they are substituted into an exact symbolic model.

Using Equation 12.2 an estimate can be made of the age of the tree from which the board in Figure 12.3 was cut. Superimpose an X, Y coordinate system on a ring that is the longest and most regular, and identify three points on it. The farther apart the points are, the more accurate the calculations will be. Placing the origin of the coordinate system on the ring simplifies calculations because one ordered pair will be (0, 0) and then $C = 0$. For this example the three points selected are (x_1, y_1) = (0, 0), (x_2, y_2) = (2.90, 0.46), and (x_3, y_3) = (4.30, 0.06).

TABLE 12.2

Tree ring data from cut samples.

Circumference of tree including bark, inches	Number of growth rings counted
$18\frac{1}{2}$	17
$6\frac{1}{2}$	14
$7\frac{7}{8}$	11
$9\frac{7}{8}$	14
$6\frac{3}{4}$	7
$7\frac{1}{2}$	10
$16\frac{1}{8}$	15
$6\frac{7}{8}$	4
$10\frac{3}{8}$	9
$7\frac{1}{4}$	9
$7\frac{1}{2}$	10
$7\frac{1}{2}$	10
$7\frac{1}{4}$	7
$8\frac{3}{4}$	8
$8\frac{1}{4}$	6
$8\frac{1}{8}$	12
$7\frac{5}{8}$	5
$6\frac{1}{2}$	12
$25\frac{1}{4}$	30
$29\frac{7}{8}$	6
$22\frac{1}{8}$	41
$48\frac{1}{2}$	18
$49\frac{1}{2}$	26
6	8
$7\frac{3}{8}$	11
$9\frac{3}{4}$	14

Substitute the values into Equation 12.2 and solve for A, B, and C.

$$0^2 + 0(A) + 0^2 + 0(B) + C = 0 \tag{1}$$
$$2.90^2 + 2.90A + 0.46^2 + 0.46B + C = 0 \tag{2}$$
$$4.30^2 + 4.30A + 0.06^2 + 0.06B + C = 0 \tag{3}$$

From (1), $C = 0$, and (2) and (3) become:

$$8.62 + 2.90A + 0.46B = 0 \tag{2}$$
$$18.49 + 4.30A + 0.06B = 0 \tag{3}$$

which are satisfied when $A = -4.43$ and $B = 9.13$; so $r = \sqrt{0 + (-4.43/2)^2 + (9.13/2)^2}$ = 5.07. An estimate for d can be determined from the figure, $6d = 0.68$ and $d = 0.11$, and the tree's age is approximately equal to 5.07/0.11 = 46 years.

FIGURE 12.2

Sample tree growth data.

[Graph showing Number of rings vs Circumference, inches. Data points include Pine and spruce, Cedar, Mulberry, Birch, Pine, Poplar, Silver maple. Lines indicate S = 0.06, S = 0.12, S = 0.25, S = 0.38, S = 0.50, S = 0.75. Note: "Number of growth rings counted on a sample of cut trees, and their measured circumference. S = growth ring spacing, inches"]

A mathematical model can be applied to a physical situation any time the assumptions and accuracy are within acceptable limits. One of the biggest errors an engineer can make is to believe an exact answer is obtained just because an exact model is used.

12.2 Boundary Conditions

When a system is being modeled, the replacement system often has variations created by the assumptions made. Matching boundary conditions is one method to use for model-sample compatibility.

FIGURE 12.3

Sample tree section.

[Diagram of a tree section with coordinates (x_1, y_1), (x_2, y_2), (x_3, y_3) and $6d = 0.68$, with X and Y axes.]

FIGURE 12.4

Schematic for Example 12.5.

(a) Cantilevered beam with uniform load w, lb/ft over length l_u, ft.

(b) Cantilevered beam with point load F, lb at free end over length l_s, ft.

EXAMPLE 12.5:

Matching boundary conditions.

It is desired to model a uniformly loaded cantilevered beam, Figure 12.4(a), with a cantilevered beam with a single load at the free end, Figure 12.4(b). Possible model assumptions include:

1. Equal free end (maximum) deflection
2. Equal midpoint deflection
3. Equal bending moment at the support end, or
4. Equal free end slope angle.

This example will assume equal free end deflection. The free end deflection of a uniformly loaded beam $\delta_{max} = wl_u^4/8E_uI_u$, and for a singly loaded beam $\delta_{max} = Fl_s^3/3E_sI_s$, therefore $wl_u^4/8E_uI_u = Fl_s^3/3E_sI_s$, and $F = (3\ E_sI_swl_u^4)/(8E_uI_ul_s^3)$. If the material, length, and section moduli of the two beams are equal, then $F = 3wl/8$. If modeling of scale and material are also to be considered, then the ratios E_s/E_u, I_s/I_u, l_u^4/l_s^3 all become important in determining F.

12.3 | Full-Scale Models

Some engineering projects do not lend themselves to the use of scale models. Scaling effects are often more difficult and costly to undertake than are the full-sized models. This is often the case when a component or assembly is used in a variety of ways, some of which are unknown.

Large machinery such as a bulldozer or motor grader are hard to test on a scale basis because it is difficult to scale the applied loads and component strengths. Testing a full-sized model is more reliable, and more information about the machine can be determined, including such items as performance specifications, noise levels, fits and clearances, operator comfort, alternate uses, and data for future production, such as manufacturing processes and assembly procedures. Certain specifications such as weight, digging depth, available horsepower, and traveling speeds can also be more reliably determined from a full-size model. NASA sends full-sized unmanned spacecrafts into space to test rocket components and performance. Automobile engines are tested at full size, partly to be certified for exhaust emission standards and fuel economy requirements. Automobile companies crash test full-size models to determine actual standards. Scale models don't answer the questions with the necessary reliability. *Note*: Recent advances in computer simulation have reduced the need for some real product crash testing.

12.4 Scale Models

Many products are tested at reduced size because of cost, time to build, size of test apparatus, and the data to be obtained from the model. Some products modeled to scale are dams and buildings. Part of the reason is that a full-sized model would be expensive and impractical to site. The models built in these cases are not generally used to test for product strength (see Figure 12.5). They are built to study the interrelationships of systems such as heating, electric, and plumbing, and to permit visualization by those who do not have the time or ability to study detail drawings. A scale model of a dam is easily understood, and this is particularly important when presenting a project for approval and financing. Three-dimension visualization is much easier to do with a model, even for people who can read drawings.

When a scale model is built for test, different concepts and procedures are required than if a model is being used just for conceptual analysis. Dimensional analysis plays a role in scale model testing. Dimensional analysis not only helps determine the test requirements such as size, loads, or fluid media, but it allows consolidation of experiments by controlling dimensional groups.

If a model is going to be scaled and used for visual analysis only, then any material can be used. Only the dimensions of the objects are of concern. If a product will be scaled in half, all linear dimensions are reduced by one-half. In some cases very small details are not built at all, but are painted or stenciled on the model. This is common for small details such as wire, nuts and bolts, nails, or floor tile layouts on models of plant layouts or houses.

If, however, a scale model is being tested for such things as pressure or heat transfer, care must be exercised. If a model is built to one-half size, then areas will be one-quarter size. The results obtained from mathematical models using area, such as Force = Pressure × AREA, Strength = Stress × AREA, or Heat Transfer = Rate × AREA, must be properly analyzed, as the next example illustrates.

EXAMPLE 12.6:
Scale model adjustments.

A model of a boat sail is built to one-quarter actual size. If the test force measured on the model is 50 lbf (222 N), what force would you expect on the full-size sail?

A model built to 1/4 size will have areas $(1/4)^2 = 1/16$ size, so the expected force on the full-sized sail is 16(50) = 800 lbf (3559 N).

FIGURE 12.5

Three dimensional model showing the details of a continuous caster line.

If the volume of a scale model is of concern for testing, then mathematical models using such variables as weight, material cost, and buoyancy force need to be compensated for by volume ratios. Volume ratio analysis requires the cube of a linear dimension, as the next example illustrates.

EXAMPLE 12.7:
Scale model adjustment.

A one-tenth scale model of a spherical water tank is being built. If the full-size tank is intended to hold 10,000 gallons, how much will the model hold?

The volume of a sphere is $V = (4/3)\pi r^3$, where r is the radius. Therefore, the volume of the model is $(1/10)^3$ of the full-sized tank, or 1/1000 of 10,000 gallons = 10 gallons. This reduces the load of the water to 1/1000 of the actual load for the model.

12.5 Dimensional Analysis

More complicated modeling problems can benefit from dimensional analysis. Dimensional analysis is a method used to formulate a mathematical model that can then be used as a basis for theoretical analysis, or for experimental analysis. Dimensional analysis is used to create and analyze functional relationships based on the requirement that the Newtonian dimensions of length, L, time, T, and mass, M, must be the same if terms are to be added or subtracted; and the powers of the L, T, and M factors must be the same on both sides of an equation.

Checking dimensions on existing models is not difficult, as the following example illustrates.

EXAMPLE 12.8:
Dimension and unit verification.

The formula $A = h(b_1 + b_2)/2$ is used to find the area of a trapezoid, where b_1 and b_2 are the lengths of the parallel bases, and h is the perpendicular distance between the bases. The dimension of b_1 and b_2 must be the same because they are being added; and the dimension of A must be the same as the product of the dimension of h and $(b_1 + b_2)$. In this case the dimension of both sides of the equation is L^2. The constant 1/2 does not have a dimension.

Having equal dimensions does not guarantee that the units are correct, however. Consider the case where b_1 and b_2 are measured in inches and h is measured in feet. A conversion factor is required depending on the desired units for A. If the value of A is desired in square inches, then the formula must be adjusted to $A = 12h(b_1 + b_2)/2$; if the value of A is desired in square feet, then the formula must be adjusted to $A = (h/12)(b_1 + b_2)/2$. In both cases the constant 12 is dimensionless, but it does have units of in./ft.

Constants can exist in any one of three conditions:

1. They can be dimensionless and have units, such as material strain, which is dimensionless but has units of in./in.

2. They can be dimensionless and unitless such as e, the base of the natural logarithms, and

3. They can have dimensions and have units, such as the gravitational constant, which has dimensions of LT^{-2}, and units of ft/sec² or m/sec².

Dimensional analysis is also helpful when trying to establish new mathematical models to predict or explain physical phenomena. The basic Buckingham Theory, used to analyze dimensional relationships between variables, states that there exists (Π_1, Π_2,..., Π_n) dimensionless groups such that $F(\Pi_1, \Pi_2,..., \Pi_n) = 0$. One valuable result of the theory is that if a variable is forced to appear in only one of the Π_i groups, then it can be separated and solved for as a dependent variable.

Product functions are an attempt to use dimensional analysis to create a mathematical model which can be verified or rejected by experiment. The theorized model becomes a working model used to try to explain variable interrelationships, but there are no guarantees. The basic procedure used to obtain the Π groups is explained in the next example using a familiar physical model, that of a structural beam with a single load.

EXAMPLE 12.9:
Dimensional analysis of a beam

Step #1: Identify the variables that can have an effect on the problem. This is not always an easy thing to do, because sometimes the very thing that is making experimental results fluctuate will be the variable omitted from the analysis. On the other hand, the addition of a variable that is not related to the physical situation will only confuse the calculations and possible outcomes. Table 12.3 lists the relevant variables for the beam problem.

Step #2: Specify the dimensions of each of the variables; these have been included in Table 12.3. Refer to Appendix A for a list of commonly used variables and their units. Variables that are not listed in the appendix can generally be determined from combinations of entries.

Step #3: Assume a product and exponent function that relates all the variables, including a proportionality constant: $1 = kP^a X^b I^c E^d \delta^e$ (12.4)

TABLE 12.3

Variables and dimensions for a beam analysis.

Variables	Dimensions
P—Load	ML/T^2
X—Beam length	L
I—Cross section moment of inertia	L^4
E—Material modulus of elasticity	M/LT^2
δ—Deformation	L

TABLE 12.4

Beam variables and exponent coefficients.

Variable	P	X	I	E	δ
Exponent	a	b	c	d	e
M (Mass)	1	0	0	1	0
L (Length)	1	1	4	−1	1
T (Time)	−2	0	0	−2	0

This form requires that 1 be dimensionless. The value of k will eventually be determined from experiment, mathematical theory, and/or unit conversion if the variables are measured in mismatched units, such as feet and inches. The values of the exponents must now be selected so that each of the basic dimensions, L, T, and M cancels and leaves a dimensionless equation.

Step #4: Establish a matrix showing the exponents and their respective units as shown in Table 12.4.

Step #5: Write linear equations with the exponents as variables and the matrix entries as coefficients.

$$\text{For } M: a + 0(b) + 0(c) + d + 0(e) = 0 \tag{12.5}$$

$$\text{For } L: a + b + 4c - d + e = 0 \tag{12.6}$$

$$\text{For } T: -2a + 0(b) + 0(c) - 2d + 0(e) = 0 \tag{12.7}$$

Step #6: Solve the equations simultaneously. If there are more unknowns than equations, it is necessary to solve the equations parametrically. From Equation 12.5, $a = -d$; Equation 12.7 is redundant and gives the same relationship. Substituting $a = -d$ into Equation 12.6 gives $b = -4c + 2d - e$. Substitute the values for a and b into Equation 12.4 and $1 = k(P^{-d})(X^{-4c+2d-e})(I^c)(E^d)(\delta^e)$. (12.8)
Group the bases with each of the exponents, and $1 = k(I/X^4)^c (X^2E/P)^d (\delta/X)^e$. (12.9)
This results in three dimensionless groups: $\Pi_1 = I/X^4$, $\Pi_2 = X^2E/P$, $\Pi_3 = \delta/X$. Alternate groups can be created by multiplying or dividing these groups together. For example, Π_2 can be replaced with $\Pi_2 \times \Pi_3 = XE\delta/P$.

By assigning arbitrary values to the exponents in Equation 12.9 various relationships can be obtained, some helpful, some not. For example, if $e = d = 0$ and $c = 1$, then $I = X^4/k$, not a very meaningful relationship because no functional constant k can be found that relates cross sectional moment of inertia with beam length. In general the exponents should not be set equal to zero because critical variables may disappear from the analysis. If $e = d = a = 1$, then $\delta = (1/k)(PX^3/EI)$ and this should look familiar. The value of k can be determined from experiment by varying the value of PX^3/EI and measuring δ; PX^3/EI is treated as a single variable. The values of P, X, E, and I can vary individually as long as the value of PX^3/EI is held constant. If several physical models are made from the same material, then E is constant and the variable becomes PX^3/I. For a simply supported center loaded beam the value of k turns out to be 48 when X and δ are in inches, I is in^4, and E is lb/in^2. For an end loaded cantilever beam, $k = 3$.

For consistent experimental results, relating scale model results to the full size object, the value of each Π group should be the same for the full-size real product *(p)* and the scale model *(s)*, that is: $I_p / X_p^4 = I_s / X_s^4$, $X_p^2 E_p / P_p = X_s^2 E_s / P_s$, and $\delta_p / X_p = \delta_s / X_s$.

EXAMPLE 12.10:

Use of a dimensional group from Example 12.9.

If a full size beam has a length of 10 meters and a section moment of 8 mm⁴, a model beam 2 meters long should have a section moment = $(2^4)(8)/10^4 = 0.0128$ mm⁴. The value of 0.0128 does not determine a specific shape.

A circular cross section, $I = \pi D^4/64$, will have $D = [(64)(0.0128)/\pi]^{1/4} = 0.71$ mm.

A square cross section, $I = S^4/12$, will have $S = [(12)(0.0128)]^{1/4} = 0.63$ mm.

The following case study further illustrates the dimensional analysis procedure and some of the problems encountered when it is used.

CASE STUDY #11:

Wind Force Modeling.

Consider the problem of trying to determine a mathematical model to quantify the force created on a flat plate exposed to the wind or other moving fluid.

Trial #1: Assume the related variables are A, the flat plate area, V, the fluid velocity, and F, the force on the plate, as listed in Table 12.5.

Assume a product function $1 = kF^a A^b V^c$, which implies $(MLT^{-2})^a (L^2)^b (LT^{-1})^c = M^0 L^0 T^0$. The exponent of each variable, (M, L, or T), must be the same on both sides of the equation, so:

for M: $a = 0$

for L: $a + 2b + c = 0$

for T: $-2a - c = 0$

Solve these three equations simultaneously and $a = b = c = 0$, so no function exists. This means the correct variables were not selected for the problem, if a product function is to be used.

Trial #2: Include another variable, ρ, the density of the air, and consider the new list of variables shown in Table 12.6.

Assume a product function $1 = kF^a A^b V^c \rho^d$, which implies that $(MLT^{-2})^a (L^2)^b (LT^{-1})^c (ML^{-3})^d = M^0 L^0 T^0$. Analyze exponents of the variables M, L, and T:

for M: $a + d = 0$

for L: $a + 2b + c - 3d = 0$

for T: $-2a - c = 0$

TABLE 12.5

Variables and dimensions for Case Study #11, trial 1.

Variable	Dimensions
F—Force, Newtons	MLT^{-2}
A—Area, m²	L^2
V—Wind velocity, m/sec	LT^{-1}

Solve these three equations simultaneously and $d = -a$, $c = -2a$, and $b = -a$, so

$$1 = kF^a A^b V^c \rho^d \text{ becomes } 1 = kF^a A^{-a} V^{-2a} \rho^{-a}$$

If $a = 1$, then $F = A\rho V^2/k = C_1 A\rho V^2$, where C_1 is a constant. If this model is tested experimentally, using the units of measure listed in Table 12.6, and the flat plate is perpendicular to the wind, then C_1 turns out to be 1/2.

Trial #3: Experimentation shows that if the plate is not oriented perpendicular to the wind, C_1 is not 1/2, but some other value, and if the angle with the wind is not held constant, then C_1 is not constant, and another variable should be considered.

The angle θ that the plate makes with the wind has no dimensions, and if included in the procedure of trial #2, will drop out of the analysis. Experimental data indicates that the angle does affect the force, and so it must be included. At this point the more generalized theory must be employed and θ is included as an additional Π group. Because the form of the additional group is not known, it must be included as a function, $h(\theta)$. The model for the force, extended from trial #2, now becomes $F = C_2 A\rho V^2 h(\theta)$. The constant will likely be a different value so subscripts are used to differentiate them from model to model. The function $h(\theta)$ must be estimated from experimental data, from mathematical theory, or intuition; is $h(\theta) = \cos \theta$ a good choice? Consider the area of the plate projected perpendicular to the wind as shown in Figure 12.6.

Trial #4: Consider the possibility that fluid viscosity might influence the results. Table 12.7 shows the listing of variables and dimensions necessary to include viscosity in the analysis.

TABLE 12.6

Variables and dimensions for Case Study #11, trial 2.

Variable	Dimensions
F—Force, Newtons	MLT^{-2}
A—Area, m²	L^2
V—Wind velocity, m/sec	LT^{-1}
ρ—Air density, kg/m³	ML^{-3}

FIGURE 12.6

Projected area of a flat plate exposed to the wind.

Assume a product function $1 = kF^a A^b V^c \rho^d \mu^e$, which implies that $(MLT^{-2})^a (L^2)^b (LT^{-1})^c (ML^{-3})^d (ML^{-1}T^{-1})^e = M^0 L^0 T^0$. Analyze the exponents of M, L, and T, and

$$\text{for } M: a + d + e = 0 \qquad (1)$$
$$\text{for } L: a + 2b + c - 3d - e = 0 \qquad (2)$$
$$\text{for } T: -2a - c - e = 0 \qquad (3)$$

From (1), $a = -d - e$, substitute in (3), $2d + 2e - c - e = 0$, and $c = 2d + e$. Substitute a and c in (2), and $-d - e + 2b + 2d + e - 3d - e = 0$ and $b = (e + 2d)/2 = e/2 + d$.

$1 = kF^a A^b V^c \rho^d \mu^e$ becomes $1 = kF^{-d-e} A^{e/2+d} V^{2d+e} \rho^d \mu^e = k(\rho A V^2/F)^d (V\mu(\sqrt{A})/F)^e$, so

$\Pi_1 = \rho A V^2/F$ and $\Pi_2 = V\mu\sqrt{A}/F$.

TABLE 12.7

Variables and dimensions for Case Study #11, trial 5.

Variable	Dimensions
F—Force, Newtons	MLT^{-2}
A—Area, m^2	L^2
V—Wind velocity, m/sec	LT^{-1}
ρ—Air density, Kg/m^3	ML^{-3}
μ—Air viscosity, N·sec/m^2	$ML^{-1}T^{-1}$

The problem now is that F occurs in both groups and cannot be separated unless d or e are set equal to zero. Setting an exponent equal to zero is risky because an important variable can be lost. The problem of separating F can be resolved by creating a new Π group to replace an existing group using a combination of existing groups. Combinations of groups are permissible because parameters other than d and e could have been selected for the final form, and other Π groups derived. Replace Π_1 with $\Pi_1' = \Pi_1/\Pi_2 = \rho V\sqrt{A}/\mu$, and $1 = k(\rho V\sqrt{A}/\mu)^d(V\mu(\sqrt{A})/F)^e$. Now let $d = e = 1$ and solve for F: $F = C_3 \rho V^2 A$, and viscosity drops out. If viscosity is still deemed an important factor, then it must be included as a separate product function $g(\mu)$ as was done with angle variation, and $F = C_3 \rho V^2 A\ g(\mu)h(\theta)$.

Trial #5: Suppose experimental observation still shows a variation of the force. What other variables might be included? How about temperature? If temperature is included as a variable, an additional product function, $k(t)$, must be included because temperature, whether measured in °C or °F, has no units of $M, L,$ or T, and would not be part of other Π_i groups. The model becomes $F = C_4 A \rho V^2 h(\theta)k(t)$.

If only one of the variables is associated with the temperature changes, such as density, then that variable could be made the temperature function. If that is done, the model becomes $F = C_5 A \rho(t) V^2 h(\theta)$.

Another variation that has not been considered is geometry, because the assumption was that the exposed area is flat. If the projected area perpendicular to the direction of the wind is held constant, but its shape changed, the force will vary. The problem is how to incorporate a geometry change into the model. Should A become A(geometry)? The solution, as students of fluid mechanics are aware, is to use a factor C_D, called the drag coefficient, which is a function of plate geometry, fluid viscosity and velocity, and surface smoothness. The final model becomes $F = [C_D A_p \rho(t) V^2]/2$, where A_p is the area projected perpendicular to the wind.

Some dimensionless groups common to particular fields of engineering are used regularly without needing to be derived for each problem. Table 12.8 lists some of the groups used frequently in fluid mechanics. Most of the groups were derived to permit the running of valid scale model experiments.

The Reynolds number, for example, is used commonly in fluid flow analysis. It is symbolized by the dimensionless quantity $\rho VD/\mu$, where ρ is the fluid density, V is the fluid velocity, D is the diameter of the pipe in which the fluid is flowing or the critical dimension of an object exposed to the flowing fluid, and μ is the absolute viscosity of the fluid. If the value of the calculated Reynolds number for a pipe flow model is less than 2700–4000, the flow is laminar; if the value is above that range, the flow is turbulent. The transition region is typical of physical systems that have surface roughness variation. Two systems with the same Reynolds number are said to be dynamically similar.

The Mach number is another common reference group ratio, usually used in relating the speed of an airplane or other missile to the speed of sound. Mach 2 is a speed twice the speed of sound. *Note*: The speed of sound is not constant, but usually taken as 1120 feet/sec (321.4 m/sec) at sea level for average atmospheric conditions.

TABLE 12.8

Common dimensionless groups used in fluid mechanics.

Group Name	Force Ratio	Quantity Ratio
Reynolds number	Inertia/viscosity	$\rho VD/\mu$
Pressure coefficient	Pressure/inertia	$\Delta p/(\rho V^2)/2$
Froude number	Inertia/gravity	$V\sqrt{gL}$
Weber number	Inertia/surface tension	$V\sqrt{\rho L/\sigma}$
Drag coefficient	Drag/inertia	$D/(\rho V^2 D^2)/2$
Lift coefficient	Lift/inertia	$L/(\rho V^2 D^2)/2$
Mach number	Inertia/inertia	$V\sqrt{K\rho}$
Thrust coefficient	Thrust/inertia	$F_t/(\rho V^2 D^2)/2$
g	Gravity conversion factor	
μ	Absolute fluid viscosity	
σ	Fluid surface tension	
V	Fluid velocity	
D or L	Critical dimension in direction of fluid flow	
ρ	Fluid density	
Δp	System pressure variation	
K	Fluid compressibility	

12.6 Experimental Data Analysis

When experiments are performed engineers anticipate that some consistent relationship will exist between input variable change and output variable change. If this anticipated relationship follows certain criteria, an empirical equation relating the variables can be established. The most common relationships include the following models, where X is the independent variable and Y is the dependent variable.

The model of a straight line on **linear coordinate graph paper** is:

$$Y = mX + b, \text{ where} \qquad (12.10)$$

m = the slope of the line, and

b = the Y-axis intercept, the value of Y when $X = 0$.

The models for straight lines on **semilog coordinate graph paper** are:

$$Y = B(10)^{m_{10}X}, Y = B(e)^{m_e X}, \text{ or } Y = BM^X, \text{ where} \qquad (12.11)$$

m_{10} is the slope of the line using common logarithms,

m_e is the slope of the line using natural logarithms, $M = (10)^{m_{10}} = (e)^{m_e}$, and

B = the Y-axis intercept, the value of Y when $X = 0$.

Often, to obtain a straight line on semilog paper, it is necessary to interchange the axis of the variables, and two choices are possible, either log Y versus X, or log X versus Y.

The model for a straight line on **log-log coordinate graph paper** is:

$$Y = B(X)^M, \text{ where} \tag{12.12}$$

M = the slope of the line, and

B = the Y-axis intercept, the value of Y when $X = 1$.

12.6.1 Least Squares Analysis

A mathematical method for determining the values of the parameters for slope and intercept is based on the method of least squares analysis. This method seeks a function such that the sum of the squares of the residuals is a minimum. The generalized method of least squares analysis is as follows.

Start with a function that is a linear combination of common functions that has a chance of closely fitting the experimental data. Any function can be used, but you have to decide which may be best. Some examples include:

$Y = c$

$Y = aX^3 + bX^2 + cX + d$

$Y = a \sin X$

$Y = a \cos X + b \exp(X^2)$

Write an expression for the sum of the squares of the residuals. A residual is the difference between the value of Y as calculated from the function and the experimental value of Y. If the function chosen in step 1 is $Y = mX + b$, a straight line on linear coordinate graph paper, then the sum of the residuals $R = \Sigma[(mx_i + b) - y_i]^2$, where (x_i, y_i) are the data for $i = 1, 2, \ldots, n$ number of data pairs.

Make the value of R as small as possible by taking first partial derivatives and setting them equal to zero. In this example the unknowns are m and b, so:

$\partial R/\partial m = 2\Sigma(mx_i + b - y_i)x_i = 0$ and $\partial R/\partial b = 2\Sigma(mx_i + b - y_i) = 0$.

Solve the differential equations simultaneously: $\partial R/\partial m$ becomes $m\Sigma x_i^2 + b\Sigma x_i = \Sigma x_i y_i$, and $\partial R/\partial b$ becomes $m\Sigma x_i + nb = \Sigma y_i$, so

$$m = \frac{n\Sigma(x_i y_i) - \Sigma x_i \Sigma y_i}{n\Sigma x_i^2 - (\Sigma x_i)^2}, \text{ and} \tag{12.13}$$

$$b = \frac{\Sigma y_i - m\Sigma x_i}{n} = \frac{\Sigma x_i^2 \Sigma y_i - \Sigma x_i \Sigma(x_i y_i)}{n\Sigma x_i^2 - (\Sigma x_i)^2} \tag{12.14}$$

The values of M and B for a straight line on semilog paper are found by modifying Equations 12.13 and 12.14 as follows.

$$m_{10} = \log M = \frac{n\Sigma(x_i \log y_i) - \Sigma x_i \Sigma \log y_i}{n\Sigma x_i^2 - (\Sigma x_i)^2} \tag{12.15}$$

$$\log B = \frac{\Sigma x_i^2 \Sigma \log y_i - \Sigma x_i \Sigma(x_i \log y_i)}{n\Sigma x_i^2 - (\Sigma x_i)^2} \tag{12.16}$$

The values of M and B for a straight line on log-log paper are found by modifying Equations 12.13 and 12.14 as follows.

$$M = \frac{n\Sigma(\log x_i \log y_i) - \Sigma\log x_i \, \Sigma\log y_i}{n\Sigma(\log x_i)^2 - (\Sigma\log x_i)^2} \tag{12.17}$$

$$B = \frac{\Sigma(\log x_i)^2 \Sigma\log y_i - \Sigma\log x_i \, \Sigma(\log x_i \log y_i)}{n\Sigma(\log x_i)^2 - (\Sigma\log x_i)^2} \tag{12.18}$$

12.6.2 Graphical Methods

An alternate method to the analytical method can be used if the data is plotted on appropriate graph paper and a "best" straight line is drawn through the points. "Best" is best in your judgment, so keep the line as close to as many points as possible.

On linear coordinate graph paper the value of b can be read directly from the graph where the line crosses the Y-axis. The value of m can be found in two ways:

1. Measure Δx and Δy with a scale, adjust for any differences in X-axis and Y-axis divisions, and divide: $m = \Delta y/\Delta x$; or
2. Select two points on the straight line drawn, (x_a, y_a) and (x_b, y_b), and use the slope equation $m = (y_b - y_a)/(x_b - x_a)$.

The graphical solution for a semilog "best" straight line is similar to the method for a linear coordinate graph. Draw a straight line through the data keeping the line as close to as many data points as possible. The intercept B is read from the graph where the line drawn crosses the Y-axis. The slope is determined by selecting two data points (x_a, y_a) and (x_b, y_b) on the line and calculating M. Do not measure Δx and Δy using a scale and calculate $\Delta y/\Delta x$ for slope, as this does not work for semilog paper.

$$\log M = \frac{\log y_b - \log y_a}{x_b - x_a}, \quad \text{or} \quad \ln M = \frac{\ln y_b - \ln y_a}{x_b - x_a} \tag{12.19}$$

The graphical solution for log-log paper locates B where the line drawn crosses the line $Y = 1$. The value of M can be found in two ways.

1. Select two data points on the line drawn and use the equation:

$$M = \frac{\log y_b - \log y_a}{\log x_b - \log x_a}, \text{ or}$$

2. Measure Δy and Δx with a scale and calculate $M = \Delta y/\Delta x$.

Measuring Δy and Δx is permissible as long as the length of one log cycle on both axes is the same. Commercial log-log paper is printed this way.

12.6.3 Additional Adjustments

On semilog graph paper, if the data plots close to a straight line but is still slightly curved, an adjustment can be attempted to obtain a better straight line; Equation 12.11 becomes:

$$Y - c = BM^X \tag{12.20}$$

The value of c is estimated as follows.

1. Select two points, (x_a, y_a) and (x_b, y_b), from the *curve* on semilog paper that are closest to the data. These points may be the first and last data pairs because points farther apart will improve the accuracy of the calculation.

2. Calculate $x_3 = (x_a x_b)/2$. (There's that $ab/2$ model again!)
3. Read y_3 from the *curve* at the value of x_3.
4. Calculate $c = \dfrac{y_a y_b - y_3^2}{y_a + y_b - 2y_3}$.

On log-log graph paper, if the data plot is close to a straight line but is still slightly curved, an adjustment can be attempted to obtain a better straight line; Equation 12.12 becomes:

$$Y - c = BX^M \quad (12.21)$$

The value of c is estimated as follows.

1. Select two points, (x_a, y_a) and (x_b, y_b) from the *curve* on log-log paper that comes closest to the data.
2. Calculate $x_3 = \sqrt{x_a x_b}$.
3. Read y_3 from the curve at x_3.
4. Calculate $c = \dfrac{y_a y_b - y_3^2}{y_a + y_b - 2y_3}$.

An alternate method for determining c for the log-log model is made by estimating the Y-axis intercept on the *linear coordinate plot* of the data.

For both the semilog and the log-log model, the value of c is negative if the original curve is concave down and positive if the original curve is concave up. Keep in mind that the value of c is an estimate. A better value for c, and a better "best" straight line, may be possible by trying other values of c in the neighborhood of the estimated value. An iterative computer program would be a good tool to use to improve the results. To verify that some other value for c is "best," calculate data residuals relative to the empirical equation and/or a coefficient of determination, and/or a percent error at each data point.

12.6.4 Polynomial Model

If none of the preceding models give a satisfactory fit to the experimental data, it may be worthwhile to try to fit data to the basic polynomial equation:

$$Y = a_n X^n + a_{n-1} X^{n-1} + \cdots + a_1 X + a_0.$$

The number of data pairs must be one or more larger than the degree of the equation model sought. If there are 10 pairs of data, polynomials up to degree 9 are possible. The higher degree models are rarely used. This is not because of the calculation required, for computers have simplified that step for us. The main reason is that if the degree must be high to get a good fit, the "best" fit is probably being forced, and the data are really not statistically related at all. Additionally, a polynomial model of high degree can deviate significantly from a smooth curve representative of the data between data values and if the data are extrapolated. See reference (3) for additional discussion on the errors of polynomial fits.

Using the least squares procedure from Section 12.6.1, the equations to use for n pairs of data to estimate a second degree polynomial, $Y = aX^2 + bX + c$, are:

$$a\Sigma x_i^2 + b\Sigma x_i + nc = \Sigma y_i \tag{12.22}$$
$$a\Sigma x_i^3 + b\Sigma x_i^2 + c\Sigma x_i = \Sigma x_i y_i$$
$$a\Sigma x_i^4 + b\Sigma x_i^3 + c\Sigma x_i^2 = \Sigma x_i^2 y_i, \text{ where } i = 1, 2, 3, \ldots, n$$

These equations are solved simultaneously for a, b, and c. This procedure minimizes the sum of the squares of the vertical residuals from each data point to the parabola determined.

A polynomial equation of degree $\leq n$ can be found that passes through n pairs of data using techniques such as the Newton interpolating polynomial, but there is no control of what happens to the equation between data points or outside the range of the data.

12.6.5 Coefficient of Correlation

As previously mentioned, following a precise model does not guarantee meaningful results. Models obtained using least squares analysis are no different. One indicator that helps judge the validity of the results of linear regression analysis is the coefficient of determination, r^2, or the coefficient of correlation, r. If the least squares analysis yields $y' = mx + b$, and $\mathbf{y}$ is the mean of the dependent data, then the coefficient of determination is defined as follows.

$$r^2 = \frac{\Sigma(y_i' - \mathbf{y})^2}{\Sigma(y_i - \mathbf{y})^2} = \frac{\Sigma(mx_i + b - \mathbf{y})^2}{\Sigma(y_i - \mathbf{y})^2}, \text{ where } i = 1, 2, 3, \ldots, n \tag{12.23}$$

The value of r^2 is the part of the data variation explained by the least square model; an r^2 value of 0.85 means that the least squares equation explains or accounts for 85 percent of the variation of the data. An r^2 value of 1 means that there is a perfect match between the variables in an increasing relationship (positive slope). An r^2 value of -1 represents a perfect match with a decreasing relationship (negative slope). The ability of r to determine the correlation depends on the number of data pairs used. As the number of data pairs increases, a lower value of r predicts the same correlation reliability. Appendix K is a tabulation of coefficients of correlation for two variables. If the calculated value of r is larger than the table entry, then correlation is probable to the significance level indicated. The coefficient of correlation r can be calculated for any empirical equation.

Alternate forms for calculating r^2 are available; three are given here for analysis of a straight line on linear coordinate paper.

$$r^2 = \frac{(n\Sigma x_i y_i - \Sigma x_i \Sigma y_i)^2}{[n\Sigma x_i^2 - (\Sigma x_i)^2][n\Sigma y_i^2 - (\Sigma y_i)^2]} = \frac{[\Sigma(x_i - \mathbf{x})(y_i - \mathbf{y})]^2}{\Sigma(x_i - \mathbf{x})^2 \Sigma(y_i - \mathbf{y})^2} \tag{12.24}$$

$$r^2 = \frac{(\Sigma x_i y_i - n\mathbf{xy})^2}{[\Sigma x_i^2 - n\mathbf{x}^2][\Sigma y_i^2 - n\mathbf{y}^2]}$$

For semilog and log-log analysis make the same substitutions made in Equations 12.13 and 12.14 to obtain Equations 12.15 and 12.16, and Equations 12.17 and 12.18. The next example illustrates the empirical equation method.

EXAMPLE 12.11:
Linear coordinate analysis.

Consider the following data pairs taken from a test on a coil tension spring within the elastic limit of the material. The data = (load on spring, N; spring length, mm): (0, 1.45), (1, 2.10), (2, 2.45), (3, 3.03), (4, 3.38), and (5, 4.09).

The data plots reasonably straight on linear coordinate graph paper (see Figure 12.7). The form $Y = mX + b$ is appropriate, where X represents the load and Y represents the spring length. The "best" straight line has been drawn on the graph. There is good reason to draw the line through the point (0, 1.45). This is the free length of the spring and the spring would have been manufactured to this length. A more complete test would check this length after the test to be sure that the test did not exceed the elastic limit of the material, and to see if the original spring length has been influenced by coating materials or temperature variation. Variations from linearity occur because the material is less than perfect in reacting to a load, and measuring instruments are less than perfect in recording the true length of the spring. The slope is calculated by using two points on the line drawn,

$$m = \frac{3.75 - 1.45}{4.45 - 0} = 0.52 \text{ mm/N, and } Y = 0.52X + 1.45.$$

A purely mathematical solution, using Equations 12.13 and 12.14, can also be

FIGURE 12.7

Graph for Example 12.11.

TABLE 12.9

Calculations for linear coordinate least squares fit, Example 12.11.

x	y	x^2	xy	y^2
0	1.45	0	0	2.10
1	2.10	1	2.10	4.41
2	2.45	4	4.90	6.00
3	3.03	9	9.09	9.18
4	3.38	16	13.52	11.42
5	4.09	25	20.45	16.73
Totals	15	16.50	55	50.06	49.84

determined. It is convenient to tabulate the data for ease and accuracy of calculations (see Table 12.9).

$$b = \frac{55(16.5) - 15(50.06)}{6(55) - (15)^2} = 1.49 \qquad m = \frac{6(50.06) - 15(16.5)}{6(55) - (15)^2} = 0.50$$

$Y = 0.50X + 1.49$, and the coefficient of determination is:

$$r^2 = \frac{[6(50.06) - 15(16.5)]^2}{[6(55) - 15^2][6(49.84) - 16.5^2]} = 0.993, \quad \text{and } r = 0.997$$

It is not surprising that the mathematical slope and intercept differ from the graphical values because the graph line was forced through (0, 1.45). Your judgment must decide which is "best."

For a problem with six data pairs, Appendix K shows a value of $r = 0.9172$ for a 99 percent confidence level. Remember to enter the table at $(n - 2)$. If the calculated value of r is greater than 0.9172 there is less than 1 chance in 100 that no real correlation between the variables exists. This example has good correlation because $0.997 > 0.9172$.

The following case study illustrates how analysis refinement and optimization can be performed using the empirical formula method.

CASE STUDY #12:
Empirical formula refinement.

Consider the following data representing the amount of energy consumed in certain years.

Year	1	11	16	21	26	27	28	29	30	31
Energy, Btu's $\times 10^{12}$	24	34	40	45	54	57	59	62	66	69

FIGURE 12.8

Linear coordinate graph for Case Study #12.

The data plotted on a linear coordinate graph, Figure 12.8, does not show a good straight line and would ordinarily be passed over for something better. But let's see what a mathematical analysis shows using the least squares models. Table 12.10 contains the preliminary data calculations required.

$$m = \frac{10(12{,}495) - 220(510)}{10(5710) - 220^2} = 1.47 \qquad b = \frac{510 - m(220)}{10} = 18.76$$

$$Y = 1.47X + 18.76,$$

$$r^2 = \frac{[10(12495) - 220(510)]^2}{[10(5710) - 220^2][10(27964) - 510^2]} = 0.956 \text{ and } r = 0.978$$

The coefficient of correlation might suggest that the equation $Y = 1.47X + 18.76$ is a good one. There is, however, a residual at each point as shown in Table 12.11. *Note*: Even though the sum of the residuals is small because of + and − values, the sum of the squares of the residuals is quite large. Observe in Figure 12.8 that the line that comes closest to all the data points is curved, and the end points lie on one

TABLE 12.10

Calculations for linear coordinate least squares fit, Case Study #12.

	x	y	x^2	xy	y^2
	1	24	1	24	576
	11	34	121	374	1156
	16	40	256	640	1600
	21	45	441	945	2025
	26	54	676	1404	2916
	27	57	729	1539	3249
	28	59	784	1652	3481
	29	62	841	1798	3844
	30	66	900	1980	4356
	31	69	961	2139	4761
Totals	220	510	5710	12495	27964

side of the best straight line while the middle points lie on the other side. Can this "answer" be improved?

Consider the same data plotted on semilog paper as shown in Figure 12.9, and the related calculations in Table 12.11.

$$\log M = \frac{10(384.18) - 220(16.87)}{10(5710) - 220^2} = 0.0150, m_{10} = 1.035$$

$$\log B = \frac{5710(16.87) - 220(384.18)}{10(5710) - 220^2} = 1.3573, B = 22.76, \text{ and}$$

$$Y = 22.76(1.035)^X$$

TABLE 12.11

Residual calculations, linear coordinate least squares fit, Case Study #12.

	x	y	$y_i = 1.47x + 18.76$	$\Delta y = y_i - y$	$(\Delta y)^2$
	1	24	20.23	−3.77	14.21
	11	34	34.93	0.93	0.86
	16	40	42.28	2.28	5.20
	21	45	49.63	4.63	21.44
	26	54	56.98	2.98	8.88
	27	57	58.45	1.45	2.10
	28	59	59.92	0.92	0.85
	29	62	61.39	−0.61	0.37
	30	66	62.86	−3.14	9.86
	31	69	64.33	−4.67	21.81
Totals				1.00	85.58

Experimental Data Analysis   463

FIGURE 12.9

Semilog graph for Case Study #12.

[Semilog plot: Energy consumed, BTU × 10^{12} vs Year, showing fitted line $Y = 22.76(1.035)^X$ passing through (0, 23.0).]

TABLE 12.12

Calculations for semilog least squares fit, Case Study #12.

	x	y	log y*	x log y*	(log y)²*
	1	24	1.38	1.38	1.90
	11	34	1.53	16.85	2.35
	16	40	1.60	25.63	2.57
	21	45	1.65	34.72	2.73
	26	54	1.73	45.04	3.00
	27	57	1.76	47.41	3.08
	28	59	1.77	49.58	3.14
	29	62	1.79	51.98	3.21
	30	66	1.82	54.59	3.31
	31	69	1.84	57.00	3.38
Totals	220		16.87	384.18	28.67

* Values rounded from 7 digit calculations.

The value of 22.76 agrees well with the graphical answer of 23.0.

$$r^2 = \frac{[10(384.18) - 220(16.87)]^2}{[10(5710) - 220^2][10(28.67) - 16.87^2]} = 0.929, \text{ and } r = 0.964$$

Table 12.13 summarizes the residuals and their squares. Recall that the theory requires that the sum of the squares of the residuals is a minimum; 21.02 is as small as the sum of the residuals can be with the semilog model, which is approximately one-fourth the amount determined using the linear coordinate model. The data in Table 12.13 suggests that the plotted data is still curved, however, because the residuals go from − to + to − as X increases. Can the residuals be made smaller, or the − to + to − error pattern eliminated?

To try for a better fit, an estimate of a data offset value (discussed in Section 12.6.3) for the Y variable is needed. This requires estimating a value for a new point x_3. For x_3 to be within the range $1 \leq x_3 \leq 35$, a careful selection of x_a and x_b is needed. Let $x_a = 1$ and $x_b = 31$, then $x_3 = 1(31)/2 = 15.5$. From the dashed line in Figure 12.8, y_3 is approximately equal to 39, and

$$c = \frac{24(69) - 39^2}{24 + 69 - 2(39)} = 9.0$$

TABLE 12.13

Residual calculations for semilog least squares fit, Case Study #12.

x	y	$y_i = 22.76(1.035)^x$**	$\Delta y = y - y_i$	$(\Delta y)^2$*
1	24	23.56	− 0.44	0.20
11	34	33.23	− 0.77	0.59
16	40	39.47	− 0.53	0.29
21	45	46.87	1.87	3.51
26	54	55.67	1.67	2.79
27	57	57.62	0.62	0.38
28	59	59.64	0.64	0.40
29	62	61.72	− 0.28	0.08
30	66	63.88	− 2.12	4.48
31	69	66.12	− 2.88	8.30
Totals			0.38	21.02

* Values rounded from 7 digit calculations.

Experimental Data Analysis

The curve is concave upward, so c is positive and subtracted from the Y values. Table 12.14 lists the adjusted data and preliminary calculations.

$$\log M = \frac{10(366.849) - 220(15.912)}{10(5710) - 220^2} = 0.01929, m_{10} = 1.045$$

$$\log B = \frac{5710(15.912) - 220(366.849)}{10(5710) - 220^2} = 1.16675, B = 14.68$$

$$Y - 9 = 14.68(1.045)^X$$

$$r^2 = \frac{[10(366.849) - 220(15.912)]^2}{[10(5710) - 220^2][10(25.647) - 15.912^2]} = 0.9878, \text{ and } r = 0.994$$

Additional accuracy was used on this calculation because r is approximately equal to one. The maximum value of r should be one, but round off error may make $r > 1.00$. When only two-digit accuracy is used in Table 12.14, $r = 1.054$. Additional refinement could be made by using additional decimal accuracy. Write and run a computer program with double precision accuracy to investigate the possibility of better answers.

Table 12.15 lists the residuals for this solution, and Figure 12.10 shows the resulting graph. This adjustment does produce a smaller sum of the squares of the residuals, a smaller sum of the residuals, and a better coefficient of correlation. Whether or not this is the "best" fit of the data is still a decision to make. Is the sum of the residuals or the correlation coefficient most important? A computer iteration program could investigate other values in the neighborhood of 9.0 and perhaps find a "better" answer.

TABLE 12.14

Calculations for adjusted semilog least squares fit, Case Study #12.

x	$y' = y - c$	log y'*	x log y'*	(log y')²*
1	15	1.176	1.176	1.383
11	25	1.398	15.377	1.954
16	31	1.491	23.862	2.224
21	36	1.556	32.682	2.422
26	45	1.653	42.984	2.733
27	48	1.681	45.394	2.827
28	50	1.699	47.571	2.886
29	53	1.724	50.004	2.973
30	57	1.756	52.676	3.083
31	60	1.778	55.123	3.162
Totals	220	15.912	366.849	25.647

* Values rounded from 7 digit calculations.

FIGURE 12.10

Adjusted semilog graph for Case Study #12.

$Y^1 = 14.68(1.045)^X$
where $Y^1 = Y - 9.0$

TABLE 12.15

Residual calculations for adjusted semilog least squares fit, Case Study #12.

x	y'	$y_i' = 18.80(1.039)^x$*	$\Delta y = y_i' - y'$	$(\Delta y)^2$*
1	15	15.34	0.34	0.12
11	25	23.82	−1.18	1.38
16	31	29.69	−1.31	1.72
21	36	37.00	1.00	0.99
26	45	46.11	1.11	1.22
27	48	48.18	0.18	0.03
28	50	50.35	0.35	0.12
29	53	52.61	−0.39	0.15
30	57	54.98	−2.02	4.07
31	60	57.46	−2.54	6.47
Totals			−4.46	16.27

* Values rounded from 7 digit calculations.

12.7 Experimentation

Experiments are run when information is needed that cannot be obtained using exact mathematical modeling. Either no theory exists to explain the situation being studied or the theory uses assumptions and boundary conditions that cannot be matched. Experiments may also be used to establish base line information such as material strength properties, or they may be run to find a relationship between variable quantities such as the electrical conductivity as a function of temperature. Questions often lead to experiments.

1. What happens to the shear strength of a material as the rate of loading changes?
2. How does the percent yield from a chemical process change as the temperature changes?
3. How does the traffic flow rate change as the red-green cycle time changes on a traffic light?
4. How does the life of a forging die change when a different lubricant is used?
5. How does the life of a light bulb change as the ambient temperature changes?

Engineers need to know what will happen when changes are made. Sometimes the answer to a question can be found by testing a sample or scale model of the real thing, such as the heat conductivity of a metal. Other times only a real product test will be valid, such as determining the life of a car tire. The purpose of an experiment is to control the variables, whatever they are, establish relationships that can be verified by others doing the same experiment, and obtain relationships to use to predict future events. One of the difficulties with experiments is knowing which variables are affecting the results, and then planning a scheme to keep those variables under control.

After the variables are under control, their values must be randomized so an experiment can be done without equipment or operator bias. If two products, A and B, are to be tested, the person running the experiment should not test all of product A and then test all of product B. This can result in bias and unreliable results. Uncontrolled variables may include the temperature of the test samples, the operating speed of the test machine, or the operator's skill in running the test machine. The instruments showing data values may change calibration during the experiment. Any number of events can occur over the course of the experiment and not be noticeable because of the slow change. Time differences can also affect the hardness of samples if they are subject to precipitation hardening. Time can also allow residual stress relaxation, and creep can occur. By randomizing the order of the test samples any time dependent, uncontrollable, or unknown variations are averaged out. One method for doing this is to give all the test samples a number and select them for test based on a random number table. Even testing on an alternating A-B-A-B basis is better than testing all of one type in sequence. Another method of reducing the effects of uncontrollable variables is to run all samples simultaneously. This could be done if the life of a light bulb experiment is run. If temperature, voltage, and outside vibrations are not controlled, having all the bulbs on at the same time will reduce the effect of these variations, as all the bulbs will be exposed to the same variation.

Another important experiment criterion is the accurate measuring and recording of the data. One of the first experiments run to measure the speed of light using lanterns could not succeed because of the crude measuring technique used. Experimental data with a large deviation or a large range should be suspect and used with caution. Good data can also be lost or become confusing because of poor recording. A person recording data may become bored with the repetition of the task and become careless, or read dials and meters incorrectly. Automatic recording instruments will help avoid that situation, but the instruments must be calibrated and checked from time to time during the experiment.

The following items should be considered when setting up and running an experiment. Each item will be discussed using a light bulb burn-out life experiment as an example.

1. State the problem or question attempted to be answered. Is the 750 hour rated life as advertised for brand X light bulb an average life?

2. Consider appropriate variables and response scales to use. The variable chosen to record is time-to-burnout, measured in hours. Other life criteria could be measured, such as time to reduce to a given level of illumination. A more elaborate experiment could control and vary voltage, ambient temperature, position of bulb filament, or induced vibration.

    Some experiments may require judgment evaluations of variable factors. The degree of cloudiness during a solar collector experiment, or physical comfort during a temperature-humidity experiment are examples. Always determine ahead of time how such judgments are to be handled.

    It is also possible that in some experiments the first choice of measuring units will not be appropriate. Samples may have to be run to determine the best units to use. Seconds would be too small a unit for the light bulb experiment, but minutes might be too large a unit for an experiment measuring the speed of light.

3. Select relevant factors to be varied. For a light bulb experiment, one variation is to have one set of bulbs on continuously, and a second set of bulbs cycling on-off. This can be done if equipment is available and the amount of time needed to run the experiment is short. The voltage and ambient temperature can be monitored if not controlled. Bulb position, horizontal, vertical, or at some angle, must be consistent.

4. Select levels of variation. Select an on-off cycle such as a 10 second interval, 5 seconds on, 5 seconds off. If voltage variation is desired, levels such as 110 ± 2, 115 ± 2, 120 ± 2, 125 ± 2, and 130 ± 2 volts are possibilities. If filament position is chosen as a variable, horizontal, vertical up, vertical down, or other angles can be selected. Ambient temperature can also be selected as a variable, any range of temperatures can be selected. To reduce the effect of the uncontrollable variables, both sets of bulbs must be tested at the same time.

    Be cautious about changing the level of variation during an experiment. It is better to wait out a portion of a test and start over with a different level of a variable. Of course a test could be set up as a changeable pattern on purpose. A load cycle, for example, could be one second, two seconds, three seconds,....

5. Establish combinations of variable levels. In the light bulb experiment this is not necessary because the only variation selected is the on-off cycle.

To illustrate how this *can be* done, consider an experiment with three variables, each with a range of values: $X_1 \pm \Delta X_1$, $X_2 \pm \Delta X_2$, and $X_3 \pm \Delta X_3$. Figure 12.11 shows how seven different combinations can be used to obtain a first approximation of the effects of variable changes. Figure 12.12 shows how nine different combinations can be used to obtain a first approximation of the effects of variable changes. If the center value (X_1, X_2, X_3) is not used, then the eight points remaining represent the factorial experiment in 3 variables ($2^3 = 8$ runs). Be certain that measuring instruments are capable of accuracies less than the experimental ranges established. Second approximations for increasing accuracy within any of the octants would be appropriate for a second round of experiments (see Exercise 12.52).

Past experience and judgment are important when considering variable ranges to investigate. It is also important to lay out and document the plan clearly so efficient use of time and equipment can be made. See Case Study #18, Experimental Optimization, in Chapter 13 for additional discussion on establishing experimental variable combinations.

6. Determine the number of observations to make. This is a very difficult pre-experiment judgment to make. It depends on the number of variables and the degree of accuracy desired. In some cases more data may be required after the preliminary tests are performed. Relationships that change abruptly or frequently require more data for analysis. As the desired accuracy increases, so does the need for additional experiments. Many times this is done with no additional variation but repeated for accuracy only. Additional data improves reliability confidence.

7. Analyze the data collected. In the case of the light bulb experiment the mean bulb life was calculated and compared with the advertised 750 hours. In both cases,

FIGURE 12.11

Variations around a point, scheme #1.

1. X_1, X_2, X_3
2. $X_1 + \Delta X_1, X_2, X_3$
3. $X_1 - \Delta X_1, X_2, X_3$
4. $X_1, X_2 + \Delta X_2, X_3$
5. $X_1, X_2 - \Delta X_2, X_3$
6. $X_1, X_2, X_3 + \Delta X_3$
7. $X_1, X_2, X_3 - \Delta X_3$

FIGURE 12.12

Variations around a point, scheme #2.

1. X_1, X_2, X_3
2. $X_1 + \Delta X_1, X_2 + \Delta X_2, X_3 + \Delta X_3$
3. $X_1 + \Delta X_1, X_2 + \Delta X_2, X_3 - \Delta X_3$
4. $X_1 + \Delta X_1, X_2 - \Delta X_2, X_3 + \Delta X_3$
5. $X_1 + \Delta X_1, X_2 - \Delta X_2, X_3 - \Delta X_3$
6. $X_1 - \Delta X_1, X_2 + \Delta X_2, X_3 + \Delta X_3$
7. $X_1 - \Delta X_1, X_2 + \Delta X_2, X_3 - \Delta X_3$
8. $X_1 - \Delta X_1, X_2 - \Delta X_2, X_3 + \Delta X_3$
9. $X_1 - \Delta X_1, X_2 - \Delta X_2, X_3 - \Delta X_3$

steady on and five second on-off, the average life was not 750 hours, but 560.

If the purpose of an experiment is to determine an empirical relationship then Equations 12.10, 12.11, or 12.12 might be used.

8. Interpret the results. In the case of the light bulb experiment a positive conclusion was difficult to make. This was because the voltage level, which was monitored during the test, was always higher than 120, ranging from 123 to 126 volts. Research on the design of light bulbs indicates that high voltage does reduce life (see Exercise 12.56). The only conclusion made from the samples tested was that at the mean recorded voltage level of 124.5 volts, the mean life was 560 hours with a standard deviation of 115 hours. There was no statistical difference in life to burn out between the bulbs that were on continually and those that were subjected to the on-off cycle.

Occasionally an experiment will be run and no conclusion can be made other than the variables chosen have no effect on each other. If that occurs a different hypothesis must be considered. Do not consider such an experiment a failure.

For all reversible experiments record data going in both directions. This will help uncover warm-up errors or hysteresis phenomenon. In material strength measurements, for example, reverse loading will indicate if the elastic limit was exceeded.

Study a system with and without a component if possible. Then a before and after comparison can be made. This can be done only when a system will function with and without a component. A street intersection, for example, will function with or without a stop sign, and another level of variation would be a yield sign. These three levels of variation are fixed, and provide an example of a step function experiment. Experiments can also be run to optimize a desired output of a process or system, as discussed in Section 13.5.

12.8 Additional Study Topics

1. Absolute measuring uncertainties
2. Relative measuring uncertainties
3. Newton-Raphson root location method
4. Lagrange interpolating polynomial
5. Newton interpolating polynomial
6. Hyperbolic curve fitting
7. Cubic polynomial spline curve fitting

12.9 References

1. Baird, D.C. *An Introduction to Measurement Theory and Experimental Design.* Englewood Cliffs, N.J.: Prentice Hall, 1962.
2. Cheney, Ward and David Kincaid. *Numerical Mathematics and Computing,* 2nd ed. Monterey: Brooks/Cole, 1985.
3. Clothier, Walter. "Design Models Are Naturals For Managing Information." *Design Graphics World,* May 1987, 40-41.
4. Clothier, Walter "Good Models Never Die." *Design Graphics World,* Dec. 1986, 28-30.
5. Dana, Forest C. and L.R. Hillyard. *Engineering Problems Manual*, 5th ed. New York: McGraw-Hill, 1958.
6. David F.W. *Experimental Modeling in Engineering.* Boston: Butterworth, 1982.
7. Dally J. W., W. F. Riley, and K. G. McConnell. *Instrumentation for Engineering Measurement.* New York: John Wiley and Sons, 1984.
8. Gajda, Walter J., Jr. and William E. Biles. *Engineering: Modeling and Computation.* Boston: Houghton Mifflin, 1978.
9. Gallin, Daniel. *Finite Mathematics.* Glencoe, Ill.: Scott Foresman and Co., 1983. Chapters 2 and 5.
10. Hillier, Frederick S. and Gerald J. Lieberman. *Operations Research* , 2nd ed. Oakland, Calif.: Holden-Day, Inc. 1974.
11. John, J.A. *Experiments: Design and Analysis.* New York: Macmillan, 1977.
12. Kapur, J. N. *Mathematical Modeling.* India: Wiley Eastern Limited, New York: John Wiley & Sons 1988.
13. Meredith, Dale D., et al. *Design and Planning of Engineering Systems.* Englewood Cliffs, N.J.: Prentice Hall, 1973.
14. Mischke, Charles R. *Mathematical Model Building*, 2nd ed. Ames, Iowa: Iowa State University Press, 1980.
15. Montgomery, Douglas C. *Design and Analysis of Experiments,* 2nd ed. New York: John Wiley and Sons, 1984.

15. Piascik, Chester. *College Mathematics with Applications*. Columbus: Charles E. Merrill, 1984. Chapter 6.
16. Steidel, Robert F. and Jerald M. Henderson. *The Graphic Language of Engineering*. New York: John Wiley & Sons, 1983. Chapter 8.
17. Tufte, Edward R. *The Visual Display of Quantitative Information*. Cheshire, Conn.: Graphics Press, 1983.

12.10 | Exercises

1. Discuss the following models. What kind of a model is each? What is being modeled? What cautions must be observed when using each one?
   a) Strain gage
   b) Flight simulator
   c) Lacquer stress coating
   d) Electricity closed circuit laws (Kirchoff's and others) when analyzing a water system
   e) Current flow equations when analyzing heat flow

2. An alternate method to calculate the radius of a circle when three points on the circle are known is to find the point of intersection of the perpendicular bisectors of lines drawn between the points. Only two bisectors are required. The distance between the intersection point and any one of the given points is the radius. Determine the models required for this method of analysis.

3. Another mathematical model for determining the radius of a circle when only a portion of it is shown involves the length of an arc chord, L, and the maximum height of the arc above the chord, h.
   a) Derive the formula for the radius of a circle as a function of L and h.
   b) Use the model from (a) on the sample in Figure 12.3.

4. The grooves on a phonograph record are not a series of concentric circles but a tightly wound spiral. Create a model to calculate the length of a groove on a record and discuss how to collect the data needed to use the model.

5. The model $y = 4L^3W/Ebh^3$ determines the maximum deflection of a rectangular cantilevered beam with weight W in pounds on the free end. Identify the other variables including their units and discuss the model limitations.

6. The maximum deflection of a simply supported beam is given as: $y = WL^3/4Ebh^3$ where y = maximum deflection, at the center; W = load in the center; L = length of the beam; E = modulus of elasticity of the beam material; b = width of the beam; h = height of the beam.
   a) Establish a consistent set of units assuming that E will be given as lbf/in^2.
   b) If you made a model of this beam, and made the model geometrically 1/10 as large as the original, used 1/10 of the load, but used the same material, what would you expect the deflection of the model to be compared to the original?

7. Verify Equation 12.3.

8. If the velocity of sound in a gas is considered to be a function of gas density, pressure, and dynamic viscosity, determine the relevant control groups using dimensional analysis.

9. Find the dimensionless groups for the following situation. A hole is drilled in the side of an open tank, a fluid is placed in the container, and the fluid leaves the hole with some velocity.

10. Find a dimensionless groups relationship for a pump using pressure change through the pump, the pipe diameter, and fluid properties density and viscosity.

11. Find the dimensionless groups for the fluid velocity flow over an inclined roof assuming the relative variables are fluid density, roof slope angle, fluid viscosity, surface tension, length of roof in the direction of flow, and gravity.

12. Discuss the following test that was performed on a nuclear fuel rod shipment container. A one-quarter scale model was dropped 30 feet onto concrete. The real tank weighs 50 tons, and holds 200 spent fuel rods and about 200 gallons of water.

13. What is the function of modeling when investigating such concepts as crack growth and stress concentration behavior?

14. Prove the statement, "One inch of circumference change in a circle is associated with a 0.32 inch change in diameter."

15. Prove the statement, "Each half inch of bark will change the circumference of the tree by π inches" mentioned in Case Study #10.

16. Write the models for the coefficient of correlation for
    a) semilog line of regression, and
    b) log-log line of regression.

17. Occasionally it is efficient to offset X or Y by a fixed amount when the data values are very large or very small. Consider the following data representing oil production in the United States in billions of barrels of oil per year.

Year:	1934	1935	1936	1937	1938	1939	1940	1941
Production:	0.9	1.0	1.1	1.3	1.2	1.3	1.4	1.4

    Determine the "best" fit of the data by using the axis translation models $P = m(Y - 1934) + b$ and $P = B(M)^{Y-1934}$.

18. Extrapolate the answer from Exercise 12.17 to current years and determine if the growth of oil production has actually followed the trend established in the years 1934–1941. Library research is required.

19. Repeat the analysis method of Exercises 12.17 and 12.18 for the following data representing the population of the United States in millions of people.

Year:	1880	1900	1910	1920	1930	1940
Population:	62.9	76.0	92.0	105.7	122.8	131.7

    For Exercises 12.20–12.24, determine an appropriate empirical equation for the data given. Also determine a coefficient of determination, the sum of the residuals, and the sum of the squares of the residuals.

20. Temperature and resistance of an electronic resistor.

   Temperature, °F: 73.2   83.3   87.8   95.0   98.8   105.3   106.2
   Resistance, ohms: 24.3   24.9   25.1   25.5   25.9   26.1   26.4

21. Pressure and volume of a gas.

   Pressure, psi: 106.0   89.0   70.8   49.0   40.0   30.0   24.5   19.5
   Volume, in^3:   1.5   1.7   2.0   2.6   3.0   3.6   4.2   5.0

22. The length of a pendulum and its period of oscillation.

   Length, inches: 25.6   31.5   39.4   49.2   59.1   78.7
   Period, seconds: 1.60   1.80   2.00   2.24   2.44   2.84

23. Fuel consumption of a particular vehicle.

   Speed, mph: 20   22   24   26   28   30   32   34
   Gal/mile: 0.183   0.188   0.191   0.193   0.195   0.197   0.198   0.200

24. Altitude and pressure above the earth's surface.

   Altitude above sea level, meters: 0   500   1000   1500   2000   2500
   Pressure in mm of mercury: 762   716   672   632   593   561

25. Determine an appropriate empirical equation and related coefficient of correlation for the following data.

   X: 1.19   2.00   2.32   3.10   3.42   5.30   5.47   6.91   7.33   7.54
   Y: 7.50   12.87   15.09   18.63   19.16   21.37   23.89   17.52   17.42   14.75

   X: 7.98   8.26   8.72
   Y: 14.29   12.52   9.83

26. Determine an appropriate empirical equation and related coefficient of correlation for the following data.

   X: 1.27   2.21   2.38   3.21   4.32   5.39   6.09   7.30   7.35   7.69
   Y: 10.75   14.97   17.61   19.06   22.69   22.41   22.33   17.36   16.63   16.77

   X: 8.18   8.41
   Y: 11.15   9.40

27. Compare the data from Exercises 12.25 and 12.26. Is it probable that the data came from the same universe?

28. An analysis of a set of n data pairs resulted in the equation $Y = 1.1X + 4.2$ with an $r = 0.80$. What is the minimum number of data pairs required so that the equation is a good representation of the data at least 95 percent of the time? At least 99 percent of the time?

29. An analysis of 10 pair of data resulted in the equation $Y = -1.2X + 15.1$ and $r = 0.70$. Is this reasonable?

30. An analysis of 30 pair of data resulted in the equation $Y = -0.9X + 28.2$ and $r = 0.400$. Is this a good representation for the data?

31. Show that a reduction in the degree of the polynomial model by one in Equation 12.22, leads to Equations 12.13 and 12.14.

32. Expand Equation 12.22 used to determine a, b, and c, for the second degree equation

    $Y = aX^2 + bX + c$, and determine a, b, c, and d for the third degree equation
    $Y = aX^3 + bX^2 + cX + d$.

33. The following model is used to evaluate flow capacity of a gas valve. $F = GA^{0.82}$, where

    F = required flow rate in m³/minute at standard pressure and temperature
    A = total outside surface area of the storage tank in m^2
    G = 10.66 for propane, and G = 8.98 for butane.

    a) What kind of graph paper will show this model as a straight line?
    b) For a given value of flow capacity, which tank will be the largest, one for propane or butane?

34. What is the significance of the 2.35 in the following model used for a design stress calculation? Maximum stress = $S_{ut}/2.35$.

35. You are going to drive a friend home who lives 9 miles away if you drive the freeway at 55 miles per hour, but who lives only 6.5 miles away if you drive the country road at 35 miles per hour. The cost of driving is estimated at $0.01V^2$ cents per mile, where V is the car speed in miles per hour. Which route will you take under each of the following circumstances?
    a) Your time is worth $0.15 per minute.
    b) Your friend is already late getting home.
    c) Your time is worth $3.00 per minute.
    d) Your friend is pleasant, a good listener, and you enjoy talking.
    e) What additional information would you like to have before answering the questions and making a choice?

36. Suppose you are working on a project designing a control device that could be analyzed with the model: $V = S^{0.5}D^2/WC^3$, where
    S = device size
    D = the distance between the control device and the object being controlled
    W = the weight of the device
    C = the cost of the device, and
    V = an indicator that relates to product acceptability and profitability, high numbers are best.
    a) Which variable of the project seems to be the most important to analyze, and would you want it to have a high or low value? Why?
    b) Which variable would you begin to work with to change if the product is already in production and you have the job of making it more profitable for your company? Why?
    c) Does this model seem realistic? Why?

37. A house has windows in it for the convenience of the people who live in it. Windows allow outside light to shine inside, allow outside visibility, provide emergency exits, and provide solar heat collecting. A single thickness of regular window glass has an R-value of 1.0, adding a storm window doubles the value to 2.0, and a triple glazed window may increase the value to 3.5.

Various products are sold to cover windows that increase the effective R-value. Make a list of the items that need to be considered to evaluate whether or not a product should be purchased and added to the windows of a house.

38. The following table lists Iowa health care expenditures per capita per year (Source: *Des Moines Register*, October 26, 1986). The data was extrapolated to arrive at the 1988 and 1990 figures.

   Year: 1966 1968 1970 1972 1974 1976 1978 1980 1982 1984
   Cost: $196   242   293   351   453   556   724   931 1182 1358

   Year: 1986 1988 1990
   Cost: $1588 1877 2206

   a) Determine a probable method of extrapolation and judge its validity.
   b) Consider alternate methods of extrapolation and compare the resulting figures for 1988 and 1990 with those given.

39. Under what circumstances would it be important to know the sequence of data recordings as well as the values, when performing an experiment?

40. List some experiments that contrast the difference between step function and continuous function control.

41. A particular liquid was tested for its boiling temperature in °C at five different pressures. Analyze the 21 data values collected and listed below.

Pressure, psig	67	71	75	79	83
Boiling temperature @ given pressure	90	89	86	84	83
	92	90	87	85	85
		90	87	85	
		91	87	86	
			88	86	
			88	87	
			88		

42. Refer to Figure 12.11 and determine the seven values of the three variables for the octant bounded by 4-1-2-7 using the same system of point selection. How many new points are needed?

43. If it is not known that the deflection of a beam is a function of the section modulus (moment of inertia), a guess might be made that deflection is a function of cross-sectional area.
   a) Use this assumption and determine the Π_i groups in Example 12.9.
   b) What would happen when the value of k is being determined experimentally if this Π group is used?

44. Model a simply supported uniformly loaded beam with a simply supported beam and a center load, assuming equal maximum deflection.

45. Repeat Exercise 12.44 assuming equal end of beam slope angle. Would this model have more or less deflection than the answer to Exercise 12.44?

46. What happens to maximum bending moment in Exercise 12.45 after the model substitution?

47. What happens to the maximum shear stress in Exercise 12.45 after the model substitution?

48. A cylindrical storage tank is being designed to hold 10,000 gallons. The design guideline calls for a diameter: height ratio of 1 : 2.
    a) Determine the dimensions of a scale model of the tank built to a scale of 1/10.
    b) How many gallons will the model hold?

49. A concrete or cement block wall is thin compared to its length and height. A freestanding wall loaded from above can be modeled as a column. Mathematical modeling is difficult when the effects of openings for doors and windows are considered. Discuss the variables and experimental control you would use to empirically determine the effect of wall openings on P_c, the load before buckling.

50. Attached walls also have an effect on the stability of walls and the value of P_c. Discuss the variables and experimental control you would use to empirically determine the effect of attached walls on P_c.

51. An engineer measured the diameter and length of 12 manufactured bolts and calculated a coefficient of correlation between the data of 0.945. The conclusion made was that by controlling the manufacture of only diameter or length, the length or diameter would also be controlled. The engineer is 99 percent confident of this conclusion. Do you agree? Explain.
    Data, measured in inches, (Diameter, length): (0.990, 2.988), (1.002, 3.005), (1.010, 3.012), (0.995, 2.993), (0.998, 2.997), (1.005, 3.008), (1.003, 2.999), (0.997, 2.994), (1.001, 3.001), (1.003, 3.004), (0.997, 2.999), (0.999, 3.000).

52. Consider a light bulb experiment with three bulb positions: vertical up, vertical down, and horizontal; three voltages: 120 ± 1, 125 ± 1, and 115 ± 1; and three ambient temperatures: 70°F, 100°F, and 150°F. Discuss a method of experimentation considering the methods illustrated in Figures 12.12 and 12.13.

53. Prove that Equation 12.10 can also be expressed as $Y - \mathbf{y} = m(X - \mathbf{x})$ where $\mathbf{y}$ is the mean of the y data, and $\mathbf{x}$ is the mean of the x data.

54. The data in Table 12.16 represents federal guidelines for maximum noise exposure to avoid hearing loss.
    a) Explain the rapid percent decline in the time exposure with a corresponding small percent increase in noise level.
    b) Find an empirical equation that best represents the data.

55. While doing research on sewage particulate settling, an engineer developed the model $V_s = \sqrt{[4gd/3C_D][(\rho_s - \rho_l)/\rho_l]}$, where
    V_s = terminal settling velocity, cm/sec
    g = gravitational constant, cm/sec^2
    d = particle diameter, cm
    ρ_s = particle density, g/cm^3
    ρ_l = liquid density, g/cm^3, and
    C_D = coefficient of drag.
    Discuss problems that you might encounter if you were to run experiments to determine the settling velocity of various particulate matter.

TABLE 12.16

OSHA sound level exposure limits.

Noise level, dbA	Maximum exposure, hours per day.
90	8.00
92	6.00
95	4.00
97	3.00
100	2.00
102	1.50
105	1.00
110	0.50
115	0.25

56. Experimental work with light bulbs has resulted in a predictor of actual life based on actual voltage: actual life = rated life(rated voltage/actual voltage)$^{13.1}$. Does the voltage level recorded in the experiment discussed in Section 12.7 account for all the reduction in bulb life?

57. The range of data collected in the experiment of Section 12.7 was 338 to 829 hours. What actual voltage existed for the data collected if voltage was the only reason for reduced life?

58. Repeat Exercise 12.57 for data with a range of 441 to 798 hours and a mean of 605 hours.

13 Optimization

"Why is it we never have time to optimize the design the first time, but always have time to revise the design when something fails?"
◆ AN ANONYMOUS ENGINEER

Optimization methods and analysis techniques give engineers the capacity to manipulate and fine tune their design development work so the results are as efficient, functional, and cost effective as possible. This does not mean that the final design will be the most efficient and most functional and have the lowest cost, because all constraints usually cannot be optimized at the same time. The engineer must select the criteria that are most important, maximize or minimize as desired, but without forcing other constraints to reach unacceptable levels. All discussions of problem solving and specific design considerations in this book have optimization as a primary objective, to select a value for a primary variable that does not create an unacceptable value for a related variable. The engineer's goal is to obtain the best results for a given problem situation within the imposed restrictions. The physically imposed restrictions, including size, weight, material available, processes available, cost, and a due date, are real constraints that must be satisfied. Engineers must also understand that an optimum determined today may not be the optimum of tomorrow, because physical constraints and technology may change as time passes. Another reason for the changing status of selected optimum values is the result of system variances. System variations include changes in material prices, interest rates, future scrap value, available material compositions and sizes, shipping costs, ability of workers, condition of machines used to manufacture products, and consistency of material purchased; as well as changes in the environment such as temperature, humidity, composition of air, and atmospheric pressure. Because of the range of variable values that can occur, assumptions about the status of the variables are made prior to using optimization techniques. A temperature range may be assumed; market potential and sales price estimated; and material availability and cost established. It is important to establish values (or ranges of values) for variables prior to searching for optimum solutions, because solutions are only optimum within the restrictions and the assumptions made. If the assumptions are changed, a different optimum may occur. This is one of the reasons why engineers run many configurations (different sets of assumptions) while developing a product. The goal is to find an optimum that has low sensitivity to changes in boundary conditions. Low slopes on function curves, when a change in an independent variable causes a small change in the dependent variable, indicate low sensitivity.

Even though many optimization techniques are available to an engineer, some are more general in nature and their principles can be extended to a wide variety of problem situations. This chapter will concentrate on several of the more general techniques, both calculus and non-calculus, and their applications.

13.1 Calculus Techniques

Engineers learn early in their education that calculus provides methods of optimization if a mathematical model can be established and is differentiable. This is not always possible, particularly in a totally new situation. Often, however, a mathematical model can be estimated from experimental or observational data as was done in Chapter 12. The specific mathematical model depends on the nature of the data collected. Fitting data to known models (such as Equations 12.10, 12.11, 12.12, 12.20, and 12.21) to determine the best empirical model can itself be an optimization process, as illustrated in Case Study #12, Empirical Formula Refinement.

13.1.1 First Derivative Analysis

If a differentiable mathematical model can be developed, then standard derivative techniques are available to find extreme (maximum or minimum) points. If a function $f(x)$ has a first derivative such that $f'(x_i) = 0$, three possibilities exist.

1. If $f^{II}(x_i) < 0$, or if $f^{II}(x_i) = 0$ and $f^{III}(x_i) = 0$ and $f^{IV}(x_i) < 0$, then $(x_i, f(x_i))$ represents a local maximum extreme value.
2. If $f^{II}(x_i) > 0$, or if $f^{II}(x_i) = 0$ and $f^{III}(x_i) = 0$ and $f^{IV}(x_i) > 0$, then $(x_i, f(x_i))$ represents a local minimum extreme value.
3. If $f^{II}(x_i) = 0$ and $f^{III}(x_i) \neq 0$, then $(x_i, f(x_i))$ represents a point of inflection.

If more than one x_i exists within an interval $[x_a, x_b]$ such that $(x_i, f(x_i))$ are local extreme values, then the global extreme within the interval is determined by comparing the values of all local extreme values. The function may also have a value at the end of an interval that is larger or smaller than local extreme values located using $f'(x_i) = 0$. Be sure to examine the value of the function at the interval endpoints.

Keep in mind that the extreme values found are valid only if the assumptions, usually the constant values in the function, are stable. An engineer should always investigate the effect of changes in the assumptions. The following example illustrates the first derivative method and a variable change sensitivity check.

EXAMPLE 13.1:
Optimize a one variable function.

The cost, in dollars per hour, of running a particular vehicle is expressed as $C_{hour}(v) = 0.05v^2 + 72$, where v = the speed of the vehicle in kilometers/hour. The 0.05 is a cost associated with the speed, and 72 is a fixed cost that applies even if the vehicle is stationary. Determine the speed for minimum cost per kilometer, and determine whether or not the relationship is more sensitive to a change in fixed cost or variable cost.

The cost per kilometer is obtained by dividing the cost per hour by the speed in kilometers per hour: $C_{dist}(v) = 0.05v + 72/v$

Take the first derivative with respect to v, set it equal to zero, and solve for v:

$$dC_{dist}/dv = 0.05 - 72/v^2 = 0$$

$v^2 = 72/0.05$, so $v = 37.94$ kilometers per hour for minimum cost, and the minimum operating cost = \$3.79 per kilometer.

Intuitively, this answer represents a minimum rather than a maximum, because maximum cost per kilometer will occur when v approaches zero and $72/v$ become increasingly large; however, checking the second derivative will verify this. The second derivative is $d^2C_{dist}/dv^2 = 216/v^3$, which is greater than zero for all positive velocities and so a minimum is assured. The graph of $C_{dist}(v) = 0.05v + 72/v$, Figure 13.1, also shows that a minimum occurs.

FIGURE 13.1

Cost as a function of velocity, Example 13.1.

If a percentage variation in the values of 0.05 and 72 is assumed (±1 percent, ±5 percent, or ±10 percent), then the sensitivity of the 0.05 and 72 can be studied. If the 72 varies ±10 percent then the minimum speed varies ±5 percent, and if the 0.05 varies ±10 percent the minimum speed also varies ±5 percent, so the relationship has low sensitivity to variations in both constants. The reader should verify these results.

13.1.2 Partial Derivative Analysis

If it is necessary to optimize functions in more than one variable, derivative techniques can be used, but partial derivatives are required, and the tests for extreme values are different. If a function $Z = f(x, y)$ is a model of the problem at hand, take partial derivatives and set them equal to zero, $\partial Z/\partial x = 0$ and $\partial Z/\partial y = 0$. Solve the derivative equations simultaneously, obtain pairs of values (x_i, y_i) that satisfy the equations, and test for extreme values.

The test for extremes requires taking three, second partial derivatives, $\partial^2 Z/\partial x^2$, $\partial^2 Z/\partial y^2$, and $\partial^2 Z/\partial x \partial y$, and defines $\Delta(x, y) = (\partial^2 Z/\partial x^2)(\partial^2 Z/\partial y^2) - (\partial^2 Z/\partial x \partial y)^2$. The value of $\Delta(x, y)$ is calculated at each of the (x_i, y_i) values which are determined by solving the first derivative equations simultaneously. If $\Delta(x, y) > 0$ then $[(x_i, y_i), f(x_i, y_i)]$ represents a maximum if $\partial^2 Z/\partial x^2 < 0$ or if $\partial^2 Z/\partial y^2 < 0$, or a minimum if $\partial^2 Z/\partial x^2 > 0$ or if $\partial^2 Z/\partial y^2 > 0$. If $\Delta(x, y) < 0$ then a saddle point (neither a maximum or minimum) exists, and if $\Delta(x, y) = 0$ the test does not allow a conclusion about an extreme to be made. If $\Delta(x, y) = 0$ then other methods, such as a computer search, may be appropriate. The following example illustrates the partial derivative technique.

EXAMPLE 13.2:
Optimize a two variable function.

Find any extreme values of the function $Z = f(x, y)$, where Z is defined as:

$$Z = xy^2 - 6y^2 + 72y + x^2y - 18xy - 6x^2 + 72x - 216$$

Take partial derivatives, set them equal to zero, and solve the resulting equations simultaneously:

$$\partial Z/\partial x = y^2 + 2xy - 18y - 12x + 72 = 0$$
$$\partial Z/\partial y = 2xy - 12y + 72 + x^2 - 18x = 0$$

The derivatives are the equations of parabolas and can have 0, 1, 2, 3, or 4 distinct pairs of answers. Exercise 13.2 asks for verification of the solutions (6, 6), (6, 0), (0, 6), and (4, 4).

Take second derivatives to check for minimum and maximum: $\partial^2 Z/\partial x^2 = 2y - 12$, $\partial^2 Z/\partial y^2 = 2x - 12$, and $\partial^2 Z/\partial x \partial y = 2y + 2x - 18$, so $\Delta(x, y) = (2y - 12)(2x - 12) - (2y + 2x - 18)^2$:

$\Delta(6, 6) = -36$, so (6, 6) is a saddle point
$\Delta(6, 0) = -36$, so (6, 0) is a saddle point
$\Delta(4, 4) = 12$, $\partial^2 Z/\partial x^2 = -4$, so (4, 4) is a maximum and $f(4, 4) = 8$.

Some extreme value optimization problems can only be solved when restrictions are placed on the variables. Restrictions are necessary when the preliminary partial derivative equations are solved and the only solution is a zero value for all variables, a point without much significance. The following example illustrates this predicament.

EXAMPLE 13.3:

Optimize a two variable function with a restriction on the variables.

Maximize the moment of inertia of a rectangular cross section. The moment of inertia of a rectangle is a function of the height, h, and base, b; $I(b, h) = bh^3/12$. The partial

FIGURE 13.2

Graph of $I = bh^3/12$.

derivatives are $\partial I/\partial b = h^3/12$ and $\partial I/\partial h = 3bh^2/12$. If they are set equal to zero the only solution is (0,0), which results in $I = 0$, a trivial minimum, and no information of a maximum is found. Figure 13.2 is a graph of $I(b, h) = bh^3/12$ with the domains $0 < b < 12$, and $0 < h < 12$.

If a restriction is placed on the variables the triviality of a zero extreme value can be overcome and a useful extreme determined. Continue the example with the restriction added that the beam is to be cut from a 12 inch (0.30 meter) diameter log (see Figure 13.3). This changes the problem to one of maximizing $I = bh^3/12$ subject to the restrictions $b^2 + h^2 = 12^2$, $0 < h < 12$, and $0 < b < 12$.

The problem can be reduced to a one variable problem by solving $b^2 + h^2 = 12^2$ for b, $b = \sqrt{144 - h^2}$, and substituting in $I(b, h)$. Then $I(b, h)$ becomes $I(h) = h^3[\sqrt{144 - h^2}]/12$.

Now take a non-partial first derivative, set it equal to zero, and solve for h.

$dI/dh = \{h^3(-2h)/[(12)(2)\sqrt{144 - h^2}]\} + \{(3h^2)\sqrt{144 - h^2}/12\}$
$dI/dh = [-h^4 + 432h^2 - 3h^4]/[12\sqrt{144 - h^2}]$
$= [-4h^4 + 432h^2]/[12\sqrt{144 - h^2}] = 0$
Multiply by $[12\sqrt{144 - h^2}]$ (permissible as long as $h \neq 12$),
and $-4h^4 + 432h^2 = 0$
$-4h^2(h^2 - 108) = 0$, and $h_1 = h_2 = 0$, $h_3 = -10.39$, and $h_4 = 10.39$.

The values of h_1, h_2, and h_3 are outside the physical reality of the problem, so if a maximum exists, it must be when $h = 10.39$. When $h = 10.39$ inches (263.9 mm), then $b = 6.00$ inches (152.4 mm), and $I(6.00, 10.39) = 560.81$ in^4 ($2.33 \cdot 10^8$ mm^4). Exercise 13.3 asks for a second derivative verification that $h = 10.39$ produces a maximum value for I, but again the graph, Figure 13.4, shows that $I = bh^3/12$ has a maximum under the restriction $b^2 + h^2 = 144$.

FIGURE 13.3

Beam section to cut from a 12 inch diameter.

FIGURE 13.4

Graph of $b^2 + h^2 = 144$ superimposed on $I = bh^3/12$.

Many times conditions exist such that a single answer is not appropriate even if derivative methods are used. This occurs when the range of limiting values is discrete rather than continuous. The following case study further illustrates the use of derivative analysis, and is an example of a situation where the optimum calculus answer may not be able to be implemented because of physical constraints.

CASE STUDY #13:
Tool Life Optimization

Machine tool speeds and feeds are critical in controlling manufactured part tolerances, surface finish, production rates, and part cost. Total cost to manufacture a part includes the purchase cost of perishable tools such as drills, reamers, lathe turning tools, taps, honing stones, broaches, and milling cutters; and the cost of resharpening them when they become worn through use.

Experiments have been run to determine the effect of cutting speed on tool life so economic decisions can be made about machining processes. A good correlation between cutting speed and tool life has been demonstrated using the model $VT^n = k$ (2,486-490), where: (13.1)

V = Cutting speed at the tool-material interface, feet/minute (meters/minute)

T = Tool life in minutes, and

k and n are parameters determined by such factors as:

1. The material being machined
2. The tool material
3. The surface coating of the cutting tool
4. The tool form and shape
5. The size and shape of the cut, and
6. The cutting fluid used.

The model $VT^n = k$ is a straight line on a log-log graph where n is the slope and k is the value of V when $T = 1$ (see Figure 13.5). This makes k the cutting

FIGURE 13.5

Typical tool-life curves, based on cutting speeds.

Tool material	Work material	n
High speed steel	Steel	0.125
	Cast iron	0.14
Cemented carbide	Steel	0.20
	Cast iron	0.25
Sintered oxide	Steel	0.70

Average values of n

$$n = \frac{\log V_1 - \log V_2}{\log T_2 - \log T_1}$$

	m/min	fpm
C_1	14	47
C_2	46	150
C_3	122	400

Sintered oxide $VT^{0.7} = C_3$

Cemented carbide $VT^{0.2} = C_2$

High speed steel $VT^{0.13} = C_1$

From Doyle/Keyser/Leach/Schrader/Singer, *Manufacturing Processes & Materials For Engineers*, 3rd ed., p 487. © 1985. Reprinted by permission of Prentice-Hall, Inc., Englewood Cliffs, New Jersey.

speed for a tool life of one minute. Combinations of values for n and k can be found in various resources and are available for many common cutting conditions. The relationship $VT^n = k$ can be used to determine the optimum cutting conditions for minimum cost using the following analysis.

Consider the cost, C_t, to supply a cutting edge to a machining process:

C_t = tool change cost + tool sharpening cost + tool cost

$C_t = (t_1 R_c) + (t_2 R_s/N_1) + [C/(N_1 N_2)]$, where

t_1 = time to change a tool, minutes

t_2 = time to grind a tool, minutes

R_c = machine operator cutting cost including overhead, $/minute

R_s = tool grinding cost including overhead, $/minute

N_1 = number of cutting edges on the tool

C = original tool cost, $, and

N_2 = number of times a tool can be sharpened, $N_2 \geq 1$. If a tool is purchased sharp and cannot be resharpened, then $N_2 = 1$.

The cost added to a part while it is being machined is $C_m = TR_c$, and the material removed during machining time is $Q = KVTfd$, where

Q = volume of material removed, in^3 (mm^3)

f = width of cut, inches (millimeters)

d = depth of cut, inches (millimeters)

K = 12 in./ft (1 mm/mm), and

T = tool life, minutes.

From Equation 13.1, $V = k/T^n$, so $Q = Kfdk/T^{n-1}$.

The total cost, C_T, of removing material in $/unit^3 is:

$$C_T = \frac{C_t + C_m}{Q} = \frac{C_t + TR_c}{Q} = \frac{C_t T^{n-1} + R_c T^n}{Kfdk} \tag{13.2}$$

Take the first derivative of C_T relative to tool life T, set it equal to zero, and solve for the value of tool life for minimum material removal cost, T_m. (Exercise 13.4 asks for this calculation.) The result is:

$$T_m = (C_t/R_c)(1 - n)/n. \tag{13.3}$$

Substitute this value of T_m into Equation 13.1 to obtain the velocity required for minimum material removal cost:

$$V = k/T_m^n = k/[(C_t/R_c)(1 - n)/n]^n$$

The cutting speed for maximum production rate can also be determined. The production rate, P, is defined as material removed per minute, and is expressed as:

$$P = Q/(T + t_1) \text{ in}^3/\text{minute (mm}^3/\text{minute)} \tag{13.4}$$

Substitute $Kfdk/T^{n-1}$ for Q, and $P = (KfdkT^{1-n})/(T + t_1)$.

Take the first derivative relative to tool life T, set the derivative equal to zero, and solve for the value of tool life that results in maximum production rate, T_M.

(Exercise 13.6 asks for this calculation.) The result is Equation 13.5, which holds as long as $T \neq -t_1$.

$$T_M = t_1(1-n)/n \tag{13.5}$$

Substitute this value of T_M into Equation 13.1 to obtain the velocity required for maximum material removal (maximum production):

$$V = k/T_M^n = k/[t_1(1-n)/n]^n$$

The comparison of T_M with T_m shows that the tool life for maximum production will be less than the tool life for minimum material removal cost if $t_1 < C_t/R_c$.

Note that an implied condition in this analysis is that the supply of cutting tools is unlimited. There is no point in getting high production for six hours and then sitting and waiting for two hours for tools to be resharpened.

Equations 13.3 and 13.5 give different answers for tool life and cutting speed, and in general the practical machine setting is between the two values, because both optimums cannot be satisfied at the same time. The following numeric details illustrate this situation.

$t_1 = 1$ minute	$R_c = \$0.20$/minute
$t_2 = 20$ minutes	$R_s = \$0.25$/minute
$C = \$12.00$	$N_2 = 10$ times
$N_1 = 8$ edges	$VT^{0.2} = 180$

$C_t = (1)(0.20) + (20)(0.25/8) + 12.00/(10)(8) = \$0.98$

$T_m = (0.98/0.20)(1 - 0.2)/0.2 = 20$ minutes

V for minimum tool cost $= 180/20^{0.2} = 99$ feet/minute (30.2 m/minute)

$T_M = 1(1 - 0.2)/0.2 = 4$ minutes

V for maximum production $= 180/4^{0.2} = 136$ feet/minute (41.5 m/minute)

Now what? An optimum range from 99 to 136 feet/minute (30.2 to 41.5 m/minute) has been determined. When the machine selected for the operation is set up, only one machine setting may exist in this range. In that case no special decision is required; use the setting available. If there is more than one setting available in the range, choose the one closest to the preferred results, minimum material removal cost or maximum production rate. If there are no settings available in the range calculated, then the decision must be made if low tool cost or high production is most important. For higher production use high values of cutting speed, for longer tool life use lower values. Some modern machines are designed with speeds and feeds infinitely variable over a fixed range so that settings at a calculated value are possible. This setting flexibility is made possible by using electronic control, programmable controllers, or direct computer control. Keep in mind that this analysis requires several pieces of data, the most elusive being the values of n and k to use in Equation 13.1.

The results of the maximum tool life/maximum production rate analysis may not result in the tolerances or surface finish specified on the part drawing, or the machine may run out of power when combined with the recommended feed rates. Compromise (there's that word again) is usually necessary. This analysis also does

13.1.3 Lagrange Multipliers

The use of Lagrange multipliers is an alternate calculus method for determining an extreme value of a function with *any* number of variables when the function is subjected to equality constraints. The object is to find an extreme of a function $F(x_i)$, $i = 1, 2,\ldots, n$, subject to restrictions $R_j(x_i)$, $j = 1, 2,\ldots, m$. The procedure is as follows.

Define a new function that is a linear combination of the given function and the restrictions, $F(x_i, \lambda_j) = F(x_i) - \Sigma \lambda_j R_j(x_i)$, where the λ_j are unknown Lagrange multipliers. Take partial derivatives $\partial F/\partial x_i$ and $\partial F/\partial \lambda_j$ and set each of them equal to zero. If the resulting $(m + n)$ simultaneous equations are independent, they will yield values of the variables for extreme function evaluation. If only one extreme is located, then deciding if it is a maximum or minimum is based on intuition, or the geometry, or physical restrictions of the problem. A single extreme, calculated purely mathematically, has no interpretation as a maximum or minimum. If several extreme values are found, then direct comparison of function values will determine if the extremes are maximum or minimum. The extreme values found using Lagrange multipliers are not likely to be the extreme of the function without the constraints. The next example illustrates the technique of Lagrange multipliers using the same situation as Example 13.3.

EXAMPLE 13.4:

Use of the Lagrange multiplier method.

Find the extreme value of $I(b, h) = bh^3/12$ subject to the constraint $R_1 = 12^2 - b^2 - h^2$. The new function is $F(b, h, \lambda_1) = bh^3/12 - \lambda_1(12^2 - b^2 - h^2)$. The three derivatives are:

$\partial F/\partial b = h^3/12 + 2b\lambda_1 = 0$ (1)
$\partial F/\partial h = bh^2/4 + 2h\lambda_1 = 0$ (2)
$\partial F/\partial \lambda_1 = -12^2 + b^2 + h^2 = 0$ (3)

Subtract b times (2) from h times (1) and

$h^4/12 - b^2h^2/4 = 0$ (4)

Solve (3) for b^2 and substitute in (4), eliminate fractions, combine terms and

$4h^4 - 432h^2 = 0$

Factor, set each factor equal to zero, and $h_1 = h_2 = 0$, $h_3 = -10.39$, and $h_4 = 10.39$. Note that the value of λ_1 did not have to be determined in this case.

The method of Lagrange multipliers can be extended to problems with any number of constraints. The next example illustrates how additional restrictions are included in the procedure.

EXAMPLE 13.5:

Use of the Lagrange multiplier method with more than one constraint.

Find any extreme values of $G(x, y, z) = -2x^2 - y^2 - 3z^2$ subject to the constraints $R_1 = x + 2y + z - 1 = 0$, and $R_2 = 4x + 3y + 2z - 2 = 0$. Note that without restrictions $G(x, y, z)_{max} = 0$ when $x = y = z = 0$, and that $F(x, y, z)_{min}$ does not exist because there is no limit to how small $G(x, y, z)$ can be, as each term of the function is always negative.

Establish $F(x, y, z, \lambda_1, \lambda_2) = -2x^2 - y^2 - 3z^2 - \lambda_1(x + 2y + z - 1) - \lambda_2 (4x + 3y + 2z - 2)$ and take partial derivatives:

$\partial F/\partial x = -4x - \lambda_1 - 4\lambda_2 = 0$
$\partial F/\partial y = -2y - 2\lambda_1 - 3\lambda_2 = 0$
$\partial F/\partial z = -6z - \lambda_1 - 2\lambda_2 = 0$
$\partial F/\partial \lambda_1 = -(x + 2y + z - 1) = 0$
$\partial F/\partial \lambda_2 = -(4x + 3y + 2z - 2) = 0$

Solve simultaneously, Exercise 13.8, and $x = 5/27$, $y = 10/27$, $z = 2/27$, $\lambda_1 = -4/27$, and $\lambda_2 = -4/27$, resulting in the extreme value of $G(5/27, 10/27, 2/27) = -2/9$. Depending on the application to a physical problem, the value $-2/9$ could be a maximum or minimum.

An engineer must never assume that mathematical or computer optimization techniques give answers that can be implemented, as Case Study #13, Tool Life Optimization, showed. Numerical solutions that come from mathematical analysis may not be realistic in ways other than the lack of machine settings. Mathematical solutions of 2.3 people, 4.1 lanes of traffic, or 7.1 floors of a building are also not meaningful. The following case study illustrates a situation where the mathematics is straightforward enough, but the mathematical answer has several physical implementation problems.

CASE STUDY #14:
Pipe Packing Optimization

In the design of fluid flow heat exchangers, the length of pipe in the exchanger influences the rate of heat exchange. The cost of the pipe is a governing factor in the optimization process, as well as the cost of the tank shell within which the pipes are located and the temperature-altering fluid flow. The shell can be fabricated in a variety of shapes: a cylinder, which encloses the most area (most pipes) for a given perimeter; a rectangle, which may occupy less floor space than a cylinder, or some other shape that is easy to fabricate. The manufacture of a cylindrical tank, for example, requires a roll machine to form the metal into shape, and the corners of a rectangular tank require hydraulic or mechanical forming machines.

Consider the cost of a cylindrical heat exchanger tank of diameter D in feet (meters) and length L in feet/pipe-length (meters/pipe-length), that requires a minimum of 300 feet (91.5 m) of pipe to meet heat exchange requirements. The requirement of 300 feet (91.5 m) is itself an optimization process dependent upon heat flow rates, fluid volume flow rates, material heat capacities, incoming fluid temperature, required outgoing temperature, and cost and availability of pipe material. Assume that the cost of the tank shell has been estimated as the sum of the pipe cost of $700, the shell cost (based on other cylindrical tanks that have been built) of $25D^{2.5}L$, and the floor space cost of $20DL$; $C = 700 + 25D^{2.5}L + 20DL$. Additional data given with the tank design is that 20 pipes will fit in 1 ft² (0.093 m²) of tank cross section. The 20 pipes/ft² translates into the following analytic restriction of the problem:

$$[\pi D^2/4][(L)(20 \text{ pipes/ft}^2)] \geq 300 \text{ ft}([\pi D^2/4][(L)(20 \text{ pipes}/0.093 \text{ m}^2)]$$
$$\geq 91.5 \text{ m}), \text{ or } \pi D^2 L \geq 60 \text{ ft } (\pi D^2 L \geq 1.70 \text{ m})$$

A minimum cost of $1538 is obtained using the information given, and occurs when $D = 1.37$ ft (0.418 m) and $L = 10.18$ ft/pipe-length (3.104 m/pipe-length) (see Exercises 13.15 and 13.16).

Now analyze the results and determine if the calculations result in a realistic solution. If the length of each pipe is 10.18 ft (3.104 m), and 300 ft (91.5 m) are required, then 300 ft/10.18 ft/pipe = 29.47 pipes are required. That is impossible. Either 29 or 30 pipes must be used. Twenty-nine pipes require that $L = 300/29 = 10.34$ ft (3.152 m), $D = \sqrt{[60/(\pi)(10.34)]} = 1.36$ ft (0.415 m), and $C = \$1539$. Thirty pipes require that $L = 300/30 = 10.00$ ft (3.049 m), $D = \sqrt{60/(\pi)(10)} = 1.38$ ft (0.421 m) and $C = \$1535$. The three costs are approximately the same and reflect a system that is insensitive to small variations in the value of the variables. Keep in mind that no variable can be consistently controlled to an exact value, and the small variation range ($1535–$1538) is likely within the accuracies of calculation and tolerances of the rest of the system.

The commercial availability of the length of pipe is also a factor affecting the design. If material comes in 20.00 ft (6.097 m) lengths, then $L = 10$ ft (3.049 m) is a good choice. If pipe comes in 24 ft (7.317 m) lengths, then additional analysis is required to determine the overall effect if $L = 12.00$ ft (3.659 m). The length of each pipe also affects the design of the tank shell, for its length is the same as the pipe lengths. If the material for the shell comes 10 ft (3.049 m) wide, then again $L = 10$ ft (3.049 m) is a good choice. The widest material the roll machine can handle, perhaps only 8 ft (2.44 m), may also be the limiting factor. But there is a larger difficulty with the heat exchanger analysis than available material and machines. Did you recognize it?

The larger problem is the assumption that the ratio of the number of pipes per square unit is constant regardless of the area. That assumption requires closer analysis. Consider enclosing a number of pipes of diameter two units, in a cylindrical shell, sample cross sections of which are shown in Figure 13.6. If one pipe is enclosed in a cylindrical shell, the area of the cross section of the shell is π unit², and the pipe density is 1 pipe/π unit² = 0.32 pipes/unit². The density of 0.32 pipes/unit² is the largest pipe density possible because there is no wasted space with one pipe. If two pipes are enclosed in a cylindrical shell the shell cross section area is 4π units², and the pipe density is 2 pipes/4π units² = 0.16 pipes/unit², half as efficient as using one pipe. If three pipes are enclosed in a cylindrical shell the shell cross section area is $\{[(2\sqrt{3})/3] + 1\}^2 \pi = 4.64\pi$ units², and the pipe density = 0.21 pipes/units². This analysis can be continued, and in general as the number of pipes

FIGURE 13.6

Typical pipe shell cross sections.

One pipe shell

Two pipe shell

Three pipe shell

increases, the pipe density increases, but never exceeds 0.32. There are, however, some configurations that are more efficient than others on a local level. It is left as an exercise, 13.17, to show that the number of pipes, n, enclosed in a circular cross section is most efficient on a local level when $n = 1, 7, 13, 19, 31, 37,\ldots$. If this is so, then the number of pipes in the shell under study should be 31 and $L = 300/31 = 9.68$ ft (2.95 m). Perhaps the optimization compromise is to use thirty-one 10 ft (3.05 m) pipes and increase the margin of safety, by using 310 ft (94.51 m) of heat transfer pipe rather than 300 ft (91.5 m).

Optimization techniques do not always result in a single answer because of the number of variables connected with the problem and uncertainty about their values. In cases when uncertainty exists, optimization decisions are often made that reflect a combination of variable values. The following case study illustrates such a situation.

CASE STUDY #15:
Cost Ratio Analysis

Consider the cost of a cylindrical food container can of height h, diameter D, volume $V = \pi D^2 h/4$, and $t = $ the material thickness. Then the volume of the material in the can $= \pi D h t + 2(\pi D^2 t/4)$, and the seam joining length $= 2\pi D + h$. Assume that a rectangular blank is rolled into a cylinder and two circular blanks are used for the ends.

Let $C_1 = $ the cost of the material in \$/unit³, $C_2 = $ the cost of seaming in \$/unit, and the total cost (material and seaming) $= C_1 t(\pi D h + \pi D^2/2) + C_2(2\pi D + h)$. The cost of rolling the cylinder and blanking the ends from sheet metal are considered constant for a wide range of dimensions and are not included in the relationship. Some may prefer to use the volume of material as $t(\pi D h + 2D^2)$ because the circular ends are probably cut from square blanks. In some manufacturing companies, however, the cost of material includes a credit for the value of scrap sold for recycling. Again a choice must be made (see Exercise 13.24).

Any of the six variables, C_1, C_2, D, h, t, or V, could be considered the main variable, but D and h are probably the two that have the most flexibility. The volume will be governed by the serving size determined by marketing; the material thickness will be determined by required strength and material availability; and the costs C_1 and C_2 will also be governed by market conditions. To reduce the total cost equation to a more manageable relationship, determine the value of h in terms of V and D, substitute in the cost equation, and limit the analysis to D: $h = 4V/\pi D^2$, and $C_t = C_1 t(4V/D + \pi D^2/2) + C_2(2\pi D + 4V/\pi D^2)$.

Take the first derivative relative to D (all other variables are assumed constant) and set the derivative equal to zero in an attempt to find a minimum cost:

$$dC_t/dD = C_1(-4Vt/D^2 + \pi Dt) + C_2(2\pi - 8V/\pi D^3) = 0$$

To reduce the number of parameters let $k = C_2/C_1$, so

$$-4Vt/D^2 + \pi Dt + k(2\pi - 8V/\pi D^3) = 0$$

To eliminate fractions multiply by πD^3, and

$$\pi^2 D^4 t + 2\pi^2 D^3 k - 4\pi VtD - 8Vk = 0 \tag{13.6}$$

This expression can be analyzed for minimum cost in different ways.

1. Determine D based on an assumed volume, material thickness, and values for C_2 and C_1.
2. Consider a range of parameter values that may be desirable and determine their effect on each other. This permits the consideration of variable sensitivity, and may suggest alternatives that have not been considered.

For example, the diameter may be restricted to $2.00 \leq D \leq 4.00$ inches (0.051 $\leq D \leq 0.102$ m) by existing manufacturing capability, and thickness is governed by normally available materials: 30 gage [0.0120 inches (0.3048 mm)], 28 gage [0.0149 inches (0.3785 mm)], 26 gage [0.0179 inches (0.4567 mm)], 24 gage [0.0239 inches (0.6071 mm)], 22 gage [0.0299 inches (0.7595 mm)], and 20 gage [0.0359 inches (0.9119 mm)].

Equation 13.6 has one sign change, so there is one positive real root regardless of the positive values given to t, V, and k. If a value for volume is assumed, then D can be studied as a function of k and t. Figure 13.7 shows the result of letting $V = 20$ in^3 (327.7 cm^3), $t =$ the six gage thickness values listed above, and solving for D. The corresponding value of h is calculated using $h = 4V/\pi D^2$. Figure 13.7 can be used in different ways.

1. Select a value for t, a value for k, and determine the corresponding D and calculate h for minimum cost. Example: If $t = 0.030$ in. and $k = 0.10$, then $D = 2.3$ in. and $h = 4.8$ in.
2. Select a value of t and D and determine the necessary cost ratio for minimum cost. Example: If $t = 0.012$ and $D = 2.15$, then k is approximately equal to 0.11.

Figure 13.7 also shows that as the value of k increases there is less sensitivity in the associated values of D and t. Low sensitivity in a manufacturing operation is good; as k gets larger, however, so does the cost. Total cost can be qualified by changing the expression for C_t to $C_t/C_1 = 4Vt/D + \pi D^2 t/2 + k(2\pi D + 4V/\pi D^2)$. Figure 13.8 illustrates the variance of C_t using the same values of t as Figure 13.7. Compromise is again required, low cost and high sensitivity, high cost and low sensitivity, or someplace in between.

FIGURE 13.7

Analysis summary, Case Study #15.

FIGURE 13.8

Cost ratio summary, Case Study #15.

The derivative method of locating extreme values of functions is an easy to use straightforward procedure that can be generalized and extended to problems of more than two variables. The test is to determine the values of the x_i that make $\partial F/\partial x_i = 0$ or nonexistent, for $i = 1, 2, 3,\ldots, n$. The second derivative test to check for maximum, minimum, or saddle point, however, does not generalize to multiple dimensions, and physical interpretations are usually used for making decisions about extremes. This can be done by evaluating the function in the neighborhood of the extreme value in question. Mathematical analysis of second derivatives of functions of more than two variables is possible, but is beyond the scope of this text.

13.2 Non-Calculus Techniques

Non-calculus techniques are required for optimization analysis when an analytical function is difficult or impossible to differentiate, or if an analytic function is not available. When analytic functions are not available, analysis and optimization must be done experimentally. The non-calculus techniques are often called search techniques because the methods search for extreme values using systematic evaluation. Search techniques are valid for analytical or experimental evaluation. The plan for search must be determined prior to performing calculations or experiments so time and resources are used efficiently. This is particularly important when calculations or experiments take a long time and/or are costly in other ways. Four search methods will be discussed here, the equal division search, the Fibonacci search, the golden section search, and the slope search.

13.2.1 Equal Division Search

The most elementary search technique is performed by dividing the domain of the variable(s) into some number of equal segments and evaluating the system at each of the division values. The technique does not place a high premium on finding *the* best, only the best at the ends of the interval selected. The equal division method can be used to fine tune a system beyond the first stage optimization if the domain is narrowed to the segment most likely to contain the desired extreme (minimum or maximum), and additional subdivision and calculation is performed. If the cost of each evaluation is large, as it is when performing a complex experiment, then the number of subdivisions will be small because of cost considerations. If the cost and time to implement each evaluation is small, then a large number of evaluations can be justified, such as the case when computer evaluation is performed.

One use of the equal division search technique is to find the zeros of a function. A function, $f(x)$, is evaluated over a closed interval $[a, b]$, or an open interval (a, b), using a fixed increment Δx, and whenever the values of $f(x_i)$ and $f(x_i + \Delta x)$ have opposite signs, at least one root of the equation is isolated. Further refinement of the root is possible by using an increment smaller than Δx within the interval $[x_i, x_i + \Delta x]$. An exact root of the equation (exact to the accuracy of the calculation) will be found if at some evaluation $f(x_i + \Delta x) = 0$. The equal division search for roots is usually done on a computer because calculations are done rapidly, and good accuracy is possible.

The equal division method can also be used for finding function extremes within an interval. There are two approaches to consider. The first approach determines only the global extreme value within an interval. Evaluation is performed across an interval $[a, b]$ using an increment Δx. The search process is started by letting the location and the value of the local extreme of the function equal $(a, f(a))$. The value at each interval of the domain $f(a + n\Delta x)$, where n = the number of segments used over the domain, is compared to the previous local extreme value. If any $f(a + n\Delta x)$ is larger (or smaller) than the previous extreme, then the new value becomes the local extreme. The local extreme after the interval $[a, b]$ has been searched becomes the global extreme for the interval. If the interval is open, (a, b), then the first local limit is $(a + \Delta x, f(a + \Delta x))$.

A second approach locates *all* local extreme values on an interval, but it requires that the function be differentiated, and the derivative must exist in the interval. The derivative of the function, $f'(x)$, is evaluated over an interval $[a, b]$ using a fixed increment Δx. Whenever the value of $f'(x_i)$ and $f'(x_i + \Delta x)$ have opposite signs, an extreme of the function is isolated. If the value of $f'(x_i) > 0$, then when $f'(x_i)$ and $f'(x_i + \Delta x)$ have opposite signs a local maximum is found. If the value of $f'(x_i) < 0$, then when $f'(x_i)$ and $f'(x_i + \Delta x)$ have opposite signs a local minimum is found. Save the value of $[(x_i + x_i + \Delta x)/2, f(x_i + x_i + \Delta x)/2)]$ as a local extreme each time the signs of $f'(x_i)$ and $f'(x_i + \Delta x)$ are different. A local extreme will also be located if $f'(x_i) = 0$ and $f'(x_i - \Delta x)$ and $f'(x_i + \Delta x)$ have the opposite sign. The global extreme will be the largest (or smallest) of the individual values saved.

The conditions of Example 13.3 will be used to illustrate the equal division search so comparison of the methods can be made.

EXAMPLE 13.6:

The equal division search used on Example 13.3.

Find the maximum value of $I = bh^3/12$ subject to the restrictions $b^2 + h^2 = 12^2$, $0 < h < 12$, and $0 < b < 12$. Divide the range $0 < h < 12$ into 7 equal segments and make evaluations at 6 points. The segments are: (0, 1.71], (1.71, 3.43], (3.43, 5.14], (5.14, 6.86], (6.86, 8.57], (8.57, 10.29], (10.29, 12.00). Table 13.1 summarizes the calculations, and Figure 13.9 illustrates the equal division search in graph form. The maximum value of I is 560.21, only slightly less than the derivative value of 560.81.

TABLE 13.1

Summary of equal division search for I_{max}.

Evaluation #	h	b	I	I_{max}
1	1.71	11.88	4.95	4.95
2	3.43	11.50	38.67	38.67
3	5.14	10.84	122.67	122.67
4	6.86	9.85	264.99	264.99
5	8.57	8.40	440.60	440.60
6	10.29	6.17	560.21	560.21

FIGURE 13.9

I_{max} for the equal division search method. Example 13.6.

13.2.2 Fibonacci Search

The steps of the Fibonacci search method are similar to those of the equal division search except the segments of the search domain are not equal. The segment length is a function of the number of intervals and the ratio of numbers in the Fibonacci series, F_n, and gets smaller at each successive calculation. The Fibonacci series is an infinite sequence of integers that starts with $F_0 = 1, F_1 = 1$ and continues with each successive term equal to the sum of the preceding two terms; $F_2 = 2, F_3 = 3, F_4 = 5, F_5 = 8, \ldots$.

The Fibonacci search is a procedure that discards domain segments that are least likely to contain an extreme value. For computer Fibonacci searches, the number of evaluations can be high with little change in cost. For physical experiments, however, the number of evaluations may be small because of cost or time constraints. To use the Fibonacci search method on an interval $[a, b]$ (or (a, b)), the number of evaluations to make must be specified before the search is started.

The subdivision segment determination, and the evaluation procedure for an open, (a, b), or closed interval, $[a, b]$, of length $(b - a)$ is performed as follows. Specify n, the number of evaluations to make. The first two evaluations are made at a distance (F_{n-2}/F_n) $(b - a)$ from each end of the interval; $x_1 = a + (F_{n-2}/F_n)(b - a)$ and $x_2 = b - (F_{n-2}/F_n)$

$(b - a)$. At each step of the search process the domain is divided into three parts (left, center, and right) and the left or right segment is discarded.

If a maximum function value is being sought and $f(x_1) > f(x_2)$, discard the region of the domain $[x_2, b]$, redefine the search domain $[a, b] = [a, x_2]$, and make the next evaluation at $x_3 = a + (F_{n-3}/F_{n-1})(b - a)$. If a minimum function value is being sought and $f(x_1) > f(x_2)$, discard the region of the domain $[a, x_1]$, redefine the search domain $[a, b] = [x_1, b]$, and make the next evaluation at $x_3 = b - (F_{n-3}/F_{n-1})(b - a)$. Continue to eliminate the right or left segment of the shortened interval, redefine the interval length $[a, b]$, and make another calculation at a distance $(F_{i-2}/F_i)(b - a)$ from the end of the interval that was not changed. When $(n - 1)$ calculations have been made, select a domain value slightly greater or slightly less than the domain value that is presently giving the extreme function value. Evaluate the function at this selected domain value to determine on which side of the present extreme the global extreme lies. This identifies the interval where the global extreme is likely to occur. The next example uses the Fibonacci search technique on the problem from Example 13.3.

EXAMPLE 13.7:

The Fibonacci search used on Example 13.3.

Find the maximum value of $I = bh^3/12$ subject to the restrictions $b^2 + h^2 = 12^2$, $0 < h < 12$, and $0 < b < 12$. Set the number of evaluations to six to match the number used in the equal division search. The first two evaluations performed on the interval $(0, 12)$ are performed at $h_1 = 0 + (5/13)(12) = 4.62$ and $h_2 = 12 - (5/13)(12) = 7.38$ (see Figure 13.10).

For $h_1 = 4.62$, $b_1 = 11.07$, and $I_1 = 90.97$.

For $h_2 = 7.38$, $b_2 = 9.46$, and $I_2 = 316.87$.

The second interval is $(4.62, 12)$; $h_3 = 12 - (3/8)(12 - 4.62) = 9.23$, $b_3 = 7.67$, and $I_3 = 502.60$.
The third interval is $(7.38, 12)$; $h_4 = 12 - (2/5)(12 - 7.38) = 10.15$, $b_4 = 6.40$, and $I_4 = 557.70$.
The fourth interval is $(9.23, 12)$; $h_5 = 12 - (1/3)(12 - 9.23) = 11.08$, $b_5 = 4.61$, and $I_5 = 522.56$.
The sixth evaluation is done on one side of 11.08. Choose $I_6 < 11.08$ or $I_6 > 11.08$. If $h_6 = 11.00$, $b_6 = 4.79$, and $I_6 = 531.29$. This indicates that h for maximum I lies in the interval $(10.15, 11.08)$. Notice that this method did not produce a maximum as large as the equal division search. In general it is not predictable which method will result in the value closest to the global extreme; however, the Fibonacci search will narrow to the segment that contains the solution faster than the equal segment search unless the extreme value lies on an equal segment boundary.

13.2.3 Golden Section Search

The golden section search technique is similar to the equal division and the Fibonacci search methods, with one important difference, it is not necessary to determine the number of divisions and calculations to use before starting the search. The method can

FIGURE 13.10

I_{max} for the Fibonacci search method, Example 13.7.

be continued until either the size of the division reaches a preset value, or the difference in the value of the function reaches a preset value. This is particularly advantageous when using the computer because a search loop can be continued until a predetermined level of accuracy is reached. The interval divisions are calculated using the golden section ratio value 0.618. The value of 0.618 is the ratio of the width of a rectangle to its length which is considered most pleasing to geometricians and architects. The rectangle ratio of 0.618 is equivalent to dividing a line into a golden section. This is done by starting with a line AB and locating a point C on it such that $AC/AB = BC/AC$ (see Exercise 13.28).

The golden section search for the extreme of a function, $f(x)$, on an interval $[a, b]$ (or (a, b)) begins in a manner similar to the Fibonacci search. Two evaluations are performed, one at $x_1 = a + (0.618)(b - a)$ and one at $x_2 = b - (0.618)(b - a)$. The left or right interval is discarded as was done in the Fibonacci search and a new interval established, either (a, x_2) or (x_1, b). At this point the second difference between the Fibonacci and the golden section searches becomes important. In the Fibonacci search the new evaluation was performed at a calculated distance measured from the end of the interval that did not change. In the golden section search the next evaluation is performed at a distance from the end of the interval that did change; so the next evaluation is performed at either $[x_2 - (0.618)(x_2 - a)]$ or $[x_1 + (0.618)(b - x_1)]$. The internal increment selection-evaluation-discard-new interval iteration is continued until one of three preset conditions is reached:

1. Until a fixed number of evaluations are performed. This is still a real restriction if the cost of each evaluation is high.
2. Until the interval reaches a predetermined small value.
3. Until the difference in function value, $f(x_{n+1}) - f(x_n)$, reaches a predetermined value.

The next example uses the golden section search on the conditions from Example 13.3 so the methods can be compared.

EXAMPLE 13.8:

Golden section search used on Example 13.3.

Find the maximum value of $I = bh^3/12$ subject to the restrictions $b^2 + h^2 = 12^2$, $0 < h < 12$, and $0 < b < 12$. On the interval $(0, 12)$ the first two calculations are performed at $h_1 = 12 - (0.618)(12) = 4.58$, and $h_2 = 0 + (0.618)(12) = 7.42$ (see Figure 13.11).

For $h_1 = 4.58$, $b_1 = 11.09$, and $I_1 = 88.79$.
For $h_2 = 7.42$, $b_2 = 9.43$, and $I_2 = 321.03$.

The second interval is $(4.58, 12)$:

$h_3 = 4.58 + (0.618)(12 - 4.58) = 9.17$, $b_3 = 7.74$, and $I_3 = 497.36$.

The third interval is $(7.42, 12)$:

$h_4 = 7.42 + (0.618)(12 - 7.42) = 10.25$, $b_4 = 6.24$, and $I_4 = 559.98$.

The fourth interval is $(9.17, 12)$:

$h_5 = 9.17 + (0.618)(12 - 9.17) = 10.92$, $b_5 = 4.98$, and $I_5 = 540.40$.

The fifth interval is $(9.17, 10.92)$:

$h_6 = 10.92 - (0.618)(10.92 - 9.17) = 9.84$, $b_6 = 6.87$, and $I_6 = 545.46$.

The maximum value of I found with the golden section search, using the same number of evaluations, is nearly equal to the maximum found using the equal division search. The optimum values determined in general depend on the nature of the function and the number of calculations performed.

Using the equal division, Fibonacci, or golden section search techniques is no guarantee that the global extreme within the interval will be located. There is also no reason to believe that any one of the methods is more likely than the others to produce the highest maximum, or lowest minimum. Any number of things can happen to a function within a sub-interval, and each method uses a slightly different sub-interval length. High

FIGURE 13.11

I_{max} for the golden section search method, Example 13.8.

slope values cause rapid changes in function values, and a spike condition may occur within an interval and be missed if the interval is too large.

A second problem arises with the Fibonacci and golden section searches if the evaluation at any two consecutive calculations results in the same function value. If this happens it is not obvious which sub-interval to discard. The usual solution to this problem is to make an evaluation assuming the left segment is discarded, then make an evaluation assuming the right segment is discarded, and select the best of the two choices. If neither calculation improves the extreme value already located, then the midpoint of the interval indicating equal extreme values can be used in a final attempt at improving the extreme value, or the search can be discontinued. An alternative solution to the problem is to reduce the size of the segments and use a larger number of calculations. For function evaluation this solution is satisfactory, particularly if a computer is available; but for experimentation it is not usually practical to increase the number of experiments arbitrarily.

If a function is being evaluated, and a computer is available, then the increment used on the independent variable(s) can be small enough to avoid the problem of missing narrow spikes (or valleys). A second advantage of using the computer is that all local extreme values can be found if a sequential equal division search method is used.

13.2.4 Slope Search

The slope search method is similar to the equal division search method but it permits beginning the search at any value within the interval under study. The equal division search started evaluation at the low end of an interval and progressed incrementally toward the upper end. By selecting a starting point within the interval, at the midpoint for example, it is likely that a local extreme will be located faster. This increase in speed and reduction in the number of evaluations is particularly important if each evaluation requires an experiment rather than analytical evaluation. The speed of analytical evaluation, by hand or by computer, can also be increased by the slope search, but only local extreme values will be located.

Start the slope search method by selecting the midpoint of the interval, $x_1 = (b - a)/2$, or a point where it is believed the extreme to be located, and a search increment, Δx. The function is evaluated at three points: the base point x_1, $x_1 + \Delta x$, and $x_1 - \Delta x$. If a maximum is being sought and $f(x_1 + \Delta x) > f(x_1 - \Delta x)$ the second base point $= x_2 = x_1 + \Delta x$; if $f(x_1 - \Delta x) > f(x_1 + \Delta x)$, then $x_2 = x_1 - \Delta x$. If the search is for a minimum, then the direction of the inequalities is reversed. Once a direction of function increase is determined, then advancement toward a maximum is continued for a fixed number of evaluations, or until $f(x_{i+1}) < f(x_i)$ at which time $(x_i, f(x_i))$ is a local maximum. In a similar manner, advancement toward a minimum is continued for a fixed number of evaluations, or until $f(x_{i+1}) > f(x_i)$ at which time $(x_i, f(x_i))$ is a local minimum. If the function is analytic, and a computer is available, then Δx can be small and the answer can be made quite accurate. If the problem is not analytic and experimentation is required, then the economics of the size of Δx will limit the accuracy. The method does not give all extreme values if the search is stopped at the first maximum or minimum located. The slope search method is illustrated in the next example using the conditions from Example 13.3 so comparison with the other methods can be made.

EXAMPLE 13.9:

Slope search method used on Example 13.3.

Find the maximum value of $I = bh^3/12$ subject to the restrictions $b^2 + h^2 = 12^2$, $0 < h < 12$, and $0 < b < 12$. To be consistent with the other methods, six evaluations will be used. If the midpoint of the interval is selected as the starting value, then $h_1 = (12 - 0)/2 = 6$, and by letting $\Delta h = 1.2$, then five evaluations will divide the half interval equally. One evaluation is necessary in the other half of the interval to determine in which direction to increment. All six evaluations will not be required if the maximum is found early in the search.

FIGURE 13.12

I_{max} for the slope search method, Example 13.9.

Start evaluation at the base point $h_1 = 6$, then $b_1 = 10.39$, $I_1 = 187.02$. The next two calculations are done on either side of the base point (see Figure 13.12).

$h_2 = h_1 + \Delta h = 7.20$, $b_2 = 9.60$, $I_2 = 298.60$, and

$h_3 = h_1 - \Delta h = 4.80$, $b_3 = 11.00$, $I_3 = 101.38$

The new base point becomes $h = 7.20$ because it resulted in the largest output, and the next evaluation is made at an increment larger than the base point because that is the direction of increasing output:

$h_4 = 7.20 + 1.20 = 8.40$, $b_4 = 8.57$, and $I_4 = 423.29$.

Continue increasing h in that direction because $423.29 > 298.60$:

$h_5 = 8.40 + 1.20 = 9.60$, $b_5 = 7.20$, and $I_5 = 530.84$.

Continue increasing h in that direction because $530.84 > 423.29$:

$h_6 = 9.60 + 1.20 = 10.80$, $b_6 = 5.23$, and $I_6 = 549.02$.

13.2.5 Method Comparison

Two factors are important when comparing the efficiency of search methods.

1. The speed of the method to reach a given limit of accuracy, evaluated by noting the interval size that likely contains the desired extreme;
2. The extreme value obtained.

The final interval of the search methods, and the numerical value obtained, taken from Examples 13.3, 13.6, 13.7, 13.8, and 13.9 are summarized in Table 13.2.

The Fibonacci search ends with the smallest interval when the methods all use the same number of evaluations. This is always true; the Fibonacci search converges to an extreme value faster than the other search methods. There is no reason to believe, however, that there is a correlation between the final interval size and the maximum (or minimum) value obtained. Any of the search methods may result in the most extreme value. In this example the smallest final interval size produced the third largest extreme.

13.2.6 Multi-dimensional Search

All of the search methods can be extended and used on multi-variable problems, but the number of evaluations increases rapidly. If six evaluations per variable are being considered then an n variable problem requires 6^n evaluations. This is not much of a problem for analytic function evaluation, particularly if a computer is used, but for experimental evaluation it can be an important factor as it increases the time and cost of data collection. The following example illustrates the equal division search on a multi-variable problem.

EXAMPLE 13.10:
Multi-variable, equal division search.

Consider the total heating cost per year for a special purpose, 1000 cubic unit rectangular storage cubicle, which is estimated at $4/year per square unit of ceiling area, $3/year per square unit for the east, north, and west walls, and $2/year per

TABLE 13.2

Comparison of optimization methods.

Method	Example #	Final interval size	I_{max}
First derivative	13.3	0	560.81
Fibonacci search	13.7	11.08 − 10.15 = 0.93	557.20
Equal division search	13.6	12.00 − 10.29 = 1.71	560.21
Golden section search	13.8	10.92 − 9.17 = 1.75	559.98
Slope search	13.9	12.00 − 9.60 = 2.40	549.02

TABLE 13.3

Cost in dollars for incremental values of y and z for Example 13.10.

		y				
		3	6	9	12	15
	3	3036	2275	2051	1966	1937
	6	2441	1716	1546	1515	1540
z	9	2273	1602	**1486**	1509	1588
	12	2216	1599	1537	1614	1747
	15	2203	1640	1632	1763	1950

square unit for the south wall. Find the dimensions of the cubicle for minimum heating cost, and determine the annual heating cost.

Let x = the south and north wall length dimension, y = the east and west wall length dimension, and z = the height of the building. Then the annual heating cost is $C = 4xy + 3[2(zy) + zx] + 2(zx) = 4xy + 6zy + 5zx$, and $xyz = 1000$ is the volume restriction.

Solve $1000 = xyz$ for any one of the variables, substitute in the cost equation, determine a reasonable domain for the other two variables, select a sub-interval for each domain, substitute sequentially in the cost equation, and select the minimum value calculated.

Solve $xyz = 1000$ for x, $x = 1000/yz$, substitute for x in $C = 4xy + 6zy + 5zx$, and $C = 4000/z + 6zy + 5000/y$. The domain for y and z, and the increment to use are arbitrary, but it seems logical to start with small values and work up rather than trying to determine an upper limit and work down; let $\Delta y = 3$ and $\Delta z = 3$. Table 13.3 summarizes the search calculations, and Figure 13.13 shows the graph of the data.

The results of the search for minimum cost in the intervals $0 < y < 18$ and $0 < z < 18$ are $y = 9$, $z = 9$, and cost = \$1486. Substitute $y = 9$, and $z = 9$ into $x = 1000/yz$, and $x = 12.3$. If it is desirable to find a more accurate solution, then the intervals $6 < y < 12$ and $6 < z < 12$ can be divided into smaller increments and calculations continued, for example let $\Delta y = \Delta z = 1$.

The Fibonacci search and the golden section search are applied to multi-variable functions in a manner similar to the equal division search. Each variable is assigned values based on the interval division criteria of the method selected, and the function is evaluated at each variable combination.

The slope search method can also be extended to n dimensions, but the number of slopes to calculate increases to at least 2^n at each point selected. In a two-variable problem ($n = 2$) the four incremental slopes from a base point (x_1, y_1) are calculated to $(x_1, y_1 + \Delta y)$, $(x_1, y_1 - \Delta y)$, $(x_1 + \Delta x, y_1)$, and $(x_1 - \Delta x, y_1)$. If diagonal slopes are considered, then the number of calculations is 2^{n+1}, and additional slopes are calculated from (x_1, y_1) to $(x_1 + \Delta x, y_1 + \Delta y)$, $(x_1 + \Delta x, y_1 - \Delta y)$, $(x_1 - \Delta x, y_1 + \Delta y)$, and

FIGURE 13.13

Graph of multi-variable search, Example 13.10.

$(x_1 - \Delta x, y_1 - \Delta y)$. After the incremental evaluations are made from the first base point, the base point moves to the location where the function has the largest value if a maximum is being sought, or the smallest value if a minimum is being sought. The procedure is continued until the four calculations from the current base point all result in smaller (or larger) values of the function.

If the slope search method is used for experimental situations, then often the base point is moved more than one increment in the direction of largest (or smallest) output. The hope is that fewer experiments can be run, and that the global extreme will not be skipped over in the process.

13.3 Linear Programming

Linear programming is a non-calculus optimization tool used to maximize or minimize a function, $F(x_i)$, which is dependent upon a variety of *inequality* constraints, $R_j(x_i)$, rather than *equality* constraints. The inequalities may be < 0, ≤ 0, > 0, or ≥ 0. The function to be optimized, and the constraints are usually continuous on the interval under consideration. If the assumption of continuity is not valid, then the optimum solution may have to be modified to fit the physical environment as was the situation in Case Study #13, Tool Life Optimization.

Linear programming is an analysis method used in business and industry to help make decisions on questions of product mix or resource distribution planning. As with other analysis methods, linear programming results must be adjusted for the practicalities of using resources such as time, money, and personnel. The following examples illustrate the general technique of linear programming, and practical implementation concerns that should follow.

EXAMPLE 13.11:

Optimize the output of a chemical process.

The maximum production of a particular chemical process is 1000 units per day. The chemical process is capable of making two product grades, A and B. A minimum of 500 units of grade A and 200 units of grade B are required each day to meet customer commitments. The profit on the sale of grade A is $2 per unit and $1 per unit on grade B. If all the output from the process can be sold, what is the maximum profit possible from the process and what output ratio will produce this result?

Let A = Units of grade A to be produced per day
Let B = Units of grade B to be produced per day
Let P = Total profit per day, dollars

The objective is to maximize the profit function $P(A, B) = 1B + 2A$, subject to the constraints: $A \geq 500$ units, $B \geq 200$ units, and $A + B \leq 1000$ units

The shaded area of Figure 13.14 shows the region of the A-B plane that satisfies all the constraints. The coordinates of the vertices of the shaded area can be found by carefully reading the graph or by solving simultaneous equations. The object is to determine the location of the profit function on the graph that results in highest profit. Because the profit function is linear, only the vertices of the shaded area need be checked in the profit function to determine a maximum profit. When the function to maximize or minimize is not linear, then its location on the graph must be determined using the equations of the lines between vertex points as well as the vertices.

$P(200, 500) = 200 + 1000 = \$1200$

$P(500, 500) = 500 + 1000 = \$1500$

$P(200, 800) = 200 + 1600 = \$1800$, which is maximum. The graph shows this profit line as a dashed line.

FIGURE 13.14

Graph for Example 13.11.

[Graph showing feasible region with vertices (200, 800), (200, 500), (500, 500), and objective line 2A + B = 1800. Axes labeled A (vertical) and B (horizontal), both scaled to 1000.]

The required production for a maximum profit of $1800 per day is 200 units of grade B per day and 800 units of grade A per day. The solution to this problem would not have to have been integral values; decimal parts of a unit are acceptable because production is continuous.

This is a simple straightforward example of the linear programming procedure. Though many students feel at this point they are done, the analytic solution is only the start of a complete analysis. In a real example additional questions should be asked and further analysis conducted considering such items as included in the following list.

1. The difference in profit between grade A and B indicates a preference to sell grade A. Why not make only grade A?
2. Perhaps there is too much capacity if just grade A is produced. Was grade B added to the product line just to use up excess capacity?
3. Perhaps the difference in profit indicates inefficiencies in the production of grade B. What can be done to improve the profitability of grade B?
4. Perhaps the profit on grade B is low because of market competition and this is a temporary situation. Should grade B be advertised more? Can additional markets be found for grade B?
5. It is assumed that all of the product produced can be sold. Should there be a limit on the daily output of either or both grades?

6. Is this profit worth the investment in the machinery needed to produce the product?
7. Is the market potential great enough to warrant investment in more productive equipment?
8. Are there waste products that occur that need special treatment? Can they be used to improve the profitability of the whole process?
9. Are there economies of production mix ratios that should be investigated?

As an engineer, always go beyond the first question asked and consider the future of your activities in relation to the overall objectives of your company. Proper planning and appraisal of activities can make the difference between a "flash in the pan" and a sound, stable operation.

Notice also in Figure 13.14 that as the profits for grades A and B become more equal, that is, a profit function slope closer to -1, it will make little difference which grade is made, and the limiting factor is the process limit of 1000 units. Notice also that if the minimum commitment for grade B is gone, there would be no reason to make any grade B unless it is done to use personnel and machines efficiently. It is often inefficient for processes to be shut off and started up except for major overhaul or maintenance. Reducing the production of one product may raise the cost of the other beyond economic operation.

EXAMPLE 13.12:
Optimize a manufacturing mix.

A company manufactures two products, each of which is manufactured using the same three production machines. Product A requires one hour, one hour, and two hours on machines #1, #2, and #3, respectively. Product B requires three hours, one hour, and one hour on each of the machines, respectively. Each product A sold yields a profit of $20, and each product B sold yields a profit of $30. Machine #1 is available 24 hours a day, machine #2 is available only 10 hours a day, and machine #3 is available 16 hours a day due to other product line usage. Find the most profitable product mix under these circumstances.

Let A = the number of product A's to make each day
Let B = the number of product B's to make each day
Let P = the profit for a day's production, dollars

Maximize the profit function, $P(A, B) = 20A + 30B$, subject to the following constraints:

$A \geq 0, B \geq 0$,
$1A + 3B \leq 24$, hours, the limit on machine #1
$1A + 1B \leq 10$, hours, the limit on machine #2
$2A + 1B \leq 16$, hours, the limit on machine #3

FIGURE 13.15

Graph for Example 13.12.

[Graph showing shaded feasible region with vertices labeled (0, 8), (3, 7), (4.8, 9.6), (6, 4), and (8, 0), with axes A (horizontal) and B (vertical).]

Figure 13.15 shows the region that satisfies the constraints. The vertices of the acceptable area are found by carefully reading the graph or by solving simultaneous equations. This example illustrates an error that can occur in linear programming analysis. If the graph is not drawn, an error will occur if the intersection point (4.8, 9.6) is used. This point is the intersection of two of the lines, but it lies outside the region satisfying all the constraints.

Evaluation of the profit function for the vertices of the shaded region results in $P(0, 0) = \$0$, $P(0, 8) = \$240$, $P(3, 7) = \$270$, $P(6, 4) = \$240$, and $P(8, 0) = \$160$.

The answer to the original question is to make three product A's and seven product B's for a daily profit of $270. The optimum answer does not have to be an integer because the process is assumed continuous and products that are partially complete on one day can be completed the next day. Notice that two other vertices give answers close to $270 and a change in profit or customer demand might influence the product mix decision. Suppose, for example, that the estimates for profit could be off by ±10 percent. The profit on A could range from $18 to $22, and the profit on B could range from $27 to $33. Would this make a difference? There are now four profit functions to use with the limits of product profit potential. Table 13.4 lists the four functions and the possibilities to consider. In each case the product mix for maximum profit stays the same, but maximum profit ranges from $243 to $297.

TABLE 13.4

Profit variations for Example 13.12.

$P(A, B)$	$P(0, 8)$	$P(3, 7)$	$P(6, 4)$
$18A + 27B$	$216	$243	$216
$18A + 33B$	264	285	240
$22A + 27B$	216	255	240
$22A + 33B$	264	297	264

This indicates that the system as modeled is "robust" (not sensitive to minor fluctuations in the values of the parameters.) This is good because minor changes are impossible to avoid. Other less robust systems could result in a different product mix as the profit ratio changes.

Linear *integer* programming is required when the values of the variables must be integers, such as the number of pipes in a heat exchanger, the number of people assigned to a task, or the number of vehicles needed to satisfy a transportation problem. Linear integer programming analysis is done using the same techniques as non-integer linear programming; however, the check for optimum may require moving away from the calculated extreme value if the calculation does not produce integers. See Exercise 13.45 for such a situation.

13.4 Additional Concerns

Engineers must use optimization techniques with care during the design process, just like all other methods that are available to them. No one method will always be the best to use, so flexibility and combinations of methods often make the search for the best more efficient. The following two case studies illustrate the flexibility and adaptability of optimization procedures, and they focus attention on additional optimization concerns. The first study illustrates a problem whose solution is facilitated by using the computer.

CASE STUDY #16:
Balcony Seating Optimization

Figure 13.16 is a layout of an older style field house basketball court with a balcony above. The balcony was once a jogging track but is now partially occupied with bleacher seating for basketball games and other activities held in the field house. The main problem with the seating in the balcony is visibility. The seats at the side of the balcony are so close to the edge of the basketball court that only first row spectators can see the whole court, and then only by leaning over a bit. The areas at

Additional Concerns 513

FIGURE 13.16

Court layout for Case Study #16.

29.00'
Radius of balcony rail

31.10'
Radius of eye positions

50.00'
Court width

Base line

17.60'

Seat positions 1 – 19

PLAN VIEW

Position of eye

ELEVATION SECTION

29.00' radius

Balcony floor

17.37'
Eye height

13.37'
Rail height

9.80'

Sight point on floor

the ends of the court presently do not have seating, but offer potential for better visibility because the edge of the balcony is farther back from the end of the court. In this design problem the input is not variable, the basic restriction is that no changes can be made to the existing structure, only the addition of seats is permitted.

Optimization situations such as this require a decision to be made on the output criteria, because the input is fixed. The criteria to optimize could be:

1. Place the maximum number of seats to have 100 percent visibility of the basketball court without regard to how many seats are available.
2. Maximize the number of seats that can see a defined percent of the court.
3. Maximize the number of seats without regard to visibility.

In this case one row of seats will be placed for optimum (maximum) visibility, close to the edge of the balcony and as high as practical, and the visibility determined. If each seat has 100 percent visibility, then a second row of seats will be considered. The analysis, therefore, must determine the visibility of the edge of the court from each seat.

Figure 13.17 shows the details of one-half of the balcony area end zone behind one basket, the other end of the court being the same. Visibility will be

FIGURE 13.17

Court details for computer program.

improved when each seat is close to the edge of the balcony, and as high as practical. The limit of 17.37 foot (5.30 m) maximum height for the eyes of a seated spectator is controlled by the ceiling height. The spectator must have head clearance when standing up. The 31.10 foot (9.48 m) radius of the eyes of a seated spectator allows 1.30 foot (0.40 m) clearance between the edge of the spectator's seat and the balcony rail. The 1.30 foot (0.40 m) clearance is judged the minimum acceptable for the average spectator. Room between adjacent seats on the 31.10 foot (9.48 m) radius is set at 1.50 feet (0.46 m). This allows 19 seats on the 31.10 foot (9.48 m) arc before reaching the sideline. *Note*: If S = the number of seats, then $1.50S/31.10 \leq$ arctangent $(25.00/17.60)$ and $S \leq 19$.

How much of the floor can be seen from the eye position of the spectator in each of the 19 seats? A layout can be made from each eye position: P_i, $0 \leq i \leq 19$, toward various points on the court baseline; S_j, $0 \leq j \leq 25$ feet, to locate the sight points on the floor. The locus of all such points describes the visibility range for each seat. This locus will be an ellipse because it is the intersection of a cone and a flat plane.

For 19 seats and 25 vision lines, 475 separate layouts need be made to determine the visibility patterns, and that is just for half the baseline; 475 are also required for the other half of the base line. Perhaps there is an easier way.

If an equation can be developed to define the geometry of the situation, then a computer program can be written to solve the problem. Following the graphical method, the computer algorithm is as follows.

1. Determine the coordinates of point P_i using the equations $XP = 31.10 \cos(1.50i/31.10)$, $YP = 31.10 \sin(1.50i/31.10)$, and $ZP = 17.37$. The argument of the sine and cosine are in radians, the center of the balcony radius is the origin of the coordinate system, and $0 \leq i \leq 19$.

2. Determine the equation of the line, $Y = mX + b$, on the floor from directly under P_i (XP, YP) to S_j (XS, YS). This changes for each vision line. For a complete analysis, $-25 \leq j \leq +25$. Only when $i = 0$ is the pattern symmetric and $0 \leq j \leq 25$ adequate.

3. Solve the equation from step #2 with the equation of the projection of the balcony edge on the floor, $XC^2 + YC^2 = 29.00^2$. The results are the coordinates of point $L(XL, YL, 13.37)$.

4. Calculate $AISLE = \sqrt{(XP - XL)^2 + (YP - YL)^2}$. $AISLE$ should always be greater than or equal to 2.1.

5. Calculate $A = 13.37[AISLE/(17.37 - 13.37)]$.

6. Calculate $HY = \sqrt{(XS - XP)^2 + (YS - YP)^2}$, and if $A < HY$, then the baseline is visible.

7. Calculate $YA = YP - A(YP - YS)/HY$.

8. Calculate $XA = (YA - b)/m$.

9. Increment j until $j = 25$ and check again.

10. Reset $j = 0$, increment i, and start again.

11. Quit when $i = 19$.

FIGURE 13.18

Line of sight from seat #5.

With graphics added, the vision limits for each seat can be generated. Figure 13.18 illustrates the vision limit for seat #5 ($i = 5$). When the data from the computer for each seat are analyzed, it is discovered that at least 82 percent of the baseline can be seen from each of the 19 seats, but only from the first 5 seats from the court center line can the entire baseline be seen.

The case study solution is a compromise on the mode of optimization, for it allows only one row of seats to be placed around the curved end, and the question was how much of the court could be seen from each seat.

The next case study illustrates how existing material sizes, properties, and costs can influence the selection of an optimum material for a design requiring a minimum capacity, a restricted search problem.

CASE STUDY #17:
Spherical Storage Tank Optimization

Consider a spherical storage tank to be designed to withstand a maximum internal pressure of 100 psi (0.689 MPa).

The stress in the shell, if $t \leq 0.05D/2$, is $\sigma = Dp/4t$, where (13.7)

σ = allowable material tensile stress, psi (MPa)

D = shell internal diameter, inches (meters)

p = internal pressure, psi (MPa), and

t = shell thickness, inches (meters).

TABLE 13.5

Materials available for Case Study #17.

Material	k $/lb	k $/kg	ρ lb/in³	ρ g/cm³	σ lb/in²	σ MPa
1	0.26	0.57	0.282	7.81	13,000	89.6
2	0.29	0.64	0.282	7.81	14,000	96.5
3	0.31	0.68	0.282	7.81	15,000	103.4
4	0.75	1.65	0.282	7.81	30,000	206.8
5	0.70	1.54	0.098	2.71	10,000	68.9

The cost of the shell material is: $C = k\rho\pi D^2 t$, where: (13.8)

C = material cost, dollars

ρ = material weight density, lb/in³(kg/m³), and

k = material cost, dollars/lb (dollars/kg).

Choose a material from the list in Table 13.5 that will give lowest cost for this application.

Some algebraic manipulation may help the analysis. Begin by solving Equation 13.7 for D and substituting in Equation 13.8, then

$$C = 16k\rho\pi t^3 \sigma^2/p^2.$$ (13.9)

Some calculus students might take the first derivative of Equation 13.9 and set the result equal to zero to get a minimum cost, but this gives no practical value for t. Try it.

As a beginning, it might be of value to keep the numerator of the cost function a minimum by making the product $k\sigma^2\rho$ a minimum. Table 13.6 gives these results. This preliminary analysis does not give a conclusive answer to the question because no overall tank size has been given. How big does the tank have to be?

TABLE 13.6

Material property products for Equation 13.9 and materials from Table 13.5.

Material	$k\sigma^2\rho$ English	$k\sigma^2\rho$ Metric	Rank (1 = smallest)
1	1.24·10⁷	3.57·10⁴	2
2	1.60·10⁷	4.65·10⁴	3
3	1.97·10⁷	5.68·10⁴	4
4	19.04·10⁷	55.11·10⁴	5
5	0.69·10⁷	1.98·10⁴	1

TABLE 13.7

Minimum material thickness, Case Study #17.

Material	Calculated minimum t inches	mm	Standard available t inches	mm	Rank (1 = thinnest)
1	0.617	15.67	0.625	15.88	3,4
2	0.573	14.55	0.625	15.88	3,4
3	0.535	13.59	0.5625	14.29	2
4	0.267	6.78	0.3125	7.94	1
5	0.802	20.38	0.8125	20.64	5

More data is required. Assume the tank will have to hold a minimum of 10,000 ft^3 (283.2 m^3).

The volume of the tank is $V = (4\pi/3)(D/2)^3$ and the corresponding value of D for a minimum volume of 10,000 ft^3 (283.2 m^3) is $D \geq [(10,000 \text{ ft}^3)(1728 \text{ in}^3/\text{ft}^3)(6/\pi)]^{1/3} = 321$ inches (8.15 m). From $D = 4t\sigma/\rho$, $t\sigma \geq (321 \text{ in.})(100 \text{ lb/in}^2)/4 = 8020 \text{ in.} \cdot \text{lb/in}^2$, and $t \geq 8020/\sigma$ inches (1.403/σ m). For each material a limiting value of t can be determined (see Table 13.7). Commercially available sizes are also included in Table 13.7, based on the assumption that material is available in 0.0625 inch (1.588 mm) increments.

Table 13.8 lists the tank size available for each preliminary answer from Table 13.7. Volumes are compared by calculating D from $D = 4t\sigma/\rho$ and then the volume from $V = (4\pi/3)(D/2)^3$. The increased thickness over the minimum allows the tank to be built larger than the minimum of 10,000 ft^3 (283.2 m^3). Material #1 or #5 appears to be "best," but cost still must be checked. Table 13.9 lists costs based on the value of purchased t listed in Table 13.8.

In this particular case, material #1 has the lowest total cost and the lowest cost per unit volume, and results in the optimum solution. If a safety allowance of 10 percent is used for storage volume, then materials #2 and #3 become

TABLE 13.8

Actual tank diameter and volume, Case Study #17.

Material	Purchased thickness, t inches	mm	Diameter based on purchased t inches	mm	Tank volume ft^3	m^3	Rank (1 = closest to objective)
1	0.625	15.88	325	8.26	10,400	294.5	1,2
2	0.625	15.88	350	8.89	12,990	367.9	4
3	0.5625	14.29	338	8.59	11,700	331.4	3
4	0.3125	7.94	375	9.53	15,980	450.0	5
5	0.8125	20.64	325	8.26	10,400	294.5	1,2

TABLE 13.9

Cost of storage tank material, Case Study #17.

Material	Cost based on purchased t	Cost of storage $/ft^3$	$/m^3$
1	$15,200	1.46	51.55
2	$19,670	1.51	53.32
3	$17,600	1.50	52.97
4	$29,200	1.83	64.62
5	$18,500	1.78	62.85

more attractive. If the manufacturing processes can only handle a maximum of 0.500 inch (12.7 mm) thickness, then material #4 would be the optimum selection.

This analysis has been simplified and many other factors need to be considered before a tank can be fabricated. Other factors to be considered include:

1. Total weight and the cost of the supporting structure
2. Land area required, its availability, and cost
3. Annual maintenance cost for painting, a function of surface area and type of paint used
4. Delivery schedule of the preferred material compared to the project schedule
5. Weldability of the material selected
6. Fracture toughness of the selected material
7. Anticipated environment (salt water, acid rain).

13.5 Experimental Optimization

Several times in this chapter reference has been made to experimental optimization, and for many engineering problems it is the only solution method available. Experimental techniques are necessary when the physical situation cannot be modeled with an analytic function, or when variation of an input is observed to have an effect on the output, but the mathematical relationships are not known. This is a common occurrence in many engineering design situations. Frequently it is not known which variables have an effect on the outcome, or, if the outcome is affected, by how much. Experiments are conducted to answer these questions. Investigations are conducted to study relationships such as:

1. Material hardness as a function of annealing time with temperature held constant
2. Material hardness as a function of annealing temperature with time held constant
3. Melting temperature of a two element alloy as a function of element ratios.

Perhaps *after* an experiment is run and data collected a mathematical relationship may be developed. Then analytic methods can be used for additional analysis. If only one variable is under investigation, any of the search methods of Section 13.2 may be used to study the experimental input-output relationship.

The basic search methods can also be used if more than one variable is being considered if each variable is assigned incremental values within its domain, but this increases the number of experiments exponentially. If the cost of each experiment is high, or the time to conduct the experiment is long, multiple variable experiments can prolong an investigation.

An alternate to the basic search methods is to select and test various combinations of the variables and carefully study the results. Pre-experiment decisions are made to select the domain of values in the experiment and the increment used to cover the domain. For a two variable experiment four combinations are used to start the process:

1. Assign each variable the value at the low end of its domain
2. Assign each variable the value at the high end of its domain
3. Assign the first variable the value at the low end of its domain and the second variable the value at the high end of its domain
4. Assign the first variable the value at the high end of its domain and the second variable the value at the low end of its domain.

The results of these four experiments will indicate where macro changes are occurring, and indicate where subsequent experiments should be conducted to lead to the desired extreme. The following example illustrates this procedure, and also illustrates the Yates algorithm used to determine which variable(s), if any, is (are) most active. A change in the most active variable affects the output of the experiment the most.

EXAMPLE 13.13:
Experimental optimization for a two variable problem.

Consider an experiment in two variables, a and b.

Experiment #1, when both variables were at their low level, symbolized as the $(--)$ condition, had an output of 210. Experiment #2, when a was kept low and b was high, symbolized $(-+)$, had an output of 180. Experiment #3, when a was at its high value and b was at its low level, symbolized $(+-)$, had an output of 240. Experiment #4, when a and b were both at their high level, symbolized $(++)$, was an output of 200. The conclusion is that a change in a is the best thing to do to increase the output level, because $240 > 210 > 200 > 180$. The conclusion cannot be made that the condition $(+-)$ produces maximum results, but the maximum is probably in the vicinity of $(+-)$, and additional experiments can be conducted near that condition in search of a better optimization. The rectangular domain area created by conditions $(--)$, $(-+)$, $(+-)$, and $(++)$ can be subdivided into quarters, eights, or smaller partitions to narrow the region in which to conduct additional experiments.

The Yates analysis algorithm, used to determine the most active variable(s), requires that the experimental results be listed in special order by experimental code. Table 13.10 shows this order and also the calculations required for the additional

TABLE 13.10

Calculations for the Yates analysis method for the two variable analysis of Example 13.13.

Experimental code	Experimental response	(1)	(2)	Effect estimate
(− −) 1	210	450	830	208
(+ −) a	240	380	50	25
(− +) b	180	30	−70	−30
(+ +) ab	200	20	−10	−5

analysis. The required sequence of table entries is controlled by column #1, the experimental code.

Column #2 entries are the results of the experiment. The numbers in the column headed (1) are obtained from the experimental response, column #2. Entry #1 in column (1) is the sum of the first two experimental response values; entry #3 is their difference. Entry #2 in column (1) is the sum of the third and fourth experimental response values; entry #4 is their difference. The entries in column (2) are determined using the values from column (1) and performing the same operations as were performed on the experimental response entries to obtain the entries in column (1). Be sure to subtract the number lower in the column from the number higher in the column. Two iterations are required because there are two variables. The second, third, and fourth entries in column (2) indicate the influence of each of the variable combinations on the outcome of the experiment, and the improvement due to a alone is confirmed because $50 > -10 > -10$. The first entry in the effect estimate column is obtained by dividing the first entry from column (2) by 2^2 and the remaining entries are obtained by dividing the column (2) entries by 2^{2-1}. The exponent 2 in these calculations is used because there are two variables under study.

In many engineering situations, it is not only desirable to optimize the situation, such as maximum output or minimum cost, but also to determine the sensitivity of the system to variation and combinations of one or more variables. This condition can be analyzed by performing a graphical analysis, hence the need for the last three columns in Table 13.10.

The graphical analysis is performed as follows. List all entries except the first from the effect estimate column in order of magnitude, along with P, their respective percentage of the whole. The percent of the whole is determined using the formula $P = 100(i - 1/2)/n$, where $i = 1, 2,\ldots, n$, and $n =$ the number of values listed. Table 13.11 shows the results of this listing and calculation for the data from Table 13.10. After the calculations are made, the values are plotted on normal probability paper as shown in Figure 13.19. In this problem the three points lie on a straight line, so no condition appears more active than any other, and it is not clear which variable to manipulate for maximum response. Had there been a factor more active than the others, the points would not have been on a straight line. If the points are not on a straight line, the variable(s) responsible for the point(s) highest and to the right are the most active.

Optimization

TABLE 13.11

Evaluations to use for graphical analysis of variable activity.

i	1	2	3
P	16.7 percent	50 percent	83.3 percent
Effect estimate	−5	−35	25

When three variables are under consideration, the minimum-maximum variable combinations again can be used, and if properly analyzed, can also lead to the variables that are having the greatest effect on the experimental outcome. The following case study illustrates an experiment conducted to determine the most active variable of a three variable analysis, as well as progressing toward a maximum output value.

FIGURE 13.19

Effect estimate plot for Example 13.13.

CASE STUDY #18:
Experimental Optimization Performed on Three Variables.

This problem requires the maximization of the output from an alternating current electric generator being driven by a windmill. A preliminary analysis was conducted and the decision was made to limit the investigation to the following variables:

1. The number of blades on the windmill shaft
2. The shape of the blades
3. The angle that the blades made with the plane of rotation of the windmill.

All experiments used a wind speed of 8 miles/hour (12.88 kilometers/hour), and a blade diameter of 15.40 inches (0.391 m). The power output was monitored by measuring the voltage and current flow of a one resistor circuit. Preliminary experiments determined which direction to proceed for additional experiments, and which of the three variables is the most active and influential on maximizing the output.

Each variable was coded as follows. The variable a was assigned to the number of blades, with a low value (−) of two blades and a high value (+) of four blades. The variable b was assigned to the shape of the blades, with a low value (−) representing a flat blade and a high value (+) assigned to an airfoil shape. Variable c was assigned to the angle of the blade with the plane of rotation, with a low value (−) of 55° and a high value (+) of 45°. The decision as to which value is the + and which is the − is arbitrary, final analysis sorts out the variation direction.

A three variable problem requires 8 experiments to be run because there are $2^3 = 8$ combinations possible when three variables each have 2 values. The eight combinations and the experimental order used are shown in Table 13.12, generalized using the (+) (−) notation. The order of the experiments is not critical but the coding is. When all variables are at the low level the condition (− − −) exists and is coded 1. Each subsequent code uses the letter for the variable(s) that are in the high value mode, hence (+ + −) is ab, (+ − +) is ac, and (− + −) is b.

The next step in the procedure requires a special order of the experimental code as listed in Table 13.13. The first column of Table 13.13 is the three variable extension of the first column of Table 13.10. Table 13.13 also includes the experimental response and the necessary Yates analysis calculations.

Experimental response shows that increasing the number of blades from two to four is the greatest single factor affecting higher output. Changing the angle from 55° to 45° is the second most important factor affecting higher output, and changing the blade shape has the least (but some) effect on increasing output. This observation is based on the data relationships 8.2 > 7.1 > 4.7 > 4.5. The highest output occurred when all three changes (four blades, airfoil shape, and 45°) were made at the same time, 15.8 being higher than any other output.

The additional calculations required for the Yates analysis for a three variable experiment are extensions of the calculations performed for the two variable experiment, but the addition-subtraction iteration calculation is performed three

TABLE 13.12

Combination codes and output for eight experiments in a three variable experiment.

Experiment number	Variable A: Number of blades	Variable B: Shape	Variable C: Angle	Experiment code	Output: milliwatts
#1	− (2 blades)	− (flat blades)	+ (45°)	c	7.1
#2	+ (4 blades)	−	+	ac	14.1
#3	−	+ (airfoil blades)	+	bc	5.7
#4	+	+	+	abc	15.8
#5	−	−	− (55°)	1	4.5
#6	+	−	−	a	8.2
#7	−	+	−	b	4.7
#8	+	+	−	ab	11.0

times instead of two. The first four entries in column (1) are formed by adding consecutive entries in the response column; for example, the first entry in column (1) is 12.7, the sum of the first two entries (4.5 and 8.2) in the response column. The second entry in column (1) is 15.7, the sum of the next two entries (4.7 and 11.0) in the response column. The last four entries in column (1) are the differences of the consecutive entries in the response column. The first four entries in column (2) are formed by adding consecutive entries in column (1), and the last four entries in column (2) are formed by subtracting the consecutive entries in column (1). Column (3) is formed from column (2) as column (2) was formed from column (1). The entries in the effect estimate column are obtained by dividing the first entry in column (3) by 2^3 and subsequent entries by 2^{3-1}.

TABLE 13.13

Calculations for the Yates method of experimental data analysis for three variables.

Experimental code	Response: milliwatts	1	2	3	Effect estimate
1	4.5	12.7	28.4	71.1	8.9
a	8.2	15.7	42.7	27.1	6.8
b	4.7	21.1	10.0	3.3	0.8
ab	11.0	21.5	17.1	5.7	1.4
c	7.1	3.7	3.0	14.3	3.6
ac	14.1	6.3	0.3	7.1	1.8
bc	5.7	7.0	2.6	−2.7	−0.7
abc	15.8	10.1	3.1	0.5	0.1

TABLE 13.14

Evaluations to use for graphical analysis of variable activity.

i	1	2	3	4	5	6	7
P	7.1%	21.4%	35.7%	50%	64.3%	78.5%	92.9%
Effect Estimate	−0.7	0.1	0.8	1.4	1.8	3.6	6.8

Continue the analysis in search of the most active factors. The effect estimate values (from the last column in Table 13.12) are placed in order, along with the corresponding P value, as shown in Table 13.14; use the formula for P from Example 13.13. The data is graphed on normal probability paper, Figure 13.20, and shows two points that are not on the straight line formed by the other points. These two points are associated with the single factor changes (the angle of the blades, and the number of blades), not combined factor changes. The conclusion is that the number of blades alone, and the blade angle alone, are the most active variables, and should be pursued independently of the other combinations.

The second half of the study focused on the effect of blade angle on power output. Four airfoil blades were used because that combination produced increases in output in the first series of experiments. The time to run each experiment in the next series of experiments was short enough so the domain was set at $0° <$ angle $< 90°$ with an increment of $10°$, even though it became clear the maximum output occurred at small angles. Table 13.15 contains the results of experiment #2, the output with $10°$ increment changes in blade angle. The highest output of 82.4 milliwatts came at $10°$ and represents an 1800 percent increase over the starting output value of 4.5 milliwatts. The experiment was repeated at $45°$ to monitor experimental consistency. The 15.8 milliwatt output compares with a 7.0 milliwatt output from the (+ + +) condition from experiment #1. This variation in output is attributed mainly to a non-constant wind source (see Exercises 13.58 and 13.59), variation in bearing tightness on the wind mill shaft, and the rigidity and induced vibration in the test apparatus. The density of the air, which is dependent on atmospheric pressure, temperature, and humidity can also contribute to output variation. The mathematical model in Exercise 13.58 shows how the density of air enters into the analysis.

Table 13.15 also contains a listing of the results of experiment #3, which used an angle increment of $2°$. The maximum output from experiment #3 occurred at $8°$. Notice again the experimental variation in power output at the monitoring values of $10°$ and $20°$: 82.4 compared to 86.5, and 46.7 compared to 41.2.

Even though the output has been increased, the study is not complete. The repeatability of the experiment must be improved by controlling wind speed, perhaps by using a wind tunnel. Structural changes in the windmill support and bearing adjustment will also help control the experimental variation. Each experiment did increase the output, but percent changes are not a reliable measure of differences because of the variance in results. Optimizing based on one condition does not guarantee a global optimum, only a local optimum. Additional experi-

FIGURE 13.20

Effect estimate plot for Case Study #18.

ments will show whether other variations will affect the output more. These experiments should investigate:

1. The effect of additional blades
2. The effect of combinations of the number of blades and blade angle
3. The effect of other blade shapes.

To add to the difficulty of this experiment, a problem was created during the last experiment that was not present at the beginning of the study. The 8° blade angle, airfoil shape, four blade configuration, which resulted in maximum power output, would not self start. The configuration did not have enough starting torque to overcome static friction; so at the benefit of higher output, the cost of needing an auxiliary start was created. More problems for the engineer.

TABLE 13.15

Results of blade angle variation on power output.

Experiment #2, Δblade angle = 10°		Experiment #3, Δblade angle = 2°	
Blade angle	Power output, milliwatts	Blade angle	Power output, milliwatts
0°	0	0°	0
10°	82.4	2°	0
20°	46.7	4°	64.8
30°	18.5	6°	78.3
40°	8.2	8°	103.1
45°	7.1	10°	86.5
50°	5.4	12°	80.8
60°	3.4	14°	64.1
70°	2.0	16°	55.2
80°	1.1	18°	45.8
90°	0.3	20°	41.2

Optimization procedures, like all engineering analysis tools, must be selected and used after both their advantages and disadvantages have been considered. The same method will not always be the most appropriate to use. Decision judgment must be tempered with concern for accuracy, cost, personnel and equipment utilization, and the health and safety consequences of unacceptable conclusions. The results of analysis must be balanced with experience, responsibility, and a regard for professional ethics.

13.6 Additional Study Topics

1. Jacobian maximization technique
2. Derivative techniques for functions of more than three dimensions
3. Additional search techniques
    a) Method of Bolzano
    b) Method of Rosenbrock
    c) Method of Powell
    d) Brent's Praxis Method
    e) Internal halving
    f) Exhaustive search
4. Nonlinear programming techniques
5. Function root refinement techniques
    a) Bisection method
    b) Newton method

13.7 References

1. Berkey, Dennis D. *Calculus*. New York: Saunders College Publishing, 1984.
2. Budnick, Frank S. *Applied Mathematics*. New York: McGraw-Hill, 1979. Chapters 6, 7, 12, 13, and 14.
3. Coleman, Hugh W. and W. Glenn Steele, Jr. *Experimentation and Uncertainty Analysis for Engineers*. New York: John Wiley & Sons, 1989.
4. Doyle, Lawrence E. *Manufacturing Processes and Materials for Engineers*, 3rd ed. Englewood Cliffs, N.J.: Prentice-Hall, 1985.
5. Foulds, L. R. *Optimization Techniques*. New York: Springer-Verlag, 1981.
6. Frazer, J. Ronald. *Applied Linear Programming*. Englewood Cliffs, N.J.: Prentice Hall, 1968.
7. Gajda, Walter J., Jr. and William E. Biles. *Engineering: Modeling and Computation*. Boston: Houghton Mifflin, 1978. Chapter 5.
8. Hicks, Charles R. *Fundamental Concepts in the Design of Experiments*, 3rd ed. New York: Holt, Rinehart and Winston, 1982.
9. McDonald, T. Marll. *Mathematical Methods for Social and Management Scientists*. Boston: Houghton Mifflin, 1974. Chapter 8.
10. Piascik, Chester. *College Mathematics: With Applications*. Columbus, Ohio: Charles E. Merrill, 1984. Chapter 6.
11. Woodson, Thomas T. *Introduction to Engineering Design*. New York: McGraw-Hill, 1966. Chapter 15.

13.8 Exercises

1. Expand the discussion of extreme function analysis for the condition when $f^I(x_i) = 0$, $f^{II}(x_i) = 0$, $f^{III}(x_i) = 0$, and $f^{IV}(x_i) = 0$.
2. Verify the intersection points of the two parabolas in Example 13.2.
3. Verify that the answer of $h = 10.39$ in Example 13.3 does give a maximum by checking the value of the second derivative.
4. Verify Equation 13.3 for minimum material removal cost: $T_m = (C_t/R_c)((1-n)/n)$.
5. Show that $T_m = (C_t/R_c)((1-n)/n)$ does give a minimum cost value for removing material by evaluating the second derivative of Equation 13.2 and showing that it is greater than 0.
6. Verify the value of T for maximum production $T_M = (t_I)(1-n)/n$, and that it does give a maximum by evaluating the second derivative of Equation 13.4.
7. The following represent some typical $VT^n = k$ relationships:
   High speed steel cutter on

Steel	$VT^{0.125} = 47$ ft/min
Cast iron	$VT^{0.14} = 47$ ft/min
Nonferrous cast alloys	$VT^{0.16} = 47$ ft/min

Cemented carbide cutter on
Steel $VT^{0.2} = 150$ ft/min
Cast iron $VT^{0.25} = 150$ ft/min
Sintered carbide on
Steel $VT^{0.7} = 400$ ft/min

Discuss the significance of the exponent change and the constant change.

8. Verify with calculations the extreme value of $-2/9$ for Example 13.5.

9. Find any minimum or maximum values of $f(x, y) = xy + 8/x + 8/y$ for real values of x and y.

10. Find any minimum or maximum values of $f(x, y) = (x - 2)^2 - y^2$.

11. Find any maximum, minimum, or inflection points of each of the following functions using calculus derivatives. Use a domain of all real numbers.
    a) $f(x) = x^4 + 6$
    b) $f(x) = x^4 - 12x^3 + 52x^2 - 96x + 64$
    c) $f(x) = x^2 - 4.25x + 4$

12. Find any maximum, minimum, or inflection points of the function $f(x) = x + \cos x$ using calculus derivatives. Use the domain $0 \leq x \leq 2\pi$.

13. A wooden I beam is made from a log 12 inches (0.305 m) in diameter shown in Figure 13.21. The center section of the I beam is made from the longest piece, the top and bottom made from the shorter pieces, and all three pieces have the same thickness. Determine the dimension for x that results in maximum area moment.

14. Will changing the thickness of the center section of the I beam in Exercise 13.13 to $x/2$ increase or decrease the maximum moment possible? By how much?

15. Verify the minimum cost of $1538 in Case Study #14 using derivative calculus techniques.

16. Verify the minimum cost of $1538 in Case Study #14 using Lagrange multipliers.

17. Verify that the optimum packing of circles within a larger circle, as used in Case Study #14, occurs on a local level when the number of small circles follows this

FIGURE 13.21

Sections to cut from a log to construct an I beam, Exercise 13.13.

sequence: $P_1 = 1, P_2 = 7, P_3 = 13, P_4 = 19, P_5 = 31, P_6 = 37,\ldots$ where the general term is $P_k = P_{k-1} + 6$ for each even numbered term in the sequence, and $P_k = P_{k-1} + 3(k-1)$ for each odd numbered term in the sequence, $k > 1$.

18. If the radius of each small circle enclosed in the larger circle is r, then verify that the radius of the large enclosing circle for each term in the sequence in Exercise 13.17 follows this sequence: $r, 3r, 2r\sqrt{3} + r, 5r, \sqrt{3(r\sqrt{3})^2 + r^2} + r, 7r,\ldots,(n+1)r$ when n is even,$\ldots$, $\sqrt{(((n+1)/2)(r\sqrt{3}))^2 + ((n-3)r/2)^2} + r$ when n is odd.

19. Case Study #14 used a cost estimate of a cylindrical shell as $25D^{2.5}L$ based on other shells manufactured. The use of a base model for estimating costs is not unusual, and is a satisfactory procedure within specified boundary conditions. Some of these conditions are the physical limits on material and manufacturing processes, and physical limits on the installation site including ceiling heights, door opening sizes, and available handling equipment. The cost of a cylindrical shell can be more closely estimated if more detail is used, and this allows a better comparison to be made with alternate designs. Determine a cost model for fabricating a welded cylindrical shell, made with a rectangular blank rolled into a cylinder and capped with two circular blanks. Let D = tank diameter, t = material thickness, and L = tank length. Comment on the simplification to the cost model $25D^{2.5}L$ used in the case study. Is the simplification justifiable?

20. Repeat Exercise 13.19 for a rectangular cross section tank and compare results. Compare the total cost of installing the rectangular tank with the total cost of installing the circular tank, $700 + 25D^{2.5}L + 20DL$.

Refer to Case Study #15 for Exercises 21-24.

21. What happens to Figure 13.7 when a larger volume is used, such as 30 in^3?

22. A 20 in^3 can is being made at minimum cost with 0.0239 thick material and a 2.16 diameter.
    a) What is the height of the can?
    b) If seaming costs $0.01/inch, what is material cost?
    c) What is the total material and seaming cost?

23. Suppose 20 in^3 cans are in production with $t = 0.0239$ inches and $D = 2.28$ inches when you discover that material with $t = 0.0179$ inches is strong enough to be used. What action would you take to reduce cost?

24. Repeat the calculation for Case Study #15, Cost Ratio Analysis, using the material volume as $t(\pi Dh + 2D^2)$.

25. Repeat Example 13.6 by using seven equal division search calculations on the variable b instead of h.

26. Illustrate that the equal division method may result in the global maximum, and that fewer divisions may be more efficient than more divisions, by redoing Example 13.6 using six divisions instead of seven.

27. Use the equal division search technique on the functions in Exercise 13.11 on the interval $-2 \le x \le 6$. Use eight equal segments and search for a minimum function value.

28. Use the Fibonacci search technique on the functions in Exercise 13.11 on the interval $-2 \leq x \leq 6$. Use nine evaluations and search for a minimum function value.

29. Prove that the golden mean ratio is 0.618 by using the ratio definition $AC/AB = BC/AC$ given in the text. Calculate AC as a percentage of AB.

30. Illustrate that the ratio of consecutive terms of the Fibonacci sequence is the same as the golden section search ratio if $n > 8$.

31. Use the golden section search technique on the functions in Exercise 13.11 on the interval $-2 \leq x \leq 6$. Use nine evaluations and search for a minimum function value.

32. Use the equal division search technique on Exercise 13.9 looking for a minimum function value. Use six intervals on the domains $0 < x < 5$, and $0 < y < 5$.

33. Use the Fibonacci search technique on Exercise 13.9 looking for a minimum function value. Use six intervals on the domains $0 < x < 5$, and $0 < y < 5$.

34. Use the golden section search technique on Exercise 13.9 looking for a minimum function value. Use six intervals on the domains $0 < x < 5$, and $0 < y < 5$.

35. Write an algorithm to search a domain of a function for the minimum and maximum value within the interval.

36. Refine the solution of Example 13.10 in search of a minimum by subdividing the intervals $6 < y < 12$ and $6 < z < 12$ using an increment equal to 1.

37. Solve Example 13.10 using the Fibonacci search and five evaluations on each of the intervals $0 < y < 18$ and $0 < z < 18$.

38. Solve Example 13.10 using the golden section search and five evaluations on each of the intervals $0 < y < 18$ and $0 < z < 18$.

39. Solve Example 13.10 using the four direction slope search method starting at $y = 4$, $z = 6$, and $\Delta y = \Delta z = 1$.

40. There are equations that can be solved algebraically, either for a root or a functional value at a given location, that are often solved by using the computer search method, because it is faster and more consistent. In these cases it is helpful to do a little pre-computer analysis so the search interval is selected wisely. This is particularly helpful when only *real* roots or functional values are meaningful. For example, the function $f(x) = \sqrt{x-5} - 6$ has real values only if $(x - 5) \geq 0$, so for real values the domain is $x \geq 5$. Analyze each of the following functions in a similar way to determine the acceptable domain for real function values search.

    a) $f(x) = \sqrt[3]{x-1} + \sqrt{x+3} - 2$
    b) $f(x) = \sqrt{2x+1} - \sqrt{4x-5} + \sqrt{6x-8}$
    c) $f(x) = \sqrt{x^2 + 2x} - 2\sqrt{x^2 - x - 6} - x$

41. The radial flow of heat between two concentric spheres can be represented as: $H(r, R, t, T) = [(t - T)4\pi k R r]/(R - r)$, where k depends on the material involved, r = the radius of the inner sphere, R = the radius of the outer sphere, t = the temperature of the inner sphere, and T = the temperature of the outer sphere.

FIGURE 13.22

Material layout for Exercise 13.42.

(diagram showing rectangular sheet of width W and length L with x by x squares cut from corners (scrap), eight cut lines and four fold lines indicated)

An engineer wants to design a two-sphere system to contain a heat producing experiment, but is concerned that the estimates of the temperature difference from inner sphere to outer sphere may be off by as much as a factor of two. The engineer does not want to change material (k is the material constant), but wants flexibility to increase the outer sphere radius to handle any variations that may arise. Analyze the mathematical model and determine if there are any particular sphere radii combinations that should be avoided if the engineer wants flexibility to change the radius of the outer sphere during the experiment.

42. A five sided rectangular box is to be made by folding the cutout from a single sheet of material as shown in Figure 13.22. Use one of the optimization methods to determine the dimensions L, W, and x for minimum surface area for a box to hold one cubic unit. What are the dimensions of the box?

43. Determine the dimensions of a rectangular box so its volume is a maximum. The only restriction is that the perimeter of the smallest right cross section plus the length perpendicular to the cross section is less than or equal to 84 inches.

44. Analyze the relationship between the three dimensions of a six sided box of a given volume for minimum cost, considering that the material for the top and bottom has a cost different from the material for the four sides. Use the cost ratio between the two as a single quantity.

45. A company makes tables and chairs. The tables require three machine hours and one hour of hand labor each. The chairs require one hour on the machine and two hours of hand labor. One machine is available nine hours per day and there are eight hours of hand labor available per day for this operation. If the profit on a chair is $4

and the profit on a table is $10, what product mix should be manufactured per day for maximum profit? Assume there is no market problem and all the products made can be sold. Can the answer legitimately be a fraction? Explain.

46. What happens to the analysis of Exercise 13.45 if the requirement is changed to requiring that one table and four chairs be sold as sets? What recommendation would you make to management concerning machinery and labor to be made available? What effect do you suppose this might have on profit?

47. To make the operation in Exercise 13.46 more profitable would you add machines or hand labor? What quantity would you add?

48. Consider the following situation. A football team has 2.50 minutes left in a game and is behind by 4 points. They have 40 yards to go for a touchdown and have been having success with running and passing plays. Running plays have been averaging 6 yards per play while taking 30 seconds, and pass plays have averaged 12 yards while taking only 20 seconds. What combinations of plays will best suit the team's situation? Ignore time outs. *Hint:* There is no function to maximize or minimize in this problem.

49. Refer to Example 13.12 and answer the following questions:
    a) How many hours per day are each of the three machines used to make maximum profit?
    b) Which of the three machines would run out of capacity first if production needed to be increased? Explain how this affects plant capacity limits.
    c) Is it reasonable to assume that machine #1 will be available 24 hours per day? Discuss your answer.

50. What minimum profit variation (percent or absolute) would change the production mix in Example 13.12 to:
    a) No A's and eight B's?   b) Six A's and four B's?

51. A chair manufacturer has on hand 1000 board feet of hardwood and 750 board feet of softwood. The cost is $0.30 and $0.15 per board foot, respectively. The company can make chairs, which use five board feet of hardwood and ten board feet of softwood, or can make tables, which use 25 board feet of hardwood and 15 board feet of softwood. The chairs can be sold for $75 each and yield a profit of $12, while tables can be sold for $150 each and yield a profit of $30. What combination of chairs and tables should be produced to make a maximum profit? How much wood is left over if this is done? What is the gross income?

52. Reevaluate Exercise 13.51 assuming that a table must be sold with six chairs.

53. A salesman working for the company in Exercise 13.51 suggests selling four chairs with one table. Is that a good suggestion?

54. Problems in linear programming often involve three variables and a three-dimensional graphical analysis and can be completed with the method presented in section 13.3. The graph can be drawn in any of the common three-dimensional methods, such as isometric or oblique. Expand the two-dimensional method to three-dimensions and solve this problem that way. A company wants to make a new alloy containing at least 25 percent of each of three special elements. The alloys do not come separated but as part of available compounds. The available compounds,

TABLE 13.16

Data for Exercise 13.54.

Compound	Element 1	Element 2	Element 3	Cost/unit
A	45 percent	20 percent	35 percent	$0.73
B	45 percent	45 percent	10 percent	0.90
C	5 percent	25 percent	35 percent	0.80

their percentages of the three elements, and their cost are shown in Table 13.16. Determine how many ounces of each compound should be used to make the new alloy with a minimum percent of each element of 25 percent, and keep cost to a minimum.

55. Select a material to minimize the cost of material for a spherical storage tank, $C = k\rho\pi D^2 t$ subject to the volume constraint $(4\pi/3)(D/2)^3 \geq 1000$ in^3. Select from the materials in Case Study #17.

56. Repeat Case Study #17 assuming the materials are only available in 0.125 inch thickness increments.

57. Expand the Yates algorithm as used in Case Study #18 and outline the table and instructions necessary to carry out a four variable experiment.

58. Derive the model for the power of wind air flow, $P_{in} = (\pi D^2)(\rho)(0.447v)^3/8$, where D = the diameter of the rotating windmill blades in meters, ρ = the density of the air in kg/m^3, v = the air flow speed in miles/hour, and 0.447 converts miles/hour to meters/second. *Hint:* Consider the kinetic energy in a column of moving air the same diameter as that of the rotating blades.

FIGURE 13.23

Diagram for Exercises 13.61–13.64.

59. Using the model from Exercise 13.58, demonstrate that at the low air speeds used in Case Study #18 a small variation in wind speed results in a large variance in experimental conditions. In particular show that a wind speed of 8 ± 1 miles/hour results in approximately 112 percent variation in power from the windmill experiment.

60. Investigate the effect of temperature and humidity on the density of air and determine if the variation is a significant factor in explaining the experimental variation of Case Study #18.

61. A pole on which a horizontal pull of F lbf is applied is to have a wire-rope guy wire attached in the plane of the pull as shown in Figure 13.23. The diameter of the guy wire depends on its strength, S_{ut}, the angle θ, and the load F. If the cost of the guy wire installation is a function of its volume, determine the angle θ for minimum wire cost.

62. Continue Exercise 13.61 by determining general expressions for:
    a) The length of the wire-rope as a function of H, the height of the pole and θ, the angle.
    b) The volume of the rope as a function of H, θ, F, the pull on the pole, and S_{ut}, the strength of the wire rope.
    c) The diameter of the rope as a function of F, θ, and S_{ut}.

63. One grade of steel wire-rope comes in diameter increments of 0.125 inches in the range $0.250 \leq D \leq 1.00$ inches. The wire-rope strength is 76,000 psi based on the nominal area of the rope, $0.38D^2$, rather than $\pi D^2/4$, and its cost is proportional to the volume used, $V = (\text{area})(\text{length}) = 0.38D^2L$. The wire-rope is to be used as guy wire support, so a safety factor of 3.5 is appropriate. Assume the pole in Figure 13.23 is 20 feet high and the pole and guy wire must support a horizontal pull of 1000 lbf. Determine the angle θ, the wire diameter, and the length to use to minimize the cost as analyzed in Exercises 13.61 and 13.62.

64. Analyze the situation and discuss the implications if the load F in Exercise 13.61 increases by 1000 lbf increments.

65. Physical limitations often exist that restrict the location of a guy wire. One restriction is where the cable attaches to the ground. The optimum location may be in the middle of a road, or the location may interfere with some other space use. The restriction may limit the location to a range of locations, or a single location as identified by C in Figure 13.24. A second restriction may be a minimum vertical clearance next to the pole, identified by h in the figure. One solution to this problem is to add a rigid member AB that creates the necessary clearance. The diameter of the guy wire depends on its strength, S_{ut}, the angle θ, the distance C, and the load F. If the cost of the guy wire installation is a function of its volume, determine the angle θ for minimum cost. Assume the cost of member AB is constant regardless of the force it must withstand, that C is constant, and that the total length of wire must be the same diameter.

66. Continue Exercise 13.65 by determining general expressions for:
    a) The length of the wire-rope as a function of H, the height of the pole, C, the horizontal clearance distance, and θ, the angle.
    b) The volume of the rope as a function of H, θ, F, and S_{ut}.
    c) The diameter of the rope as a function of F, θ, C, and S_{ut}.

FIGURE 13.24

Diagram for Exercises 13.65–13.68.

FIGURE 13.25

Circle layouts for Exercises 13.69 and 13.70.

67. Using the data from Exercise 13.63, and the restrictions that $C = 5$ feet and $h \geq 7$ feet, determine the specifics of the design for minimum cost.

68. Analyze the situation and discuss the implications if the load in Exercise 13.65 increases by 1000 lbf increments.

69. Silos are used to store grain and other foodstuffs. Silos are built round because it is well known that for a given perimeter, a circle encloses the most area, and so a circular silo will contain the most volume for a given surface area. Supposing, however, that two silos are to be built next to each other. Is there any advantage in building the silos together with a common straight wall as shown in Figure 13.25? The question that needs answering is whether or not at some angle θ, the total area of the two partial adjoining circles is greater than the area of two separate circles. To answer this question maximize the combined area of the two partial circles and compare with $2(\pi R_1^2)$ while keeping total perimeter constant.
Note: $R_2 \neq R_1$.

70. Repeat Exercise 13.69, but change the conditions so that the straight common segment between the partial circles costs half as much as the circular segments.

14 Financial Analysis

"In today's business environment, a manager must know what costs should be and what they actually are and why."
♦ DON KIRKLAND, DIRECTOR, ARTHUR YOUNG & COMPANY

Engineers need to know about financial matters because finances are the backbone of any engineering project. Financial analysis and control will not make a product function better or last longer, but if finances are not controlled, a good product design may never get produced and reach the market place. Engineers are managers of time (projects have time deadlines) and other people (project analysis and implementation involves the effort of many other people), and because projects need money for implementation, engineers must also be concerned with the management of money. A project that is financially unsound will fail regardless of how good the intentions are. Engineers must get involved in project financial analysis and not leave themselves at the mercy of others. This chapter is offered to familiarize you with some of the concerns of financial managers so you can be the exceptional engineer, one who can follow financial planning and be involved in financial decision making. Some of the topics covered may be a review for you, but they are all important when financial understanding and planning of engineering projects are required. For the derivation of the formulas given in this chapter, see reference (1) or (8).

The most important financial concern is having money available to fund the development of a project. Money can be obtained by selling stock if the company is incorporated (common or preferred), selling bonds, receiving grants, and borrowing money. Common stock sale does not burden the company with the necessity of repaying interest, but it does affect the company's management because the stockholders can influence and vote on management decisions. Preferred stock is often sold with the promise of paying a fixed dividend, and this does place a repayment burden on the company. The amount of money available by selling stock is limited, because the price per share is governed by market conditions and the success of the company's previous operations.

Selling bonds is another way to obtain money for product development but it adds the interest payments to the cost of the project. The added cost may result in a higher selling price or the need to sell more product, and the bonds must be repaid at some future date.

Grants can be obtained for product development from the government or a related company, and are nice because in general there is no obligation to repay them. The grantor usually expects a tangible result, however, such as a new product or process that will improve some present deficiency, produce a change in the state of the art, or create new jobs. Grants are not necessarily continued, and small companies may not be able to survive when a grant runs out.

Borrowing money is also a common way to obtain money for project development. Borrowing money is similar to selling bonds because there is an obligation to repay the loan principal at a future date and to pay interest. The periodic interest payments add cost to the project and place demands on cash flow. Borrowing is perhaps the most common method of obtaining money for product development, and various levels of borrowing exist, from short term loans, measured in months, or long term loans, measured in years. The next five sections will review the mechanics of borrowing and repaying money.

14.1 Simple Interest

Simple interest is used for calculating interest for the repaying of short term loans. The model for simple interest is $I = Pit$, where (14.1)

I = the amount of interest, $

P = the loan principal, $

i = interest rate expressed as a decimal, and

t = the number of time periods the money is borrowed, in the same time units as i.

EXAMPLE 14.1:

Short term loan.

On Monday morning Hal asked June if he could borrow $20. He promised to pay back the $20 plus $2 interest on the following Monday. June agreed, mainly because Hal was a friend. Later in the day she grew curious about the rate of interest the $2 represented. She solved Equation 14.1 for i using I = $2/week, P = $20, and t = 1 week:

$i = 2 / (20)(1) = 0.10 = 10$ percent/week $= 520$ percent/year.

June was surprised by the high value and decided to calculate the amount of interest she would charge for one week using a modest interest rate of 10 percent/year. Using the same formula and solving for I resulted in $I = (20)(0.10/52)(1) = \$0.04$.

June was now surprised by the small amount of interest and decided not to get further involved and charge Hal only $1.

14.2 Installment Loans

Installment loans are loans that are repaid in equal installments over a given period of time, usually ranging from one to five years.

EXAMPLE 14.2:
Installment loan.

Henry wanted to buy a stereo system that cost $1000, but did not have the money. He asked the dealer if the system could be purchased with a down payment and 12 monthly payments. The dealer responded by asking for a $200 down payment and twelve $72 payments starting in one month. Henry could afford that, so he wrote a check for $200, signed contract papers, and took the stereo home. Only later did Henry analyze his purchase.

Original cost = $1000, less down payment of $200, leaves a balance due of $800. Twelve payments of $72 each requires a total repayment of $864 for the $800 debt—$64 interest. The interest rate cannot be calculated using Equation 14.1 because the principal is not constant over the 12 month period. However, the problem can be modeled using averages to approximate the rate. The average amount of interest paid per month is $64/12 = $5.33. Using the average interest, the 12 months can be analyzed as shown in Table 14.1. The average monthly balance owed = $5199.96/12 = $433.33. Now P and I are assumed constant and $I = Pit$ can solved for i: $i = 64/(433.33)(1 \text{ year}) = 0.148 = 14.8$ percent annual rate. Other models could be used, such as assuming interest is paid in full before any principal payments are made, or assuming principal is paid up first and the interest is paid last. Different assumptions and resulting models will result in variations from the 14.8 percent obtained in the example.

14.3 Compound Interest

Compound interest is a method used to account for the fact that the loan time differs from the time specified for adding interest. An interest adding period might be per year but the loan might be for three years or one-half year.

The simplest situation to model for compound interest is to have a beginning

TABLE 14.1

Monthly payback table.

During month	Amount owed	End of month payment	Amount toward interest	Amount toward principal
1	$800.00	$72	$5.33	$66.67
2	733.33	72	5.33	66.67
3	666.66	72	5.34	66.66
4	600.00	72	5.33	66.67
5	533.33	72	5.33	66.67
6	466.66	72	5.34	66.66
7	400.00	72	5.33	66.67
8	333.33	72	5.33	66.67
9	266.66	72	5.34	66.66
10	200.00	72	5.33	66.67
11	133.33	72	5.33	66.67
12	66.66	72	5.34	66.66
Totals	$5199.96	$864	$64.00	$800.00

principal and no deposits or withdrawals for a known length of time. The model for this situation is

$S = P(1 + i)^n$, where (14.2)

P = the beginning principal amount, $

$i = r/k$, the period interest rate,

r = the annual interest rate expressed as a decimal

k = the number of compounding periods in a year

n = the total number of compounding periods of interest, and

S = the amount in the account after n periods, $.

EXAMPLE 14.3:
Compound interest, single deposit.

If Henry deposits his $200 down payment into an account paying 5.25 percent annual interest, compounded monthly, then after one month he would have $200 + 200(0.0525/12)(1) = $200.88 in his account. The $200.88 would then earn interest during the next month. A month-by-month calculation could be made for any number of months to determine any end-of-month balance. An easier method is to use Equation 14.2: $S = 200(1 + 0.0525/12)^{12} = $210.77. At the end of 12 months Henry would have $210.77 in his account. Penny differences may occur in actual practice because of rounding off calculations done at the end of each month, or if the monthly calculation is done for the number of days in the month divided by 365.

It is common for deposits to be made into accounts on a regular basis, where each deposit earns interest at the same interest rate, and each deposit and its accrued interest contribute to the account total. The model for this situation is:

$$S = \frac{R[(1+i)^n - 1]}{i}, \text{ where} \quad (14.3)$$

S = the amount in the account after n periods, $

R = the equal *end of the period* deposits

i = the period interest rate expressed as a decimal, and

n = the number of periods.

EXAMPLE 14.4:

Compound interest, multiple deposits.

Suppose Henry deposited his monthly payment of $72 in the same account into which he put the $200. If he starts making $72 deposits at the beginning of the second month, the first $72 will earn interest for 11 months, and will be worth $72(1 + 0.0525/12)^{11}$ = $75.54 at the end of the twelfth month. Money deposited at the beginning of the next month would earn interest for 10 months, and so on. The individual calculations could be done for each deposit and added up, but Equation 14.3 does this very thing. For Henry's case $i = 0.0525/12 = 0.004375$, and

$$S = \frac{72[(1 + 0.004375)^{12} - 1]}{0.004375} = \$885.10$$

Combine this with the amount resulting from the $200 deposit, and Henry has $1095.87 in his account after 12 months. If the price of the stereo did not change during the year, Henry could buy it and have $95.87 left over for records. *Note:* The last $72 deposit did not earn interest, but it is necessary to include it when using Equation 14.3.

14.4 | Mortgages

A mortgage is an example of a long term debt payment arrangement. A beginning sum of money is owed, which the borrower pays back in regular periodic payments, including interest on the unpaid balance, while paying periodic interest at a known rate. The question generally asked is, "What is the periodic payment required to repay the loan in x years?" This is accomplished by using Equation 14.4.

$$R = \frac{(A)(i)}{1 - (1 + i)^{-n}}, \text{ where} \quad (14.4)$$

A = amount of money owed at the beginning of the mortgage, also called the present value, $

R = regular periodic payments, $

n = the number of payback periods, and

i = the *period* interest rate expressed as a decimal.

The effect of Equation 14.4 is to annualize a one-time cost over the life of a project by taking the time value of money into account. See also Exercise 14.38.

EXAMPLE 14.5:
Mortgage loan repayment.

A plant facilities addition is built for a cost of $1,000,000, repayment due in regular monthly payments over a 20 year period. If the annual interest rate is 10 percent, what is the required monthly payment?

$A = \$1,000,000$, $i = 0.10/12$, and $n = 240$.

$$R = \frac{1,000,000(0.10/12)}{1 - (1 + 0.10/12)^{-240}} = \$9650.22.$$

The total interest paid over the 20 year period is $(240)(\$9650.22) - \$1,000,000 = \$1,316,053$. Long term repayment plans raise the "price" of owning. Not only does long term debt increase the cost of owning, it also puts a burden on cash flow: $9650.22 in cash is required each month to make the payment.

14.5 Annuities

An annuity is a financial arrangement where one party pays another party a fixed amount per period for a specified number of periods. The payment could be required because of a long term loan, such as a mortgage debt, or it could be a person receiving the proceeds of an insurance policy or retirement account over a period of years, receiving a check each month. The model that is needed is a relationship between the money in the account at the beginning of the time period, and the regular periodic withdrawals (payments) to be made for a fixed number of periods. The analysis is done by solving Equation 14.4 for A with the following notation:

A = amount of money in the annuity, also called the present value, $

R = regular periodic withdrawals, $

n = the number of periods, and

i = the *period* interest rate expressed as a decimal.

If the account did not draw interest, the answer would be easy, $A = Rn$. A no-interest account would not be a wise financial decision because an interest bearing account will allow higher periodic withdrawals for the same number of periods, or the same withdrawal for more periods. For an interest bearing account, where the interest compounding period and the withdrawal period coincide, Equation 14.5 is used to calculate the present value of an annuity.

$$A = \frac{R[1-(1+i)^{-n}]}{i} \tag{14.5}$$

EXAMPLE 14.6:

Annuity present value.

What initial amount is needed in an account so $200 can be withdrawn each month for three years, if the account earns $5\frac{1}{4}$ percent annual interest compounded monthly?

$i = 0.0525/12 = 0.004375$, and Equation 14.5 becomes

$$A = \frac{200[1-(1+0.004375)^{-36}]}{0.004375} = \$6648.12$$

The interest earned over the three year period is the difference between the beginning annuity amount and the withdrawals, ($36)($200) − ($6648.12) = $551.88.

There are other financial concerns a business must consider other than borrowing money. Special funds for expansion, facility replacement, or pension obligations are often required; cash flow and the tax laws must be considered; and product cost must be controlled. The next sections consider some of these concerns.

14.6 Taxes

Federal and state business income taxes are a concern when making economic judgments relative to engineering decisions. The laws governing this aspect of operating a business are arbitrary and changeable, but do place pressure on the decision process. Taxes are charged on profits, which are calculated by subtracting the cost of operations from total income derived by selling a product or service. Some forms of income are not taxable, such as municipal bond interest and life insurance proceeds. The cost of operation includes most of the expenses of doing business. These expenses include many items: the cost of material and the freight cost to deliver it, labor to produce the product, rent on buildings and equipment, interest on money borrowed, sales promotions, insurance, depreciation, maintenance, telephone, electricity, heat, water, license fees, royalty fees, transportation, advertising, postage, paid employee vacations, health and accident insurance, social security matching payments, and unemployment compensation insurance. Some expenses are not allowed as deductions from income, such as goodwill amortization, life insurance premiums where the company is the beneficiary, and fines and other legal penalties. Expenses incurred in a year that are not offset by

TABLE 14.2

Corporate income tax rates for 1990.

Federal rates:
  15% of the first $50,000 of net income
  25% of any amount over $50,000 up to $75,000
  34% of any amount over $75,000 up to $100,000
  39% of any amount over $100,000 up to $335,000
  34% of any amount over $335,000

Iowa state rates:
  6% of the first $25,000 of net income
  8% of any amount over $25,000 up to $100,000
  10% of any amount over $100,000 up to $250,000
  12% of any amount over $250,000

income can be carried forward to succeeding years. Table 14.2 lists the federal and Iowa state corporate income tax in effect in 1989. It doesn't take much analysis to realize that a variable tax rate, from 21 to 51 percent, is an important factor in a company's financial planning program. One concern is that cash must be available to make the tax payments.

Use caution when analyzing long term projects because tax laws are subject to change. Consider the investment tax credit that was allowed by the Internal Revenue Service (IRS) for many years. The investment was a deduction from taxes owed, based on a percentage of the cost of capital equipment purchases. This percent varied from 3 to 10 percent depending on the equipment. The investment tax credit law reduced the net cost of the equipment and provided additional cash for operations. It was the government's way of encouraging equipment purchases. A disadvantage of the tax credit was that it encouraged companies to go into debt to buy equipment, and the result was often a debt burden that could not be repaid. Banks and other financial institutions often encouraged the debt and they too suffered as loans could not be paid. In 1985 the investment tax credit was dropped.

14.7 Depreciation & Depletion

Depreciation is an accounting procedure that charges current operations for equipment purchased in earlier periods as the equipment wears out. A piece of equipment that is depreciable does not become part of the final product, but is used to make, service, or deliver the product. Depreciation expense is an allowable deduction for income tax purposes, but the tax law permits depreciation methods that do not necessarily match the actual wearing out of a piece of equipment. The law changes from time to time so any discussion of depreciation is subject to change without notice. The reason the law is changed is to control companies' incentives to buy more equipment when it is thought necessary to stimulate the economy, or reduce incentives when it is necessary to slow the economy down. In practice it rarely works as desired. A company still has to have money or good credit to make purchases, and if business is bad no amount of tax credit incentive will help. Borrowing money just to take advantage of tax breaks often leads to cash flow problems and financial difficulties; it is not wise management policy.

Straight line depreciation is the easiest method of depreciation to apply, and all past tax laws have permitted its use. The cost of a piece of equipment is charged off in equal increments over its estimated life. The value of the equipment at the end of the estimated life, called salvage value, *can be* considered in the calculation.

$$D = (C - S)/n, \text{ where} \tag{14.6}$$

D = Periodic depreciation charge, generally for one year

C = Original purchase price

S = Salvage or scrap value, and

n = Number of periods the equipment will last.

The difficult part of using this model is estimating a realistic scrap value for a date in the future, and estimating how long a piece of equipment will last. The IRS has published guidelines for the life of various classes of equipment. These guidelines may or may not agree with actual use but should be used unless other data are available. The straight line depreciation method assumes equal wear for each period of time. Many pieces of equipment do not wear out at a linear rate.

EXAMPLE 14.7:
Straight line depreciation.

A truck costs $35,000, has an estimated life of 5 years, and an estimated scrap value of $5000. The annual depreciation is $D = (\$35,000 - \$5000)/5 = \$6,000$.

If the truck was purchased October 1, 1988, and the company operated on a calendar year basis, the depreciation allowed for 1988 would be $(3/12)(\$6,000) = \$1500$. The depreciation allowed for each subsequent year, 1989, 1990, 1991, and 1992 would be $6,000. The depreciation for 1993 would be $(9/12)(\$6,000) = \$4500$.

If the truck operated an average of 2000 hours per year, the depreciation "cost" charged to the customer would be $3 per hour. The $3 per hour is added to the cost of a driver, fuel, and maintenance.

The **sum-of-the-digits** is another method of calculating depreciation. This method permits higher depreciation charges in the earlier years of the equipment life. The fraction of total value allowed each year is the year number, in reverse order, divided by the sum of the years. This method of depreciation encourages investment purchases, as large first year depreciation costs help reduce the income tax due.

EXAMPLE 14.8:
Sum-of-the-digits depreciation.

Calculate the depreciation by the sum-of-the-digits method for the truck in Example 14.7.

The sum of the years 1, 2, 3, 4, and 5 is 15, therefore:

Year	Fraction	Depreciation allowed
1	$\frac{5}{15}$	$10,000
2	$\frac{4}{15}$	8000
3	$\frac{3}{15}$	6000
4	$\frac{2}{15}$	4000
5	$\frac{1}{15}$	2000

Some businesses, such as mining and oil drilling, are given depletion allowances. A depletion allowance is similar to depreciation in that something is used up and is of less value to the business as time goes by. In this case the something is a natural resource: oil, lumber, or minerals.

Depreciation and depletion allowances are the result of having purchased something that is wearing out. While depreciation may not be a current year cash outlay, it is necessary to buy something to be able to claim depreciation. Only land is not depreciable, on the assumption that it does not wear out.

14.8 Overhead Charges To Operations

Businesses often try to maintain cost control by dividing operations into segments and allocating costs to those segments. For example, a company in the business of selling and repairing boats and accessories could divide the business into:

1. Selling new products
2. Selling used products, and
3. Repairing customer products.

A further breakdown is possible if new products included inboard engine boats, outboard engine boats, canoes, and sailboats.

When a product division is made, then all income and expenses are categorized to allow analysis of each phase of the business. In this way the contribution of each segment of the business can be studied relative to the whole. Decisions can be made regarding such items as expansion, new lines of products, hourly service charges to customers, and personnel assignment. Ordinarily 20 percent of a company's products and services generate 80 percent of its income. This does not imply that the other 80 percent of the products and services should be eliminated. Often the complementing products create more total customer purchases because of complete service in one location.

14.8.1 Material Overhead

Material that is purchased for use in a product has additional cost associated with its ordering, shipping, receiving, and storing. For some low cost materials, such as scrap metal for a foundry, the associated costs may be equal to or greater than the cost of the material, while in the case of high price material, such as electronic or hydraulic components, the associated costs may be a low percent of the total cost. Heavy steel sheets and plates require overhead cranes for handling, small purchased parts can be handled with a forklift, and fluids and gases may require pipes to transfer them from storage tanks to the point of use. Averaging all material receiving and distribution costs within a plant, and applying a fixed overhead rate to all operations, will not allow good control of costs because averages do not allow analysis of high and/or low cost conditions.

Overhead charges directed to a particular purchased material composition and size, or purchased component, will allow better cost analysis. Many companies have separated the detail cost of using various materials, and have made progress on reducing material overhead cost. This is accomplished by eliminating incoming inspection by certifying the supplier at the supplier's facilities, and by eliminating storage costs by receiving material as needed for production (just-in-time inventory management).

14.8.2 Process Overhead

In large manufacturing firms each process or operation is often assigned an overhead rate. This rate is based on the historic relationship between direct labor cost (money paid to employees) and support costs attributable to the process including machine tool depreciation, maintenance, tooling (fixed and perishable), power, insurance, and the percent of plant occupancy for heat, lights, and insurance. The advantages of such a system include the ability to isolate high cost areas of operation so cost reduction action can be taken, and the ability to calculate the cost of a product for market pricing. Disadvantages include the amount of record keeping required and the potential inflexibility of conducting individual analysis within a process. The inflexibility comes from the use of historical averages. If all the costs for a year are summarized by category and become averages applied to the next year, changes in current operations are not reflected until another year. This makes cost reduction efforts difficult to measure because an entire year is averaged. A real-time system allows changes in cost to be made the day that a change in operations is made. Not many systems do that. Some companies try to overcome the lack of real-time system adjustments by summarizing cost changes in variance accounts. That is adequate for summary purposes but not for individual part or process analysis.

Material and finished product storage cost is also an overhead expense that needs analysis. The cost of the interest on the money borrowed to purchase the material, the storage space, and the related utilities are all costs to charge against storage. Shouldn't the product that uses the most material be charged with the highest storage overhead?

CASE STUDY #19:
Misapplied Overhead

A company has its production cost separated by process type. Each different manufacturing process, such as shearing, welding, drilling, and grinding has its own overhead rate based on historical data. A new machine was installed in the plant that would do milling, drilling, and boring under numerical control. The machine had a tool changer and a rotary index table with a load and unload station. It was the most expensive machine in the plant and was different in operation from any other process. The engineer assigned to the development of the machine thought it would be appropriate to create a new process overhead account for the machine because of its uniqueness. This was in lieu of grouping it with other milling machines. The accounting department was consulted and agreed to the proposal. At the end of year when an updated overhead rate sheet was published, the engineer noticed that the account number for the new machine had the same overhead rate as the other milling machines in the plant. This was not anticipated, so a review with the accounting department was made. The answer, "Oh, we didn't have time to separate all the costs, so we lumped them together and gave both processes the same rate." So much for cost control analysis in that plant.

There are many costs that can be directly attributed to a particular process, including:

1. Machine and tooling depreciation and interest on borrowed money used for purchase
2. Machine and tooling maintenance and repair, parts and labor
3. The cost of machine tooling: dies, jigs, and fixtures, including interest if money is borrowed to make the purchases
4. Perishable tooling: drills, punches, reamers, broach bars; purchase cost and resharpening cost
5. Electric, gas, and steam power; water if used in a process like water spray cooling for flame cutting
6. Machine and tooling insurance: fire, flood, earthquake, and theft
7. Percent of plant occupancy for utilities; depreciation, maintenance, and insurance
8. Plant layout costs, rearranging present facilities to create manufacturing cells or flexible machining systems

Keeping track of all these expenses does take a good system, but it is not a question of whether or not it can be done. It is a question of whether or not time and money are going to be invested to control cost analysis procedures and operations.

Average overhead rates can hamper cost analysis. There are companies that group a single spindle drill press (cost of $30,000) with a numerically controlled automatic tool changer drill (cost of $300,000) in the same category for overhead purposes. It isn't just the machine cost difference that makes for a poor grouping. Consider also the differences in floor space required, the percent of lights and heat required for the space occupied, maintenance, tooling, insurance, employee training, power required, and staff support people. Should a minute of machine time on both machines be charged the same overhead rate? Machine time should include all the time the machine is used to make a part; tool cut time, load and unload time, tool change time, and setup time. Many companies charge these costs into some "support services" category, and specific costing is nearly impossible.

14.8.3 Product Overhead

There are costs of operations that are difficult to charge against a particular process or machine tool, but are chargeable against a part for a particular product. Some of these are:

1. Product advertising if attributable to one product or one line of products
2. Purchasing department costs, the cost of ordering parts and material for a particular product line
3. Plant layout costs such as laying out a new product assembly line
4. Product development costs, prototype building and testing
5. Sales and distribution costs, getting the product to the market
6. Warranty costs, parts and labor, record keeping
7. Income tax, a percent of net profit
8. Labor cost and fringe benefit cost as a percent of labor cost, such as social security, hospitalization, workman's compensation, and unemployment insurance

Many of these costs can be charged against a particular product, however, so it is convenient to establish *product* overhead charges as well as *process* overhead charges and *material* overhead charges.

14.8.4 Administrative Overhead

Another category of expense involves costs of operation charges such as:

1. Plant manager's salary
2. Company image advertising
3. Main gate security costs
4. Base telephone charges (Long distance calls could presumably be charged against a specific process or product.)
5. Accounting department costs
6. General maintenance: grounds and custodian workers, and
7. Research and development not attributable to an existing product

Research and development cost is often not appropriated against overhead but is given a separate budget that comes from general operations profit. If, however, the research and development is for a specific product or process, the cost can be charged accordingly.

One model for total product cost is:

Product cost = $\{[\Sigma(MC_i + MO_i) + \Sigma(PC_j + PO_j)][1 + POR_p]\}\{1 + AOR\}$,
where:

MC = material cost

MO = material overhead

i = each material form that requires procurement and storage, such as bars, plates, forgings, castings, fluids, powders, and purchased complete parts

PC = process cost

PO = process overhead

j = each process or machine used to manufacture the material into a product

POR = product overhead rate

p = the particular product being analyzed; and

AOR = administrative overhead rate.

There are as many variations to the procedures of costing products as there are companies. The objective of all of them should be to determine the cost of a product properly and to allow for the analysis of day-to-day operations. Unfortunately this does not happen in some companies because of macro grouping of costs, and because of the use of averages. To be a usable management tool, overhead rates should be reflective of current operations. If a product redesign or modification is made, the sooner the cost change can be reflected in the current operations, the better. An engineer should not have to wait a year to see the cost effect of changes, whether it be positive or negative. Timely adjustments permit better decisions and promote accountability.

14.9 Return On Investment

The selling price of a product must allow for a profit: selling price = cost + profit. Profit can be measured in absolute terms, as a percentage of capital investment, often referred to as Return on Investment (ROI), or as a percentage of company equity. If there is no profit, then investment money has been tied up in buildings and machines that would earn more money if simply deposited in a savings certificate. The profit measured in absolute terms can give a distorted view of a business, as the following example illustrates.

EXAMPLE 14.9:

Return on investment analysis.

Company A reports profits of $1,000,000 while company B reports profits of $1,300,000. Which is best? High profits are good, but what investment was required to make the profit? If company A had invested $10,000,000 and company B had invested $20,000,000, then:

$ROI(A) = \$1,000,000/\$10,000,000 = 0.100 = 10.0$ percent, and

$ROI(B) = \$1,300,000/\$20,000,000 = 0.065 = 6.5$ percent.

What does all this have to do with engineering design? You must understand the implications of your decisions in the context of your whole company. If five proposals are presented to management, the money for implementation will be appropriated to those with the highest likelihood of success and with the highest projected *ROI*. If there is only enough investment money for two of the five projects, yours may be rejected even though it is a soundly based proposal.

If your proposal is rejected you may have to redesign to reduce the manufacturing cost or create a higher sales appeal. Learn how your company makes financial decisions, including the costing and pricing system used. You may not be able to change the system, although it is always worth a try if it is very bad, but you must understand the system if you are going to operate efficiently within it.

14.10 Cash Flow

A business must have cash to operate on a day-to-day basis. Cash is needed to pay current bills: wages, utilities, materials and supplies, periodic estimated income tax payments, social security payments, shipping, postage, and other continuing charges. Cash flow problems have caused companies to go out of business despite some very worthwhile products. Cash is required for product research and development, and for building and testing of preliminary idea prototypes. Cash is needed for expenses before any comes back from the sale of product to pay the bills. Cash is generated by selling products, by borrowing, or by getting investors to buy shares of stock or bonds. Many good projects have been turned down by management because of a poor cash flow projection.

CASE STUDY #20:
Cost Reduction Proposal

Consider a proposal to install storm windows on all the windows in an office building. The total cost of the windows is quoted at $45,000, and the estimated fuel cost savings is $8000 per year. Assume a straight line depreciation period for the windows of 15 years, no salvage value, and no investment credit. The increase in taxable income is the difference between the money not spent on expense, $8000, and the depreciation chargeable against expenses, $3000. Table 14.3 summarizes the 15 year analysis.

It takes nearly 6 years to save the money spent the first year for the improvement, $45,000/$8000 = 5.6 years. If after-tax savings is considered, it takes over 7 years: $45,000/($8000 − $1700) = 7.14.

The rate of return on this project is not $49,500/$45,000, because the dollars of positive cash flow in future years are not worth the same as in year one. The

TABLE 14.3

Cash flow summary, Case Study #20.

Year #	Pretax cash flow change	Depreciation	Increase in taxable income	34%* income tax	Net cash flow change
1	$-37,000	$3,000	$5,000	$1,700	$-38,700
2-15	8,000	3,000	5,000	1,700	6,300
Totals	$75,000	45,000	75,000	25,500	49,500

*Variable depending on total taxable income for the year.

present value of future dollars must be considered. Money income in the future has less value today because of the interest earning potential for the length of time considered. The annual interest rate that yields a present value of $45,000 is the *ROI* for the project. An analysis of this can be done by determining what rate of interest on an investment of $45,000 will allow 15 annual payments of $5000 each—use Equation 14.4. If a before tax rate is desired, use $8000 in place of $5000.

$$45,000 = \frac{5000[1 - (1 + i)^{-15}]}{i}$$

The variable i cannot be separated, so an iterative approach must be used for the solution. A short computer program can be written to solve for i; preliminary simplification will make that easier.

$$\frac{(45,000)i}{5000} = 1 - (1 + i)^{-15},$$

$9i - 1 = -(1 + i)^{-15}$, and after computer iteration, $i = 7.2$ percent

The implication of this analysis is that if money can be invested on the open market at a rate greater than 7.2 percent, then this project should not be done. Does that mean that the project should not be done at all? No, there may be items to consider that were neglected during the first analysis.

Perhaps the maintenance cost for the existing windows will be reduced if new storm windows are installed. The workers who have to spend the day by the windows may be relieved from cold drafts and become more efficient. Maybe the heating system will have less maintenance cost because it will not have to run as much and will last longer than originally estimated. The looks of the building may be enhanced and office space will be easier to rent out, or maybe the rents can be increased. An estimate could be made on the projected increase in cost of fuel over the 15 year period, changing the annual savings.

You may be wondering why all these items were not considered during the first analysis. A first analysis is often done quickly with the best and most reliable data available. Items to include are the ones that are the most reliable, and the ones you feel confident in evaluating; secondary items are included as intangible advantages. The first analysis is often referred to as a feasibility study. Do the preliminary findings justify the project and warrant further analysis? If the project can make the grade on the preliminary analysis basis, then any savings not quantified will be frosting on the cake. Finding answers to all the secondary questions could postpone submitting a proposal, and an opportunity for implementation may be lost. Some projects are implemented solely on intangible items. It may not be practical to estimate savings on some of the secondary considerations; for example, how can extended furnace life be estimated?

Return on investment is a consistent method of analyzing and comparing project proposals. It is not, however, the only thing to consider when deciding the worth of a project. Some projects, such as the first computer controlled machine in a factory or a new CAD/CAM system, could be the beginning of a change in the company's manufacturing philosophy. The first steps of such a change must be studied in detail. Detailed cost comparison for projects like that are more difficult because in-plant data to use for analysis is lacking. The first step may be an investigative trial to determine if the change is worth pursuing.

Cash flow and return on investment are both used for project proposal acceptance. Project A, with a *ROI* of 18 percent, may receive approval over project B, with a *ROI* of 22 percent, because of a better cash flow projection. Good cash flow allows management to continue projects and developments on a steady basis. Borrowing money to finance a project will decrease the saving potential because of the interest cost.

Other criteria are available for project analysis and comparison. One of these is the cash payback period, the length of time it takes savings to replace the money spent on the project. Ordinarily this is considered on an after-tax basis. In Case Study #20, the project takes 15 years to reach a cash payback on an after-tax basis. Some rapidly growing companies in technologically advancing industries will not consider any project that requires longer than a three year cash payback period. The business they are in is too volatile to consider longer periods of time. Some projects are analyzed more on subjective criteria rather than objective criteria, as the next case study illustrates.

CASE STUDY #21:
New Process Proposal

A particular plant had no numerically controlled machinery. An engineer at that plant made a proposal to purchase a numerically controlled bar lathe to run small lots of parts, particularly repair parts. The plan called for listing all the parts presently run on the bar lathes in order of the quantities of parts required, and to run the lowest on the list on the new machine. Periodically, as needs changed, the list would be reviewed and adjustments made in the manufacturing routing. The proposal also included a similar procedure for other processes where numerically controlled machinery was available, such as machining centers and punching machines. The beginning of the plan required additional money for programming support, and some changes in tooling setup and tool sharpening procedures. New skills were needed, and plans were presented for additional training sessions. The main advantages of the plan were:

   1. Faster response to repair parts orders and increased customer satisfaction

2. Fewer changeovers and setups on regular production machines, increasing their production efficiency
3. Reduced overhead cost because of #2
4. Added flexibility in producing prototype parts.

When proposing such a project include plans for review of before and after effects. It might be difficult to estimate how much less time it would take to fill repair orders with the proposed system, but a before and after study could be made. Good follow-up is important for project review. It is doubtful that a dollar value can be placed on customer satisfaction, as that has to be analyzed subjectively.

The number of setups before and after project implementation should be easy to keep track of. The reduction in overhead cost is more difficult to quantify. Some people fill up their work day regardless of how much actual work there is to do. Others take on additional duties, occasionally performing a meaningful task that had not been done before.

Adding flexibility to prototype build is an important advantage. Being able to make parts in-house allows better scheduling and control. A prototype may be built in less time if in-house capabilities exist. The actual savings are hard to quantify because a prototype by definition is something new, and something new probably does not have a historical time standard to measure against. Another subjective decision is required.

14.11 Forecast Uncertainty

As with any method of analysis it is not good practice to make a decision based on only one criterion. Financial analysis is no different. Poor estimates of the future can cause any project to fail. No one can predict the future with certainty, but projections into the future can be based on past experience. Methods do exist for considering the variations in future possibilities. One method is to use weighted averages. One weighted average value = $[E_c + 4E_m + E_l]/6$ where E_c is the most conservative estimate, E_m is the most likely estimate, and E_l is the most liberal estimate. Ratios other than 1 : 4 : 1 can be used depending on the circumstances.

CASE STUDY #20 CONTINUED:

Suppose the $8000 per year fuel savings was a most likely estimate, but a low of $6000 is possible in a mild winter, and a high of $11,000 is possible in a cold winter. The weighted average method of calculation results in the annual savings = [$6000 + 4($8000) + $11,000]/6 = $8170. The analysis done with that value gives a slightly different numerical answer. Taxable income would increase by 15($8170 − $8000) = $2550, and net cash flow change would improve by $2550 − 0.34($2550) = $1683.

A second method for considering variations in estimates is to calculate *ROI* based on a conservative estimate and again on a liberal estimate, so the range of possibilities is known. A savings estimate of $6000 per year would change taxable income to $45,000 and after-tax cash payback to nearly 23 years, while an $11,000 estimate changes taxable income to $120,000 and an after-tax cash payback time to approximately 8.5 years.

14.12 Summary

Many engineers treat the financial matters of their business as a necessary evil rather than as a source of data to assist the decision making process. Many financial people feel that if it weren't for them the company couldn't survive and they respond, "Why are engineers always arguing that they are being hamstrung by accountants?" If you are lucky, you will work in a situation where neither of these attitudes prevails, but don't count on it. Many projects are judged solely by the short term effects on the profit and loss statement rather than by the lasting effects on the company. Some managers aim for short term gains and quick promotion rather than long term gains and continued growth.

14.13 Additional Study Topics

1. Break even analysis—linear and nonlinear
2. LIFO versus FIFO accounting methods
3. Alternate depreciation methods
4. Perpetual inventory systems
5. Bond discounting
6. Lease versus purchase of equipment
7. Annuities with unequal period or unequal amount withdrawals

14.14 References

1. Budnick, Frank S. *Applied Mathematics for Business, Economics, and the Social Sciences.* New York: McGraw-Hill, 1979. Chapter 9.
2. Chapman, C.B., D.F. Cooper, and M.J. Page. *Management for Engineers.* New York: John Wiley and Sons, 1987.
3. Cooper, Robin and Robert S. Kaplan. "Measure Costs Right: Make the Right Decision." *Harvard Business Review*, Sept./Oct., 1988, 96–103. Excellent discussion of overhead accounting.
4. Ferraro, R. A., T. E. Hunley, and O. Y. Shackney. "Banishing Management Barriers to Automation." *Manufacturing Engineering.* January 1988, 44–48.
5. Gaither, Norman. *Production and Operations Management.* Hinsdale, Ill.: The Dryden Press, 1987. Chapter 9, Inventory Systems.

6. Gallin, Daniel. *Finite Mathematics.* Glencoe, Ill.: Scott, Foresman and Company, 1984. Chapter 12.
7. Grant, Eugene L. and Richard S. Leavenworth. *Statistical Quality Control,* 5th ed. New York: McGraw-Hill, 1980. Chapter 19.
8. Hayes, Robert H. and Ramchandran Jaikumar. "Manufacturing Crisis: New Technologies, Obsolete Organizations." *Harvard Business Review,* Sept./Oct., 1988, 77–85.
9. Helfert, Erich A. *Techniques of Financial Analysis.* Chicago: Richard D. Irwin Inc., 1987.
10. Johnson, H. Thomas. "Managing Costs: An Outmoded Philosophy." *Manufacturing Engineering.* May 1989, 42–46.
11. McDonald, T. Marll. *Mathematical Methods for Social and Management Scientists.* Boston: Houghton Mifflin Co., 1974. Chapter 4.
12. Piascik, Chester. *College Mathematics with Applications.* Columbus, Ohio: Charles E. Merrill, Publishing Co., 1984. Chapter 4.
13. Worthy, Ford S. "Accounting Bore You? Wake Up." *Fortune,* October 12, 1987, 43–53.

14.15 Exercises

1. Refer to Example 14.2 and explain why an 8 percent interest rate was used in advertising the finance rate before the truth-in-lending law was passed.

2. Redo Table 14.1 assuming that interest is paid on the first of each month at an annual rate of 8 percent.

3. Estimate the true interest rate for a $5000 purchase with a 10 percent down payment and forty-eight, $125 payments. Use:
   a) The before truth-in-lending law method.
   b) The method of Example 14.3.
   c) Equation 14.4.

4. Redo Exercise 14.3 using the method from Exercise 2 and $i = 8\frac{1}{3}$ percent.

5. Modify Equation 14.3 to calculate the balance if deposits are made at the beginning of the period instead of at the end of the period.

6. Explain why the model for *continuous* interest is $S = Pe^{rt}$, where e is the base of the natural log system.

7. How much time is required for a single deposit of $500 to grow to $1000 at 8 percent annual interest if the compounding is done:
   a) Annually?
   b) Quarterly?
   c) Monthly?
   d) Daily?
   e) Continually?

8. Show why $(1 + r/k)^k - 1$ represents the effective annual interest rate for a stated interest rate of r if there are k compounding periods in the year. *Hint*: Compare interest earned in a year using Equations 14.1 and 14.2.

9. A bank advertised a 6 percent annual rate of interest with an effective yield of 6.168 percent. How many compounding periods are there?

10. How much money should a person deposit monthly in an account that earns 6 percent compounded monthly if $10,000 is needed in four years?

11. What single sum of money must be deposited into an account that pays 6 percent compounded monthly if $10,000 is needed in four years?

12. Verify Equation 14.2 by proper derivation. *Hint*: Set up a geometric series and use summation procedures.

13. How can Equation 14.3 be used if the compounding period is quarterly and payments are made monthly?

14. How can Equation 14.3 be used if the compounding period is monthly and payments are made quarterly?

15. A sinking fund is used to accumulate money to make a purchase in the future. Periodic payments are made into an interest bearing account for this purpose. What monthly payments are required so that a $1,500,000 building can be built in three years if the account earns 8 percent compounded monthly?

16. In Example 14.6 what percent of the first payment goes toward paying off the mortgage?

17. How much does a $25 product increase in cost over a three-year period if the cost follows an inflation rate of 6 percent per year?

18. A potential home buyer can make $500 per month house payments and has $15,000 for a down payment. If the current finance rate is 11 percent for a 20 year mortgage, and monthly tax escrow payments of $80 are estimated, what price house can this person afford? Escrow payments are payments made by the buyer to the lender to create a fund from which taxes are paid. The escrow is deducted from the monthly payment before interest and principal payments are made. Most lenders do not pay interest on this fund.

19. Occasionally a lender will apply the full monthly payment to interest and principal and when tax payments are made add them back onto principal. What effect would this have on the example in Exercise 14.18?

20. An insurance policy has an annuity option for the owner. At age 65, monthly payments can be taken from the annuity. The minimum annual interest on the annuity is 4 percent but has been averaging 6.2 percent. If the beginning annuity balance is $20,000, and the account compounds monthly, what minimum equal monthly withdrawals can be made if the life expectancy is 13 years and the owner wants the fund to equal zero after 13 years? How does the answer change if the current monthly average continues?

21. Prove that $[1 - (1 + i)^{-n}]/i < n$, and as a result A in Equation 14.5 is less than A in $A = Rn$ as mentioned in Section 14.5.

22. Compare the following two project proposals on the basis of cash flow. Either project will accomplish the same goal.

	Project A	Project B
Project life	10 years	10 years
First cost	$80,000	$120,000
Time to install	12 months	18 months
Annual operating cost	$4000	$3000
Savings/month after installation	$2000	$3000

23. What efforts could be made to make the poorest cash flow project in Exercise 14.22 the better cash flow?

24. An automatic sensor is available that turns the lights in a room off if no one is in the room. The sensor costs $75. Discuss how you would go about determining the feasibility of installing such a sensor. Use $0.06 per kilowatt hour and a cash payback period of three years for a goal. Determine a model relating the number of lights in a room with the number of hours the lights should be off to meet the goal.

25. A six foot long part can be bent to shape on a press brake in one minute. The charge for this operation is $8.50 per hour for labor and $2.00 per hour for overhead. A special rolling machine is proposed for the job that will roll the desired shape at the rate of 60 feet per minute. The proposed machine costs $35,000 plus $10,000 for the required tooling. The labor rate on the proposed machine is the same as the press brake. The tooling must be paid for in one year; the machine must be paid for in 5 years. Analyze this situation.

26. What is the effective federal tax rate on a business that has net profit of:
    a) $50,000?
    b) $100,000?
    c) $500,000?
    d) $1,000,000?

27. Compare these two projects on the basis of cash flow. One of the projects must be implemented because of a change in the law.

	Project A	Project B
Initial capital cost	$10,000	$100,000
Salvage value	—	$10,000
Life of equipment	5 years	10 years
Annual maintenance cost	$4000	$500
Reduced product cost/year	—	$2000

Financial Analysis

28. Compare the probable part cost from machine A and machine B, assuming each will make the part to the same specification. Which machine yields the lowest part cost?

	Machine A	Machine B
Cost	$35,000	$150,000
Life	10 years	8 years
Salvage value	$3500	$15,000
Parts required per year	10,000	10,000
Labor cost per hour	$8	$10
Time to make one part	20 minutes	10 minutes
Maintenance cost per year	$500	$1500

29. Can the decision in Exercise 14.28 be reversed if a ±10 percent variation in life, salvage value, time to make one part, and maintenance costs are considered?
30. Redo Case Study #20 on the basis of a 20 year project life.
31. What is the *ROI* before taxes in Case Study #20?
32. Inflation is always a factor to be reckoned with in financial matters. Table 14.4 lists the annual consumer price index of the United States since 1967. If the sinking fund in Exercise 14.15 is started in 1977 for a planned building project in 1981, how much short will the fund be?

TABLE 14.4

Urban consumer price index in United States, June of each year.

Year	Index	Year	Index
1967	100.0	1979	217.4
1968	104.2	1980	246.8
1969	109.8	1981	272.4
1970	116.3	1982	289.1
1971	121.3	1983	298.4 *(99.6)
1972	125.3	1984	103.9
1973	133.1	1985	107.6
1974	147.7	1986	109.6
1975	161.2	1987	113.6
1976	170.5	1988	118.3
1977	181.5	1989	123.8
1978	195.4		

*The index base was re-established at 100.
Source: United States Department of Labor, Bureau of Labor Statistics.

33. Repeat Exercise 14.32 for the 3 year period 1970–1974.

34. If you were establishing a 3 year sinking fund, how would you take inflation into account?

35. What would the price index for 1989 be if the index had not been adjusted to 100 in 1983?

36. An owner of a plating facility has equipment that needs extensive overhaul because of new government regulations. An overhaul will cost $10,000, and annual operating costs are estimated at $700 per year. The overhaul is estimated to last 10 years. First year maintenance costs are estimated at $1000 and will likely increase 15 percent per year. New equipment would cost $25,000 and have operating costs of $100 per year. Maintenance costs are estimated at $500 in the first year and will increase by 10 percent per year. The scrap value of the present equipment is $3000. Would you overhaul or buy new? Discuss thoroughly.

37. A bridge with an estimated life of 30 years costs $250,000 to build, has annual maintenance costs of $3000, and $8000 painting costs every 5 years. How much money is required at the beginning of the project so no additional funds must be added to the project over the 30 year life if the value of money is 8 percent per year?

38. Refer to Exercise 12.46. A particular house has 200 ft^2 of window area covered with storm windows. It is located where there are an average of 7000 degree days per heating season, and natural gas costs $0.55/100 ft^3. If the required payback period is 5 years, and the current interest rate is 10 percent, how much money can be spent on window covers if they are made with a material that has an R-value of 5?

39. Determine a model that can be used to equalize successive annual payments that increase each year by a constant amount considering a constant interest rate is in effect. In other words, what equal annual payments A, are equivalent to unequal annual payments $G, 2G, 3G, \ldots$. This model is also referred to as the equivalent uniform annual cost form.

40. The model from Exercise 14.39 has value when trying to compare between unequal cost sequences.
    Compare machines A and B. Machine A costs $4,000, has estimated annual maintenance costs of $220, 440, 660, \ldots$, and estimated annual product savings of $800. Machine B costs $12,000, has estimated annual maintenance costs of $60, 120, 180, \ldots$, and estimated annual production savings of $2,000. By equalizing the annual costs, determine which machine is less expensive to own if it will be kept for only a few years, and determine the break-even number of years where both machines cost the same amount per year to own. Assume a 12 percent annual interest rate. *Hint*: Combine the results of Equation 14.5 and Exercise 14.39 for each machine and compare.

41. Reanalyze Exercise 14.22 on the basis of equalized annual payments using Equation 14.4.

42. Reanalyze Exercise 14.27 on the basis of equalized annual payments using Equation 14.4.

43. Case Study #21 suggests listing the parts in quantity required order. What other order could be used for determining which parts to run on the new machine?

15 Ethics

"Example is not the main thing in influencing others. It is the only thing."
◆ ALBERT SCHWEITZER

The purpose of this chapter is to provide food for thought concerning the professional ethics of an engineer. Ethics are influenced and controlled by individuals. Ethics affect the person and the person's profession, and must be considered globally when performing engineering activities. Even though your basic moral character is probably well ingrained in your life, this chapter asks you to stop and consider if your attitudes are consistent with those expected of a professional engineer. It also raises questions to make you think about specifics that you may never have considered. Additional reading of cases that have occurred is good study material. Concern for ethical matters will prepare you to maintain a proper perspective when an unusual or unexpected situation arises. Planning ahead in ethical matters should receive no less of your thoughts than planning ahead for other problem solving activities.

15.1 Definition

Ethics is an area of study involving what is acceptable and what is not acceptable in association with moral duty and personal and professional obligation. Ethical conflicts can cause as much anxiety as other societal pressures such as obtaining food, clothing, and shelter for survival. Engineers are not immune to ethical pressures and in some ways may be more exposed to them than other professionals. Engineers are involved with using resources, many non-replaceable, to produce goods and services for the public. Engineers must be concerned about the uses of the raw materials and the uses of the end products. Are they being used in an ethical manner and for ethical purposes? Should an engineer get involved with the design and construction of a building that is going to be used for the offices of a tobacco company? What if the engineer is convinced that tobacco is harmful to the health of humans? Should an engineer help design planned obsolescence into a product knowing there are limited natural resources to use for the product?

There are no easy answers to many of the ethical questions that engineers have to face in their career. Personal conflicts arise because of the pressure triangle created by wanting to be a responsible engineer, by wanting to fulfill an obligation to the public, and by wanting to be loyal to an employer or client. It is often difficult to balance all three and not sacrifice one's self interest. Conflict of interest within oneself, or between alternatives, exists when personal desires and professional constraints interfere with an employment obligation or when two or more desires and conditions cannot be met at the same time.

15.2 Sources of Problems

Ethical and legal questions can arise from a number of sources. Generally it is a question of the action to take, or not to take, under a given set of circumstances.

Is it ethical, or legal, to use someone else's trade secret if it is used in an unrelated industry? What if your company will go bankrupt if the secret is not used?

Is it ethical to use the company's computer for your own jobs when it is not being used for company jobs? What is the "cost" to the company when someone does this?

Is it ethical to "borrow" a patented design for use if the patent owner is dead? What if your idea is "somewhat" different from the actual patent?

Is it ethical to advertise yourself as available for engineering jobs?

Is it ethical to bid competitively on jobs, thinking that you can do this job for less than standard rates to get your foot in the door? Will competitive bidding guarantee a safe and reliable product for the consumer? How can contract negotiations be impartially centered on the merits of the contract alone?

Is a kickback ethical? Legal? What is the difference between a kickback and an earned commission? Should an engineer accept a commission from a contractor or manufacturer?

Is it ethical to let others, including the public, know when another engineer is doing something that you consider to be bad engineering? Suppose you are wrong in your accusation? When can freedom of speech interfere with ethical behavior within the confines of your profession? Must you always speak out?

How ethical is it to design a product or service system when you are not licensed as an engineer?

Is it ethical, or legal, to agree with another company on the selling price of a product?

Must you go along with your employer when you are asked to perform an activity you consider unethical, such as altering test data? What rights do you have to balance the responsibility between yourself and your employer?

Engineers cause trouble for themselves and the company they work for when they have personal beliefs about their actions that in practice are not true. Engineers often rationalize their behavior by believing an action taken is really not illegal or unethical. Activities in this category include providing favors for clients and customers, and industrial spying. Some will carry this belief to the point of saying "If my actions are wrong, there would be a law against them." In many cases a law is passed only after someone has done a "bad" thing and society passes a law to make a public statement about the activity. The passage of a law, of course, does not change the attitude and actions of all people. If it did we wouldn't need police to enforce the law. Laws cannot be passed to include every possible situation in which an engineer might be involved.

A second belief engineers can have that can cause trouble is the belief they won't get caught, even though they know an activity they engage in is illegal or unethical. From copying homework from a fellow student, exceeding the speed limit while driving, exaggerating contributions for tax deductions, to falsifying product test data—if we knew we were going to get caught, we probably wouldn't do it. Ethical behavior requires actions to be taken assuming someone will find out, whether they do or not.

Another way engineers can get into trouble is to believe that the company they work for expects unethical behavior, as long as the results benefit the company. Examples include using low cost material that does not meet specifications, creating product test data without running the tests, paying for insider information, or misrepresenting a product to a customer to get a sale. A second aspect of this behavior is to believe that the company will support and protect employees who engage in these activities if they are caught. Company managers set the environment for employees to follow by their actions or inactions, but ethical engineers cannot allow themselves to be led or forced into a compromising situation. To help engineers in questionable cases, codes of ethics have been created.

15.3 Codes

Codes of ethics have been developed to help engineers answer questions that arise, to avoid potential problems, and to offer guidelines for expected behavior. Codes are established in many professional areas. Oaths of public office or of organizational leaders are a form of ethical code. Other professionals such as doctors and lawyers also have codes of ethics to guide them in their activities. Engineering societies have written and adopted codes to guide them in engineering activities, but there are over 150 different engineering societies. This presents a small problem because there is no single code comparable to the doctors' AMA or lawyers' state bar. However, all the engineering codes attempt to perform the same service, to help guide the engineer. Codes of ethics for engineers have been around for a long time: the first engineering code in the United States was adopted in 1912 by the American Institute of Electrical Engineers.

A code of ethics has several purposes. One is to give guidance and inspiration to engineers. A code will specify actions to take that engineers should consider when making decisions about their work. The code also is a support to an engineer. It specifies rationale to use for a particular decision. A code also can act as a deterrent to certain activities that otherwise might be considered. The code adds discipline to the profession.

The code can also be used to educate people about the role of the engineer. Education promotes mutual understanding and contributes to the level of professionalism of engineers. Codes also protect the status quo and promote consistent business practices.

Codes not only contain professional restrictions but also specify protections for an individual's rights. Among the rights an engineer has is the right to pursue legitimate interests freely without the fear of discrimination. An engineer has the right to receive payment for work performed. An engineer has the right to refuse to carry out illegal and unethical work.

Discussion is a key ingredient when deciding which actions to take to avoid unethical actions and behavior. The study topics and the exercises are meant for discussion and the sharing of personal perspectives. Any number of sample codes could be included for reference and discussion. The following code is the Code of Ethics of Engineers adopted by The American Society of Mechanical Engineers.

FIGURE 15.1

Code of Ethics

SOCIETY POLICY

ETHICS

ASME requires ethical practice by each of its members and has adopted the following Code of Ethics of Engineers as referenced in the ASME Constitution, Article C2.1.1.

CODE OF ETHICS OF ENGINEERS

The Fundamental Principles

Engineers uphold and advance the integrity, honor and dignity of the engineering profession by:

  I. using their knowledge and skill for the enhancement of human welfare;
  II. being honest and impartial, and serving with fidelity the public, their employers and clients, and
  III. striving to increase the competence and prestige of the engineering profession.

The Fundamental Canons

1. Engineers shall hold paramount the safety, health and welfare of the public in the performance of their professional duties.
2. Engineers shall perform services only in the areas of their competence.
3. Engineers shall continue their professional development throughout their careers and shall provide opportunities for the professional development of those engineers under their supervision.
4. Engineers shall act in professional matters for each employer or client as faithful agents or trustees, and shall avoid conflicts of interest.
5. Engineers shall build their professional reputation on the merit of their services and shall not compete unfairly with others.
6. Engineers shall associate only with reputable persons or organizations.
7. Engineers shall issue public statements only in an objective and truthful manner.

THE ASME CRITERIA FOR INTERPRETATION OF THE CANONS

The ASME criteria for interpretation of the Canons are advisory in character and represent the objectives toward which members of the engineering profession should strive. They constitute a body of principles upon which an engineer can rely for guidance in specific situations. In addition, they provide interpretive guidance to the ASME Board on Professional Practice and Ethics in applying the Code of Ethics of Engineers.

1. Engineers shall hold paramount the safety, health and welfare of the public in the performance of their professional duties.
    a) Engineers shall recognize that the lives, safety, health and welfare of the general public are dependent upon engineering judgments, decisions and practices incorporated into structures, machines, products, processes and devices.
    b) Engineers shall not approve or seal plans and/or specifications that are not of a design safe to the public health and welfare and in conformity with accepted engineering standards.
    c) Whenever the Engineers' professsional judgment is over-ruled under circumstances where the safety, health, and welfare of the public are endangered, the Engineers shall inform their clients and/or employers of the possible consequences and notify other proper authority of the situation, as may be appropriate.
        c.1 Engineers shall do whatever possible to provide published standards, test codes, and quality control procedures that will enable the public to understand the degree of safety or life expectancy associated with the use of the designs, products, or systems for which they are responsible.
        c.2 Engineers shall conduct reviews of the safety and reliability of the designs, products, or systems for which they are responsible before giving their approval to the plans for the design.
        c.3 Whenever Engineers observe conditions which they believe will endanger public safety or health, they shall inform the proper authority of the situation.
    d) If engineers have knowledge or reason to believe that another person or firm may be in violation of any of the provisions of these Canons, they shall present

such information to the proper authority in writing and shall cooperate with the proper authority in furnishing such further information or assistance as may be required.
2. Engineers shall perform services only in areas of their competence.
   a) Engineers shall undertake to perform engineering assignments only when qualified by education or experience in the specific technical field of engineering involved.
   b) Engineers may accept an assignment requiring education or experience outside of their own fields of competence, but their services shall be restricted to other phases of the project in which they are qualified. All other phases of such project shall be performed by qualified associates, consultants, or employees.
3. Engineers shall continue their professional development throughout their careers, and should provide opportunities for the professional development of those engineers under their supervision.
4. Engineers shall act in professional matters for each employer or client as faithful agents or trustees, and shall avoid conflicts of interest.
   a) Engineers shall avoid all known conflicts of interest with their employers or clients and shall promptly inform their employers or clients of any business association, interests, or circumstances which could influence their judgment or the quality of their services.
   b) Engineers shall not undertake any assignments which would knowingly create a potential conflict of interest between themselves and their clients or their employers.
   c) Engineers shall not accept compensation, financial or otherwise, from more than one party for services on the same project, or for services pertaining to the same project, unless the circumstances are fully disclosed to, and agreed to, by all interested parties.
   d) Engineers shall not solicit or accept financial or other valuable considerations, for specifying the products or material or equipment suppliers, without disclosure to their clients or employers.
   e) Engineers shall not solicit or accept gratuities, directly or indirectly, from contractors, their agents, or other parties dealing with their clients or employers in connection with work for which they are responsible.
   f) When in public service as members, advisors, or employees of a governmental body or department, Engineers shall not participate in considerations or actions with respect to services provided by them or their organization(s) in private or product engineering practice.
   g) Engineers shall not solicit an engineering contract from a governmental body on which a principal, officer, or employee of their organization serves as a member.
   h) When, as a result of their studies, Engineers believe a project(s) will not be successful, they shall so advise their employer or client.
   i) Engineers shall treat information coming to them in the course of their assignments as confidential, and shall not use such information as a means of making personal profit if such action is adverse to the interests of their clients, their employers or the public.
      i.1 They will not disclose confidential information concerning the business affairs or technical processes of any present or former employer or client or bidder under evaluation, without his consent, unless required by law.

  i.2 They shall not reveal confidential information or finding of any commission or board of which they are members unless required by law.
  i.3 Designs supplied to Engineers by clients shall not be duplicated by the Engineers for others without the express permission of the client(s).
 j) The Engineer shall act with fairness and justice to all parties when administering a construction (or other) contract.
 k) Before undertaking work for others in which the Engineer may make improvements, plans, designs, inventions, or other records which may justify copyrights or patents, the Engineer shall enter into a positive agreement regarding the rights of respective parties.
 l) Engineers shall admit and accept their own errors when proven wrong and refrain from distorting or altering the facts to justify their decisions.
 m) Engineers shall not accept professional employment outside of their regular work or interest without the knowledge of their employers.
 n) Engineers shall not attempt to attract an employee from another employer by false or misleading representations.

5. Engineers shall build their professional reputation on the merit of their services and shall not compete unfairly with others.
 a) Engineers shall negotiate contracts for professional services on the basis of demonstrated competence and qualifications for the type of professional service required and at fair and reasonable prices.
 b) Engineers shall not request, propose, or accept professional commissions on a contingent basis under circumstances under which their professional judgments may be compromised.
 c) Engineers shall not falsify or permit misrepresentation of their, or their associates, academic or professional qualification. They shall not misrepresent or exaggerate their degrees of responsibility in or for the subject matter of prior assignments. Brochures or other presentations incident to the solicitation of employment shall not misrepresent pertinent facts concerning employers, employees, associates, joint venturers, or their past accomplishments.
 d) Engineers shall prepare articles for the lay or technical press which are only factual, dignified and free from ostentations or laudatory implications. Such articles shall not imply other than their direct participation in the work described unless credit is given to others for their share of the work.
 e) Engineers shall not maliciously or falsely, directly or indirectly, injure the professional reputation, prospects, practice or employment of another engineer, nor shall they indiscriminately criticize another's work.
 f) Engineers shall not use equipment, supplies, laboratory or office facilities of their employers to carry on outside private practice without consent.

6. Engineers shall associate only with reputable persons or organizations.
 a) Engineers shall not knowingly associate with or permit the use of their names or firm names in business ventures by any person or firm which they know, or have reason to believe, are engaging in business or professional practices of a fraudulent or dishonest nature.
 b) Engineers shall not use association with non-engineers, corporations, or partnerships as "cloaks" for unethical acts.

7. Engineers shall issue public statements only in an objective and truthful manner.
 a) Engineers shall endeavor to extend public knowledge, and to prevent misunderstandings of the achievements of engineering.

> b) Engineers shall be completely objective and truthful in all professional reports, statements or testimony. They shall include all relevant and pertinent information in such reports, statements, or testimony.
> c) Engineers, when serving as expert or technical witnesses before any court, commission, or other tribunal, shall express an engineering opinion only when it is founded upon adequate knowledge of the facts in issue, upon a background of technical competence in the subject matter, and upon honest conviction of the accuracy and propriety of their testimony.
> d) Engineers shall issue no statements, criticisms, or arguments on engineering matters which are inspired or paid for by an interested party, or parties, unless they preface their comments by identifying themselves, by disclosing the identities of the party or parties on whose behalf they are speaking, and by revealing the existence of any pecuniary interest they may have in matters under discussion.
> e) Engineers shall be dignified and modest in explaining their work and merit, and shall avoid any act tending to promote their own interest at the expense of the integrity, honor and dignity of the profession or another individual.
>
> 8. Any Engineer accepting membership in The American Society of Mechanical Engineers by this action agrees to abide by this Society Policy on Ethics and procedures for implementation.
>
> Responsibility: Board on Professional Practice and Ethics/Council on Member Affairs

The American Society of Mechanical Engineers, 345 East 47th Street New York, NY 10017, Reprinted with permission.

15.4 Specific Situations

Many ethics violations are centered around money. A person might receive money, or other goods of value, for work not performed or for special favors. Or a person might receive kickbacks for awarding a contract, or tender gifts and favors before the issuing of a contract. Even if a company will receive a contract because of its competitive bid, the giving and receiving of gifts or special entertainment should not be condoned. Even seemingly trivial things such as paying for coffee and rolls, or a round of golf, cannot occur without some form of partiality taking place. If vendor A buys coffee on Monday, what feeling exists for vendor B on Tuesday when the same gesture is not made? What's fair for one must be fair for all; the easiest and most consistent policy is to avoid all such situations. Many corporations have established guidelines against receiving gifts of any kind from companies they do business with. It is against company policy to receive a gift whatever its value, regardless of whether the person receiving the gift has influence on company purchases or not. If all such giving and receiving is curtailed, then no one has to decide what a gift of "nominal value" is.

The following excerpt from an ethical business practices notice sent to employees of a Fortune 500 corporation (name withheld by request) illustrates the degree of importance that some companies attach to policies of professional ethics.

Our Company policy does not permit the acceptance of any gifts by our employees or members of their immediate families from organizations doing business with the Company, or their representatives. Further, our employees must never become involved in any situation which might place them under obligation to any organization with which the Company does business.

Acceptance of gifts, including advertising novelties, by employees or family members from suppliers or business associates is prohibited. The personal use of supplier owned or provided offices, hotels or other residences and facilities is prohibited. An employee or family member is prohibited from using any special discounts or privileges from a supplier that are not available to all employees. Planned social relationships with supplier representatives are discouraged.

It may be regretful that such policy statements are necessary, but business dealings must be conducted on an ethical basis if equity and fairness is to prevail.

An ethic that is violated most often, without regard to type of business or job status, is spending company money improperly. Unwise spending includes using company property for personal use (often called stealing), including the use of company property such as paper, pencils, and other supplies for personal use, as well as making personal phone calls on company time, photocopying personal papers, and using company mail for personal correspondence.

Another category of improper spending involves the purchase of luxury equipment for use on the job when standard equipment would do, such as a $10,000 mahogany desk, or extraordinary plush carpeting for an office.

Impropriety in expense account reporting is another abuse of company funds. If you wouldn't stay at an expensive hotel if you were spending your own money, why would it be considered proper to stay there just because the company is paying for it? Don't use the excuse that everyone else expects it. Just because someone else does something doesn't make it right. Be your own judge of ethical behavior; the real payoff is personal professional integrity. Some case studies will show how difficult a conflict may become for a particular situation, and why ethics can never be neglected.

CASE STUDY #22:
Personal Conflict

A mechanical engineering graduate took a job with one of the larger companies in the United States. Her first assignment was to try to correct product reliability problems on a plastic crash helmet design. The problems seemed to be of a manufacturing nature because of the wide variance in test results of production models. In the process of doing her investigation she learned about the company's facilities, previous helmet design problems that had been solved, how they had been solved, and the structure of the company in general. She worked with other engineers who were still making design changes to try to improve the helmet's performance, including changes to a microphone and earphone system that were part of the helmet. It seemed quite natural to ask about the market that the helmet sold in, its cost, and other marketing concerns. She learned the helmet was used by military pilots, that many were sold to countries other than the United States, and one of the countries was presently at war. TV coverage and news reports indicated that the country had been restricting the civil rights of many of its inhabitants, and this was disturbing to her.

Work continued on improving the quality of the helmet, but at the same time the situation began to add stress to her life; she had misgivings about how her work was contributing to situations that she opposed in other countries. Should she ask for a transfer of assignments? Quit her job? Was she adequately informed of the effect of her work on the civilians in the other country? Who could she talk to about her misgivings? Would she be ridiculed for her feelings just because of her sex? Could she in good conscience continue on the project? Would any action taken by just one person make any difference? She was sure that others would continue the work on improving the helmet. Did older engineers have similar misgivings? How did they rationalize their activities? What should she do?

As mentioned earlier in the chapter, each of us has our own level of tolerance to situations such as this. Some would not even question the significance of the work, responding, "Someone has to do it." Others might say, "If our company doesn't do the work some other company will," or "I can't be responsible for other people."

Can a situation like this have a single answer? No, the situation is in that grey area, not right or not wrong for the majority of people. The young engineer did talk to her superiors, and after working on the helmet project for over a year, and performing well, she was transferred to another project, one with no political overtones.

CASE STUDY #23:
Corporate Conflict

In 1987 the Justice Department filed charges against Chrysler Corporation and two of its executives for selling cars that had been driven with the odometers disconnected, and for selling damage repaired cars as new. Car companies, and other companies, often let employees try out new products to expose them to situations that may have been overlooked during normal testing procedures; Chrysler admitted this is what they were doing. The public exposure of the situation caused faithful Chrysler customers to wonder if they had been shortchanged, and some felt that Chrysler had been unethical in its procedures. Over a ten-year period 72 of the test loaner cars had been involved in accidents, and 40 of those had been repaired and sold as new cars. Thousands of cars had been driven with the odometer disconnected, but only three had been driven over 100 miles.

The company policy was not a secret plot to bilk the customers. No one said, "Let's drive cars with the odometers disconnected so we can sell used cars as new cars." But who is to judge whether the action is ethical or not? If an action violates a law, a person can be tried, convicted, and fined or sent to jail. If a building code is violated, an engineer may lose a job or lose the license to practice engineering, and the out-of-code structure may have to be torn down and rebuilt. When no law or code applies, the customer judges.

In this case the Justice Department considered a law had been broken and acted accordingly; in addition, customers considered the behavior unethical. Chrysler executives wanted to restore faith, make amends to the customers, and not go to jail. Chrysler offered to either replace the cars involved or extend the warranty at no cost to the customer, and pleaded nolo contendere to the criminal charge. The plea was accepted and the charges were dropped. No trials, no lawsuits, and no one lost their job. The consumer and the Justice Department made public an activity that was judged unethical, and the company responded, not with excuses or alibis, but with action.

CASE STUDY #24: Government Ethics.

Government is big business and it involves not only elected officials, but government agencies, contractors, and the voting public. Special interest groups apply pressure in different forms to influence the laws that are passed, the regulations established, and the contracts that are awarded. What constitutes unethical behavior? Where is the line that separates information and data gathering from prejudicial solicitation? When is what is best for the country as a whole not the best for a specific locale?

Engineers deal with the government and its agencies, and must be alert to possible ethical problems, whether on the giving or receiving end of information flow and contract awarding. An editorial in the *Des Moines Register* on March 21, 1989 summarizes problems and concerns.

JUST SAY NO

As inducements go, they aren't much: free tickets to a football game, free cable TV, free movie passes. The perquisites of office say the defenders; the little things to make up for the modest salaries Iowa pays legislators.

And rank hath privileges. If the little freebies are OK for the rank and file, there's nothing wrong with Senate Majority Leader Bill Hutchins and Senate Minority Leader Calvin Hultman accepting $500 payments to listen to a pitch by chemical pusher E.I. du Pont de Nemours and Co., which has a rather large interest in environmental legislation.

There's an old story that fits: The cocktail-hour conversation turns to the price of integrity, and the dowager announces that some things are beyond monetary value. Her honor, for instance. Not even for $1 million? she is asked. Well, perhaps. How about $100,000? What do you think I am? she asks indignantly. We've established that, she is told; now we're just arguing price.

While accepting two-bit freebies may seem far removed from the issue of bribery, recipients have established that they'll accept favors. Now the only question is the price at which such favors merit a return.

If $500 won't buy a crucial vote or a leader's influence regarding a chemical-pollution issue, will $5000? If a choice ticket to the Iowa-Iowa State game won't influence a vote on a majority appropriation, will a whole row of seats?

You're not for sale for any price, right? Then why allow the question to be raised at all by accepting favors?

The larger the favor the more serious the questions. The smaller the favor the easier it is to forego it. Those penny-ante perks may seem laughably insignificant to the recipients. But they're a conditioner, a preparation.

If legislators are so dense as to think the favors emanate solely from the goodness of a lobbyist's soul, their handicap is not shared by their constituents.

It's so easy to avoid even the appearance of a conflict of interest.

Just say no.

Copyright 1989, the *Des Moines Register*, Reprinted with permission.

15.5 Study Cases

1. Kepone, an insecticide ($C_{10}Cl_{10}O$) production in Hopewell, Virginia, Allied Chemical Corporation, 1949–1975
2. Hooker Chemical and Love Canal, 1975
3. B. Everett Jordan Lake and dam, 1967–1976
4. DC–10 airplane, 1972–1974
5. Lockheed bribe payments to Japan, 1972–1973
6. Kermit Vandiver and B.F.Goodrich, 1968
7. Bay Area Rapid Transit System, 1973
8. Ernest Fitzgerald and Lockheed, November, 1968
9. Ford Pinto 1967–1978
10. Three Mile Island nuclear power plant, March 28, 1979
11. Challenger disaster, January 1986
12. Exxon oil spill in Alaska, March 24, 1989
13. Jim Wright, Speaker of The House of Representatives, 1989
14. Robert T. Morris and computer security, 1990

15.6 References

1. Andrews, Kenneth R., ed. *Ethics In Practice*. Cambridge, Mass.: Harvard Business School Press, 1989.
2. Asch, Peter. *Consumer Safety Regulation*. New York: Oxford University Press, 1988.
3. Baum, Robert J. *Ethics and Engineering Curricula*. New York: Institute of Society, Ethics and the Life Sciences, 1980.
4. Berube, Bertrand G. "A Whistle-blower's Perspective of Ethics in Engineering." *Engineering Education*, Feb. 1988, 294–295.
5. Bick, Patricia Ann. *Business Ethics and Responsibility: An Information Sourcebook*. Phoenix, Ariz.: Oryx Press, 1988.
6. Blinn, Keith W. *Legal and Ethical Concepts in Engineering*. Englewood Cliffs, N.J.: Prentice Hall,1989. Case study approach.
7. (The) Business Round Table. *Corporate Ethics: A Prime Business Asset*. New York: 1988.
8. Constance, John Dennis. *How To Become a Professional Engineer*. New York: McGraw-Hill, 1978.
9. Davenport, William H. and Daniel Rosenthal. *Engineering: Its Role and Function in Human Society*. New York: Pergamon Press, 1967.
10. Dimond, Diane. "Know-How or No Way?" *Business and Industry,* January, 1988, 108–115.
11. Florman, Samuel. *The Civilized Engineer*. New York: St. Martin's Press, 1987.

12. Florman, Samuel. *The Existential Pleasures of Engineering.* New York: St. Martin's Press, 1976.
13. Gould, Leroy C., et al. *Perceptions of Technological Risks and Benefits.* New York: Russell Sage Foundation, 1988.
14. Gunn, Alastair S. and P. Aarne Vesilind. *Environmental Ethics for Engineers.* Chelsea, Mich.: Lewis Publishers, 1986.
15. Iacocca, Lee, with Sonny Kleinfield. *Talking Straight.* New York: Bantam Books, 1988.
16. King, W.J. "The Unwritten Laws of Engineering." *Mechanical Engineering,* Vol. 66, No. 5,6,7, 1944.
17. Martin, Mike. *Ethics in Engineering.* New York: McGraw-Hill, 1983. Many case summaries.
18. Red, W. Edward. *Engineering, The Career and the Profession.* Monterey, Cal.: Brooks/Cole, 1982.
19. Schaub, James H. *Engineering Profession and Ethics.* New York: Wiley-Interscience, 1983.
20. Straub, Joseph T. *Applied Management.* Boston: Winthrop Publishers, 1979.
21. Vesilind, P. Aarne. "Rules, Ethics and Morals In Engineering Education." *Engineering Education,* Feb. 1988, 289–293.

15.7 Exercises

1. If a person is ethically against taking the life of another person, does it follow that the person should also be against:
    a) Capital punishment for premeditated murder?
    b) Killing someone in self defense?
    c) War?
    d) Capital punishment for manslaughter?
    e) Mercy killing of the terminally ill?

2. A proposed ethic often heard is, "No one should cause physical suffering to another living thing." If you agreed with that ethic, could you participate in:
    a) Boxing?   *b*) Football?
    c) Hunting?   *d*) Golf?
    e) Baseball?   *f*) Branding of cattle?

3. If all project planners and designers were ethical, would the passage of the National Environmental Policy Act (NEPA) have been required? The NEPA requires an environmental impact statement for some projects.

4. Does the fact that NEPA only covers some, but not all, government projects seem ethical? Weapons testing, for example, is not covered.

5. Is it ethical for an engineer to ignore NEPA guidelines just because the project being worked on does not fall under its jurisdiction?

6. Is it ethical for an engineer to ignore the recommended safety factors in the steam boiler code even though they are not law?

7. An engineer signed a pre-employment agreement that gives the company the rights to all patentable ideas created on the job. During the course of performing product research, the engineer thinks of a novel solution to a problem encountered during previous employment. The previous job was in the food industry; the present job is in the steel making industry. The engineer takes the idea to the former employer and tries to sell it. Discuss the ethics of the situation.

8. A student who was uncertain of a major learned of a scholarship available only to English majors. The student had a 3.90 grade point average and recognized financial need, and felt confident about winning the scholarship. The student declared English as a major and applied for the scholarship. Discuss the ethics of the situation.

9. A computer operator runs a program for private purposes at work when the computer is not running company jobs. The computer is on all the time. Discuss the ethics of the situation.

10. A student gives a non-student friend the password to the school computer so the friend can access some programs. Discuss the ethics of the situation. Does it matter if the school is a private or public school?

11. A student breaks the campus computer security code and reviews student grades, but makes no changes. Discuss the ethics of the situation.

12. A programmer is asked to write a program that will give inaccurate information to the company's stockholders. When she discusses this with her supervisor she is told to write the program or be fired. Discuss the ethics of the situation.

13. An engineer makes personal phone calls at work and does not charge them to a home phone number. Each month the calls amount to $10–15. Discuss the ethics of the situation. Would it make a difference if the calls were charged to the home phone?

14. The recording secretary of a faculty committee that reviewed confidential faculty records and made confidential recommendations on tenure and promotion was a member of a department that had a member up for promotion. The committee recommended promotion but was overruled by the president and the board of regents. In a letter of appeal by the non-promoted faculty member, the phrase "and overruled the committee's recommendation" was used. If all deliberations and recommendations were confidential, how was this information known? Discuss possible unethical behavior.

15. Read and discuss each section of the Code of Ethics for Engineers. Discuss possible situations you might find yourself in where the code would help you decide on a course of action.

16. Find real cases that the Code would have jurisdiction over and discuss if the results were in agreement with the code.

17. A company offered an engineer a raise if the engineer agreed to contribute half the raise to the political campaign fund of a congressional candidate. Consider the ethics of the situation:
    a) If the engineer liked the candidate, or
    b) If the engineer did not like the candidate.

18. How concerned should an engineer be about using raw materials or products from countries whose governments violate "human rights?"

19. Consider an engineer who is against the use of tobacco and whose company has just been awarded a large contract to design and build a factory that will produce cigarettes. What are some personal implications when a supervisor tells the engineer "If our company doesn't do the work some other company will," or "I can't be responsible for other people."

20. Discuss the ethical pressures on an engineer who knows that poor product quality on a project may lead to customer dissatisfaction, but that if the project is brought in on schedule there will be a large bonus handed out.

21. Some business managers strive for short term gains in lieu of long term stability. Quick success often leads to promotion and bonus pay. Discuss the ethics of a manager who knowingly performs this way on the job.

22. Is maximum profit always an ethical goal?

23. Discuss situations where a decision might lead to lower annual profits and yet be good for the company. How could you handle such a situation ethically?

24. How would the answer to Exercise 13.50 change if a manager wanted maximum income rather than maximum profit? Can maximum income be used as a measure of success rather than maximum profit? How?

Appendices

APPENDIX A1

Basic quantities and units.

Quantity	Symbol	Dimensions	Units Metric	English
Acceleration, linear	a	LT^{-2}	meters/sec^2	ft/sec^2
Acceleration, angular	α	T^{-2}	radians/sec^2	radians/sec^2
Acceleration, gravity	g	LT^{-2}	9.80 m/sec^2	32.2 ft/sec^2
Area	A	L^2	meter^2	ft^2
Coefficient of compressibility	β	LT^2M^{-1}	—	—
Damping factor	C	MT^{-1}	N·sec/m	lbf·sec/inch
Displacement, linear	r, d	L	meter	ft
Displacement, angular	$\emptyset$	—	radian	radian
Energy	K, U	ML^2T^{-2}	joule = kg·m^2/sec^2	ft·lbf
Force	F	MLT^{-2}	newton	lbf = lbm·g
Frequency, linear	ν	T^{-1}	hertz	hertz
Frequency, angular	ω	T^{-1}	radians/sec	radians/sec
Length	L	L	meter	ft
Impulse	i	MLT^{-1}	N·sec	lbf·sec
Inertia, rotational	I	ML^2	kg·m^2	lbm/ft^2
Mass	m	M	kilogram	lbm = lbf/g
Mass density	ρ	ML^{-3}	kg/m^3	lbm/ft^3
Mass flow rate	dm/dt	MT^{-1}	kg/sec	lbm/sec
Momentum, linear	**p**	MLT^{-1}	kg·m/sec	lbm·ft/sec
Momentum, angular	L	ML^2T^{-1}	kg·m^2/sec	lbm·ft^2/sec
Period	T	T	second	second
Power	P	ML^2T^{-3}	watt = joule/sec	ft·lbf/sec
Pressure	p	$ML^{-1}T^{-2}$	pascal = N/m^2	lbf/ft^2
Spring constant	k	MT^{-2}	N/m	lbf/inch
Stress	σ	$ML^{-1}T^{-2}$	N/m^2	lbf/ft^2
Surface tension	σ	MT^{-2}	N/m	lbf/inch
Time	t	T	second	second
Torque	τ	ML^2T^{-2}	N·m	ft·lbf
Velocity, linear	v	LT^{-1}	meters/sec	feet/sec
Velocity, angular	ω	T^{-1}	radians/sec	radians/sec
Viscosity, dynamic		$ML^{-1}T^{-1}$	N·sec/m^2	lbm/ft·sec
Viscosity, kinematic		L^2T^{-1}	m^2/sec	ft^2/sec
Volume	V	L^3	meter^3	ft^3
Volume flow rate	Q	L^3T^{-1}	m^3/sec	ft^3/sec
Wavelength	l	L	meter	ft
Work	W	ML^2T^{-2}	joule	ft·lbf

APPENDIX A2

Basic quantities and units.

Quantity	Symbol	Dimensions	Units
Entropy	S	ML^2T^{-2}	joules/°K
Internal energy	U	ML^2T^{-2}	joule
Heat	Q	ML^2T^{-2}	joule
Temperature	T	—	°Kelvin
Capacitance	C	$M^{-1}L^{-2}T^2Q^2$	farad
Charge	q	Q	coulomb = amp·sec
Conductivity	σ	$M^{-1}L^{-3}TQ^2$	1/ohm·meter = mho
Current	i	$T^{-1}Q$	ampere
Current density	j	$L^{-2}T^{-1}Q$	amp/meter^2
Electric dipole moment	**p**	LQ	coulomb·meter
Electric displacement	**D**	$L^{-2}Q$	coulomb/meter^2
Electric polarization	**P**	$L^{-2}Q$	coulomb/meter^2
Electric field strength	**E**	$MLT^{-2}Q^{-1}$	volts/meter
Electric flux	Φ	$ML^3T^{-2}Q^{-1}$	volt·meter
Electric potential	V	$ML^2T^{-2}Q^{-1}$	volt
Electromotive force	V	$ML^2T^{-2}Q^{-1}$	volt
Inductance	L	ML^2Q^{-2}	henry
Magnetic dipole moment	μ	$L^2T^{-1}Q$	amp·meter^2
Magnetic field strength	**H**	$MT^{-1}Q$	amp·meter
Magnetic flux	Φ	$ML^2T^{-1}Q^{-1}$	weber = volt·sec
Magnetic induction	B	$MT^{-1}Q^{-1}$	tesla = webers/meter^2
Magnetization	M	$L^{-1}T^{-1}Q$	amp/meter
Permeability	μ	MLQ^{-2}	henrys/meter
Permittivity	ϵ	$M^{-1}L^{-3}T^2Q^2$	farads/meter
Resistance	R	$ML^2T^{-1}Q^{-2}$	ohm
Resistivity	ρ	$ML^3T^{-1}Q^{-2}$	ohm·meter
Voltage	V	$ML^2T^{-2}Q^{-1}$	volt

APPENDIX B

SI prefixes and multiplying factors.

Prefix Name	Symbol	Multiplying Factor
tera	T	10^{12}
giga	G	10^{9}
mega	M	10^{6}
kilo	k	10^{3}
hecto*	h	10^{2}
deca*	da	10
deci*	d	10^{-1}
centi	c	10^{-2}
milli	m	10^{-3}
micro	μ	10^{-6}
nano	n	10^{-9}
pico	p	10^{-12}
femto	f	10^{-15}
atto	a	10^{-18}

* Not commonly used

APPENDIX C

Selected conversion factors.

To Convert From	To	Multiply By
Btu	gram calorie	252
Btu	joule	1055
Btu	foot-pounds	778
cubic feet	gallon	7.48
foot-pound	newton·meter = joule	1.356
gallon, U.S. liquid	cubic feet	0.134
gallon, U.S. liquid	cubic inches	231
gallon, U.S. liquid	liter	3.79
gallon, U.S. liquid	pounds of water	8.34
gram/cm^3	pounds/ft^3	62.43
horsepower	watts	745.7
horsepower	ft·lbf/second	550
inch	millimeter	25.4
kilogram	pound	2.20
kilowatt·hour	Btu	3414
meter	inch	39.37
millimeters of Hg	atmosphere	0.001316
newton	pound force (lbf)	0.2248
pound	gram	453.6
lbf/in^2	kPascal	6.895
watt	Btu/hr	3.412

APPENDIX D

Average energy values of common fuels.

Fuel	Energy
Coal	
Anthracite	14,000 Btu/pound
Bituminous	12,000 Btu/pound
Lignite	7000 Btu/pound
Crude oil	143,000 Btu/gallon
Diesel fuel*	135-160,000 Btu/gallon
Ethyl alcohol (Ethanol)	75-84,000 Btu/gallon
Fuel oil	140,000 Btu/gallon
Gasoline*	110-126,000 Btu/gallon
Hydrogen (liquid)	34,000 Btu/gallon
Methanol	56,500 Btu/gallon
Natural gas (@ furnace burner)	1000 Btu/ft^3
Compressed @ 3000 psi	29,000 Btu/gallon
Liquid	73,500 Btu/gallon
Propane (bottled gas)	94,000 Btu/gallon

* A function of crude oil composition and distillation technique.

APPENDIX E

Table of selected elements.

Element	Melting Temp °C	Solid Density Mg/m^3 @ 20°C*	Crystal Structure @ 20°C	Common Valence
Aluminum	660.4	2.699	fcc	+3
Antimony	630.7	6.62	—	+5
Argon	−189.2	—	—	Inert
Arsenic	817	5.72	—	+3
Beryllium	1278±5	1.85	hcp	+2
Boron	2300	2.34	—	+3
Calcium	839±2	1.55	fcc	+2
Carbon	~3550	2.25	hex	—
Chlorine	−101	—	—	−1
Chromium	1857±20	7.19	bcc	+3
Cobalt	1495	8.9	hcp	+2
Copper	1083	8.96	fcc	+1
Fluorine	−220	—	—	−1
Germanium	937	5.32	diamond cubic	+4
Gold	1064	19.32	fcc	+1
Helium	−272.2	—	—	Inert
Hydrogen	−259.14	—	—	+1
Iodine	114	4.94	noncubic	−1
Iron	1535	7.87	bcc	+2, +3
Lead	327.5	11.34	fcc	+2
Lithium	180.5	0.534	bcc	+1
Magnesium	649	1.74	hcp	+2
Manganese	1244	7.4	—	+2
Mercury	−38.87	—	—	+2
Neon	−248.7	—	—	Inert
Nickel	1453	8.90	fcc	+2
Nitrogen	−210	—	—	−3
Oxygen	−218.4	—	—	−2
Phosphorus	44	1.8	—	+5
Potassium	64	0.86	bcc	+1
Silicon	1410	2.33	diamond cubic	+4
Silver	961.9	10.5	fcc	+1
Sodium	97.8	0.97	bcc	+1
Sulfur	113	2.07	—	−2
Tin	232	7.3	—	+4
Titanium	1660±10	4.51	hcp	+4
Tungsten	3410±20	19.3	bcc	+4
Uranium	1132	18.7	—	+4
Zinc	419.6	7.135	hcp	+2

* Divide by 27.68 to get lbf/in^3.

APPENDIX F1

Properties of selected materials.

Material	Specific gravity	Thermal conductivity, $\left[\dfrac{\text{Btu·in.}}{\text{ft}^2\cdot\text{sec}\cdot°\text{F}}\right]$ at 68°F*	Thermal expansion, in/in/°F at 68°F†	Electrical resistivity, ohm·cm at 68°F‡	Average modulus of elasticity, psi at 68°F§
Metals					
Aluminum (99.9+)	2.7	0.43	12.5×10^{-6}	2.7×10^{-6}	10×10^6
Aluminum alloys	2.7(+)	0.3(±)	12×10^{-6}	$3.5 \times 10^{-6}(+)$	10×10^6
Brass (70Cu-30Zn)	8.5	0.235	11×10^{-6}	6.2×10^{-6}	16×10^6
Bronze (95Cu-5Sn)	8.8	0.16	10×10^{-6}	9.5×10^{-6}	16×10^6
Cast iron (gray)	7.15	—	5.8×10^{-6}	—	$12\text{-}25 \times 10^6$
Cast iron (white)	7.7	—	5×10^{-6}	—	30×10^6
Copper (99.9+)	8.9	0.77	9×10^{-6}	1.7×10^{-6}	16×10^6
Iron (99.9+)	7.87	0.14	6.53×10^{-6}	9.7×10^{-6}	30×10^6
Lead (99+)	11.34	0.06	16×10^{-6}	20.65×10^{-6}	2×10^6
Magnesium (99+)	1.74	0.37	14×10^{-6}	4.5×10^{-6}	6.5×10^6
Monel (70Ni-30Cu)	8.8	0.05	8×10^{-6}	48.2×10^{-6}	26×10^6
Silver	10.5	0.79	10×10^{-6}	1.6×10^{-6}	11×10^6
Steel (1020)	7.86	0.096	6.5×10^{-6}	16.9×10^{-6}	30×10^6
Steel (1040)	7.85	0.093	6.3×10^{-6}	17.1×10^{-6}	30×10^6
Steel (1080)	7.84	0.089	6.0×10^{-6}	18.0×10^{-6}	30×10^6
Steel (18Cr-8Ni stainless)	7.93	0.028	5×10^{-6}	70×10^{-6}	30×10^6
Ceramics					
Al_2O_3	3.8	0.057	5×10^{-6}	—	50×10^6
Brick					
Building	2.3(±)	0.0012	5×10^{-6}	—	—
Fireclay	2.1	0.0016	2.5×10^{-6}	1.4×10^8	—
Graphite	1.5	—	3×10^{-6}	—	—
Paving	2.5	—	2×10^{-6}	—	—
Silica	1.75	0.0016	—	1.2×10^8	—
Concrete	2.4(±)	0.002	7×10^{-6}	—	2×10^6
Glass					
Plate	2.5	0.0014	5×10^{-6}	10^{14}	10×10^6
Borosilicate	2.4	0.002	1.5×10^{-6}	$>10^{17}$	10×10^6
Silica	2.2	0.0025	0.3×10^{-6}	$\sim 10^{20}$	10×10^6
Vycor	2.2	0.0025	0.35×10^{-6}	—	—
Wool	0.05	0.0005	—	—	—
Graphite (bulk)	1.9	—	3×10^{-6}	10^{-3}	1×10^6
MgO	3.6	—	5×10^{-6}	10^5(2000°F)	30×10^6
Quartz (SiO_2)	2.65	0.025	—	10^{14}	45×10^6
SiC	3.17	0.025	2.5×10^{-6}	2.5(2000°F)	—
TiC	4.5	0.06	4×10^{-6}	50×10^{-6}	50×10^6

Properties of selected materials. (*Continued*)

Material	Specific gravity	Thermal conductivity, $\left[\dfrac{\text{Btu·in.}}{\text{ft}^2\text{·sec·°F}}\right]$ at 68°F*	Thermal expansion, in/in/°F at 68°F†	Electrical resistivity, ohm·cm at 68°F‡	Average modulus of elasticity, psi at 68°F§
Polymers					
Melamine-formaldehyde	1.5	0.00057	15×10^{-6}	10^{13}	1.3×10^6
Phenol-formaldehyde	1.3	0.00032	40×10^{-6}	10^{12}	0.5×10^6
Urea-formaldehyde	1.5	0.00057	15×10^{-6}	10^{12}	1.5×10^6
Rubbers (synthetic)	1.5	0.00025	—	—	600–11,000
Rubber (vulcanized)	1.2	0.00025	45×10^{-6}	10^{14}	0.5×10^6
Polyethylene (LD)	0.92	0.00065	100×10^{-6}	10^{15}–10^{18}	14,000–50,000
Polyethylene (HD)	0.96	0.0010	70×10^{-6}	10^{16}–10^{18}	50,000–180,000
Polystyrene	1.05	0.00016	40×10^{-6}	10^{18}	0.4×10^6
Polyvinyl chloride	1.3	0.0003	75×10^{-6}	10^{14}	100,000–600,000
Polyvinylidene chloride	1.7	0.00025	105×10^{-6}	10^{13}	0.05×10^6
Polytertrafluoroethylene	2.2	0.0004	55×10^{-6}	10^{16}	50,000–100,000
Polymethyl methacrylate	1.2	0.0004	50×10^{-6}	10^{16}	0.5×10^6
Nylon	1.15	0.0005	55×10^{-6}	10^{14}	0.4×10^6

*Multiply by 5.19 to get joule · cm/cm² · sec · °C.
†Multiply by 1.8 to get cm/cm/°C.
‡Divide by 2.54 to get ohm · in.
§Multiply by 6900 to get n/m².

L. H. VanVlack, *A Textbook of Materials Technology.* Copyright © 1973, Addison-Wesley Publishing Co., Reading, Massachusetts. Appendix B pp. 313–315. Reprinted with permission.

APPENDIX F2

Average properties of selected engineering materials (English System of Units).

Materials	Specific Weight (lb/in.³)	Elastic Strength[a] Tension (ksi)	Elastic Strength[a] Comp. (ksi)	Elastic Strength[a] Shear (ksi)	Ultimate Strength Tension (ksi)	Ultimate Strength Comp. (ksi)	Ultimate Strength Shear (ksi)	Endurance Limit[c] (ksi)	Modulus of Elasticity (1000 ksi)	Modulus of Rigidity (1000 ksi)	Percent Elongation in 2 in.	Coefficient of Thermal Expansion (10^{-6}/°F)
Ferrous metals												
Wrought iron	0.278	30	[b]		48	[b]	25	23	28		30[d]	6.7
Structural steel	0.284	36	[b]		66	[b]		28	29	11.0	28[d]	6.6
Steel, 0.2% C hardened	0.284	62	[b]		90	[b]			30	11.6	22	6.6
Steel, 0.4% C hot-rolled	0.284	53	[b]		84	[b]		38	30	11.6	29	
Steel, 0.8% C hot-rolled	0.284	76	[b]		122	[b]			30	11.6	8	
Cast iron—gray	0.260				25	100		12	15		0.5	6.7
Cast iron—malleable	0.266	32	[b]		50	[b]			25		20	6.6
Cast iron—nodular	0.266	70			100				25		4	6.6
Stainless steel (18-8) annealed	0.286	36	[b]		85	[b]		40	28	12.5	55	9.6
Stainless steel (18-8) cold-rolled	0.286	165	[b]		190	[b]		90	28	12.5	8	9.6
Steel, SAE 4340, heat-treated	0.283	132	145		150		95	76	29	11.0	19	
Nonferrous metal alloys												
Aluminum, cast, 195-T6	0.100	24	25		36	[b]	30	7	10.3	3.8	5	12.5
Aluminum, wrought, 2014-T4	0.101	41	41	24	62	[b]	38	18	10.6	4.0	20	12.5
Aluminum, wrought, 2024-T4	0.100	48	48	28	68	[b]	41	18	10.6	4.0	19	12.5
Aluminum, wrought, 6061-T6	0.098	40	40	26	45	[b]	30	13.5	10.0	3.8	17	12.5
Magnesium, extrusion, AZ80X	0.066	35	26		49	[b]	21	19	6.5	2.4	12	14.4
Magnesium, sand cast, AZ63-HT	0.066	14	14		40	[b]	19	14	6.5	2.4	12	14.4
Monel, wrought, hot-rolled	0.319	50	[b]		90			40	26	9.5	35	7.8
Red brass, cold-rolled	0.316	60			75				15	5.6	4	9.8
Red brass, annealed	0.316	15	[b]		40	[b]			15	5.6	50	9.8

Average properties of selected engineering materials (English System of Units). (Continued)

Materials	Specific Weight (lb/in.³)	Elastic Strength[a] Tension (ksi)	Elastic Strength[a] Comp. (ksi)	Elastic Strength[a] Shear (ksi)	Ultimate Strength Tension (ksi)	Ultimate Strength Comp. (ksi)	Ultimate Strength Shear (ksi)	Endurance Limit[c] (ksi)	Modulus of Elasticity (1000 ksi)	Modulus of Rigidity (1000 ksi)	Percent Elongation in 2 in.	Coefficient of Thermal Expansion (10^{-6}/°F)
Bronze, cold-rolled	0.320	75			100				15	6.5	3	9.4
Bronze, annealed	0.320	20	[b]		50	[b]			15	6.5	50	9.4
Titanium alloy, annealed	0.167	135	[b]		155	[b]			14	5.3	13	
Invar, annealed	0.292	42	[b]		70	[b]			21	8.1	41	0.6
Nonmetallic materials												
Douglas fir, green[e]	0.022	4.8	3.4			3.9	0.9		1.6			
Douglas fir, air dry[e]	0.020	8.1	6.4			7.4	1.1		1.9			
Red oak, green[e]	0.037	4.4	2.6			3.5	1.2		1.4			1.9
Red oak, air dry[e]	0.025	8.4	4.6			6.9	1.8		1.8			
Concrete, medium strength	0.087		1.2			3.0			3.0			6.0
Concrete, fairly high strength	0.087		2.0			5.0			4.5			6.0

Exact values may vary widely with changes in composition, heat threatment, and mechanical working. More precise information can be obtained from manufacturers.

[a] Elastic strength may be represented by proportional limit, yield point, or yield strength at a specified offset (usually 0.2 percent for ductile metals).
[b] For ductile metals (those with an appreciable ultimate elongation), it is customary to assume the properties in compression have the same values as those in tension.
[c] Rotating beam.
[d] Elongation in 8 in.
[e] All timber properties are parallel to the grain.

A. Higdon, et al., *Mechanics of Materials*. 4th ed. Copyright © 1985, John Wiley & Sons, Inc., New York. Reprinted with permission of John Wiley & Sons, Inc.

APPENDIX F3

Mechanical properties of selected cast irons.

Grey Irons

ASTM Class	S_{ut} ksi	S_{uc} ksi	S_{us} ksi	S_e ksi	Brinell Hardness	E Mpsi	G Mpsi
20	22	83	26	10	156	9.6–14.0	3.9–5.6
25	26	97	32	11.5	174	11.5–14.8	4.6–6.0
30	31	109	40	14	210	13.0–16.4	5.2–6.6
35	36.5	124	48.5	16	212	14.5–17.2	5.8–6.9
40	42.5	140	57	18.5	235	16.0–20.0	6.4–7.8
50	52.5	164	73	21.5	262	18.8–22.8	7.2–8.0
60	62.5	188	88.5	24.5	302	20.4–23.5	7.8–8.5

Nodular Irons

Grade	S_{yt} ksi	S_{ut} ksi	E Mpsi	Brinell Hardness	Elongation, min. % in 2 inches
Tensile Properties—0.2% offset yield					
60-40-18	47.7	66.9	24.5	167	15.0
65-45-12	48.2	67.3	24.4	167	15.0
80-55-06	52.5	81.1	24.4	192	11.2
120-90-02	125.3	141.3	23.8	331	1.5
Compressive Properties—0.2% offset yield					
60-40-18	52.0	—	23.8	—	—
65-45-12	52.5	—	23.6	—	—
80-55-06	56.0	—	23.9	—	—
120-90-02	133.5	—	23.8	—	—
Torsional Properties—0.0375% offset yield					
60-40-18	28.3	68.5	9.1	—	—
65-45-12	30.0	68.9	9.3	—	—
80-55-06	28.0	73.1	9.0	—	—
120-90-02	71.3	126.9	9.2	—	—

Mechanical properties of selected cast irons. *(Continued)*

Class	S_{yt} ksi	S_{ut} ksi	Brinell Hardness	Elongation, min. % in 2 inches
Malleable Irons				
Ferritic: ASTM A47, A338; ANSI G48.1; FED QQ-I-666C				
32510	32	50	156 max	10
35018	35	53	156 max	18
Ferritic: ASTM A197				
—	30	40	156 max	5
Pearlitic and Martensitic: ASTM A220; ANSI G48.2; MIL-I-11444B				
40010	40	60	149–197	10
45008	45	65	156–197	8
45006	45	65	156–207	6
50005	50	70	179–229	5
60004	60	80	197–241	4
70003	70	85	217–269	3
80002	80	95	241–285	2
90001	90	105	269–321	1
Automotive: ASTM A602; SAE J158				
M3210, Annealed	32	50	156 max	10
M4504, Air quenched & tempered	45	65	163–217	4
M5003, Air quenched & tempered	50	75	187–241	3
M5503, Liquid quenched & tempered	55	75	187–241	3
M7002, Liquid quenched & tempered	70	90	229–269	2
M8501, Liquid quenched & tempered	85	105	269–302	1

Source: *Metals Handbook, Desk Edition*. Copyright © 1985, American Society for Metals, Metals Park, OH.

APPENDIX F4

Properties of selected aluminum alloys.

Alloy designation	Principal alloying elements	Hardening process*	Tensile strength, 10^7 N/m²	Yield strength, 10^7 N/m²	Elongation in 50 mm, %	Endurance strength (5×10^8 cycles), 10^7 N/m²	Typical applications
Wrought alloys							
1100	Commercial purity	Cold-working	9–16	3–15	45–15	3–6	Cooking utensils
5052	2.5% Mg	Cold-working	19–29	9–26	30–8	11–14	Bus and truck bodies
Alclad 2024	4.5% Cu, 1.5% Mg (with protective sheet of pure aluminum)	Precipitation	19–47	8–32	22–19	9–14	Aircraft
6061	1.5% Mg$_2$Si	Precipitation	12–31	6–28	30–17	6–10	General structural
7075	5.6% Zn, 2.5% Mg, 1.6% Cu	Precipitation	23–57	10–50	16–11	12–16	Aircraft
Cast alloys							
195	4.5% Cu	Precipitation	22–28	11–22	8.5–2	5–6	Sand castings
319	3.5% Cu, 6.3% Si	Precipitation	19–25	12–18	2–1.5	7–8	Sand castings
356	7% Si, 0.3% Mg	Precipitation	23–26	15–19	5–4	8–9	Permanent mold castings

* In addition to alloy hardening.
To convert from N/m² to lb/in², multiply by 1.450×10^{-4}.

Guy, A.G. and J.J. Hren, *Elements of Physical Metallurgy*, 3rd ed. Copyright © 1974, Addison-Wesley Publishing Co., Reading, Mass. Table 10.4, p. 495. Reprinted with permission.

APPENDIX F5

Properties of selected magnesium alloys.

Use	Am. Soc. of Testing Materials designation	Composition, % Al	Zn	Mn	Zr	Condition	Tensile strength, 10^7 N/m²	Yield strength, 10^7 N/m²	Elongation in 50 mm, %
Sand and permanent mold casting	AZ92A	9.0	2.0	0.1		As-cast	17	10	2
						Solution treated	27	10	10
						Aged	27	14	3
Die casting	AZ91C	9.0	0.7	0.2		As-cast	23	14	3
Sheet	AZ31B	3.0	1.0	0.45		Annealed	26	15	15
						Hard	29	21	8
						Extruded	26	20	15
Structural shapes	AZ80A	8.5	0.5	0.2		Extruded	34	25	12
						Extruded and aged	38	28	8
	ZK60A		5.5		0.5	Extruded	34	26	12
						Extruded and aged	37	30	10

To convert from N/m² to lb/in², multiply by 1.450×10^{-4}.

Guy, A.G. and J.J. Hren, *Elements of Physical Metallurgy*, 3rd ed. Copyright © 1974, Addison-Wesley Publishing Co., Reading, Mass. Table 7.7, p. 353. Reprinted with permission.

APPENDIX F6

Properties of selected brasses.

Alloy	Cu	Zn	Sn	Pb	Mn	Tensile strength, 10^7 N/m²	Yield strength, 10^7 N/m²	Elongation in 50 mm, %	Typical applications
Wrought alloys (annealed)									
Gilding metal	95	5				21	7	40	Coins, jewelry base for gold plate
Commercial bronze	90	10				23	7	45	Grillwork, costume jewelry
Red brass	85	15				25	7	48	Weatherstrip, plumbing lines
Low brass	80	20				26	8	52	Musical instruments, pump lines
Cartridge brass	70	30				28	7	65	Radiator cores, lamp fixtures
Yellow brass	65	35				30	11	55	Reflectors, fasteners, springs
Muntz metal	60	40				37	14	40	Large nuts and bolts, architectural panels
Low-leaded brass	64.5	35		0.5		30	11	55	Ammunition primers, plumbing
Medium-leaded brass	64	35		1.0		30	11	52	Hardware, gears, screws
Free-cutting brass	62	35		3.0		30	12	51	Automatic screw machine stock
Admiralty metal	71	28	1			37	15	65	Condenser tubes, heat exchangers
Naval brass	60	39	1			40	19	45	Marine hardware, valve stems
Manganese bronze	58.5	39.2	1	1 Fe	0.3	45	21	33	Pump rods, shafting rods
Cast alloys									
Cast red brass	85	5	5	5		23	12	25	Pipe fittings, small gears, bearings
Cast yellow brass	60	38	1	1		28	10	25	Hardware fittings, ornamental castings
Cast manganese bronze	58	39.7	1 Al	1 Fe	0.3	48	19	30	Propeller hubs and blades

Guy, A.G. and J.J. Hren, *Elements of Physical Metallurgy*, 3rd ed. Copyright © 1974, Addison-Wesley Publishing Co., Reading, Mass. Table 7.4, p. 348. Reprinted with permission.

APPENDIX F7
Properties of selected bronzes.

Alloy	Condition	Cu	Zn	Sn	Pb	(other)	Tensile strength, 10^7 N/m²	Yield strength, 10^7 N/m²	Elongation in 50 mm, %	Typical applications
Tin-Bronzes										
5% phosphor-bronze	Wrought, cold-worked	94.8	5			0.2 P	63	61	5	Diaphragms, springs, switch parts
10% phosphor-bronze	Wrought, cold-worked	89.8	10			0.2 P	74	68	10	Bridge bearing plates, special springs
Leaded tin-bronze	Sand cast	88	6	4.5	1.5		26	11	35	Valves, gears, bearings
Gun metal	Sand cast	88	10	2			28	11	30	Fittings, bolts, pump parts
Aluminum-Bronzes (eutectoid hardening)										
5% aluminum-bronze	Wrought, annealed	95				5 Al	41	17	66	Corrosion-resistant tubing
10% aluminum-bronze	Sand cast	86			3.5 Fe	10.5 Al	52	24	20	
Same	Sand cast and hardened						69	31	12	Gears, bearings, bushings
Silicon-Bronze										
Silicon-bronze, type A	Wrought, annealed	95			1 Mn	3 Si	39	14	63	Chemical equipment, hot water tanks
Nickel-Bronzes										
30% cupro-nickel	Wrought, annealed	70				30 Ni	41	17	45	Condenser, distiller tubes
18% nickel-silver	Wrought, annealed	65		17		18 Ni	39	17	42	Table flatware, zippers
Nickel-silver	Sand cast	64	4	8	4	20 Ni	28	17	15	Marine castings, valves
Beryllium-Bronzes (precipitation hardening)										
Beryllium-copper	Wrought, annealed	98			0.3 Co	1.7 Be	48	21	50	Springs, nonsparking tools
Same	Wrought, hardened						121	90	5	

Guy, A.G. and J.J. Hren, *Elements of Physical Metallurgy*, 3rd ed. Copyright © 1974, Addison-Wesley Publishing Co., Reading, Mass. Table 7.6, p. 351. Reprinted with permission.

APPENDIX F8

Properties of selected nickel alloys.

Alloy	Principal alloying elements	Tensile strength, lb/in²	Yield strength, lb/in²	Elongation in 2 inches, %	Endurance strength (10^8 cycles), lb/in²	Typical applications
"A" nickel, hot-rolled	none	65,000	30,000	50	30,000	Chemical industry, nickel plating
Duranickel, cold-drawn and precipitation hardened	4.4% aluminum 0.4% titanium	190,000	150,000	20	60,000	Springs, plastics, extrusion equipment
Monel, hot-rolled	30% copper	90,000	55,000	35	42,000	Oil refinery parts
"K" Monel, cold-drawn and precipitation hardened	29% copper 3% aluminum	180,000	160,000	5	46,000	Pump rods, valve stems
Inconel, hot-rolled	14% chromium 6% iron	120,000	65,000	35	45,000	Gas turbine parts
Inconel "X" hot-rolled and aged at 1300°F (704°C)	15% chromium 7% iron 2.5% titanium	180,000	130,000	25	60,000	Springs and bolts subjected to corrosion

Guy, A.G., *Physical Metallurgy For Engineers*. Copyright © 1962, Addison-Wesley Publishing Co., Reading, Mass. Table 9.2, p. 284. Reprinted with permission.

APPENDIX F9

Properties of selected titanium alloys.

Alloy	Heat treatment	Yield strength, lb/in²	Tensile strength, lb/in²	Elongation in 2 inches, %	Reduction of area, %
Commercial titanium	Annealed at 1100 to 1350°F (593 to 732°C)	80,000	95,000	20	42
Ti-6Al-4V	Water quenched from 1750°F (954°C); aged at 1000°F (538°C) for 2 hr	150,000	170,000	13	14
Ti-4Al-4Mn	Water quenched from 1450°F (788°C); aged at 900°F (482°C) for 8 hr	170,000	185,000	13	37
Ti-5Al-2.75Cr-1.25Fe	Water quenched from 1450°F; aged at 900°F for 6 hr	175,000	185,000	10	20
Ti-5Al-1.5Fe-1.4Cr-1.2Mo	Water quenched from 1625°F (885°C); aged at 1000°F for 6 hr	170,000	185,000	15	25

Guy, A.G., *Physical Metallurgy For Engineers*, Copyright © 1962, Addison-Wesley Publishing Co., Reading, Mass. Table 5.8, p. 163. Reprinted with permission.

APPENDIX F10

Properties of selected ceramics @ 20°C.

	Melting Point, °C	Tensile or Bend Strength (σ_f), MPa	Young's Modulus (E), GPa	Fracture Toughness (R), J/m²	Coefficient of Thermal Expansion (α), °C^{-1} × 10^{-6}	Thermal Conductivity (k), W/mK	Spalling Index ($k\sigma_f/E\alpha$), kW/m	Toughness/Strength Ratio (K_c/σ_f) = $\sqrt{ER/\sigma_f}$, $\sqrt{m}$
Typical porous insulating firebrick	?	~15	20	? 30	7	1	0.1	0.05
Alumina (Al$_2$O$_3$)	2050	300 (700 whisker)	380	80	8	38	3.8	5.5
Magnesia (MgO)	2850	100	315	70	14.8	44	1	0.5
Silicon carbide (SiC)	2300 (decomposes)	200	350	50	4.5	209	26	0.02
Silicon nitride (Si$_3$N$_4$)	1900 (sublimes)	600 (1400 whisker)	320	80	2.5	8	6	0.01
Titanium carbide (TiC), Ni/Mo binder	3250	200	350	400	7	17	1.4	0.6
Tungsten carbide (WC), Co binder	2620	350	700	1000	1	34	1.4	0.8
Common bulk glass	Fuses at 1600	70 (up to 4000 filaments)	70	20	10	1	0.1	0.2
Pyroceram glass ceramic	Fuses at 1600	190	126	? 40	10	3	0.5	0.1
Concrete, reinforced	?	20	35	30	7	0.5	0.1	0.1
Carbon graphite	3600 (sublimes)	30	6	100	3	~100	170	0.3

NOTE: There is much more variation in properties of these types of solid than metals, variations occurring with production methods. Properties also vary with temperature and rate, for example, k for Al$_2$O$_3$ is 4 W/mK at 1000°C and is 13 W/mK at 1000°C for SiC.

David K. Felbeck and Anthony G. Atkins, *Strength & Fracture of Engineering Solids*. Copyright © 1984, p. 527. Reprinted by permission of Prentice-Hall, Inc., Englewood Cliffs, N.J.

APPENDIX F11

Properties of concrete.

Property	Value*
Compressive strength	3000–7000 psi
Tensile strength	400 psi
Modulus of elasticity	$3.5–5.0 \cdot 10^6$ psi
Modulus of rupture	700 in·lb/in^3
Density	
Wet	145 lb/ft^3
Dry	96 lb/ft^3

* Values depend on water content, cement : sand : gravel ratio, and cure time.

APPENDIX F12

Selected values of modulus of elasticity @ 20°C (Chapter 8: 1, 11, 20, and 36).

Material	Mpsi	GPa
Aluminum	10.2	70.5
Beryllium	37.7	260
Brick	1.5–2.5	10.5–17.2
Brass	14.6	101
Ceramics—See Appendix F-10		
Cobalt	30.4	210
Concrete—See Appendix F-1 & F-11		
Chromium	40.5	280
Copper	18.8	130
Glass	9–10	62.0–68.9
Graphite	1.3	9
Irons—See Appendix F-3		
Lead	2.34	16
Magnesium	6.48	45
Molybdenum	46.4	320
Nickel	30.4	210
Nylon	0.17	1.2
Osmium	81.2	560
Polyethylene Low density	0.015–0.040	0.2–0.6
High density	0.060–0.180	0.9–2.6
Rubber	(150 psi)	0.001
Silicon Carbide	68.0	470
Silver	12.0	83
Steel—See Appendix F-2		
Titanium	17.4	120
Tungsten	59.4	410
Silica Glass	10.5	72
Sintered Carbide, 94% WC, 6% Co	100	690
Elinvar	24.6	170
36% Ni, 12% Cr & W, 52% Fe: S_{yt} = 65,000 psi		
Invar	20.3	140
36% Ni, 64% Fe: S_{yt} = 51,000 psi		
Ni-Span C	27.6	190
42% Ni, 5.5% Cr, 2.5% Ti, 50% Fe: S_{yt} = 180,000 psi		

APPENDIX G

Areas under the normal curve.

Proportion of total area under the curve that is under the portion of the curve from $-\infty$ to $(X_i - \overline{X'})/\sigma'$. ($X_i$ represents any desired value of the variable X)

$\dfrac{X_i - \overline{X'}}{\sigma'}$	0.09	0.08	0.07	0.06	0.05	0.04	0.03	0.02	0.01	0.00
−3.5	0.00017	0.00017	0.00018	0.00019	0.00019	0.00020	0.00021	0.00022	0.00022	0.00023
−3.4	0.00024	0.00025	0.00026	0.00027	0.00028	0.00029	0.00030	0.00031	0.00033	0.00034
−3.3	0.00035	0.00036	0.00038	0.00039	0.00040	0.00042	0.00043	0.00045	0.00047	0.00048
−3.2	0.00050	0.00052	0.00054	0.00056	0.00058	0.00060	0.00062	0.00064	0.00066	0.00069
−3.1	0.00071	0.00074	0.00076	0.00079	0.00082	0.00085	0.00087	0.00090	0.00094	0.00097
−3.0	0.00100	0.00104	0.00107	0.00111	0.00114	0.00118	0.00122	0.00126	0.00131	0.00135
−2.9	0.0014	0.0014	0.0015	0.0015	0.0016	0.0016	0.0017	0.0017	0.0018	0.0019
−2.8	0.0019	0.0020	0.0021	0.0021	0.0022	0.0023	0.0023	0.0024	0.0025	0.0026
−2.7	0.0026	0.0027	0.0028	0.0029	0.0030	0.0031	0.0032	0.0033	0.0034	0.0035
−2.6	0.0036	0.0037	0.0038	0.0039	0.0040	0.0041	0.0043	0.0044	0.0045	0.0047
−2.5	0.0048	0.0049	0.0051	0.0052	0.0054	0.0055	0.0057	0.0059	0.0060	0.0062
−2.4	0.0064	0.0066	0.0068	0.0069	0.0071	0.0073	0.0075	0.0078	0.0080	0.0082
−2.3	0.0084	0.0087	0.0089	0.0091	0.0094	0.0096	0.0099	0.0102	0.0104	0.0107
−2.2	0.0110	0.0113	0.0116	0.0119	0.0122	0.0125	0.0129	0.0132	0.0136	0.0139
−2.1	0.0143	0.0146	0.0150	0.0154	0.0158	0.0162	0.0166	0.0170	0.0174	0.0179
−2.0	0.0183	0.0188	0.0192	0.0197	0.0202	0.0207	0.0212	0.0217	0.0222	0.0228
−1.9	0.0233	0.0239	0.0244	0.0250	0.0256	0.0262	0.0268	0.0274	0.0281	0.0287
−1.8	0.0294	0.0301	0.0307	0.0314	0.0322	0.0329	0.0336	0.0344	0.0351	0.0359
−1.7	0.0367	0.0375	0.0384	0.0392	0.0401	0.0409	0.0418	0.0427	0.0436	0.0446
−1.6	0.0455	0.0465	0.0475	0.0485	0.0495	0.0505	0.0516	0.0526	0.0537	0.0548
−1.5	0.0559	0.0571	0.0582	0.0594	0.0606	0.0618	0.0630	0.0643	0.0655	0.0668
−1.4	0.0681	0.0694	0.0708	0.0721	0.0735	0.0749	0.0764	0.0778	0.0793	0.0808
−1.3	0.0823	0.0838	0.0853	0.0869	0.0885	0.0901	0.0918	0.0934	0.0951	0.0968
−1.2	0.0985	0.1003	0.1020	0.1038	0.1057	0.1075	0.1093	0.1112	0.1131	0.1151
−1.1	0.1170	0.1190	0.1210	0.1230	0.1251	0.1271	0.1292	0.1314	0.1335	0.1357
−1.0	0.1379	0.1401	0.1423	0.1446	0.1469	0.1492	0.1515	0.1539	0.1562	0.1587
−0.9	0.1611	0.1635	0.1660	0.1685	0.1711	0.1736	0.1762	0.1788	0.1814	0.1841
−0.8	0.1867	0.1894	0.1922	0.1949	0.1977	0.2005	0.2033	0.2061	0.2090	0.2119
−0.7	0.2148	0.2177	0.2207	0.2236	0.2266	0.2297	0.2327	0.2358	0.2389	0.2420
−0.6	0.2451	0.2483	0.2514	0.2546	0.2578	0.2611	0.2643	0.2676	0.2709	0.2743
−0.5	0.2776	0.2810	0.2843	0.2877	0.2912	0.2946	0.2981	0.3015	0.3050	0.3085
−0.4	0.3121	0.3156	0.3192	0.3228	0.3264	0.3300	0.3336	0.3372	0.3409	0.3446
−0.3	0.3483	0.3520	0.3557	0.3594	0.3632	0.3669	0.3707	0.3745	0.3783	0.3821
−0.2	0.3859	0.3897	0.3936	0.3974	0.4013	0.4052	0.4090	0.4129	0.4168	0.4207
−0.1	0.4247	0.4286	0.4325	0.4364	0.4404	0.4443	0.4483	0.4522	0.4562	0.4602
−0.0	0.4641	0.4681	0.4721	0.4761	0.4801	0.4840	0.4880	0.4920	0.4960	0.5000

APPENDIX G

Areas under the normal curve. *(Continued)*

$\dfrac{X_i - \overline{X}'}{\sigma'}$	0.00	0.01	0.02	0.03	0.04	0.05	0.06	0.07	0.08	0.09
+0.0	0.5000	0.5040	0.5080	0.5120	0.5160	0.5199	0.5239	0.5279	0.5319	0.5359
+0.1	0.5398	0.5438	0.5478	0.5517	0.5557	0.5596	0.5636	0.5675	0.5714	0.5753
+0.2	0.5793	0.5832	0.5871	0.5910	0.5948	0.5987	0.6026	0.6064	0.6103	0.6141
+0.3	0.6179	0.6217	0.6255	0.6293	0.6331	0.6368	0.6406	0.6443	0.6480	0.6517
+0.4	0.6554	0.6591	0.6628	0.6664	0.6700	0.6736	0.6772	0.6808	0.6844	0.6879
+0.5	0.6915	0.6950	0.6985	0.7019	0.7054	0.7088	0.7123	0.7157	0.7190	0.7224
+0.6	0.7257	0.7291	0.7324	0.7357	0.7389	0.7422	0.7454	0.7486	0.7517	0.7549
+0.7	0.7580	0.7611	0.7642	0.7673	0.7704	0.7734	0.7764	0.7794	0.7823	0.7852
+0.8	0.7881	0.7910	0.7939	0.7967	0.7995	0.8023	0.8051	0.8079	0.8106	0.8133
+0.9	0.8159	0.8186	0.8212	0.8238	0.8264	0.8289	0.8315	0.8340	0.8365	0.8389
+1.0	0.8413	0.8438	0.8461	0.8485	0.8508	0.8531	0.8554	0.8577	0.8599	0.8621
+1.1	0.8643	0.8665	0.8686	0.8708	0.8729	0.8749	0.8770	0.8790	0.8810	0.8830
+1.2	0.8849	0.8869	0.8888	0.8907	0.8925	0.8944	0.8962	0.8980	0.8997	0.9015
+1.3	0.9032	0.9049	0.9066	0.9082	0.9099	0.9115	0.9131	0.9147	0.9162	0.9177
+1.4	0.9192	0.9207	0.9222	0.9236	0.9251	0.9265	0.9279	0.9292	0.9306	0.9319
+1.5	0.9332	0.9345	0.9357	0.9370	0.9382	0.9394	0.9406	0.9418	0.9429	0.9441
+1.6	0.9452	0.9463	0.9474	0.9484	0.9495	0.9505	0.9515	0.9525	0.9535	0.9545
+1.7	0.9554	0.9564	0.9573	0.9582	0.9591	0.9599	0.9608	0.9616	0.9625	0.9633
+1.8	0.9641	0.9649	0.9656	0.9664	0.9671	0.9678	0.9686	0.9693	0.9699	0.9706
+1.9	0.9713	0.9719	0.9726	0.9732	0.9738	0.9744	0.9750	0.9756	0.9761	0.9767
+2.0	0.9773	0.9778	0.9783	0.9788	0.9793	0.9798	0.9803	0.9808	0.9812	0.9817
+2.1	0.9821	0.9826	0.9830	0.9834	0.9838	0.9842	0.9846	0.9850	0.9854	0.9857
+2.2	0.9861	0.9864	0.9868	0.9871	0.9875	0.9878	0.9881	0.9884	0.9887	0.9890
+2.3	0.9893	0.9896	0.9898	0.9901	0.9904	0.9906	0.9909	0.9911	0.9913	0.9916
+2.4	0.9918	0.9920	0.9922	0.9925	0.9927	0.9929	0.9931	0.9932	0.9934	0.9936
+2.5	0.9938	0.9940	0.9941	0.9943	0.9945	0.9946	0.9948	0.9949	0.9951	0.9952
+2.6	0.9953	0.9955	0.9956	0.9957	0.9959	0.9960	0.9961	0.9962	0.9963	0.9964
+2.7	0.9965	0.9966	0.9967	0.9968	0.9969	0.9970	0.9971	0.9972	0.9973	0.9974
+2.8	0.9974	0.9975	0.9976	0.9977	0.9977	0.9978	0.9979	0.9979	0.9980	0.9981
+2.9	0.9981	0.9982	0.9983	0.9983	0.9984	0.9984	0.9985	0.9985	0.9986	0.9986
+3.0	0.99865	0.99869	0.99874	0.99878	0.99882	0.99886	0.99889	0.99893	0.99896	0.99900
+3.1	0.99903	0.99906	0.99910	0.99913	0.99915	0.99918	0.99921	0.99924	0.99926	0.99929
+3.2	0.99931	0.99934	0.99936	0.99938	0.99940	0.99942	0.99944	0.99946	0.99948	0.99950
+3.3	0.99952	0.99953	0.99955	0.99957	0.99958	0.99960	0.99961	0.99962	0.99964	0.99965
+3.4	0.99966	0.99967	0.99969	0.99970	0.99971	0.99972	0.99973	0.99974	0.99975	0.99976
+3.5	0.99977	0.99978	0.99978	0.99979	0.99980	0.99981	0.99981	0.99982	0.99983	0.99983

E. L. Grant and R. S. Leavenworth, *Statistical Quality Control*, 5th ed. Copyright © 1980 by McGraw-Hill Publishing Co., New York. Reprinted with permission.

APPENDIX H

The t distribution.

(a) One-tail α

(b) Two-tail α

Given v, the table gives (a) the one-tail t_0 value with α of the area above it, that is, $P(t \geq t_0) = \alpha$, or (b) the two-tail $+t_0$ and $-t_0$ values with $\alpha/2$ in each tail, that is, $P(t \leq t_0) + P(t \geq +t_0) = \alpha$

	\multicolumn{6}{c	}{One-tail α}
	0.10	0.05	0.025	0.01	0.005	0.001
	\multicolumn{6}{c	}{Two-tail α}
v	0.20	0.10	0.05	0.02	0.01	0.002
1	3.078	6.314	12.706	31.821	63.657	318.300
2	1.886	2.920	4.303	6.965	9.925	22.327
3	1.638	2.353	3.182	4.541	5.841	10.214
4	1.533	2.132	2.776	3.747	4.604	7.173
5	1.476	2.015	2.571	3.305	4.032	5.893
6	1.440	1.943	2.447	3.143	3.707	5.208
7	1.415	1.895	2.365	2.998	3.499	4.785
8	1.397	1.860	2.306	2.896	3.355	4.501
9	1.383	1.833	2.262	2.821	3.250	4.297
10	1.372	1.812	2.228	2.764	3.169	4.144
11	1.363	1.796	2.201	2.718	3.106	4.025
12	1.356	1.782	2.179	2.681	3.055	3.930
13	1.350	1.771	2.160	2.650	3.012	3.852
14	1.345	1.761	2.145	2.624	2.977	3.787

APPENDIX H

The t distribution. (*Continued*)

	\multicolumn{6}{c}{One-tail α}
	0.10	0.05	0.025	0.01	0.005	0.001
ν	0.20	0.10	0.05	0.02	0.01	0.002
	\multicolumn{6}{c}{Two-tail α}

ν	0.20	0.10	0.05	0.02	0.01	0.002
15	1.341	1.753	2.131	2.602	2.947	3.733
16	1.337	1.746	2.120	2.583	2.921	3.686
17	1.333	1.740	2.110	2.567	2.898	3.646
18	1.330	1.734	2.101	2.552	2.878	3.611
19	1.328	1.729	2.093	2.539	2.861	3.579
20	1.325	1.725	2.086	2.528	2.845	3.552
21	1.323	1.721	2.080	2.518	2.831	3.527
22	1.321	1.717	2.074	2.508	2.819	3.505
23	1.319	1.714	2.069	2.500	2.807	3.485
24	1.318	1.711	2.064	2.492	2.797	3.467
25	1.316	1.708	2.060	2.485	2.787	3.450
26	1.315	1.706	2.056	2.479	2.779	3.435
27	1.314	1.703	2.052	2.473	2.771	3.421
28	1.313	1.701	2.048	2.467	2.763	3.408
29	1.311	1.699	2.045	2.462	2.756	3.396
30	1.310	1.697	2.042	2.457	2.750	3.385
40	1.303	1.684	2.021	2.423	2.704	3.307
60	1.296	1.671	2.000	2.390	2.660	3.232
80	1.292	1.664	1.990	2.374	2.639	3.195
100	1.290	1.660	1.984	2.365	2.626	3.174
∞	1.282	1.645	1.960	2.326	2.576	3.090

L. Blank, *Statistical Procedures for Engineering, Management, and Science.* Copyright © 1980 by McGraw-Hill Publishing Co., New York. Reprinted with permission.

APPENDIX I

The F distribution ($\alpha = 0.10$, 0.05, and 0.01).

Given v_1 and v_2, the table gives the F_0 value with α of the area above it, that is
$$P(F \geq F_0) = \alpha$$

The F distribution (α = 0.10, 0.05, and 0.01). (Continued)

v_2	α	1	2	3	4	5	6	7	8	9	10	11	12	14	15	19	20	24	30	50	100	500	∞
1	.10	39.9	49.5	53.6	55.8	57.2	58.2	58.9	59.4	59.9	60.2	60.5	60.7	61.1	61.2	61.6	61.7	62.0	62.3	62.7	63.0	63.3	63.3
	.05	161	200	216	225	230	234	237	239	241	242	243	244	245	246	248	248	249	250	252	253	254	254
2	.10	8.53	9.00	9.16	9.24	9.29	9.33	9.35	9.37	9.38	9.39	9.40	9.41	9.42	9.42	9.44	9.44	9.45	9.46	9.47	9.48	9.49	9.49
	.05	18.5	19.0	19.2	19.2	19.3	19.3	19.4	19.4	19.4	19.4	19.4	19.4	19.4	19.4	19.4	19.4	19.5	19.5	19.5	19.5	19.5	19.5
	.01	98.5	99.0	99.2	99.2	99.3	99.3	99.4	99.4	99.4	99.4	99.4	99.4	99.4	99.4	99.4	99.4	99.5	99.5	99.5	99.5	99.5	99.5
3	.10	5.54	5.46	5.39	5.34	5.31	5.28	5.27	5.25	5.24	5.23	5.22	5.22	5.20	5.20	5.18	5.18	5.18	5.17	5.15	5.14	5.14	5.13
	.05	10.1	9.55	9.28	9.12	9.10	8.94	8.89	8.85	8.81	8.79	8.76	8.74	8.71	8.70	8.67	8.66	8.64	8.62	8.58	8.55	8.53	8.53
	.01	34.1	30.8	29.5	28.7	28.2	27.9	27.7	27.5	27.3	27.2	27.1	27.1	26.9	26.9	26.7	26.7	26.6	26.5	26.4	26.2	26.1	26.1
4	.10	4.54	4.32	4.19	4.11	4.05	4.01	3.98	3.95	3.94	3.92	3.91	3.90	3.88	3.87	3.84	3.84	3.83	3.82	3.80	3.78	3.76	3.76
	.05	7.71	6.94	6.59	6.39	6.26	6.16	6.09	6.04	6.00	5.96	5.94	5.91	5.87	5.86	5.81	5.80	5.77	5.75	5.70	5.66	5.64	5.63
	.01	21.2	18.0	16.7	16.0	15.5	15.2	15.0	14.8	14.7	14.5	14.4	14.4	14.2	14.2	14.0	14.0	13.9	13.8	13.7	13.6	13.5	13.5
5	.10	4.06	3.78	3.62	3.52	3.45	3.40	3.37	3.34	3.32	3.30	3.28	3.27	3.25	3.24	3.21	3.21	3.19	3.17	3.15	3.13	3.11	3.10
	.05	6.61	5.79	5.41	5.19	5.05	4.95	4.88	4.82	4.77	4.74	4.71	4.68	4.64	4.62	4.57	4.56	4.53	4.50	4.44	4.41	4.37	4.36
	.01	16.26	13.27	12.06	11.39	10.97	10.67	10.46	10.29	10.16	10.05	9.96	9.89	9.77	9.72	9.58	9.55	9.47	9.38	9.24	9.13	9.04	9.02
6	.10	3.78	3.46	3.29	3.18	3.11	3.05	3.01	2.98	2.96	2.94	2.92	2.90	2.88	2.87	2.84	2.84	2.82	2.80	2.77	2.75	2.73	2.72
	.05	5.99	5.14	4.76	4.53	4.39	4.28	4.21	4.15	4.10	4.06	4.03	4.00	3.96	3.94	3.88	3.87	3.84	3.81	3.75	3.71	3.68	3.67
	.01	13.74	10.92	9.78	9.15	8.75	8.47	8.26	8.10	7.98	7.87	7.79	7.72	7.60	7.56	7.42	7.40	7.31	7.23	7.09	6.99	6.90	6.88
7	.10	3.59	3.26	3.07	2.96	2.88	2.83	2.78	2.75	2.72	2.70	2.68	2.67	2.64	2.63	2.60	2.59	2.58	2.56	2.52	2.50	2.48	2.47
	.05	5.59	4.74	4.35	4.12	3.97	3.87	3.79	3.73	3.68	3.64	3.60	3.57	3.53	3.51	3.46	3.44	3.41	3.38	3.32	3.27	3.24	3.23
	.01	12.25	9.55	8.45	7.85	7.46	7.19	6.99	6.84	6.72	6.62	6.54	6.47	6.36	6.31	6.18	6.16	6.07	5.99	5.86	5.75	5.67	5.65
8	.10	3.46	3.11	2.92	2.81	2.73	2.67	2.62	2.59	2.56	2.54	2.52	2.50	2.47	2.46	2.43	2.42	2.40	2.38	2.35	2.32	2.30	2.29
	.05	5.32	4.46	4.07	3.84	3.69	3.58	3.50	3.44	3.39	3.35	3.31	3.28	3.24	3.22	3.16	3.15	3.12	3.06	3.02	2.97	2.94	2.93
	.01	11.26	8.65	7.59	7.01	6.63	6.37	6.18	6.03	5.91	5.81	5.73	5.67	5.56	5.52	5.38	5.36	5.28	5.20	5.07	4.96	4.88	4.86
	.10	3.36	3.01	2.81	2.69	2.61	2.55	2.51	2.47	2.44	2.42	2.40	2.38	2.35	2.34	2.31	2.30	2.28	2.25	2.22	2.19	2.17	2.16

APPENDIX I

The F distribution ($\alpha = 0.10$, 0.05, and 0.01). (Continued)

v_2	α	1	2	3	4	5	6	7	8	9	10	11	12	14	15	19	20	24	30	50	100	500	∞
9	.05	5.12	4.26	3.86	3.63	3.48	3.37	3.29	3.23	3.18	3.14	3.10	3.07	3.03	3.01	2.95	2.94	2.90	2.86	2.80	2.76	2.72	2.71
	.01	10.56	8.02	6.99	6.42	6.06	5.80	5.61	5.47	5.35	5.26	5.18	5.11	5.00	4.96	4.83	4.81	4.73	4.65	4.52	4.42	4.33	4.31
10	.10	3.28	2.92	2.73	2.61	2.52	2.46	2.41	2.38	2.35	2.32	2.30	2.28	2.25	2.24	2.21	2.20	2.18	2.16	2.12	2.09	2.06	2.06
	.05	4.96	4.10	3.71	3.48	3.33	3.22	3.14	3.07	3.02	2.98	2.94	2.91	2.86	2.85	2.78	2.77	2.74	2.70	2.64	2.59	2.55	2.54
	.01	10.04	7.56	6.55	5.99	5.64	5.39	5.20	5.06	4.94	4.85	4.77	4.71	4.60	4.56	4.43	4.41	4.33	4.25	4.12	4.01	3.93	3.91
11	.10	3.23	2.86	2.66	2.54	2.45	2.39	2.34	2.30	2.27	2.25	2.23	2.21	2.18	2.17	2.13	2.12	2.10	2.06	2.04	2.00	1.98	1.97
	.05	4.84	3.98	3.59	3.36	3.20	3.09	3.01	2.95	2.90	2.85	2.82	2.79	2.74	2.72	2.66	2.65	2.61	2.57	2.51	2.46	2.42	2.40
	.01	9.65	7.21	6.22	5.67	5.32	5.07	4.89	4.74	4.63	4.54	4.46	4.40	4.29	4.25	4.12	4.10	4.02	3.94	3.81	3.71	3.62	3.60
12	.10	3.18	2.81	2.61	2.48	2.39	2.33	2.28	2.24	2.21	2.19	2.17	2.15	2.11	2.10	2.07	2.06	2.04	2.01	1.97	1.94	1.91	1.90
	.05	4.75	3.89	3.49	3.26	3.11	3.00	2.91	2.85	2.80	2.75	2.72	2.69	2.64	2.62	2.56	2.54	2.51	2.47	2.40	2.35	2.31	2.30
	.01	9.33	6.93	5.95	5.41	5.06	4.82	4.64	4.50	4.39	4.30	4.22	4.16	4.05	4.01	3.88	3.86	3.78	3.70	3.57	3.47	3.38	3.36
14	.10	3.10	2.73	2.52	2.39	2.31	2.24	2.19	2.15	2.12	2.10	2.06	2.05	2.02	2.01	1.97	1.96	1.94	1.91	1.87	1.83	1.80	1.80
	.05	4.60	3.74	3.34	3.11	2.96	2.85	2.76	2.70	2.65	2.60	2.57	2.53	2.48	2.46	2.40	2.39	2.35	2.31	2.24	2.19	2.14	2.13
	.01	8.86	6.51	5.56	5.04	4.69	4.46	4.28	4.14	4.03	3.94	3.86	3.80	3.70	3.66	3.53	3.51	3.43	3.35	3.22	3.11	3.03	3.00
15	.10	3.07	2.70	2.49	2.36	2.27	2.21	2.16	2.12	2.09	2.06	2.04	2.02	1.98	1.97	1.93	1.92	1.90	1.87	1.83	1.79	1.76	1.76
	.05	4.54	3.68	3.29	3.06	2.90	2.79	2.71	2.64	2.59	2.54	2.51	2.48	2.42	2.40	2.34	2.33	2.29	2.25	2.18	2.12	2.08	2.07
	.01	8.68	6.36	5.42	4.89	4.56	4.32	4.14	4.00	3.89	3.80	3.73	3.67	3.56	3.52	3.40	3.37	3.29	3.21	3.08	2.98	2.89	2.87
16	.10	3.05	2.67	2.46	2.33	2.24	2.18	2.13	2.09	2.06	2.03	2.01	1.99	1.95	1.94	1.90	1.89	1.87	1.84	1.79	1.76	1.73	1.72
	.05	4.49	3.63	3.24	3.01	2.85	2.74	2.66	2.59	2.54	2.49	2.46	2.42	2.37	2.35	2.29	2.28	2.24	2.19	2.12	2.07	2.02	2.01
	.01	8.53	6.23	5.29	4.77	4.44	4.20	4.03	3.89	3.78	3.69	3.62	3.55	3.45	3.41	3.28	3.26	3.18	3.10	2.97	2.86	2.78	2.75

v_1 (numerator)

The degrees of freedom are v_1 for the numerator and v_2 for the denominator.

The F distribution (α = 0.10, 0.05, and 0.01). (Continued)

ν_2	α	1	2	3	4	5	6	7	8	9	10	11	12	14	15	19	20	24	30	50	100	500	∞
18	.10	3.01	2.62	2.42	2.29	2.20	2.13	2.08	2.04	2.00	1.98	1.96	1.93	1.90	1.89	1.85	1.84	1.81	1.78	1.74	1.70	1.67	1.66
	.05	4.41	3.55	3.16	2.93	2.77	2.66	2.58	2.51	2.46	2.41	2.37	2.34	2.29	2.27	2.20	2.19	2.15	2.11	2.04	1.98	1.93	1.92
	.01	8.29	6.01	5.09	4.58	4.25	4.01	3.84	3.71	3.60	3.51	3.43	3.37	3.27	3.23	3.10	3.08	3.00	2.92	2.78	2.68	2.59	2.57
19	.10	2.99	2.61	2.40	2.27	2.18	2.11	2.06	2.02	1.98	1.96	1.94	1.91	1.87	1.86	1.82	1.81	1.79	1.76	1.71	1.67	1.64	1.63
	.05	4.38	3.52	3.13	2.90	2.74	2.63	2.54	2.48	2.42	2.38	2.34	2.31	2.26	2.23	2.17	2.16	2.11	2.07	2.00	1.94	1.89	1.88
	.01	8.18	5.93	5.01	4.50	4.17	3.94	3.77	3.63	3.52	3.43	3.36	3.30	3.19	3.15	3.03	3.00	2.92	2.84	2.71	2.60	2.51	2.49
20	.10	2.97	2.59	2.38	2.25	2.16	2.09	2.04	2.00	1.96	1.94	1.92	1.89	1.85	1.84	1.80	1.79	1.77	1.74	1.69	1.65	1.62	1.61
	.05	4.35	3.49	3.10	2.87	2.71	2.60	2.51	2.45	2.39	2.35	2.31	2.28	2.22	2.20	2.14	2.12	2.08	2.04	1.97	1.91	1.86	1.84
	.01	8.10	5.85	4.94	4.43	4.10	3.87	3.70	3.56	3.46	3.37	3.29	3.23	3.13	3.09	2.96	2.94	2.86	2.78	2.64	2.54	2.44	2.42
24	.10	2.93	2.54	2.33	2.19	2.10	2.04	1.98	1.94	1.91	1.88	1.85	1.83	1.79	1.78	1.74	1.73	1.70	1.67	1.62	1.58	1.54	1.53
	.05	4.26	3.40	3.01	2.78	2.62	2.51	2.42	2.36	2.30	2.25	2.21	2.18	2.13	2.11	2.04	2.03	1.98	1.94	1.86	1.80	1.75	1.73
	.01	7.82	5.61	4.72	4.22	3.90	3.67	3.50	3.36	3.26	3.17	3.09	3.03	2.93	2.89	2.76	2.74	2.66	2.58	2.44	2.33	2.24	2.21
30	.10	2.88	2.49	2.28	2.14	2.05	1.98	1.93	1.88	1.85	1.82	1.79	1.77	1.73	1.72	1.68	1.67	1.64	1.61	1.55	1.51	1.47	1.46
	.05	4.17	3.32	2.92	2.69	2.53	2.42	2.33	2.27	2.21	2.16	2.13	2.09	2.04	2.01	1.95	1.93	1.89	1.84	1.76	1.70	1.64	1.62
	.01	7.56	5.39	4.51	4.02	3.70	3.47	3.30	3.17	3.07	2.98	2.91	2.84	2.74	2.70	2.57	2.55	2.47	2.39	2.25	2.13	2.03	2.01
50	.10	2.81	2.41	2.20	2.06	1.97	1.90	1.84	1.80	1.76	1.73	1.70	1.68	1.64	1.63	1.58	1.57	1.54	1.50	1.44	1.39	1.34	1.33
	.05	4.03	3.18	2.79	2.56	2.40	2.29	2.20	2.13	2.07	2.03	1.99	1.95	1.89	1.87	1.80	1.78	1.74	1.69	1.60	1.52	1.46	1.44
	.01	7.17	5.06	4.20	3.72	3.41	3.19	3.02	2.89	2.79	2.70	2.63	2.56	2.46	2.42	2.29	2.27	2.18	2.10	1.95	1.82	1.71	1.68
100	.10	2.76	2.36	2.14	2.00	1.91	1.83	1.78	1.73	1.70	1.66	1.63	1.61	1.57	1.56	1.50	1.49	1.46	1.42	1.35	1.29	1.23	1.21
	.05	3.94	3.09	2.70	2.46	2.31	2.19	2.10	2.03	1.97	1.93	1.89	1.85	1.79	1.77	1.69	1.68	1.63	1.57	1.48	1.39	1.31	1.28
	.01	6.90	4.82	3.98	3.51	3.21	2.99	2.82	2.69	2.59	2.50	2.43	2.37	2.26	2.22	2.09	2.07	1.98	1.89	1.73	1.60	1.47	1.43
500	.10	2.72	2.31	2.10	1.96	1.86	1.79	1.73	1.68	1.64	1.61	1.58	1.56	1.52	1.50	1.45	1.44	1.41	1.36	1.28	1.21	1.12	1.09
	.05	3.86	3.01	2.62	2.39	2.23	2.12	2.03	1.96	1.90	1.85	1.81	1.77	1.71	1.69	1.61	1.59	1.54	1.48	1.38	1.28	1.16	1.11
	.01	6.69	4.65	3.82	3.36	3.05	2.84	2.68	2.55	2.44	2.36	2.28	2.22	2.12	2.07	1.94	1.92	1.83	1.74	1.56	1.41	1.23	1.16
∞	.10	2.71	2.30	2.08	1.94	1.85	1.77	1.72	1.67	1.63	1.60	1.57	1.55	1.51	1.49	1.43	1.42	1.38	1.34	1.26	1.18	1.08	1.00
	.05	3.84	3.00	2.60	2.37	2.21	2.10	2.01	1.94	1.88	1.83	1.79	1.75	1.69	1.67	1.59	1.57	1.52	1.46	1.35	1.24	1.11	1.00
	.01	6.63	4.61	3.78	3.32	3.02	2.80	2.64	2.51	2.41	2.32	2.25	2.18	2.08	2.04	1.90	1.88	1.79	1.70	1.52	1.36	1.15	1.00

ν_1 (numerator)

L. Blank, *Statistical Procedures For Engineering, Management, and Science.* Copyright © 1980 by McGraw-Hill Publishing Co., New York. Reprinted with permission.

APPENDIX J

The χ^2 distribution.

Given ν, the table gives the χ_0^2 value with α of the area above it; that is

$$P(\chi^2 \geq \chi_0^2) = \alpha$$

ν \ α	0.995	0.990	0.975	0.950	0.900	0.500	0.100	0.050	0.025	0.010	0.005
1	0.00	0.00	0.00	0.00	0.02	0.45	2.71	3.84	5.02	6.63	7.88
2	0.01	0.02	0.05	0.10	0.21	1.39	4.61	5.99	7.38	9.21	10.60
3	0.07	0.11	0.22	0.35	0.58	2.37	6.25	7.81	9.35	11.34	12.84
4	0.21	0.30	0.48	0.71	1.06	3.36	7.78	9.49	11.14	13.28	14.86
5	0.41	0.55	0.83	1.15	1.61	4.35	9.24	11.07	12.83	15.09	16.75
6	0.68	0.87	1.24	1.64	2.20	5.35	10.65	12.59	14.45	16.81	18.55
7	0.99	1.24	1.69	2.17	2.83	6.35	12.02	14.07	16.01	18.48	20.28
8	1.34	1.65	2.18	2.73	3.49	7.34	13.36	15.51	17.53	20.09	21.96
9	1.73	2.09	2.70	3.33	4.17	8.34	14.68	16.92	19.02	21.67	23.59
10	2.16	2.56	3.25	3.94	4.87	9.34	15.99	18.31	20.48	23.21	25.19
11	2.60	3.05	3.82	4.57	5.50	10.34	17.28	19.68	21.92	24.72	26.76
12	3.07	3.57	4.40	5.23	6.30	11.34	18.55	21.03	23.34	26.22	28.30
13	3.57	4.11	5.01	5.89	7.04	12.34	19.81	22.36	24.74	27.69	29.82
14	4.07	4.66	5.63	6.57	7.79	13.34	21.06	23.68	26.12	29.14	31.32
15	4.60	5.23	6.27	7.26	8.55	14.34	22.31	25.00	27.49	30.58	32.80
16	5.14	5.81	6.91	7.96	9.31	15.34	23.54	26.30	28.85	32.00	34.27
17	5.70	6.41	7.56	8.67	10.09	16.34	24.77	27.59	30.19	33.41	35.72
18	6.26	7.01	8.23	9.39	10.87	17.34	25.99	28.87	31.53	34.81	37.16
19	6.84	7.63	8.91	10.12	11.65	18.34	27.20	30.14	32.85	36.19	38.58
20	7.43	8.26	9.59	10.85	12.44	19.38	28.41	31.41	34.17	37.57	40.00
21	8.03	8.90	10.28	11.50	13.24	20.38	29.62	32.67	35.48	38.93	41.40
22	8.64	9.54	10.98	12.34	14.04	21.34	30.81	33.92	36.78	40.29	42.80
23	9.26	10.20	11.69	13.09	14.85	22.34	32.01	35.17	38.08	41.64	44.18
24	9.89	10.86	12.40	13.85	15.66	23.34	33.20	36.42	39.36	42.98	45.56
25	10.52	11.52	13.12	14.61	16.47	24.34	34.38	37.65	40.65	44.31	46.93
26	11.16	12.20	13.84	15.38	17.29	25.34	35.56	38.89	41.92	45.64	48.29
27	11.81	12.88	14.57	16.15	18.11	26.34	36.74	40.11	43.19	46.96	49.65

The χ^2 distribution. (*Continued*)

ν \ α	0.995	0.990	0.975	0.950	0.900	0.500	0.100	0.050	0.025	0.010	0.005
28	12.46	13.57	15.31	16.93	18.94	27.34	37.92	41.34	44.46	48.28	50.99
29	13.12	14.26	16.05	17.71	19.77	28.34	39.09	42.56	45.72	49.59	52.34
30	13.79	14.95	16.79	18.49	20.60	29.34	40.26	43.77	46.98	50.89	53.67
40	20.71	22.16	24.43	26.51	29.05	39.34	51.80	55.76	59.34	63.69	66.77
50	27.99	29.71	32.36	34.76	37.69	49.33	63.17	67.50	71.42	76.15	79.49
70	43.28	45.44	48.76	51.74	55.33	69.33	85.53	90.53	95.02	100.42	104.22
100	67.33	70.06	74.22	77.93	82.36	99.33	118.50	124.34	129.56	135.81	140.17

L. Blank, *Statistical Procedures For Engineering, Management, and Science*. Copyright © 1980 by McGraw-Hill Publishing Co., New York. Reprinted with permission.

APPENDIX K

The correlation coefficient.

Values of the Correlation Coefficient for Different Levels of Significance

n	0.1	0.05	0.02	0.01	0.001	n	0.1	0.05	0.02	0.01	0.001
1	0.98769	0.99692	0.999507	0.999877	0.9999988	16	0.4000	0.4683	0.5425	0.5897	0.7084
2	0.90000	0.95000	0.98000	0.990000	0.99900	17	0.3887	0.4555	0.5285	0.5751	0.6932
3	0.8054	0.8783	0.93433	0.95873	0.99116	18	0.3783	0.4438	0.5155	0.5614	0.6787
4	0.7293	0.8114	0.8822	0.91720	0.97406	19	0.3687	0.4329	0.5034	0.5487	0.6652
5	0.6694	0.7545	0.8329	0.8745	0.95074	20	0.3598	0.4227	0.4921	0.5368	0.6524
6	0.6215	0.7067	0.7887	0.8343	0.92493	25	0.3233	0.3809	0.4451	0.4869	0.5974
7	0.5822	0.6664	0.7498	0.7977	0.8982	30	0.2960	0.3494	0.4093	0.4487	0.5541
8	0.5494	0.6319	0.7155	0.7646	0.8721	35	0.2746	0.3246	0.3810	0.4182	0.5189
9	0.5214	0.6021	0.6851	0.7348	0.8471	40	0.2573	0.3044	0.3578	0.3932	0.4896
10	0.4973	0.5760	0.6581	0.7079	0.8233	45	0.2428	0.2875	0.3384	0.3721	0.4648
11	0.4762	0.5529	0.6339	0.6835	0.8010	50	0.2306	0.2732	0.3218	0.3541	0.4433
12	0.4575	0.5324	0.6120	0.6614	0.7800	60	0.2108	0.2500	0.2948	0.3248	0.4078
13	0.4409	0.5139	0.5923	0.6411	0.7603	70	0.1954	0.2319	0.2737	0.3017	0.3799
14	0.4259	0.4973	0.5742	0.6226	0.7420	80	0.1829	0.2172	0.2565	0.2830	0.3568
15	0.4124	0.4821	0.5577	0.6055	0.7246	90	0.1726	0.2050	0.2422	0.2673	0.3375
						100	0.1638	0.1946	0.2301	0.2540	0.3211

Appendix K is taken from Table VII of Fisher and Yates: *Statistical Tables for Biological, Agricultural and Medical Research*, published by Longman Group UK Ltd. London (previously published by Oliver and Boyd Ltd. Edinburgh) and by permission of the authors and publishers.

If the computed value of r at a given level of significance exceeds the table value, then the null hypothesis that there is no relationship between the variables is rejected. The values given are for a two-tail, two variable test, n = the degrees of freedom.

APPENDIX L

Stress concentration factors.

Top chart: Axial loading of a stepped round bar with fillet radius r, larger diameter D, smaller diameter d.

$$\sigma_{nom} = \frac{P}{A} = \frac{4P}{\pi d^2}$$

K_t vs r/d, curves for $D/d \geq 2$, 1.1, 1.03, 1.01.

Bottom chart: Bending of a stepped round bar with fillet radius r.

$$\sigma_{nom} = \frac{Mc}{I} = \frac{32M}{\pi d^3}$$

K_t vs r/d, curves for $D/d \geq 2$, 1.1, 1.03, 1.01.

Stress concentration factors. (*Continued*)

$$\tau_{nom} = \frac{Tc}{J} = \frac{16T}{\pi d^3}$$

Curves: $D/d \geq 2$, 1.1, 1.01 vs r/d

$$\sigma_{nom} = \frac{P}{A} = \frac{P}{bh}$$

Curves: $H/h = 3$, 2, 1.5, 1.15, 1.05, 1.01 vs r/h

APPENDIX L

Stress concentration factors. (*Continued*)

$$\sigma_{nom} = \frac{P}{A} = \frac{4P}{\pi d^2}$$

$$\sigma_{nom} = \frac{Mc}{I} = \frac{6M}{bh^2}$$

Stress concentration factors. (*Continued*)

$$\tau_{nom} = \frac{Tc}{J} = \frac{16T}{\pi d^3}$$

$D/d = 2$
1.2
1.09

Torsion:
$$\tau_{nom} = \frac{Tc}{J} \approx \frac{T}{(\pi D^3/16) - (dD^2/6)}$$

Axial load:
$$\sigma_{nom} = \frac{P}{A} \approx \frac{P}{(\pi D^2/4) - Dd}$$

Bending (in this plane):
$$\sigma_{nom} = \frac{Mc}{I} \approx \frac{M}{(\pi D^3/32) - (dD^2/6)}$$

R. E. Peterson, *Stress Concentration Factors*. Copyright © 1974 by John Wiley & Sons, Inc., New York. Reprinted with permission of John Wiley & Sons, Inc.

APPENDIX M

Stress intensity factors.

Stress intensity factors. (*Continued*)

APPENDIX M

Stress intensity factors. (*Continued*)

Stress intensity factors. (*Continued*)

APPENDIX M

Stress intensity factors. (*Continued*)

APPENDIX N

Copper wire data.

Wire Gauge	Diameter, mm	Resistance, ohms/km
0000	11.6840	0.1608
000	10.4038	0.2028
00	9.2659	0.2557
0	8.2525	0.3224
1	7.3482	0.4065
2	6.5430	0.5128
3	5.8268	0.6463
4	5.1892	0.8153
5	4.6203	1.0279
6	4.1148	1.2963
7	3.6652	1.6345
8	3.2639	2.0610
9	2.9058	2.5988
10	2.5883	3.2772
12	2.0526	5.2100
14	1.6276	8.2841
16	1.2908	13.1759
18	1.0236	20.9482
20	0.8118	33.3005
22	0.6438	52.9661
24	0.5106	84.2460
26	0.4049	133.9047

Reprinted with permission from *Handbook of Chemistry and Physics*, 67th ed. R. C. Weast, ed. Copyright 1986, CRC Press, Inc. Boca Raton, Fla.

APPENDIX O

Shear, moment, and deflection equations for cantilever beams.

	Slope at Free End	Maximum Deflection	Deflection at Any Point x
1. Concentrated load at end	$\theta = \dfrac{PL^2}{2EI}$	$\delta_{max} = \dfrac{PL^3}{3EI}$	$\delta = \dfrac{Px^2}{6EI}(3L - x)$

$V = P$

$M = -PL$

2. Concentrated load at any point	$\theta = \dfrac{Pa^2}{2EI}$	$\delta_{max} = \dfrac{Pa^2}{6EI}(3L - a)$	For $0 \leq x \leq a$: $\delta = \dfrac{Px^2}{6EI}(3a - x)$ For $a \leq x \leq L$: $\delta = \dfrac{Pa^2}{6EI}(3x - a)$

$V = P$

$M = -Pa$

Shear, moment, and deflection equations for cantilever beams *(Continued)*.

	Slope at Free End	Maximum Deflection	Deflection at Any Point x
3. Uniform load	$\theta = \dfrac{wL^3}{6EI}$	$\delta_{max} = \dfrac{wL^4}{8EI}$	$\delta = \dfrac{wx^2}{24EI}(x^2 + 6L^2 - 4Lx)$

$V = wL$

$M = -wL^2/2$

	Slope at Free End	Maximum Deflection	Deflection at Any Point x
4. Moment load at free end	$\theta = \dfrac{M_b L}{EI}$	$\delta_{max} = \dfrac{M_b L^2}{2EI}$	$\delta = \dfrac{M_b x^2}{2EI}$

Fundamentals of Machine Component Design, by Robert C. Juvinall. Copyright © 1983, John Wiley & Sons, Inc. Reprinted with permission of John Wiley & Sons, Inc.

Shear, moment, and deflection equations for simply supported beams *(Continued)*.

	Slope at Ends, θ	Maximum Deflection, δ_{max}	Deflection at Any Point x, δ
1. Concentrated center load	$\dfrac{PL^2}{16EI}$	At center: $\dfrac{PL^3}{48EI}$	For $0 \leq x \leq L/2$: $\dfrac{Px}{12EI}\left(\dfrac{3L^2}{4} - x^2\right)$
2. Concentrated load at any point	At left end: $\dfrac{Pb(L^2 - b^2)}{6LEI}$	At $x = \sqrt{\dfrac{L^2 - b^2}{3}}$: $\dfrac{Pb(L^2 - b^2)^{3/2}}{9\sqrt{3}LEI}$	For $0 \leq x \leq a$: $\dfrac{Pbx}{6LEI}(L^2 - x^2 - b^2)$

Shear, moment, and deflection equations for simply supported beams *(Continued)*.

	Slope at Ends, θ	*Maximum Deflection,* δ_{max}	*Deflection at Any Point x,* δ
3. Uniform load	$\dfrac{wL^3}{24EI}$	$\dfrac{5wL^4}{384EI}$	$\dfrac{wx}{24EI}(L^3 - 2Lx^2 + x^3)$

4. Overhung load	At left supt: $\dfrac{Pab}{6EI}$ At right supt: $\dfrac{Pab}{3EI}$ At load: $\dfrac{Pb}{6EI}(2L+b)$	$\delta_{max} = \dfrac{Pb^2L}{3EI}$	For $0 \leq x \leq a$: $\dfrac{Pbx}{6aEI}(x^2 - a^2)$ For $0 \leq z \leq b$: $\dfrac{P}{6EI}[z^3 - b(2L+b)z + 2b^2L]$

Shear, moment, and deflection equations for simply supported beams *(Continued).*

	Slope at Ends, θ	Maximum Deflection, δ_{max}	Deflection at Any Point x, δ

5. Moment load between supports

At left supt:
$$\frac{-M_0}{6EIL}(2L^2 - 6aL + 3a^2)$$

At load:
$$\frac{M_0}{EI}\left(\frac{L}{3} + \frac{a^2}{L} - a\right)$$

At right supt:
$$\frac{M_0}{6EIL}(L^2 - 3a^2)$$

At load:
$$\frac{M_0 a}{3EIL}(2a^2 - 3aL + L^2)$$

For $0 \leq x \leq a$:
$$\frac{M_0 x}{6EIL}(x^2 + 3a^2 - 6aL + 2L^2)$$

6. Overhung moment load

At left supt:
$$\frac{M_0 a}{6EI}$$

At right supt:
$$\frac{M_0 a}{3EI}$$

At load:
$$\frac{M_0(a + 3b)}{3EI}$$

$$\delta_{max} = \frac{M_0 b}{6EI}(2L + b)$$

For $0 \leq x \leq a$:
$$-\frac{M_0 x}{6aEI}(a^2 - x^2)$$

For $0 \leq x' \leq b$:
$$\frac{M_0}{6EI}(2ax' + 3x'^2)$$

Shear, moment, and deflection equations for beams with fixed ends *(Continued)*.

	Deflection, δ	*Deflection, δ, at Any Point, x*

1. Concentrated center load

At center:
$$\delta_{max} = \frac{PL^3}{192EI}$$

For $0 \leq x \leq L/2$:
$$\frac{Px^2}{48EI}(3L - 4x)$$

2. Concentrated load at any point

At load:
$$\delta = \frac{Pb^3 a^3}{3EIL^3}$$

For $0 \leq a$:
$$\frac{Pb^2 x^2}{6EIL^3} - [3aL - (3a + b)x]$$

Shear, moment, and deflection equations for beams with fixed ends *(Continued).*

	Deflection, δ	*Deflection,* δ, *at Any Point, x*
3. Uniform load	At center: $\delta_{max} = \dfrac{wL^4}{384EI}$	$\dfrac{wx^2}{24EI}(L-x)^2$

Diagram values:
- End moments: $-\dfrac{wL^2}{12}$
- Reactions: $\dfrac{wL}{2}$
- Shear: $+\dfrac{wL}{2}$ to $-\dfrac{wL}{2}$
- Max positive moment at center: $\dfrac{wL^2}{24}$
- End moments on M diagram: $-\dfrac{wL^2}{12}$

APPENDIX P

Properties of selected geometric shapes.

Rectangle
Area: $A = bh$
Centroid location: $x = 0, y = 0$
Area moment of inertia: $I_{xx} = bh^3/12$, $I_{yy} = hb^3/12$
Radius of gyration: $r_x = h/\sqrt{12}$, $r_y = b/\sqrt{12}$

Triangle
Area: $A = bh/2$
Centroid location: $y = h/3$
Area moment of inertia: $I_{xx} = bh^3/36$
Radius of gyration: $r_x = h/\sqrt{18}$

Circle
Area: $A = \pi D^2/4$
Centroid location: $x = 0, y = 0$
Area moment of inertia: $I_{xx} = I_{yy} = \pi D^4/64$
Radius of gyration: $r_x = r_y = D/4$

Hollow Circle
Area: $A = \pi(D^2 - D_i^2)/4$
Centroid location: $x = 0, y = 0$
Area moment of inertia: $I_{xx} = I_{yy} = \pi(D^4 - D_i^4)/64$
Radius of gyration: $r_x = r_y = \sqrt{(D^2 + D_i^2)/16}$

Semicircle
Area: $A = \pi R^2/4$
Centroid location: $x = 0, y = 4R/3\pi$
Area moment of inertia: $I_{xx} = \pi R^4/8$, $I_{yy} = \pi R^4[1/8 - 8/9\pi^2]$
Radius of gyration: $r_x = R/\sqrt{2}$, $r_y = (R/2)\sqrt{1/8 - 8/9\pi^2}$

APPENDIX Q1

Wide flange structural sections.

Wide flange structural sections. (*Continued*)

Nominal depth × lbs/ft	t in.	b in.	d in.	Area in^2	I_{xx} in^4	I_{yy} in^4
W 36 × 300	0.945	16.655	36.74	88.3	20300	1300
Ten sizes exist between 36 × 300 and 36 × 135.						
W 36 × 135	0.600	11.950	35.55	39.7	7800	225
W 33 × 241	0.830	15.860	34.18	70.9	14200	932
Five sizes exist between 33 × 241 and 33 × 118.						
W 33 × 118	0.550	11.480	32.86	34.7	5900	187
W 30 × 211	0.775	15.105	30.94	62.0	10300	757
Six sizes exist between 30 × 211 and 30 × 99.						
W 30 × 99	0.520	10.450	29.65	29.1	3990	128
W 27 × 178	0.725	14.085	27.81	52.3	6990	555
Five sizes exist between 27 × 178 and 27 × 84.						
W 27 × 84	0.460	9.960	26.71	24.8	2850	106
W 24 × 162	0.705	12.955	25.00	47.7	5170	443
Nine sizes exist between 24 × 162 and 24 × 55.						
W 24 × 55	0.395	7.005	23.57	16.2	1350	29.1
W 21 × 147	0.720	12.510	22.06	43.2	3630	376
Eleven sizes exist between 21 × 147 and 21 × 44.						
W 21 × 44	0.350	6.500	20.66	13.0	843	20.7
W 18 × 119	0.655	11.265	18.97	35.1	2190	253
Eleven sizes exist between 18 × 119 and 18 × 35.						
W 18 × 35	0.300	6.000	17.70	10.3	510	15.3
W 16 × 100	0.585	10.425	16.97	29.4	1490	186
Nine sizes exist between 16 × 100 and 16 × 26.						
W 16 × 26	0.250	5.500	15.69	7.68	301	9.6
W 14 × 730	3.070	17.890	22.42	215.0	14300	4720
Thirty-four sizes exist between 14 × 730 and 14 × 22.						
W 14 × 22	0.230	5.000	13.74	6.49	199	7.00
W 12 × 336	1.775	13.385	16.82	98.8	4060	1190
Twenty-seven sizes exist between 12 × 336 and 12 × 14.						
W 12 × 14	0.200	3.970	11.91	4.16	88.6	2.36
W 10 × 112	0.755	10.415	11.36	32.9	716	236
Sixteen sizes exist between 10 × 112 and 10 × 12.						
W 10 × 12	0.190	3.960	9.87	3.54	53.8	2.18
W 8 × 67	0.570	8.280	9.00	19.7	272	88.6
Eleven sizes exist between 8 × 67 and 8 × 10.						
W 8 × 10	0.170	3.940	7.89	2.96	30.8	2.09
W 6 × 25	0.320	6.080	6.38	7.34	53.4	17.1
Seven sizes exist between 6 × 25 and 4 × 13.						
W 4 × 13	0.280	4.060	4.16	3.83	11.3	3.86

APPENDIX Q2

Standard flange structural sections.

Nominal depth × lbs/foot	t in.	b in.	d in.	Area in^2	I_{xx} in^4	I_{yy} in^4
S 24 × 121	0.800	8.050	24.50	35.6	3160	83.3
Four sizes exist between 24 × 121 and 24 × 80.						
S 24 × 80	0.500	7.000	24.00	23.5	2100	42.2
S 20 × 96	0.800	7.200	20.30	28.2	1670	50.2
Two sizes exist between 20 × 96 and 20 × 66.						
S 20 × 66	0.505	6.255	20.00	19.4	1190	27.7
S 18 × 70*	0.711	6.251	18.00	20.6	926	24.1
S 15 × 50*	0.550	5.640	15.00	14.7	486	15.7
S 12 × 50	0.687	5.477	12.00	14.7	305	15.7
Two sizes exist between 12 × 50 and 12 × 31.8.						
S 12 × 31.8	0.350	5.000	12.00	9.35	218	9.36
S 10 × 35*	0.594	4.944	10.00	10.3	147	8.36
S 8 × 23*	0.441	4.171	8.00	6.77	64.9	4.31
S 7 × 20*	0.450	3.860	7.00	5.88	42.4	3.17
S 6 × 17.25*	0.465	3.565	6.00	5.07	26.3	2.31
S 5 × 14.75*	0.494	3.284	5.00	4.34	15.2	1.67
S 4 × 9.5*	0.326	2.796	4.00	2.79	6.79	0.903
S 3 × 7.5*	0.349	2.509	3.00	2.21	2.93	0.586

* A lighter weight section exists for each of these depth sizes.

APPENDIX Q3

Standard channel structural sections.

Actual depth in inches × lbs/ft	t in.	b in.	Area in²	I_{xx} in⁴	I_{yy} in⁴	x in.
C 15 × 50 (3)	0.716	3.716	14.7	404	11.0	0.798
C 12 × 30 (2)	0.510	3.170	8.82	162	5.14	0.674
C 10 × 30 (3)	0.673	3.033	8.82	103	3.94	0.649
C 9 × 20 (2)	0.448	2.648	5.88	60.9	2.42	0.583
C 8 × 18.75 (2)	0.487	2.527	5.51	44.0	1.98	0.565
C 7 × 14.75 (2)	0.419	2.299	4.33	27.2	1.38	0.532
C 6 × 13 (2)	0.437	2.157	3.83	17.4	1.05	0.514
C 5 × 9 (1)	0.325	1.885	2.64	8.90	0.632	0.478
C 4 × 7.25 (1)	0.321	1.721	2.13	4.59	0.433	0.459
C 3 × 6 (2)	0.356	1.596	1.76	2.07	0.305	0.455

( ) The number of additional smaller weights available for each depth size category.

APPENDIX Q4

Standard angle structural sections.

Leg lengths × thickness, inches	Area in²	Weight lbs/ft	I_{xx} in⁴	y in.	I_{yy} in⁴	x in.	I_{zz}* in⁴	Ø deg
L 9 × 4 × 5/8 (2)	7.73	26.3	64.9	3.36	8.32	0.858	5.54	12.2
L 8 × 8 × 1 1/8 (6)	16.7	56.9	98.0	2.41	98.0	2.41	40.64	45.0
L 8 × 6 × 1 (6)	13.0	44.2	80.8	2.65	38.8	1.65	21.30	28.5
L 8 × 4 × 1 (3)	11.0	37.4	69.6	3.05	11.6	1.05	7.87	13.9
L 7 × 4 × 3/4 (3)	7.69	26.2	37.8	2.51	9.05	1.01	5.69	18.0
L 6 × 6 × 1 (8)	11.0	37.4	35.5	1.86	35.5	1.86	15.06	45.0
L 6 × 4 × 7/8 (7)	7.98	27.2	27.7	2.12	9.75	1.12	5.86	22.8
L 6 × 3.5 × 1/2 (2)	4.50	15.3	16.6	2.08	4.25	0.833	2.59	19.0
L 5 × 5 × 7/8 (6)	7.98	27.2	17.8	1.57	5.17	1.57	7.55	45.0
L 5 × 3.5 × 3/4 (6)	5.81	19.8	13.9	1.75	5.55	0.996	3.25	24.9
L 5 × 3 × 5/8 (5)	4.61	15.7	11.4	1.80	3.06	0.796	1.91	19.2
L 4 × 4 × 3/4 (6)	5.44	18.5	7.67	1.27	7.67	1.27	3.29	45.0
L 4 × 3.5 × 5/8 (5)	4.30	14.7	6.37	1.29	4.52	1.04	2.22	36.7
L 4 × 3 × 5/8 (5)	3.98	13.6	6.03	1.37	2.87	0.871	1.61	28.1
L 3.5 × 3.5 × 1/2 (4)	3.25	11.1	3.64	1.06	3.64	1.06	1.52	45.0
L 3.5 × 3 × 1/2 (4)	3.00	10.2	3.45	1.13	2.33	0.875	1.16	35.5
L 3.5 × 2.5 × 1/2 (4)	2.75	9.4	3.24	1.20	1.36	0.705	0.78	25.9
L 3 × 3 × 1/2 (5)	2.75	9.4	2.22	0.932	2.22	0.932	0.94	45.0
L 3 × 2.5 × 1/2 (5)	2.50	8.5	2.08	1.00	1.30	0.750	0.68	33.7
L 3 × 2 × 1/2 (5)	2.25	7.7	1.92	1.08	0.67	0.583	0.41	22.5
L 2.5 × 2.5 × 1/2 (4)	2.25	7.7	1.23	0.806	1.23	0.806	0.53	45.0
L 2.5 × 2 × 3/8 (3)	1.55	5.3	0.91	0.831	0.51	0.581	0.27	31.5
L 2 × 2 × 3/8 (4)	1.36	4.7	0.48	0.636	0.48	0.636	0.21	45.0

( ) Number of additional thinner angles available at the given leg sizes.
* Minimum moment of inertia; maximum moment axis is perpendicular to ZZ through the same center.
$Z = I/c = r^2 A/c$, where Z = section modulus, in³, r = radius of gyration, in., A = section area, in², and c = the distance to the extreme fiber from the centroid, in.
If I_e = moment of inertia relative to an axis parallel to the c axis and e units away, then $I_e = I_c + Ae^2$, where I_c = moment of inertia relative to an axis passing through the centroid of the area A.
Courtesy American Institute of Steel Construction, reprinted with permission.

APPENDIX R

Insulation values of common building materials.

Material	Thickness, inches	R-value
Air space	0.75 or more	0.91
Exterior surface resistance		0.17
Interior surface resistance		0.68
Sand and gravel concrete block	8	1.11
Filled with concrete	8	1.93
Sand and gravel concrete block	12	1.28
Lightweight concrete block	8	2.00
Lightweight concrete block	12	2.27
Face brick	4	0.44
Common brick	4	0.80
Concrete cast in place	1	0.08
Wood sheathing	0.75	0.94
Fiber board insulating sheathing	25/32	2.06
Polyisocyanurate rigid foam board	0.75	5.4
Plywood	1	1.24
Bevel lapped wood siding	0.50 × 8	0.81
Bevel lapped wood siding	0.75 × 10	1.05
Vertical tongue and groove siding	0.75	0.94
Asbestos board	0.25	0.21
3/8 gypsum lath and 3/8 plaster	0.75	0.42
Gypsum board	1	0.90
Interior wood paneling	0.25	0.31
Building paper		0.06

APPENDIX R

Insulation values of common building materials *(Continued)*.

Material	Thickness, inches	R-value
Vapor barrier		0.00
Asphalt shingles	0.125	0.44
Wood shingles	0.375	0.87
Wood 2 × 4 (1.625 × 3.50)	1.625	2.03
	3.50	4.38
Aluminum siding	0.375	0.61
Aluminum siding (insulated)	0.375	1.82
Rock wool batts	1	3.70
Rock wool, loose	1	2.90
Fiberglass batts	1	3.15
Cellulose fiber	1	3.70
Pearlite	1	2.70
Vermiculite	1	2.08
Urethane, aged	1	6.25
Polystyrene, extruded low density	1	4.00
Polystyrene, extruded high density	1	5.00
Polystyrene, beadboard	1	3.57
Acoustical ceiling tile	0.75	1.78
Carpet and fiber pad		2.08
Carpet and rubber pad		1.23
Linoleum and floor tile	0.125	0.05
Stone	1	0.08

Answers To Selected Exercises

1-6. *a*) Bones in the arms and legs

b) Muscles in the arms making the forearm move. The elbow joint is the fulcrum.

c) The heart

d) Swimming fish, flying birds

e) Duck feathers, some plant leaves

f) Bats in flight, sounds of whales and other fish, dog whistles that humans cannot hear.

g) Lightning bugs, fluorescent fish, luminescent bacteria, fluorescent minerals under ultraviolet light, chlorophyll fluoresces red under certain conditions.

h) Chameleon skin, insects and bacteria that are attracted to light, plants grow toward the light.

1-7. 13°, 6° 21', 3° 1'

1-10. Approximately one ton per day

1-12. 0.432 inches

1-14. The design as specified will not work.

1-15. *a*) Approximately 12 feet *c*) 41 psi *d*) 6.11 inches

2-1. *c*) Ships have progressed from wooden, sail and oar powered, to fiberglass, steel, and other metal alloy hulls powered by steam, diesel, gasoline, and nuclear power. Designs have progressed to include submarines and hydrofoil hull designs that skim the water rather than cutting a wake. Ship design changes have reduced land travel and opened frontiers, but have not made anything obsolete.

f) Firearms started with muzzle loading, single shot pistols and rifles. Powder was carried separately from the bullets which were nearly spherical in shape. The powder was ignited with steel and flint. The powder and bullet became a single unit loaded from the breach ignited with a pressure firing pin. Single shot weapons became multiple shot, and had automatic shell ejection and new bullet loading. Barrels became rifled to make the bullet rotate for better accuracy and distance. The bow and arrow, and sling shot were made less important as weapons by firearms.

    i) Airplanes have progressed from lightweight, single engine, single person, low altitude, short flight machines to multiple engine, multiple passenger, long distance machines with pressurized cabins for high altitude flight. Jet engines have replaced internal combustion engines. Materials have changed to keep weight down and strength up. Navigational devices have been added to fly at night and in the fog. Airplanes reduced long distance travel by automobile and ocean travel by boat.

2-2.   *a*) Mopeds, carnival ride cars, farm tractors, golf carts

    b) Highway information signs by law, bricks and other building materials by convention and use efficiency

    d) Products sold by weight (1 pound of butter, 5 pounds of sugar, 50 pounds of fertilizer)

    e) Gas can (red), traffic signal lights (red, yellow, green), highway mile markers (green)

2-4. Use of automatic seat belts in cars (get in the way of passengers entering and leaving auto), kickstand on motorcycles (digs into soft surfaces), vending machines (jam and do not return money), the storage and removal of car jacks and spare tires from the trunks of cars (awkward to remove from storage).

2-6. See Figure 2.14 for sample.

2-11. A knife, football, tennis ball, golf club, bicycle rack, gasoline, oil, kitchen utensils, calculator

2-14. Most companies purchase items that have been standardized such as nuts, bolts, washers, and other hardware components; light bulbs and other electrical devices; rubber O-rings, tires, and other formed rubber parts; and specialty operations they are not equipped to handle such as chrome plating, die casting, and galvanizing.

2-16. Reduced weight from mean = 360 lbs

2-18. Fire extinguisher, match, firecracker, camera flash bulb, CO_2 cartridge, bullet, garbage bag

2-20. Light bulb, grease or oil surface rust protection, food, medicine, paint

2-22. Electric fuse, shear pin, collapsible crash barrier

2-24. Manual can opener, put crank on top; scissors, must have cutting action force on center line; manual pencil sharpener, place crank on top; TV remote control, place most used buttons—channel and volume—in center.

2-26. The interior of a car is not an easily defined geometric shape so the volume is hard to calculate. A room has perpendicular walls and the volume is easier to calculate.

2-27. Range is from $0.30/pound to $2.50/pound.

Answers To Selected Exercises

2-29. Suitcases can hold enough to become too heavy to be easily carried, weights used in exercise machines, door sizes related to people size, clothes must be sized to people dimensions, pole vault pole diameter must be sized to the hand grip of people.

2-31. Oil and air filters must be changed regularly or they will become clogged and not allow the passage of oil or air. No economical alternate method exists to remove the unwanted material from the fluid. *Note:* Electrostatic air filters are available, but most air furnace systems are not using them, perhaps because they are too expensive or noisy.

2-33. Rubber tires, clothes washers, driers, television sets, electric batteries, plastic containers, Styrofoam cups

2-35. One subjective comparison.

	Metal	Glass	Plastic	Coated paper	Uncoated paper
Raw material use	5	1	4	3	2
Fabrication	5	4	3	2	1
Disposal	4	5	3	2	1
Reuse	2	1	5	5	5
Litter life	5	5	5	2	1
Cleanup ease	2	4	2	4	4
By-product effect	3	3	3	2	2
Ease of use	2	2	1	1	1
Shelf life	2	1	3	4	5

1 = good, 5 = poor

2-37. *a)* The critical path is $A-D-F-G-H-I-J-K$ which has a total time of 30 weeks. The project cannot be completed in 27 weeks.

b) Event B has an eight week slack time, event C has a one week slack time, and event E has a ten week slack time.

c) Use four people for the first 15 weeks and two people the second 15 weeks.

2-39. *a)* This analysis has a critical path of: $A-B-C-E-F-G-H-I-L-M-Q-N-R-T-V$, with an estimated completion time of 26.5 working days depending on whether T, the three-day concrete cure time, can be scheduled for a week end.

b) An additional 15–17 days are required depending on whether or not the rain starts on the first or second day of outside work, and whether or not concrete curing time falls on the weekend.

2-40. A computer program that accounts for all the variations of user and environment cannot be written because all variations cannot be identified qualitatively and quantitatively.

2-41. *a)* Implies that other things are hard to obtain, such as technical help, maintenance manuals.

   b) Other problems exist such as light weight, damaged in shipment.

2-43 and 44.

   a) What does "best" mean? Best taste? Best price?

   b) What does "very hard" mean? How many hours? What work was done, number of problems completed?

   c) Car engine stops running, car stops moving. Did the engine run out of fuel? Did the car run into a wall? Did all the tires go flat?

   d) What is "overworked?" Sixty hours a week? Twelve hours a day? A hundred gear boxes assembled before going home?

   e) How heavy is "too heavy?" What weight can a person lift? How many times a day can that weight be lifted and carried a given distance?

   f) "Unbearable?" Backlog can't be filled in "x" weeks? Is the weight of the paper for the orders making something collapse?

   g) "Good?" What size part can be machined? What dimension tolerances can be held?

   h) "Crowded?" Crowded may just mean filled. Does the warehouse contain more than it was designed to hold? How much was it designed to hold? Volume? Weight?

   i) "Dangerous?" Are the flames apt to burn someone? Are the sheets of metal too heavy for the table and it may collapse?

   j) "Beautiful?" Beauty is in the eye of the beholder. More beautiful because of color? Compared to a Van Gogh painting? To another car?

   k) "Better than?" What tested better? Torque-in torque-out ratio? Temperature of operation?

2-46. If the pipes are truly inaccessible, then no amount of cost or time can get them repaired. A more accurate description is probably "Not easily accessed."

3-1. Walk, run, car, taxi, skateboard, bicycle, helicopter, hitch hike

3-3. *a)* How can you dry your hands?

   b) We need a loaf of bread for supper in 30 minutes.

   c) I need to remember the message I am about to receive.

   d) Where can we be comfortable to eat our lunch?

   e) How can we get across this river?

3-5. Trial and error means test, learn from the test, and iterate to improve the design. This cycle may be necessary several times in engineering design projects.

3-6. Design an automobile that carries six people, but will fold up so it can be carried to an office.

3-7.   c) Chernobyl is located approximately 30°E and 50°N on the Earth, so directly opposite would be 150°W and 50°S, somewhere in the Pacific Ocean east of New Zealand.

3-9.   The next letter is F. The pattern is the first letter of the counting numbers *o*ne, *t*wo, *t*hree, *f*our, *f*ive,...,*t*hirteen, *f*ourteen, *f*ifteen.

3-10. a) He is short and he cannot reach a floor button higher than 10.

b) The elevator was being repaired in the morning and could only go to the tenth floor.

c) He wanted some exercise in the morning walking up stairs, but not at night walking down.

d) He forgot where he was going in the morning, his office used to be on the tenth floor.

e) The coffee shop is on the tenth floor and he stopped to get a cup on the way to work and then walked up the last flights.

3-11. Contingency plans in a preliminary schedule might include extra lead time and money, or plans to add more people, time, and material to particular activities if schedule is not maintained.

3-14. Development projects such as medicines and hybrid plants, that need a lot of test time, cannot be speeded up by adding people and money as much as projects such as timber cutting, road building, and brick laying.

3-16. Motor rated too low. Too much friction between belt and rollers. Not enough friction between drive belts and pulleys. Weight rating of system doesn't include all possible product combinations.

3-17. Need curb and gutter to control water flow. Collect the water at low points for sewer openings. Need to know the highest rainfall rate in inches per hour; one-half inch per hour is equivalent to 13,577 gallons per acre per hour; system must be capable of handling the runoff. Consider sewer openings to be safe so children will not be washed into them. How can the system be cleaned out if plugged with debris? What is the shortest route for the system?

3-20. Measure amount of fuel used to produce the same output of heat, or measure the amount of heat output based on a constant input.

3-23. Testing a part outside its normal environment requires a great deal of confidence in knowing what the actual loads are. A test of the shaft all by itself is advantageous if only one aspect of the design is being studied, such as rotating fatigue. Assembled in the transmission will be a better test of all the required functions of the shaft. Tested in an automobile will be best, because outside forces coming from vibration and mounting stress will also be involved.

3-26. A traffic counter could be replaced by a person counting cars. An electronic door opener could be replaced by a door-person. A lane presence device could be replaced with a "traffic-cop." An escalator could be replaced by a stairway. A windmill used to grind grain could be replaced with an animal driven grinder.

3-30. Start on the inside and work out rather than starting on the outside and working in.

3-32. Even a simple job like changing an automobile tire has many variations. Is a bumper jack available, an axle jack, no jack? Is the tire flat, or just being rotated? Flat tires often cause the car to settle so low that there is no room for standard jacks. What kind of wrench is available? Impact? Ratchet? Spinner? Is the tire on a compact car, a full size car, a truck?

3-33. Case 1: Flip either switch up. If a switch is flipped up and the lights go out, that switch is a three-way switch. If a switch is flipped up and more lights go on, that switch is a two-way switch.

Case 2: Flipping one switch will not allow an identification.

Case 3: Flipping one switch will not allow an identification.

Case 4: Flip either switch down. If the lights go off the switch is a two-way switch. If the other lights go on it is the three-way.

4-1. Was the hood release accidentally pulled? How fast was the car moving? How far did the hood open? Does the hood hinge at the windshield, at the side, or at the front? Was there any vibration before the hood came open? Has this ever happened before?

4-3. Is it windy? Does the door have a lock open position? Were you carrying a large load? Did someone keep closing the door as you were carrying packages? Does the door swing both ways? Can you disable the automatic closer if there is one? Will someone hold the door open for you? Is there another entrance you can use?

4-5. Do you own land? Is there land available to rent? What do you want to grow? Vegetables? Flowers? What size garden do you anticipate having? Is water available at the garden site? Do you have any books on gardening? Any other reference material? Have you ever gardened before? Do you own any gardening tools?

4-7. How much will you get? Do you have any outstanding bills to pay? Have you considered investing in stocks? Bonds? Certificates of deposit? Regular savings account? Have you considered donating the money to charity? Do you have a personal need or want?

4-9. Is there a kitchen nearby? What color smoke? Do you see the smoke or smell it? Do you feel heat? Have you looked for the source yet? Is there a fireplace in the house? What time of the year is it? Is the furnace running?

4-13. The order of ranking is from easy-to-check to hard-to-check.

Is the bulb screwed in all the way? Is the bulb burned out? Is this the right switch for that light? Is the fuse blown? Is power out on our street? Has power been disconnected from the house? Is the switch defective? Is the house wiring defective?

4-14. Is the door locked? Do I have the right key? Do I have the right car? Is the lock frozen? (If wintertime.) Is the lock defective?

4-15. Did I pull fast enough? Is there fuel in the fuel tank? Is the key turned on? (If a key ignition.) Is the spark plug wire attached? Is the mower in heavy grass? Is the choke open or closed as necessary for the conditions? (Hot engine or cold engine.) Is the engine flooded with fuel? Is the spark plug bad? Is the magneto or another ignition part operating improperly?

4-18. The hitch on the two units must be compatible, same style and same height. The power of the pulling vehicle must be adequate for the additional load. The units must be compatible for stop and signal light connection. The pulling vehicle must have proper rear view mirrors. Heavy duty rear tires and shock absorbers are also critical if the trailer is very large.

4-20. Public buildings are more likely to have multiple uses, and anticipated equipment and people loads are more difficult to control. Residential homes usually have fixed usage and do not have the high volume traffic that a public building would have.

4-22. Automobiles to carry six passengers and yet be efficient enough to get 50 miles per gallon. A toaster that would not get hot enough to burn a user. A hammer that would not hurt a user's thumb if hit. A lawn mower that would not cut a user's hand even if the user cleaned grass from under the mower while it was running.

4-25. Metals were heated and hammered together. Inertia welding heats two pieces of metal by friction and then forces them together with pressure. Parts are rotated at high speeds in a lathe-like machine.

4-27. *a*) Student should leave earlier or choose a different mode of transportation.

b) Rider should wear pants that are tighter to the leg and cannot get tangled with the chain.

c) Student should learn to type better or review work to correct errors before running the program.

d) The problem could be poor assembly procedure or improper use by customer.

5-2. *a*) Flooding and hard to start, or no starting.

b) A choke that is stuck closed. Dirt in the gasoline.

c) The float was too heavy. The sound of sloshing fluid in the float. A puddle of gas by the float as it sat on the workbench.

d) (1) Drill a hole in the float. Fast, but the hole will need repair.
(2) Heat the float and boil the gas out. Effective, but hot gas fumes might explode.
(3) Wait for the gas to evaporate. Effective, but slow.
(4) Suck the gas out with a vacuum. Effective, but slow; need a vacuum source.
(5) Separate halves and dry out and resolder. Effective, but costly in time. Resolder job is critical.

e) Ranking is subject to facilities available and skill of repair person.

f) Repair as indicated in part (d). Replace float with a new one. Use a material for the float that does not absorb, or is attacked by, gasoline. Fill the existing float with a substance less dense than gasoline.

Answers To Selected Exercises

5-4. Connecting belt, spline shaft, mating gears, friction wheel, chain.

5-7. Alarm signals for fire, computer access to bank accounts, fixed response to TV questions, order products.

5-9. Issue bonus credit if turned in early. Require step by step checking and credit. Decrease credit if turned in late.

5-10. Compass, stars, prevailing winds, position of sun, aurora borealis.

5-12. Traffic person, signal lights, stop and yield signs, gate barriers, overpass, underpass.

5-14. Record license plate numbers and send bills. Prepay and only prepayers can travel road. A coin collector that matches the car speed so paying can be done without slowing down.

5-15. Number the pages. Letter the pages. Notch the sides of the pages. Use different size paper for each page. Place a diagonal line on the edges of the stack of pages.

5-17. Playground apparatus, sandbox, line driveways so cars cannot drive on grass, boat dock bumpers, truck loading dock bumpers, tree protectors, car bumpers, edge of gardens.

5-19.
   a) A buzzer sounding if the head drops, the eyelids stay shut for more than two seconds, or the tires leave the pavement.

   b) Crawl on floor to locate. Vacuum area and sift through debris. Make contacts infrared or ultraviolet sensitive and use special light to locate.

   c) Car monitoring system that keeps track of both miles traveled and gallons of fuel used. Less exact system could calculate based on miles since last fill up; miles per gallon would continue to increase until the next fill up.

   d) Remote control solenoid plungers, pull a wire that allows gravity plungers to operate.

   e) Have a person walk the aisles and check each seat belt. Use an electric circuit that uses buckle connections as switches; check only seats with weight on them. Use seat belt units that beep if not buckled whenever takeoff and landing is in progress. Install automatic bucklers, those that would buckle whenever a person sits in the seat.

5-21. Need dense material: lead, gold, concrete, stone, brick.

5-22. Place a net over the area and pull the corners in like a fish net. Make the area a moving surface like people conveyors at large airports. Train dogs to collect the balls. Make special balls with metal centers and use a magnet. Slope area to a trough. Pay someone to walk the area picking up balls. Use a machine with tires that picks up balls as it rolls over them. Make customers pick up all the balls they have hit.

5-24. Horizontal: bubble level, non-rolling of a ball, pressure sensitive altimeter, water or other fluid level.

Vertical: perpendicular to horizontal, hanging weight, falling weight, fluid over a ledge.

5-27. Make A'' an overpass. Place a yield sign at intersection of E' and A'' for A'' traffic. Reroute traffic and close A''. Erect a barrier and eliminate E'. Place a yield sign at E'. Don't allow E' and A'' and make B a one way street.

5-28. Use a fluid in a tube, see Figure 6.10. Use a hanging weight and measure the angle from vertical, see Figure 6.11. Use a mass and a spring to be compressed or stretched in the direction of acceleration, see Figure 6.12.

5-30. Use an ice cube, a lighted candle, a bowl of any solid or liquid that will evaporate. Use any substance that will absorb water such as granular fertilizer, salt, or silica gel.

5-33. Use a Styrofoam container, parachute, hard boil and drop into water, pack egg in carton with "lots" of cushion material, use tube of fluid with density just less than that of an egg and let egg settle slowly in fluid.

5-35. Keep pouring and hope enough hits the bottle until it is full. Use an extension of straws. Use a garden hose.

5-36. Rubber tipped rod attached to head. Wind from a fan. Use your nose. Tip the book and let gravity turn pages.

5-39. Long handled hammer. Throw rocks at the nail. Shoot bullets at the nail. Catapult rocks at the nail.

6-2. See Exercise 6.32, Figure 6.18.

6-4.

Favorable	Unfavorable
a. Stereo system	Car without a muffler
b. Spotlight in theater	Glare from setting sun
c. Sound speakers	Car with bad shocks
d. Night light	Changing a car tire at night
e. Chemotherapy	Nuclear waste
f. Refrigeration of food	Winter weather
g. Balloons	Low water pressure in house
h. Cooking food	Danger of burning skin
i. Cattle prod	Explosions
j. Cooling fan	Hurricane

6-6. *b)* 44 cars per minute

c) 6.9 mph

6-8. *a)* acceleration = $\Delta h/L$.

b) The area of the vertical tubes does not matter, only the difference in heights.

6-13. Two units mounted in perpendicular planes could be used, and the actual acceleration would be the vector sum of the two readings.

6-14. Force in Gs = 0.40

6-16. Approximately 5420 feet

6-17. While accelerating, 0.45 Gs; while braking, −0.95 Gs

6-20. The lowest cost = $16,825P

6-24.   *a)* Common available material thicknesses are 1.250, 1.375, or 1.500, not 1.40.

    b) Flame cut is appropriate. Cut variations from mean profile will give size variations. Matching plates for good looking assembly will be difficult.

    c) The only rivets needed are to hold the laminates together. The entire shear load should not occur at any laminate boundary. Twenty-six 1-inch holes will reduce the effective cross section of material. Drilling holes is added labor cost that should be minimized.

    d) Welding makes a one piece assembly as far as crack propagation is concerned.

    e) Hook cannot hang straight unless mass on both sides of centerline are the same.

    f) There are three critical sections to consider for possible maximum stress, at the upper hole, at the shaft cross section, and at the curved beam section at the bottom.
      (1) At the top hole, section A-A, SF is approximately 6.4.
      (2) At the mid section B-B, SF is approximately 10.
      (3) At section C-C, SF is approximately 4.9.

6-27. The cost of required Btu's at 65% efficiency is still less than 100% electric efficiency. The inefficiency of using electricity is the loss when oil, gas, or coal is converted to electric power at the power generating plant, and the power loss during distribution through the wires.

6-29.   *a)* Shortened

6-30. The cross sectional area of any component is proportional to F for a given material, so the material cost is proportional to the product of the force F and length L.

    a) Take first moments around point C: the minimum $(F_{AE} L_{AE})$ product = 40,000.

    b) Take first moments around point C: the minimum $(F_{DF} L_{DF})$ product = 50,000.

    c) Take first moments around point C: the minimum $(F_{DF} L_{DF})$ product = 40,000.

The most widely used solution is *DF* or *BD*. The members are in tension and cables can be used. Cable is easier to deliver to the job site than rigid members which are needed for compression loads. The actual answer will depend on the cable sizes available for the tension loads applied.

7-1.   *a)* Material, type of bottom (V or flat), how many passengers, weight, dimensions, maximum load, price, safety record, oars, comfort.

642 Answers To Selected Exercises

- b) Size of books it will hold, required strength, mounting, style, how many books it will hold, other items it will hold, backed or open through, book ends built in, adjustable shelves, number of shelves, material, height.
- c) Material, table surface, weight, leg stability, dimensions, style, ease of setup and take-down.
- d) Capacity, color, dimensions, weight, type of loading, price.
- e) Available software, type of disk used, type of disk drives, weight, dimensions, price, available printers, memory capacity, delivery, service.
- f) Type of seat, type of brakes, color, height, number of speeds, type of chain guard, price.
- g) Weight, capacity in number of people, material used, waterproofing, cost, appearance, ease of setup and take-down, durability, compactness for storage, fire resistance, screen doors/windows, type of tie down anchors, odor containment.
- h) Location, number and type of rooms, number of levels, type of foundation, condition of plumbing, heating, and electrical systems, appearance, cost, size of yard.

7-5. Sales, warranty claims, safety records, percent of market penetration.

7-7. Have the machine tool vendor machine parts. Have an outside vendor machine parts. Change the processing of the part to make it some other way.

7-10.
- a) *D* is the highest value choice.
- b) *D* is still the choice.
- c) *C* or *D* are equal in value.

7-12.
- a) The evaluation is somewhat subjective; high density would be important if a building is being built as a fallout shelter, thermal resistance is critical in a very hot area for keeping cool and in a cold area for keeping warm, and cost is always an important item. Use values of 6, 9, and 10.
- b) If the best value in each category is given a 10, and there is no cutoff value that is restrictive, then all other evaluations can be calculated. If the value for density of 169 and 170 is 10, then for 120: $(120/170)10 = 7.1$, for 140: $(140/170)10 = 8.2$; for 62: $(62/170)10 = 3.6$. If the value for heat resistivity of 0.25 is 10, for 0.20: $(0.20/0.25)10 = 8$, for 0.10: $(0.10/0.25)10 = 4$, for 0.02: $(0.02/0.25)10 = 0.8$, and for 0.0007 a value of zero. If the value for cost of 0.50 is 10, then for 3: $(0.50/3)10 = 1.7$, for 5: $(0.50/5)10 = 1$, for 17: $(0.50/17)10 = 0.3$, and for 30: $(0.50/30)10 = 0.2$. With these numbers the ranking chart looks like this.

Item	Value	Normalized	Material 1	2	3	4	5
Density	6	0.24	10/2.4	7.1/1.7	8.2/2.0	10/2.4	3.6/0.9
Heat *R*	9	0.36	0/0	8/2.9	4/1.4	0.8/0.3	10/3.6
Cost	10	0.40	0.3/0.1	1/0.4	1.7/0.7	0.2/0.1	10/4.0
Totals			2.5	5.0	4.1	2.8	8.5

c) Material #5 is judged the best.

d) Durability, color, ease of installation, availability

7-14. For Case Study #6:

	1 vs					
2	1	2 vs				
3	3	3	3 vs			
4	4	4	4	4 vs		
5	5	5	5	5		
Idea		1	2	3	4	5
Number of "wins"		1	0	2	3	4

Proposal 5 is the "best."

7-17.

	Advantages	**Disadvantages**
Wooden pencil	Less expensive. Always know if there is lead available.	Wears out with each sharpening. Handling changes as size and weight changes. Need access to a sharpener.
Mechanical pencil	Lead sharpness is consistent. Handling is constant. Erasers are generally changeable.	More expensive If lead advance does not work, pencil is scrap.

7-18. An example of a checklist: Items to check when buying a used house.

a) Attic insulation, affects heating and cooling costs.

b) Roof, age and general condition, 20 year "normal" life.

c) Chimney, if brick check mortar joints, does it have a clay liner? If metal check for rust and tight joints.

d) Roof flashing, check for rust and tight joints, caulking may be required.

e) Plaster, any cracks, lose or crumbly plaster. Check source of problem: sagging timbers, settling foundation.

f) Bath walls, any wet plaster around pipes and drains?

g) Wall insulation, affects cooling and heating costs.

h) Storm doors and windows, are they well sealed, do any moving sashes not slide or swing as they should? Test them.

i) Wiring, is the service entrance large enough for all appliances and future need? Has updating to circuit breakers been done?

j) Rain gutters and downspouts, any rust? Are they working properly? Check by running water off roof with a hose.

k) Siding
Brick: Check mortar joints and caulking around doors and windows.
Wood: Check for cracked or water soaked boards. Does it need paint? Check caulking around doors and windows.

l) Exterior doors, check weather stripping and latch mechanisms.

m) Foundation, look for recent cracks, water stains, and out of plumb walls.

n) Water supply, if well and pump, have water tested for contaminants. Check for rusty joints. Check all shutoff valves. Turn more than one faucet on and check water supply.

o) Furnace, check for rust. If questionable have inspected.

p) Water heater, check for rust, check age.

q) Floors, check for sagging and squeaks.

r) Check surface water runoff for potential basement problems.

8-3. $(3.12 - 2.00)/2.00 = 56\%$

$(5.12 - 4.00)/4.00 = 28\%$

$(9.12 - 8.00)/8.00 = 14\%$

An example of why gage length is important as a standard for ductility calculations.

8-5. 31,200 psi

8-8. Assuming no change in material volume, then $L = 256$ ft.

8-11. 0.0008 in./in.

8-12. Casting, grinding, plasma arc cutting; any process that does not create tensile stress.

8-15. Using the 12 materials lead, magnesium, aluminum, silver, copper, beryllium, nickel, cobalt, iron, titanium, chromium, and tungsten, $E = -1.2339 + 0.019\,T$ with a coefficient of correlation of 0.91.

8-17. BHN = force/spherical contact area = force per unit area

8-19. $G = E/[2(4/3)] = 3E/8$

8-22. The bar will be curved concave downward when it reaches room temperature.

8-25. The higher yield strength steel will have the most springback because for a constant E, a higher yield strength results in a higher elastic deformation.

8-27. Because yield strength goes down as temperature goes up.

8-29. Bronze is more expensive than aluminum, but for moving parts in a water environment, the best bronze alloy is stronger and more corrosion resistant than the best aluminum alloy. Salt will affect more grades of aluminum than grades of bronze.

8-31. *a*) $(0.96 - 0.92)/0.92 = 4.3\%$, not significant

b) $(0.001 - 0.00065)/0.00065 = 54\%$, significant

c) 150 psi $\leq \sigma(LD) \leq$ 400 psi, 420 psi $\leq \sigma(HD) \leq$ 1260 psi

8-34. *Hint*, assume $\epsilon_{yt} \lll \epsilon_{ut}$.

8-35. Using an *allowable tensile* strength for 1020 HRS of 20,000 psi, then $\Phi = 0.0010$ radians.

8-41. The preference for 4340 would be when toughness, durability, and homogeneous properties are required; 4340 can be through hardened; 1060 would be preferred when wear, abrasion resistance, and case hardening are required. The 1060 has a shallower hardenability than does the 4340.

8-43. a, b, and c are all tungsten, refer to Figure 8.13.

8-45. (a) is the lowest, (c) is the highest.

8-48. The material could be a polymer with "plastic memory."

8-56. Low density, high ultimate strength, approximately five times the strength/density ratio of steel, modulus of elasticity approximately 3.6 that of steel, corrosion resistant.

8-59. $P = 16,878$ lbf

8-61. No. The shear stress on the bond interface is approximately 98 psi.

9-3. a) Yes, as long as $t > 0$, or $t > 1/[(Vn/2v)^2 Dd/(d+D)]$.

b) Increase n 10 percent and increase C by 21 percent.
Increase t 10 percent and increase C by 21 percent.
Increase V 10 percent and increase C by 21 percent.
Increase v 10 percent and decrease C by 17.4 percent.
Increase D or d 10 percent and decrease C by 9.1 percent.

c) Keep $Dd/(d+D)$ constant.

9-5. Plating can be used to create a harder and more wear resistant surface than can be obtained with the base material. The plated material may also have electrical characteristics that are different from the base material. Improved surface finish might also be an objective of plating.

9-7. a) Reinforcing bars do not fit the exact description of bars because they do not have a constant cross section.

b) Reinforcing bars do have a regular pattern of cross section and are formed by the same process, rolling.

c) A grinding spark test would help determine the amount of carbon. A hardness test might also help. A sample could be heated and quenched, and maximum hardness compared with standards for a range of carbon content (see Figure 9.22).

9-9. b) The reasons for variations in tolerance in any process include: material properties are not constant, material hardness or melted viscosity are not constant; operators' performances are not constant; atmospheric pressure, temperature, and humidity are not constant. These and other normal variations make exact constant output unlikely. Consider performing a process capability study to determine how to get the best surface finish.

9-10. A rectangular cross-sectional part being bent is modeled as a simple supported beam loaded in the center: $\sigma = Mc/I$, $I = wt^3/12$, $c = t/2$, and $M = (F/2)(L/2)$, so $\sigma = 3FL/2wt^2$ and $F = 2\sigma wt^2/3L = 0.67\sigma wt^2/L$. The bend angle is assumed small and is not part of the calculation. This holds as long as $\sigma \leq S_{yt}$, when the entire bend is elastic. The extension to plastic bending is an empirical one, not theoretical.

9-12. *a*) Smaller   *c*) Larger

9-14. Relative cost would probably not change much unless a plateau of technology is reached.

9-17. The two guidelines are incompatible. The material from the centerline of the bend to the edge of the material $= 1.5t + 0.5[2\pi(t + xt)(0.25)]$ where $0.33 \leq x \leq 0.50$, reference Equation 9.1. Half the die opening $= 4t$.
$1.5t + 0.5[2\pi(t + xt)(0.25)] \geq 4t$; $2.285t + 0.78xt \geq 4t$; $0.78xt \geq 1.715t$; $x \geq 2.20$, which is outside the range $0.33 \leq x \leq 0.50$.

9-18. Use method #1 for 75 pieces unless the low cost tooling is not capable of maintaining required tolerances or material specification.

9-19. *b*) Part cost includes the amortization of die cost but also includes material and labor cost. Brass has the highest melting temperature of the four materials, so melt time and equipment cost is higher, and the material cost itself is higher.

9-21. Spinning is clearly the choice if lead time is short (3 weeks compared to 10 weeks). The tooling cost is less which is an advantage if the design is not finalized and if quantities required are low. Drawing is best when large quantities of a stable design are required.

9-23. *a*) lathe cost = $0.033; saw cost = $0.038.   *b*) $3.16 per hour.

c) The $3.16 does not seem reasonable, a saw operator would be paid more than that per hour. But if the hourly wage rate were higher then it would be less expensive to make bosses on a lathe. That would be poor use of a lathe. The solution often employed is to have the saw operator service more than one saw. If two saws were serviced by one operator then the hourly rate could be $6.32 and the boss cost would be equal on the two processes. If three saws were serviced then the hourly rate could be $9.48 and the boss cost would be equal.

9-24. Drilling, reaming, boring, grooving, trepanning, spotfacing, countersinking, burnishing, face milling, end milling, tapping, threading, chamfering. Honing and grinding could be done but are not recommended because of the abrasive particles generated. The machine bearings are not protected well enough for these operations.

9-26. Use a hydraulic cylinder, a conveyor, a ball screw, a linear motor, or a rack and pinion. Each could be controlled by an impulse motor, a motor with induction feedback control, limit switches, or shot pin locators.

9-28. Hot forging does not change grain size (unless very slowly cooled in which case grains can grow) and does not increase strength, cold working reduces grain size and increases strength.

a) Fatigue resistance increases as strength increases, favored by cold working provided cold working cracks or residual tensile stresses are not created.

Answers To Selected Exercises

      b) Creep is resisted by larger grains, favored by hot forging.

      c) Stress corrosion is resisted by large grains, favored by hot forging.

9-29. If the radius of bend is the same then the neutral axis would be closer to the punch side of the bend and more material would be in tension than compression. This should cause more springback. If, however, the punch radius was increased then the neutral axis could be shifted so the amount of springback should remain constant. This would make a good manufacturing experiment for a student project.

9-31. Approximately 6 in./min.

9-35.   *a*) With material the correct width, a two station punch and shear cutoff operation will suffice.

      d) The material is too thick to punch the holes efficiently; shear or saw the part to length and drill the holes, or hot punch the holes.

      e) Saw or shear to a length longer than finished length; mill to length for squareness and tolerance control. Drill the holes and ream or bore the holes to final size.

9-37. If made from bar stock the profile could be flame cut, the step milled, and the holes drilled. The 1.00 inch hole needs a flat bottom; boring, flat bottomed drill, or end mill would work.

9-39. Other critical dimensions include the width of the slot, and the two widths indicated with finish marks. Machining would best be handled with milling. The top slot and width can be milled with a three cutter setup with the middle cutter the width of the slot. The bottom width can be straddle milled. The hole should be located from the inside of the slot and the side of the web. They should be done in one operation to insure parallelism.

9-42.   *b*) The part is a little long for its diameter to be machined in one operation from a bar, unless a two stage operation is used. Machine up to the 1.00 diameter, chucking the part on the 1.00 diameter and then advance the bar out to finish the turning and cut off.

      c) Centers would have to be added to the ends of the part to hold it in a center grinder.

9-43. Welding is a casting process, metal is melted, mixed, and solidified. When the weld joint solidifies, it shrinks. The shrinking applies a tensile force to the two pieces being joined. If the pieces are held to prevent the movement toward each other, then the weld joint will be in tension. A tensile load on hot, low tensile strength metal can cause cracks to form. Pieces being welded together should be allowed to move so that tensile loads in the weld joint are avoided.

9-44. Sand cast, add material on surfaces to be machined, cast holes solid. Mill surfaces. Drill holes.

9-48. Die cast, adding material to the surfaces to be machined, 24 hole is solid, 34 radius is cast to something smaller, approximately 26. Mill bottom using three cutters, maintain 28 thickness; the middle cutter is 104 wide. Straddle mill sides to 164, hold thickness of one side to 20. Make one 20 dimension a reference so the other side becomes the datum side. Mill back side to 82, make 32 a reference. Mill top and bevels, use special cutters or two operations, one with a single flat cutter, the second with two 30° cutters. Drill hole.

9-50. Sand cast iron, add material on surfaces to be machined. Bearing caps often are match machined with the other half of the bearing, such as an automotive connecting rod. The rod and cap become an assembly and are not sold as individual pieces. Bearing caps for larger applications are often "machined in" after assembly to match an existing shaft. Whether machined as one part or in assembly, the flat surfaces and the 11 diameter, hole should be machined first. It then could be assembled and match machined. The hole in assembly is often broached because the 35 radius is cast slightly smaller, and has good clearance for the broach. If the part was to be machined by itself a mill cutter with a 35 radius could be used.

9-52. Die cast, core large hole. If corners must be square then add machining stock on casting and mill to size. Finish bore large hole and drill small holes. This part has other interesting possibilities. It could be formed from flat material, milled to shape, and finished by drilling and boring the 1.688 hole. Depending on the part's function the square corners may not be necessary. The part could be made from extruded bar and sawed to length, finished with milling, and by drilling and boring the 1.688 hole.

10-1. The combined $x = 581$ and $S = 111$.

10-3. As n increases, σ/S approaches 1.

10-5. Possible solutions include: a backup generator system, more than one operator trained to run the machines, known material substitutions available, or run some other job.

10-7. Reliability/cost ratios are:

a) 0.00046.   b) 0.00028.   c) 0.00036.   d) 0.00031.
e) 0.00028.   f) 0.00041.   g) 0.00037.

System (f) is "best." Choice (a) is not the best even though it has a higher ratio. It has a lower reliability than either component standing alone. This is another example of not just taking the highest number without thorough analysis.

10-9. a) $P(0) = 0.135$

b) $P(3 \text{ or higher}) = 0.323$

10-11. a) Checks were made every 48 min.

b) $P(\text{no terminals being available}) = 0.2$

10-13. F test: $F = 1.13$, $\alpha \sim 0.17$, and the parts are probably the "same." A second test is to use model #7 from Table 10.3.
$k = 0.575$, °f $= 8.4$, $t = 0.394 \Rightarrow \alpha \sim 0.7$, and again the parts are more likely the "same."

10-15. $P(\text{of a part being checked as good}) = 0.78$
$P(\text{of a good part being checked as good}) = 0.765$
$P(\text{a part being checked as good and is good}) = 0.98$

Answers To Selected Exercises

10-17. a) Parallel the 0.80 component with itself and the system reliability = 0.96, an increase of 0.16

10-22. $t = 9.54$

10-23. 74% of the samples will last more than 10 days.

10-24. $P(t) = 1 - \exp(-0.022\, t^{5.93})$

10-28. a) $P(6 \text{ good blocks}) = 0.17780$

b) $P(2, 3, 4, 5, \text{ or } 6 \text{ bad blocks}) = 0.46606$

c) $P(1, 2, 3, 4, 5, \text{ or } 6 \text{ bad blocks}) = 0.82202$

d) $P(6 \text{ good from } 7) = 0.0445$

10-32. Meter #1 is the more reliable meter and just needs calibration to get to 1000. Meter #2 has a good average but is not reliable.

10-34. a) log-log

b) $b = 0.125$ and $a = 13.33$

c) 138 hp

10-36. a) 0.62

c) 0.65

10-38. Approximately 0.150 inches of wear.

10-40. Use model #4 from Table 10.3, $\delta = -22.1$, and the product performance is probably not the same.

10-43. Approximately $11,350

10-44. a) 0.856

b) 0.137

c) 0.006

d) 0.00008

10-47. $\chi^2 = 7.208$, $n = 10$, α is approximately 0.70, so the dice are probably fair.

10-49. a) 0.22

b) 0.78

10-50. $R(1000 \text{ hours}) = 1 - (1 - \exp(-0.00002)(1000))(1 - \exp(-0.0000143)(1000)) = 0.9997$

10-52. Use model #1 Table 10.3, and $n \geq 87$.

10-54. Use model #9 Table 10.3, and 11 samples will be sufficient.

10-58. Approximately 0.90% of all parts are bad.

10-59. 0.7638 ± 0.0022

11-3. a) 0.04

b) 0.196

c) 0.0016

d) 0.020

11-5. Two *C*'s give a reliability = 0.995 @ $200. All other combinations are more expensive.

11-7. Overall reliability = 0.68 < 0.70. One solution is to have a second $25,000 machine on standby, which increases the reliability to 0.74. A better solution is to discover the problems keeping the machines from running and get them more efficient, 0.68 is very close to 0.70 and within reach of the present situation. If all machines were at 93.1% then overall reliability will equal 70%. Much more work needs to be done to get to 0.80 reliability. Getting all four machines up to 0.96 would be required. Duplicating the $25,000 and the $30,000 machines would increase reliability to 0.794.

11-9. $S_z = S_x n X^{n-1}$

11-11. Equation 11.6 stays the same, but the −2.33 used for Equation 11.7 becomes −3.09. This changes Equation 11.7 to $-8.45x^8 + 7.290x^5 - 1.476x^2 + 0.0005074 = 0$, and $x = 0.814$ (use 0.875 standard size), so the design is sensitive to changes in reliability.

11-16. $W = 800$ lb and $\delta = 1.73$ in.

11-18. Weld cable to end of shaft. Drill a hole in the end of the shaft, insert cable, and weld or crimp.

11-20. $\tau_{max} = 15{,}280$ psi, and reliability is approximately 95%.

11-22. $F = 438$ lb

11-24. Weld, or drill and tap and use screws.

11-26. Yes, 1650 lbf ⩾ 612.5 lbf

11-28. 12.6 lb

11-31. The minimum impact factor = 2 when $h = 0$.

11-33. Ordinarily failure will occur at the support, $x = 0$, but if the velocity of impact is large enough failure can occur at $x > 0$. The velocity caused by gravity may not be large enough, a propulsion mechanism may be required.

11-35. The effect of impact is reduced as δ_{st} increases, δ in case (a) is less than δ in case (b).

   a) $\delta_a = 4(LEF/\pi D_1^2)$, and $\delta_b = 2.5(LEF/\pi D_1^2)$.

11-38. Control groups would be of major concern with this experiment so analysis of differences could be made, such as values of E, temperature, moment of inertia and grain direction. Careful preplanning would be critical because of the long term nature of creep tests.

11-42. Controlled vibration must be induced into a chair or platform on which a person can attempt various activities; writing, reading, sorting, or assembling parts.

11-44.   a) $l_e/r\,(1) = 71$, $l_e/r\,(2) = 61$

11-46. $l_e/r = 2\pi^2 E^2/(2\pi E S_{yc} - S_{yc}^2)$

11-48. $P_{cx} = 6450$ lb, 6450 > 6300, so failure by column buckling would not be the primary mode of failure. If, however, the strength of a given 2 × 4 was > 1200 psi then column buckling would be likely, because 6300 lb is so close to 6450 lb.

11-50. $\sigma_{mean} = 375$ psi and $\sigma_{amplitude} = 11{,}622$ psi. $S_e = 0.5(90{,}000)(0.68$ for reliability$)(0.9$ for diameter$)(0.9$ for axial loading$)(0.6$ for plating$) = 14{,}870$ psi. For a life $\geq 10^6$ cycles $\sigma_{amp} \leq -14{,}870/90{,}000\, \sigma_{mean} + 14{,}870$; at $\sigma_{mean} = 375$, $\sigma_{amp} \leq 14{,}810$, and actual $\sigma_{amp} = 11{,}622 < 14{,}810$, so life is greater than 10^6 cycles.

11-52. Two channels placed back to back have $I_{xx} = 72.2$ in^4 and $I_{yy} = 8.6$ in^4, a ratio of 8.4 compared to a single section ratio of 23.6.

11-55. a) Probably (b) because it has the highest initial crack growth rate.

   b) Material (b) because $2.25 < 3.00 < 3.25$.

11-57. A washer in a home water faucet; a leak is not fatal.

11-58.

a/2b	C_a	C_b	C_c	C_d
0.01	1.002	1.000	1.000	1.000
0.03	1.007	1.002	1.002	1.001
0.05	1.011	1.006	1.005	1.004
0.07	1.016	1.012	1.010	1.008
0.09	1.023	1.020	1.017	1.014
0.11	1.030	1.031	1.025	1.021
0.13	1.041	1.044	1.036	1.029
0.15	1.054	1.059	1.048	1.040
0.17	1.070	1.078	1.063	1.052
0.19	1.091	1.100	1.081	1.067
0.21	1.116	1.125	1.114	1.094
0.23	1.146	1.155	1.126	1.105
0.25	1.183	1.189	1.155	1.128

Small is when $a/2b < 0.23$.

11-59. Small is when $a/2b < 0.01$.

11-61. a) The normal stress $= 75{,}000$ psi. Using a value of $c = 1.04$ from the results of Exercise 11.58, then $K = 80{,}650 < 90{,}000$, so the center crack is not critical.

   b) Using a value of $c = 1.22$ from Exercise 11.60, and $K = 94{,}600 > 90{,}000$ and is unsafe. This analysis does not consider crack growth, which can change a safe condition to an unsafe one.

11-63. Yes: $123{,}000/218{,}000 = 0.56$, which is in the neighborhood of the 0.50 guideline; $141{,}000/218{,}000 = 0.65$, which is in the neighborhood of the 0.63 guideline.

11-65. A keyed shaft would likely be in torsion load. The square corners in keyway (a) are in a higher stress region than the square corner of keyway (b). Keyway (b) should resist fatigue the best.

11-67. a) $> 490°C$

   b) $> 231°C$

11-68. S_e is approximately $(0.50)(500)(300) = 75{,}000$ psi.

11-70. *a*) and *b*) This design criteria allows for a uniformly distributed load of 25 lb/ft².

  c) $P_c = 1200$ psi

  d) $P_c(\text{min}) = 800$ psi

  e) $P_c(\text{max}) = 781$ psi

11-74. $\sigma_{max} = 13{,}200$ psi. From Appendix F-2, $S_{yt} = 132{,}000$ psi and $S_{ut} = 150{,}000$ psi. The component is safe from yielding, $13{,}200 < 132{,}000$, and also probably safe from fatigue, $13{,}200 \ll (0.50)(150{,}000) = 75{,}000$; geometry and surface quality would determine the actual S_e.

11-76. *a*) If the cable doesn't break, the cable will support a maximum pull of 12,760 pounds. If the cable does not stay stretched beyond its original length the cable is capable of a maximum pull of 7850 pounds.

  b) The cable will break.

11-78. Use Equation 11.18 and let $\sigma_1 = -5000$, $\sigma_2 = 10{,}000$, and $\sigma_3 = 40{,}000$. $\sigma_e = 39{,}700 > 25{,}000$ so the element will yield, but will not fracture.

11-81. Between 680 and 690 hours

11-83. Cold worked brass because the grain boundaries are already at a high stress level due to the cold working, and less additional stress is needed to cause fracture.

11-84. The material is probably a nonferrous alloy of some kind; S_{ut} is approximately equal to 107,200 psi.

11-86. Heavy machines and machines that release a large amount of mechanical energy when they operate need to have a high mass foundation to reduce the transmission of vibration to other parts of the building. This would be hard to accomplish unless the machine was at ground level.

11-88. For the sample of 15 bulbs, $P(\text{surviving } 515 \text{ hours}) = 0.95$. Use $x = 550$ and $S = 27$ from Example 11.4 and the number of standard deviations for a reliability of 0.95 from Appendix G. Let L = the life for a 95 percent reliability, and $L = 497$ hours, close to the n test, but a lower value. One would expect the complete test to be more accurate than the single failure test.

11-90. *a*) $k = 1.24$ hertz

  b) $k = 1.08$ hertz. Three springs would help if the transmitted frequency is exactly 1.24.

  c) Increase or decrease the spring stiffness to get away from the 1.24 hertz level.

11-92. The data does not appear to be consistent, more complete testing is advised.

11-94. $F = 32PL/\pi D^3$

11-96. $\sigma_{\#1} = 32PL/\pi D^3$ and $\sigma_{\#2} = 3PL/4D^3$, so beam #2 will resist fatigue failure the best.

11-99. The addition of a torsion load increases the maximum compressive load, and will make column buckling more likely.

11-100.   *a)* 9.1

         b) 3.6 for no yield design; 5.9 for no failure design

         c) 1.32

11-102. Representative wood properties taken from Table 8.7.
The velocity of impact stays the same: 5.18 ft/sec.
The velocity of the stress wave = 15,220 ft/sec.

δ_{st} = 0.00015 in., F = 12,960 lbf , and σ = 12,960 psi, σ_o = 680 psi , σ_{max} = 13,590 psi

12-3.   *a)* $R = (L^2 + 4h^2)/8h$

        b) $R = 4.87$

12-6.   *a)* y in inches, W in pounds, L in inches, b in inches, and h in inches

        b) No difference

12-8. $1 = kv^{(-2a_3)}\rho^{(-a_3)}p^{(a_3)}$, $\Pi_1 = p/(v^2\rho)$ and $v = c\sqrt{p/\rho}$

12-10. $1 = kf^{(2a_4-a_3)}\Delta p^{(a_4)}D^{(a_3)}\rho^{(a_4)}v^{(2a_4-a_3)}$, $\Pi_1 = f^2\Delta p \rho v^2$, $\Pi_2 = D/fv$

12-12. The model dropping through 30 feet has the same amount of energy to dissipate as the real object falling 7.5 feet.

12-14. Start with $C = 2\pi R$. Let C' and R' be the new circumference and radius, and X = the change in circumference. Then $C' = C + X = 2\pi(R + 0.50'')$ and $X = 2\pi R + \pi - 2\pi R = \pi$.

12-17. The offset data is:

Year	0	1	2	3	4	5	6	7
Production	0.9	1.0	1.1	1.3	1.2	1.3	1.4	1.4

Linear coordinate regression results in $P = 0.07(Y - 1934) + 0.950$, and $r^2 = 0.893$.

12-19. The offset data is:

Year	0	20	30	40	50	60
Population	62.9	76.0	92.0	105.7	122.8	131.7

Linear coordinate regression results in $P = 1.22(Y - 1880) + 57.85$, and $r^2 = 0.975$.
Semilog regression gives the model $P = 61.65(1.013^{Y-1880})$, and $r^2 = 0.987$.
Using the semilog model, $P(1950) = 152.3$, compared to 151.3 actual; $P(1960) = 173.3$ compared to 179.3 actual; $P(1970) = 197.1$ compared to 203.2 actual; and $P(1980) = 224.3$ compared to 226.5 actual.

12-21. Log-log regression gives the "best" fit with the model: $V = 40.14(P^{-0.7045})$, $r^2 = 0.9997$, a sum of the residuals $= -0.997$ and a sum of the squares of the residuals $= 0.006$.

12-23. Log-log regression gives the "best" fit with the model: $G = 0.1153(S^{0.1571})$, $r^2 = 0.975$, a sum of the residuals $= -1.0000$ and a sum of the squares of the residuals $= 0.000006$.

12-25. Use Equations 12.22 for a quadratic least squares analysis, and
$Y = -0.817X^2 + 8.39X - 0.0000003$ with a sum of the squares of the residuals = 32.3.

12-27. Yes

12-29. A value of $r = 0.70$ is acceptable for a 95 percent confidence, $0.70 > 0.632$. A value of 0.70 is not acceptable for a 99 percent confidence level, $0.70 < 0.765$.

12-32. The expanded equations are:

$$a\sum x_i^3 + b\sum x_i^2 + c\sum x_i + nd = \sum y_i$$
$$a\sum x_i^4 + b\sum x_i^3 + c\sum x_i^2 + d\sum x_i = \sum y_i^2$$
$$a\sum x_i^5 + b\sum x_i^4 + c\sum x_i^3 + d\sum x_i^2 = \sum y_i^3$$
$$a\sum x_i^6 + b\sum x_i^5 + c\sum x_i^4 + d\sum x_i^3 = \sum y_i^4$$

12-33. a) log-log

b) propane

12-36. a) Cost is the highest importance because it is to the third power. Keep cost low for a high V.

b) Look at cost as the most important and distance of control as the second most important.

c) Everything except the $S^{0.5}$. How can increased size make a control device more attractive to the customer?

12-37. Cost of purchase. Cost of installation. Cost of maintenance. Appearance and styling. Ease of operation if moving parts are required. Delivery time. Effective R-value. Product life.

12-40. Step function: Use of a three speed fan. Lights on or off. A manual five-speed transmission.

Continuous function: Rheostat control of lights. Throttle control on an auto. Pitch settings on an airplane propeller.

12-41. Calculate the average temperature at each pressure and the data becomes: (67,91), (71,90), (75,87.5), (79,85.5), and (83,84). This models well with linear coordinate regression as $T = -0.46P + 122.23$, $r^2 = 0.983$, and a sum of the squares of the residuals = 0.607.

12-43. a) $1 = kP^{(-a_4)}L^{[a_5/(a_5+2a_3-2a_4)]}A^{(a_5/a_3)}E^{(a_5/a_4)}\delta^{a_5}$

$\Pi_1 = L^2E/P$, $\Pi_2 = \delta/L$, $\Pi_3 = A/L^2$

b) The value of k will not be constant over the range of the experiment.

12-45. For the uniformly loaded beam $\emptyset = wL^3/24EI$.
For the single center load $\emptyset = PL^2/16EI$.
For slope to be the same, $wL^4/24EI = PL^2/16EI$, and $P = 2wL/3$.

12-47. The maximum shear on the uniformly loaded beam = $wL^2/2$. For the model, the maximum shear = $P/2 = (5wL/8)/2 = 5wL/16$, less than the uniformly loaded beam.

12-49. Information required to conduct an experiment should include: length of the wall; thickness of the wall; height of the wall; properties of the wall material (E, S_{yt}, S_{ut}, S_{yc}, S_{uc}); height and width of door openings, and their location from the end of the wall; height, width, distance off the ground for window openings, and their location from the end of the wall.

12-51. Even though the data shows a high correlation, whether the diameter and length are actually related depends on the processes used in the manufacture. If the length is created on a lathe and the diameter created on a grinder it is doubtful if the dimensions are truly related. If the length and diameter are being created at the same time in the same process then perhaps there is some degree of correlation. A good statistical correlation does not prove there is a real correlation between the data. Don't fall into the statistical trap.

12-53. $Y = mX + b$, $Y = mX + [\Sigma y_i - m\Sigma x_i]/n$, $Y = mX + \Sigma y_i/n - m\Sigma x_i/n$; but $\Sigma y_i/n = \mathbf{y}$ and $\Sigma x_i/n = \mathbf{x}$, so $Y = mX + \mathbf{y} - m\mathbf{x}$, and $Y - \mathbf{y} = m(X - \mathbf{x})$.

12-56. Actual voltage = 122.8

13-3. d^2I/dh^2 is approximately equal to -124, so $I(10.39)$ is a maximum.

13-7. Other variables being equal, the higher the exponent value, the lower the tool life.

13-9. $F(2,2) = 12$ is a minimum.

13-10. No maximum or minimum

13-11. a) $(0, f(0)) = (0,6)$ is a minimum.

b) $(2, f(2)) = (2,0)$ is a minimum; $(3, f(3)) = (3,1)$ is a maximum; $(4, f(4)) = (4,0)$ is a minimum.

c) $(2.125, f(2.125)) = (2.125, -0.515625)$ is a minimum.

13-12. $(\pi/2, f(\pi/2)) = (\pi/2, \pi/2)$ is an inflection point.

13-13. The maximum value of $I = 2978$ in^4 when $x = 3.078$ in.

13-14. The maximum value of $I = 3843$ in^4 when $x = 3.760$ in.

13-20. The cylindrical tank model is lower in cost than the rectangular model.

13-21. The curves move up and show a slight change in spacing.

13-22. a) $h = 5.46$ inches

b) Material cost = \$0.01/0.20 = \$0.05/in^3

c) Total cost = 28(\$0.05) = \$1.40

13-24. The diameter for lowest cost decreases compared to Case Study #15 because more material is attributed to the diameter D.

13-25. $I_{max} = 546.33$ when $b = 6.86$ and $h = 9.85$

13-26. $I_{max} = 560.81$ when $b = 6.00$ and $h = 10.39$

13-31. a) $f(x) = x^4 + 6$: Eighth interval = $[-0.1114, 0.1642]$
Eighth interval division = $(0.618)(0.1642 + 0.1114) = 0.1703$
$x_9 = 0.1642 - 0.1703 = -0.0061$, $f(-0.0061) = 6.0000$

b) $f(x) = x^4 - 12x^3 + 52x^2 - 96x + 64$: Eighth interval = $[3.9409, 4.2174]$
Eighth interval division = $(0.618)(4.2174 - 3.9409) = 0.1703$
$x_9 = 4.2174 - 0.1703 = 4.0461$, $f(4.0461) = 0.0089$

c) $f(x) = x^2 - 4.25x + 4$: Eighth interval = [1.9476, 2.2232]
Eighth interval division = $(0.618)(2.2232 - 1.9476) = 0.1703$
$x_9 = 1.9476 + 0.1703 = 2.1179, f(2.1179) = -0.5156$

13-32. A minimum value of 12.16 occurs at (1.67, 2.5) and (2.5, 1.67).

13-33. Fibonacci search. $x_6 = y_6 = 2.381 - 0.238 = 2.101$
$f(1.905, 1.905) = 12.208, f(1.905, 2.101) = 12.010, f(2.101, 1.905) = 12.010$, and
$f(2.101, 2.101) = 12.030$
Estimate for $x_7 = y_7$ = the middle of the intervals = $(1/2)(1.905 + 2.101) = 2.003$
$f(2.003, 2.003) = 12.000$, so the minimum is $f(2.003, 2.003) = 12.000$.

13-34. Golden section search. $x_6 = y_6 = 1.636 + 0.447 = 2.083$
$f(1.910, 1.910) = 12.025, f(1.910, 2.083) = 12.008, f(2.083, 1.910) = 12.008$, and
$f(2.083, 2.083) = 12.020$
Estimate for $x_7 = y_7$ = the middle of the interval = $(1/2)(2.083 + 1.910) = 1.996$
$f(1.996, 1.996) = 12.000$, so the minimum is $f(1.996, 1.996) = 12.000$.

13-36. A minimum value of C is 1484 when $y = 10$ and $z = 9$.

13-37. $C(9.692, 8.308) = 1480.5$ represents the minimum.

13-38. $C(9.502, 8.491) = 1481.4$ represents the minimum.

13-40. c) $x \leq -2$ or $x \geq 3$

13-41. $R_1 < 2r$

13-42. The box dimensions are $1.26 \times 1.26 \times 0.63$, and $A = 4.763$.

13-44. $H = \sqrt{KV/2L}$

13-46. $P_{max}(8/9, 32/9) = \$23.11$, and leaves 2.78 hours of machine time available.

13-48. The possible winning combinations (R, P) are: (3,2), (3,3), (2,3), (2,4), (1,3), (1,4), (1,5), (1,6), (0,4), (0,5), (0,6), and (0,7).

13-51. $P(22, 35) = \$1314$

13-53. $P(52, 13) = \$1014$

13-54. Use 18.7 ounces of alloy A, 21.3 ounces of alloy B, and 46.7 ounces of alloy C.

Answers To Selected Exercises

13-57. The four variable Yates algorithm format is:

Experimental code					Experimental response	(1)	(2)	(3)	(4)	Effect estimate
(1)	−	−	−	−						
a	+	−	−	−						
b	−	+	−	−						
ab	+	+	−	−						
c	−	−	+	−						
ac	+	−	+	−						
bc	−	+	+	−						
abc	+	+	+	−						
d	−	−	−	+						
ad	+	−	−	+						
bd	−	+	−	+						
abd	+	+	−	+						
cd	−	−	+	+						
acd	+	−	+	+						
bcd	−	+	+	+						
abcd	+	+	+	+						

Four columns of addition-subtraction are required before the effect estimate column can be completed. The rest of the analysis follows the three variable format.

13-60. The density of air = $1.2929[(273.13/T)(B - 0.3783e)/760]$ kg/m³, where
T = temperature in degrees centigrade plus 273.13, (°C +273.13)
B = atmospheric pressure in mm
e = vapor pressure of air

13-62. a) $L = H/\sin\theta$

b) $V = 2FH/S_{ut} \sin 2\theta$

c) $D = \sqrt{4F/\pi\, S_{ut} \cos\theta}$

13-63. 0.375 diameter is the most economical choice.

13-64. At 5000 lbf use 1.000 inch diameter.

At 6000 lbf, not possible under the restrictions of the problem.

13-65. $\sin^4\theta\,(C^2 + H^2) - 4C^2 \sin^3\theta + \sin^2\theta\,(6C^2 - H^2) - 4C^2 \sin\theta + C^2 = 0$

This equation is not practical to solve unless C and H are known, or at least the ratio between them known, and even then computer search is most likely the preferred solution method.

13-70. $\theta = 2.637 = 151.1°$

14-1. Because $(\$800)(0.08)(1) = \$64$, the amount of the interest charge.

Answers To Selected Exercises

14-3.   *a)* 8.3 percent

       b) 16.3 percent

       c) 15 percent

14-5. The final amount equals $S(1 + i) = \{R[(1 + i)^n - 1]\}\{1 + i\}$.

14-7.   *a)* 10 years

       b) 8.75 years

       c) 8.69 years

       d) 8.664 years

       e) 8.664 years

14-9. 12

14-11. $7870.98

14-13. Use the longer period of time and assume the monthly deposits made within the quarter do not earn interest. This will give a low value if the interim payments do earn interest.

14-15. $7872.55

14-17. $4.78

14-19. Assume that the $960 paid out once a year for taxes and seven fewer payments are required, saving $3500.

14-20. 13.6 percent more

14-22. Project *B* has the best cash flow, $74,370 compared to $63,280.

14-26.   *a)* 15 percent

        b) 22.25 percent

        c) 26.2 percent

        d) 30.1 percent

14-28. Machine *A*; $3.04/part   Machine *B*; $3.51/part

14-30. $i = 9.0$ percent

14-32. $751,240

14-34. Must estimate an inflation rate. Could use regression analysis on all data in Table 14.4, or just over the last 3 or 4 years.

14-37. $298,330

14-38. Five year saving = $184.80

14-39. $A = G\{(1/i) - [n/((1 + i)^n - 1)]\}$

14-41. If the project life was 3 years then machine *A* would cost $6339 per year; machine *B* would cost $16,962 per year. If the project life was 6 years then machine *A* earns $2471 per year while machine *B* earns $3813 per year.

15-4. It would seem that if NEPA was developed to protect the land and the people, that it should have jurisdiction and control over all matters. Why should any activity be exempt? The overused excuse of national security is often a copout.

15-6. Many codes exist to guide project design. They are not law in the sense that they have passed through legislation. If, however, the code has not been followed and a failure occurs, violation of the code is considered a negligent act and those responsible can be successfully sued and their licenses lost. Some cases exists where following the code was not good enough because the state of the art had advanced beyond code requirements. The engineer was successfully sued even though the code was followed. Codes are guidelines. Sound decisions must be made whether they are followed or not.

15-8. If the scholarships are offered to encourage students into a particular field, then the student reacted exactly as desired. If a scholarship is offered as a reward for past performance then the student was not responding as desired. The action would be unethical if the student only declared English as a major for the time it took to get the scholarship, and then changed to another major. At this point the school could take steps to rescind the scholarship.

15-10. Another form of stealing. The student has paid for the right to use the computer through tuition or lab fees, not the right to give the privelege to someone else. To allow someone else to use the computer is not ethical.

15-12. It is unfortunate when a person is threatened with the loss of a job to become an accomplice to a wrongdoing. It is a case of the supervisor being guilty of two counts of unethical behavior. The first is wanting to mislead the stockholders. The second is to ask someone else to join in the deception. The programmer has the choice of joining the charade and being equally guilty of deception, or of writing the program correctly and risking the loss of the job. One of the problems of lying is that usually more lies are required to keep the first lie from being found out.

15-14. Several conditions could exist. 1. The recording secretary could have told the person. 2. Someone else on the committee could have told the person. 3. The person may have assumed the actions were taken and written the statement for effect. In all three cases a breach of ethics occurred.

15-17. If the engineer liked the candidate the temptation to go along with the scheme is more enticing, but it is still wrong. Raises should be related to the worth of the engineer to the company objectives. If the engineer did not like the candidate, refusal would be easier, but alienation of a supervisor might occur. The person who makes such a request is being unethical, and to draw someone else into the picture just makes it worse.

15-20. The pressure is highest if others are involved in the bonus because the pressure comes not only from oneself but from fellow workers. Pressure for a short term gain at the expense of covering a problem is not unusual. Sometimes the engineer is the only one who knows the problem. Sometimes it is known that someone else is aware of a problem. The incentive system (the reward for high output) without a check on quality encourages poor judgement and action for short term gain over long term sound reputation.

15-23. Money might be spent on product improvement or new product research. If an improved product can be sold in later years for a higher price, or more units can be sold, then money spent now will help future profits.

Index

Numbers in ( ) are exercise numbers. Bold face entries are Appendices.

Accelerometer, 132 (5.28), 156 (6.8), 157 (6.11, 12, 13)
Activity chart, 53-57
Airplanes, 29
  Boeing 737-200, 8
  DC-10, 7
  de Havilland Comet, 6
  Voyager, 27
Alexander L. Kieland oil rig, 7
Alpha error, 331-332
Aluminum, 201-202
  coding system, 202
  property values of, 225, 226, 227, 231, **F-1, F-2, F-4**
Analysis
  beam, **O**
  column, 402-407
  data
    experimental, 454-466
    graphical, 456-457
  fatigue life, 392-393
  financial, 538-556
  force, 142-143
  functional, 135
  human factors, 136-137
  least squares, 455-456
  linear coordinate, 459-460
  idea, 81-82
  market, 96, 138, 176
  safety, 136
  service requirements, 137-138
  table data, 352-353
  using drawings, 140-144
Annuities, 543-544
Aswan Dam, 32-33
Automobile(s), 27-29
  springs, 35
Aviation, 13, 17, 27, 29

Beam analysis, **O**
Bend
  allowance, 269-272
  force, 301 (9.10)
  tolerances, 250
Beta error, 331-332
Bill of material, 43
Binomial distribution, 340-342
Blanking, 249, 261, 272
Boring, 249, 251, 274-275
Boundary conditions, 444

Brainstorming, 119-120
Brass, property values of, **F-1, F-2, F-6**
Bridges, 5
  arch, 36
  arch truss, 38
  Brooklyn, 10
  draw, 35
  Golden Gate, 5, 35
  Hackensack River, 5
  Mackinac suspension, 10
  Mianus River, 7
  Queensboro truss, 34
  South Carolina Canadian, 7
  suspension, 22 (1.14)
  Tacoma Narrows, 5, 35
  Walt Whitman suspension, 37
Brinell hardness number, 214, 215, 241 (8.17)
Brittle
  failure theory, 381-383
  material data, 215, **F-10**
Broaching, 249, 251, 279
Bronze, property values of, **F-1, F-2, F-7**
Buckingham Pi Theorem, 448-450
Buckling. *See* Column analysis
Buildings, 5, 6, 7
Burnishing, 249

Case Study #1: Door Holder-opener, 79-80
Case Study #2: Parts Moving Problem, 89-91, 120-121, 148-153
Case Study #3: Idea Generation, 121-123, 176
Case Study #4: Drawing Board Table Adjustment, 141-145
Case Study #5: Room Cooling, 147-148
Case Study #6: Ball Field Revision, 171-174
Case Study #7: Special Machine, 287-288
Case Study #8: Probabilistic Design Analysis, 376-377
Case Study #9: Hydraulic Cylinder, 415-418
Case Study #10: Tree Age Model, 440-443

Case Study #11: Wind Force Model, 450-453
Case Study #12: Empirical Formula Refinement, 460-466
Case Study #13: Tool Life Optimization, 486-490
Case Study #14: Pipe Packing Optimization, 491-493, 530 (13.18)
Case Study #15: Cost Ratio Analysis, 493-495
Case Study #16: Balcony Seating, 512-516
Case Study #17: Spherical Storage Tank, 516-519
Case Study #18: Experimental Optimization, 523-526
Case Study #19: Misapplied Overhead, 549-550
Case Study #20: Cost Reduction Proposal, 552-554, 555-556
Case Study #21: New Process Proposal, 554-555
Case Study #22: Personal Conflict, 570-571
Case Study #23: Corporate Conflict, 571-572
Case Study #24: Government Ethics, 572
Cash flow, 552-554
Casting, 253, 263-267
  centrifugal, 265, 266, 267
  ceramic mold, 266, 267
  die, 250, 251, 265, 266, 267
  investment, 251, 266, 267
  permanent mold, 265, 266, 267
  plaster mold, 251, 265, 266, 267
  sand, 251, 264-265, 266, 267
  shell mold, 266, 267
Cast iron, 5, 31, 199
  property values of, 212, 215, 221, 222, **F-1, F-2, F-3**
Ceramics, 206-207, 215
  property values of, 226, 227, 231, **F-1, F-10**
Chain saw, 30
Chi-squared test (statistics), 335-337, **J**
Code
  building, 48-50, 106 (4.20)

Index

of ethics, 564-569
Coefficient
  of correlation, 458, **K**
  of determination, 458
Column analysis, 402-407
  eccentricity, 404-405
  Euler model, 402-404
  end conditions, 403
  Johnson model, 404
  secant model, 404-407
Combinations, 322
Communication, 45, 57-64, 174. *See also* Reports
  oral, 61-64
  written, 58-61
Composite materials. *See* Materials, composite
Compromise. *See* Design, compromise
Computer(s), 7, 86, 147, 406-407, 512-516
  software, MetSel/2, 234
Concrete, 6
  property values of, **F-1, F-2, F-11**
Conductivity
  electric, 224
  thermal, 228
Consumer, 12, 136, 167, 176
Conversion factors, **C**
Copper, 201
  property values of, 225, 226, 227, 231, **F-1**
  wire, **N**
Corrosion, 412
Cost, 42, 43, 146
Coulomb-Mohr failure theory, 382
Counting methods, 321-323
Creativity, 76-80, 108
  roadblocks, 78-79
Crack(s), 7
  growth, 410-411, 428 (11.55)
Creep, 219, 221, 411
Crystal structure, **E**

Dams
  and locks, 10
  Aswan, 32-33
  Vaiont, 6
Data
  analysis, 11, 352-353, 454-466
  collection, 8, 11
  error, 11
Databases, 99
Decision, 166-173. *See also* Engineering, decisions
  error, 331
Density, 231

material values of, 208, 221, 226, **E**
Depletion, 545
Depreciation, 454-547
  straight line, 546
  sum of the digits, 547
Design
  compromise, 4, 7, 45, 74, 135
  considerations, 290-297
  constraints, 46-52
  definition, 24, 155 (6.1)
  guidelines
    codes, 47-50, 106 (4.20)
    general, 252, 284, 290-291, 398-399, 401, 414
    welding, 401-402
    wood, 430 (11.70)
  failure, 10-13
  for assembly, 290
  interferences, 368
  human aspects, 136-137
  probabilistic, 374-377
  problems, 182-195
  side effects, 31
Dielectric constant, 226
  values of, 228
Dimensional analysis, 447-454
Dimensionless groups, 453-454
Distortion energy theory, 379-381
Dixon, 428 (11.58)
Drawing(s), 174
  errors, 11-12
  manufacturing process, 251, 268, 272
Drilling, 250, 251, 277-279
Ductility, 218-219
Dynamic damper, 131 (5.26)

Earthquake, 10
Edison, 30, 37
Electric discharge, 249, 251
Electrical conductivity, 224
Electrical permeability, 226
Electrical resistivity, 223
  values of, 227
Electricity, 30
Energy, 15-16, 21
  nuclear, 10
  solar, 16
  wind, 16
Energy values of fuel, **D**
Engineer(ing)
  activities, 24-65
  advancements, 26-37
  constraints, 46-52
  day-to-day concerns, 44-46
  decisions, 5, 45, 83

design. *See* Design
developments, 26-37
experiences, 5-10
failures. *See* Failure
mistakes, 44
necessary skills, 25-26
personal characteristics and abilities, 64-65
primary job, 24-25
responsibilities, 38-44
Environment. *See* Problems, pollution
Epoxy, 206, 245
  graphite, 221
Equal division search technique, 496-498
Ethics, 45, 562-572
  code of, 564-569
  definition of, 563
  sources of problems, 563-564
Euler model for columns, 402-404
Expansion joint, 7
Experimentation, 11, 37, 467-470, 519-527
Experimental data analysis, 454-466, 519-527
Extruding, 251, 253, 269

F test (statistics), 334-335, **I**
Failure. *See also* specific modes of failure
  analysis, 366-418
  building, 5-7
  bridge, 5-7
  engineering, 5
  roof, 7
Fatigue, 8, 389-402
  characteristics, 217, 389
  factors affecting, 218, 220, 393-399
  Goodman diagram, 391-392
  part design, 401
  testing, 390-391, 400-401
  welding, 401-402
Feasibility study, 125-127
Fedderson, 428 (11.58)
Ferrous alloys, 198-201
  property values of, 212, 221, 222, 225, 226, 227, **F-1, F-2**
Fiberglass, 205
Fibonacci search technique, 498-499, 505, 506
Financial analysis, 538-556
Fire, 6, 8
Fits, classes of, 247-248
Flame cutting, 250, 251, 257-258, 261

Index

Flexible manufacturing system, 276
Forecast uncertainty, 56, 555-556
Forging, 267-273
    bend allowance, 269-270, 301 (9.17)
    drawing, 251, 268, 272
    extrusion, 251, 253, 269
    press drives, 304
    rolling, 253, 268
    shearing, 250, 255-257, 260, 272
    stretching, 272, 273
    surface finish, 251
Four-color map problem, 86 (3.30)
Fracture mechanics, 6, 34, 407-411
    design criteria, 407-408
    plane stress-plane strain, 408-409
    crack growth, 410-411, 428 (11.55)
    stress intensity factors, 408, 428 (11.58, 59, 60), **M**
Fracture toughness, 218
    values of, 221, 222
Fresnel's equations, 232
Fretting, 412
F test (statistics), 334-335, **I**

Gamma rays, 130 (5.21)
Gantt chart, 54-55
Glass transition temperature, 230
Golden section search technique, 499-503, 505, 506
Goodman diagram, 391-392
Government
    agencies, 99-100
    regulations, 47-50
Graphical analytical methods, 456-457
Graphite, 245
Grinding, 249, 251, 279-281, 300 (9.3)
Grumman bus, 7

Half-life, 233
Hardenability, 216-217, 218, 219
Hardness, 214-215
Hartford Civic Center, 7
Heat flow, 531 (13.41)
Heat treatment, 283-284
Heat of fusion, 230
Histogram, 320-321
Honing, 249, 251, 281
Hooke's law, 210
Hyatt Regency Hotel skywalk collapse, 7
Hydrogen embrittlement, 35, 413

Idea(s)
    alternate word, 113-114

analysis, 81-82
combination, 112-113
comparison, 170-173, 180 (7.14)
evaluation, 138
generation, 80-81, 108-127
implementation, 82-83, 174-176
question and answer, 114-116
recording, 114
refinement, 81
selection, 168-171
Impact, 383-389
    factor, 385
    linear, 383-386
    stress wave propagation, 387-389
    torsion, 386
Inertia formulas, **P**
Information, 10, 16
    sources of, 96-101, 113
Installment loans, 540
Interest
    compound, 540-544
    continuous, 557 (14.6)
    effective annual, 540
    simple, 539-540
Irwin, 428 (11.58)

John Hancock Center, 6
Johnson model for columns, 404

Kevlar™, 205, 226

Lagrange multipliers, 490
Lambert's law, 233
Lapping, 249, 251, 281
Laser cutting, 251, 258, 261
Lathes, 258-259, 273-274
Laws
    legal, 47
    physical, 117-119
Lead, property values of, 212, **F-1**
Least square analysis, 455-456
Liberty Ships, 6, 34
Light absorption, 233
Light bulbs, 37, 467-470, 478 (12.56)
Linear programming, 508-517
Lock and dam, 10

Machinability, 222
    values of, 227
Machine drives, 304
Machining operations, 273-281
Machines, special, 284-288
Machining center, 275
Magnesium, property values of, 225, **F-1, F-2, F-5**
Magnetic hysteresis, 226
Maintenance, 7, 51, 176

Manufacturing, 27, 47, 246-299
    design, 290-297
    process
        selection, 144-145, 246-299
        surface finishes, 251
        tolerances, 248-250, 368
Materials, 144, 198-237. *See also* specific material name or material category.
    composite, 235-237
        mechanical property values of, 236
        mechanical property formulas, 235, 244 (8.52, 53)
    density, 231
        values, 208, 221, 226, **E**
    crystal structure, 230, **E**
    forms, 252-254
        bars, 252-254
            reinforcing, 300 (9.7)
        casting, 263-267
        coils, 262-263
        plates and sheets, 259-260
    properties, 208-233
        electrical, 223-228
        mechanical, 208-227
        thermal, 228-230, **E**
    selection, 234
    strength
        theories, 377-383
            distortion energy, 379-381
            maximum normal stress, 377-379
            maximum shear stress, 379
            variations on, 381-383
        variations, 368
Mean, arithmetic, 327
Median, 327
Medicine, 17
Melting temperature, 230
    values of, **E**
Metals, 198-202
    ferrous, 198-201
    magnetic, 201
    nonferrous, 201-202
Metallurgy, 30
    powder, 273
Methane, 5
Metric
    preferred sizes, 253
    prefixes, **B**
Milling, 249, 252, 274
Mode, 327
Model(s), 146, 436-454
    analog, 437
    boundary conditions, 443-444
    full size, 444-445
    graphical, 456

Index

iconic, 437
mathematical, 454-458
  log-log, 455, 456
  polynomial, 457-458
  linear coordinate, 454, 455, 459-460, 461
  semilog, 454, 455, 462-466
scale, 445-447
symbolic, 438
testing, 135
Modulus of elasticity, 209
  values of, 208, 221, 222, 226, **F-12**. *See also* specific material or material category.
Modulus of resilience, 210
Modulus of rigidity, 212
Modulus of rupture, 212, 243 (8.34)
Mortgages, 542-543

Natural events, 18
Natural gas, 5
Neutron radiation, 9, 34
Nibbling, 262, 272
Nickel, 199
  property values of, 226, 227, **F-8**
Nonferrous alloys, 201-202
Normal distribution, 327-340, **G**
  data test, 338-340
  Chi-squared test, 335-337, **J**
  effect of sample size, 333-334
  F test, 334-335, **I**
  t test, 332-333, **H**
  Z test, 330-331
Nuclear power, 9, 34
Nylon, 205
  property values of, 226, 227, 228, **F-1**

Optimization, 46, 480-527
  calculus techniques, 481-496
    Lagrange multipliers, 490-491
  experimental, 519-527
  linear programming, 508-512
  search techniques, 496-507
    equal division, 496-498, 505-507
    Fibonacci, 498-500, 505, 506
    golden section, 499-503, 505, 506
    multi-dimensional, 505-507
    slope, 503-504, 505
Overhead charges, 547-551
  administrative, 550
  material, 548
  process, 548-550
  product, 550

Palmgren-Miner theory, 399-400
Panama Canal, 10
Paris, 410
Particulate settling, 477 (12.55)
Patent records, 97
Pendulum, 163 (6.29)
Permeability, 252
Permutations, 322
PERT, 56-57
Petroleum, 15, 27
Planing, 249, 251, 277
Plating, 281, 399
Plexiglass™, 136, 204
Plutonium oxide, 10
Pollution, 8, 13
  waterway, 18
Polyethylene, 203-204
  property values of, 203, 243 (8.31, 8.32), **F-1**
Polymers, 202-206
  property values of, 231, **F-1**
Poisson distribution, 342-343
Poisson's ratio, 211
  values of, 205, 212
Porosity, 231
Powder metallurgy, 273
Probabilistic design, 374-377
Probability (and statistics), 318-354
  definition, 318-319
  multiple event, 319-320
  parallel, 326-327
    constant rate, 327
  series, 325-326
    constant rate, 326
Problem solving activities, 74-83
Problem definition, 80, 88-104
  cautions, 102-103
  methods of, 88-96
  related activities, 96-101
Problems
  information, 16
  medicine, 17
  pollution, 13, 21 (1.4), 111-112
  sources of, 74-75
  system, 103-104
  transportation, 17
  unsolved, 13-18
Process
  capability, 346-352
  control, 333, 344
  overhead, 548-550
Product
  cost, 42, 43, 547-551
  distribution, 18
  life, 43, 50
  servicing, 44, 137-138
  testing, 45, 101, 175
Professional societies, 101

Project planning, 52-57, 175
  critical path network, 54, 56
  Gantt chart, 54-55
  implementation, 82-83
  PERT, 56-57
Proportional limit, 210
Punching, 272, 349

R value of materials, **R**
Range, 328
Reaming, 249, 251
Refinement, 138-147, 151-154
Reflectivity, 232
Refraction index, 232
  values of, 233
Reliability, 135, 319
  determining, 371-373
  measure of, 369-371
  parallel, 326-327, 370
  series, 325-326, 370, 371
Reports
  written, 58-61
    feasibility, 125-127
    project proposal, 59
    project status, 59-60
  oral, 61-64
    visual aids, 62-63
Resistivity, electrical, 223
Return on investment, 551-552
Rolling, 253, 268

Safety, 17, 167
  analysis, 136
  factors, 373-374
  features, 12, 44
Sawing, 250, 251, 254, 256, 260, 303 (9.20)
Secant model (column buckling) 405-407
Shaping, 249, 251, 277
Shearing, 250, 255-257, 260, 272
Shipping, 12, 51
Shot peening, 401
Slope search technique, 503-505
Sound exposure levels, 478
Specifications, 40-41
Spring rate, 386
Stamping, 249, 261, 272
Standard deviation, 328
  of a sum function, 375
State of the art, 5, 7, 167
Statistics (and probability), 318-354
Statistical tests, 337-338. *See also* Z test, t test, F test, Chi-squared test
Statistical scatter of data, 395-396

Steel, 35, 199
  alloy
    effect of, 199
    numbering system, 199-200
  magnetic, 201
  property values of, 212, 221, 222, 225, 226, 227, **F-1, F-2**
  stainless, 9, 225
Storage, 12, 138
Strain
  elastic, 209
  hardening, 213, 214
  normal, 209
  plastic, 209
  shear, 211, 213
  2% offset, 210
Strength
  compressive values, 208, 215
  elastic, 209-210
  fatigue, 8, 217, 389-402
    values of, 220, **F-1**
  impact, 219, 383-389
  shear values, 208
  specific, 222
    values of, 226
  tensile, values of, 208, 215, 221
  ultimate, values of, 225, 226
  yield, 209
    values of, 222
Stress
  concentration, 5, 6
    factors, 395, **L**
  fatigue, 389-402
  impact, 383-389
  intensity factors, **M**
  normal, 209

  shear, 211
  wave propagation, 387-389
Structural sections
  angles, **Q-4**
  channels, **Q-3**
  standard flange beams, **Q-2**
  wide flange beams, **Q-1**
Styling, 50
Surface coatings, 281-282
  plating, 281, 399
Survey
  land, 10
  market, 96
Suspension. *See also* Bridges
  formula, 22 (1.14)
Systems, 103-104, 367-368

t test (statistics), 332-333, **H**
Taxes (corporate), 544-545
Teflon™, **F-9**
Thermal
  conductivity, 228
  expansion, 229
  shock resistance, 244 (8.49)
Titanium, property values of, **F-2, F-9**
Tolerances, 247-252. *See also* specific machining operations.
Tooling, 288-289
  life, 486-490, 528 (13.7)
Transistors, 36
Transparency, 233
Transportation, 17
Tree diagram, 323-324
Turning, 249, 258-259

Vaiont Dam, 6
Value analysis, 110-112
Vapor pressure, 231
Variance, 328
Vector analysis, 142-143
Vibration, 413, 415
Viscosity, 231
  values of, 231
Visual aids, 62-63
Voyager (airplane), 27
Voyager 2 (spacecraft), 10

Weibull distribution, 344-347
Welding, 282-283
  design guidelines, 401-402
  fatigue, 401-402
  tolerances, 250
Wind
  testing, 5
  modeling, 450-453, 534 (13.58)
Wood, 5, 185 (7.42), 207-208
  design guidelines, 430 (11.70)
  material properties of, 208, **F-2**
  plywood, 260
  sizes, 253
Work simplification, 124-125

Xerox, 167

Yates algorithm, 520-526
Yield strength, 209-210
  yield point, 210
  proportional limit, 210
  values of, 222, 225, **F2-F10**

Z test (statistics), 330-331